Advanced Microfluidics-Based Point-of-Care Diagnostics

Advanced Microfluidics-Based Point-of-Care Diagnostics

A Bridge Between Microfluidics and Biomedical Applications

Edited by
Raju Khan, Chetna Dhand, S. K. Sanghi,
Shabi Thankaraj Salammal, and A. B. P. Mishra

CRC Press
Taylor & Francis Group
Boca Raton London New York

CRC Press is an imprint of the
Taylor & Francis Group, an **informa** business

MATLAB® is a trademark of The MathWorks, Inc. and is used with permission. The MathWorks does not warrant the accuracy of the text or exercises in this book. This book's use or discussion of MATLAB® software or related products does not constitute endorsement or sponsorship by The MathWorks of a particular pedagogical approach or particular use of the MATLAB® software.

First edition published 2022
by CRC Press
6000 Broken Sound Parkway NW, Suite 300, Boca Raton, FL 33487-2742

and by CRC Press
2 Park Square, Milton Park, Abingdon, Oxon, OX14 4RN

Library of Congress Cataloging-in-Publication Data

Names: Khan, Raju, Dr., editor. | Dhand, Chetna, editor. | Sanghi, Sunil K., editor. | Shabi, Thankaraj S. (Thankaraj Salammal), 1984- editor. | Mishra, A. B. P. (Ashtbhuja B. Prasad), editor.
Title: Advanced microfluidics based point-of-care diagnostics : a bridge between microfluidics and biomedical applications / edited by Raju Khan, Chetna Dhand, SK Sanghi, Thankaraj S Shabi, ABP Mishra.
Description: First edition. | Boca Raton : CRC Press, 2022. | Includes bibliographical references and index. | Summary: "This book provides a well-focused and comprehensive overview of novel technologies involved in advanced microfluidics based diagnosis via various types of prognostic and diagnostic biomarkers. Moreover, i also contains detailed descriptions on the diagnosis of novel techniques"-- Provided by publisher.
Identifiers: LCCN 2021048705 (print) | LCCN 2021048706 (ebook) | ISBN 9780367461607 (hardback) | ISBN 9781032218120 (paperback) | ISBN 9781003033479 (ebook)
Subjects: MESH: Microfluidic Analytical Techniques | Point-of-Care Testing |Biosensing Techniques
Classification: LCC R857.B54 (print) | LCC R857.B54 (ebook) | NLM WX 162 |DDC 610.28/4--dc23/eng/20211101
LC record available at https://lccn.loc.gov/2021048705
LC ebook record available at https://lccn.loc.gov/2021048706

ISBN: 978-0-367-46160-7 (hbk)
ISBN: 978-1-032-21812-0 (pbk)
ISBN: 978-1-003-03347-9 (ebk)

DOI: 10.1201/9781003033479

Typeset in Times
by Deanta Global Publishing Services, Chennai, India

Contents

Editor Biographies

Dr Raju Khan is a Principal Scientist & Associate Professor at the Council of Scientific and Industrial Research – Advanced Materials and Processes Research Institute (CSIR–AMPRI), Bhopal, MP, India. He has more than 15 years of experiences in the field of advanced materials for antibacterial/antiviral and electrochemical/fluorescence-based biosensors integrated with microfluidics.

Dr Chetna Dhand is a Senior Scientist & Assistant Professor at the Council of Scientific and Industrial Research – Advanced Materials and Processes Research Institute (CSIR–AMPRI), Bhopal, MP, India. She has more than 10 years of working experience in the field of nanomaterials, biomaterials, antimicrobial materials, nanobiointerfacial chemsitry and biosensors.

Dr S. K. Sanghi is a Senior Principal Scientist at Council of Scientific and Industrial Research – Advanced Materials and Processes Research Institute (CSIR–AMPRI). He has more than 35 years' experience in the field of microchip-based separation under the concept of lab-on-a-chip.

Dr Shabi Thankaraj Salammal is a Scientist & Assistant Professor at the Council of Scientific and Industrial Research – Advanced Materials and Processes Research Institute (CSIR–AMPRI), Bhopal, MP, India. He has more than 13 years of research experience in the field of organic electronics, radiation shielding materials and nanomaterials.

Dr A. B. P. Mishra is a Principal Scientist at the Department of Science and Technology, Technology Bhavan, New Delhi, India. He has more than 25 years' experience in the field of sensors.

List of Contributors

F. Alam
Khalifa University
Abu Dhabi, UAE

A. A. S. J. Alghailani
Khalifa University
Abu Dhabi, UAE

Wejdan S. AlGhamdi
KAUST Solar Centre
King Abdullah University of Science and Technology
Thuwal, Saudi Arabia.

B. Alqattan
University of Birmingham
Birmingham, UK

Nirupama Guru Aribam
School of Studies in Environmental Chemistry
Jiwaji University
Gwalior, India

Thomas D. Anthopoulos
KAUST Solar Centre
King Abdullah University of Science and Technology (KAUST)
Thuwal, Saudi Arabia.

T. G. Satheesh Babu
Amrita School of Engineering
Amrita Vishwa Vidyapeetham
Coimbatore, India.
School of Engineering
Amrita Vishwa Vidyapeetham
Coimbatore, India.

U. Banerjee
Indian Institute of Technology
Madras, India

Rinkumoni Barman
Department of Water Resources,
Ministry of Jal Shakti
Guwahati, India

H. Butt
Khalifa University
Abu Dhabi, UAE

Pranjal Chandra
School of Biochemical Engineering
Indian Institute of Technology (Banaras Hindu University)
Varanasi, India

Parul Chaurasia
School of Biomedical Engineering
Indian Institute of Technology (Banaras Hindu University)
Varanasi, India

Suparna Datta
Department of Water Resources
Ministry of Jal Shakti
Kolkata, India

Chetna Dhand
Council of Scientific and Industrial Research (CSIR)
Advanced Materials and Processes Research Institute (AMPRI)
Bhopal, India
Academy of Scientific and Innovative Research (AcSIR)
Ghaziabad, India

Snigdha Dutta
Department of Water Resources
Ministry of Jal Shakti
Guwahati, India

Neeraj Dwivedi
Council of Scientific and Industrial Research (CSIR)
Advanced Materials and Processes Research Institute (AMPRI)
Bhopal, India
Academy of Scientific and Innovative Research (AcSIR)
Ghaziabad, India

Hendrik Faber
KAUST Solar Centre
King Abdullah University of Science and Technology (KAUST)
Thuwal, Saudi Arabia.

M. Elsherif
Khalifa University
Abu Dhabi, UAE

Mayank Garg
Council of Scientific and Industrial Research (CSIR)
Central Scientific Instruments Organisation
Chandigarh, India
Academy of Scientific and Innovative Research (AcSIR)
Ghaziabad, India

Shristi Handa
Institute of Nano Science & Technology (INST)
Mohali, India

S. K. Jain
Indian Institute of Technology
Madras, India

T. Kaneta
Okayama University
Japan

Vibhav Katoch
Institute of Nano Science & Technology (INST)
Mohali, India

A. A. Khan
Lahore University of Management Sciences
Pakistan

Raju Khan
Council of Scientific and Industrial Research (CSIR)
Advanced Materials and Processes Research Institute (AMPRI)
Bhopal, India
Academy of Scientific and Innovative Research (AcSIR)
Bhopal, India

Sunil Kumar
Department of Chemistry
Indian Institute of Technology (Indian School of Mines)
Jharkhand, India

Rashmi Madhuri
Department of Chemistry
Indian Institute of Technology (Indian School of Mines)
Jharkhand, India

Sanjeev Kumar Mahto
Centre for Advanced Biomaterials and Tissue Engineering
Indian Institute of Technology (Banaras Hindu University)
Varanasi, India

Alka Mishra
Council of Scientific and Industrial Research (CSIR)
Advanced Materials and Processes Research Institute (AMPRI)
Bhopal, India
Academy of Scientific and Innovative Research (AcSIR)
Ghaziabad, India

Satanand Mishra
Council of Scientific and Industrial Research (CSIR)
Advanced Materials and Processes Research Institute (AMPRI)
Bhopal, India
Academy of Scientific and Innovative Research (AcSIR)
Ghaziabad, India

A. Nath
Indian Institute of Technology
Madras, India

Arpana Parihar
Council of Scientific and Industrial Research (CSIR)
Advanced Materials and Processes Research Institute (AMPRI)
Bhopal, India

Dipesh Singh Parihar
Engineering College Tuwa
Godhra, India

Priyanka Prabhakar
Council of Scientific and Industrial Research (CSIR)
Advanced Materials and Processes Research Institute (AMPRI)
Bhopal, India
Academy of Scientific and Innovative Research (AcSIR)
Ghaziabad, India

Aarathi Pradeep
Amrita School of Engineering
Amrita Vishwa Vidyapeetham
Coimbatore, India

Bhanu Prakash
Institute of Nano Science & Technology (INST)
Mohali, India

Priyadarsini S
Cerner Healthcare Solutions India Pvt Ltd

Keisham Radhapyari
Department of Water Resources
Ministry of Jal Shakti
Guwahati, India

Reetu Rani
Council of Scientific and Industrial Research (CSIR)
Central Scientific Instruments Organisation
Chandigarh, India
Academy of Scientific and Innovative Research (AcSIR)
Ghaziabad, India

Pushpesh Ranjan
Council of Scientific and Industrial Research (CSIR)
Advanced Materials and Processes Research Institute (AMPRI)
Bhopal, India
Academy of Scientific and Innovative Research (AcSIR)
Bhopal, India

Mohd. Abubakar Sadique
Council of Scientific and Industrial Research (CSIR)
Advanced Materials and Processes Research Institute (AMPRI)
Bhopal, India

Dennyson Savariraj
Khalifa University
Abu Dhabi, UAE

S. Seetasang
Okayama University
Japan

A. K. Sen
Indian Institute of Technology
Madras, India

Raj Kumar Sen
Council of Scientific and Industrial Research (CSIR)
Advanced Materials and Processes Research Institute (AMPRI)
Bhopal, India
Academy of Scientific and Innovative Research (AcSIR)
Ghaziabad, India

Abhinav Sharma
KAUST Solar Centre
King Abdullah University of Science and Technology (KAUST)
Thuwal, Saudi Arabia.

Amit L. Sharma
Council of Scientific and Industrial Research (CSIR)
Central Scientific Instruments Organisation
Chandigarh, India
Academy of Scientific and Innovative Research (AcSIR)
Ghaziabad, India

Suman Singh
Council of Scientific and Industrial Research (CSIR)
Central Scientific Instruments Organisation
Chandigarh, India
Academy of Scientific and Innovative Research (AcSIR)
Ghaziabad, India

Pratima R. Solanki
Special Centre for Nanoscience
Jawaharlal Nehru University
New Delhi, India

Avanish Kumar Srivastava
Council of Scientific and Industrial Research (CSIR)
Advanced Materials and Processes Research Institute (AMPRI)
Bhopal, India
Academy of Scientific and Innovative Research (AcSIR)
Ghaziabad, India

A. Sudeepthi
Indian Institute of Technology
Madras, India

Lynda Velutheril Thomas
Sree Chitra Tirunal Institute for Medical Sciences and Technology

Niraj K. Vishwakarma
School of Biomedical Engineering
Indian Institute of Technology (Banaras Hindu University)
Varanasi, India

A. K. Yetisen
Imperial College London
London, UK

1 The Basic Concept for Microfluidics-Based Devices

Vibhav Katoch and Bhanu Prakash

CONTENTS

DOI: 10.1201/9781003033479-1

LIST OF ABBREVIATIONS

CNC, 34 computer numerical control
COC, 34 cyclic olefin copolymers
COP, 34 cyclic olefin polymers
DARPA, 2 Defense Advanced Research Projects Agency
GC, 2 gas chromatography
ICs, 2 integrated circuits
LOC, 2 laboratory-on-a-chip
MEMs, 2 microelectromechanical systems
PDMS, 3 poly(dimethylsiloxane)
PC, 4 polycarbonate

PMMA, 4 polymethylmethacrylate
PS, 4 polystyrene
R_e, 11 Reynolds number
μ-TAS, 2 total analysis system

1.1 WHAT IS MICROFLUIDICS?

Microfluidics refers to the behavior and manipulation of fluids corresponding to the dimension of channels in the order of tens to hundreds of microns. This technology of fluid manipulation is described by fundamental equations used to study fluid physics at the macroscale. Therefore, microfluidics is usually categorized under fluid mechanics. In microfluidics systems, the height or width (or both) of the channel are in the range of few micrometers while it can hold the volume of liquid in the range milliliters (10^{-3}) to nanoliters (10^{-9}) or smaller as depicted in Figure 1.1. In recent years, it has emerged as a diverse field of research with emphasis on physical sciences, medicine, chemistry, and biology.

1.1.1 Evolution of Microfluidics

The evolution of microfluidics can be directly correlated right from the discovery of materials and techniques that are intensively used in recent times for the fabrication of microfluidics devices. To establish the fundamental rules or equations governing microfluidics systems, concepts are taken from diverse areas like fluid flow physics and fluid mechanics which ultimately provide a basic understanding about the flow phenomenon at microscale. Figure 1.2 depicts the invention of components of microfluidics systems in chronological order.

The origin of studies that have a relationship with the field of microfluidics is the derivation of the Hagen–Poiseuille equation given by French physicist Jean Leonard Marie Poiseuille and German civil engineer Gotthilf Henrich Ludwig Hangen in 1838. Later, Poiseuille published it and it was subsequently termed the Poiseuille law

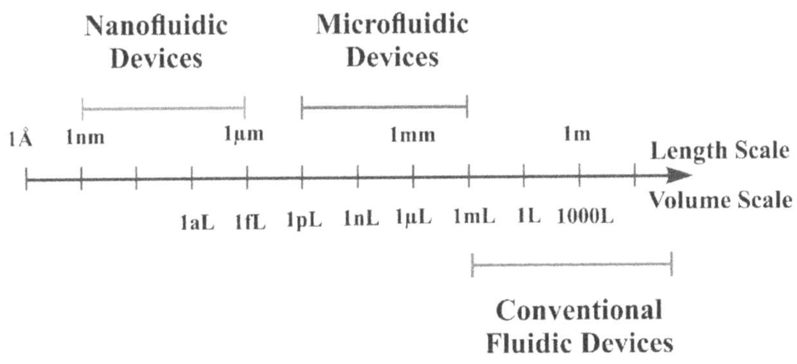

FIGURE 1.1 Microfluidics devices: Dimensions and volume scale.

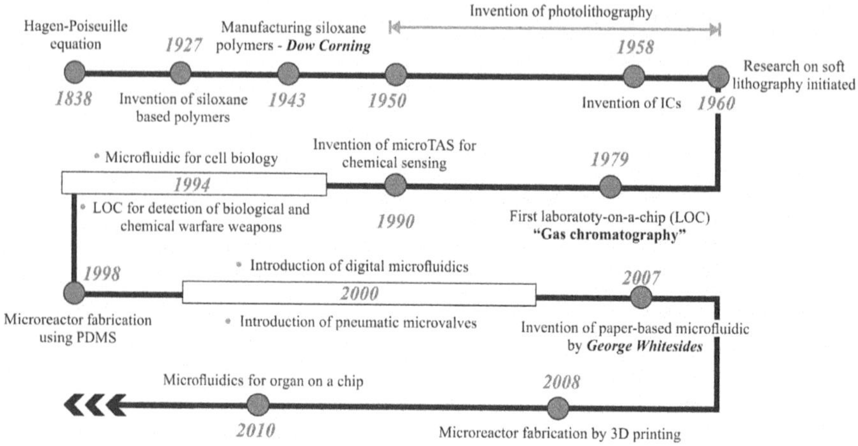

FIGURE 1.2 Evolution of microfluidics.

(Sutera 1993). This law determines the relationship between the resistance arising during fluid flow and the channel length which is given by Equation (1.1)

$$\Delta P = -\frac{8\mu L}{\pi r^4} Q \tag{1.1}$$

Here, (ΔP) is the pressure difference developed across the inlet and outlet. Also, μ is the dynamic viscosity, Q is the volumetric flow rate of the fluid, L and r are the length and radius, respectively.

In 1927, Frederic Stanley Kipping, an English chemist, was credited with the invention of siloxane ((C_2H_6OSi)n)-based polymers. As the outcome of this invention, in 1943 Dow Corning set up the large-scale manufacture of silicone-based polymers, which at present are a major contribution to the fabrication of microfluidics devices (Duffy et al. 1998, Hyde 1965).

In the 1950s, the quest for size reduction of electronic circuits was taken as a challenge by two Americans, Jay W. Lathrop and his co-worker James R. Nall working in the US National Bureau of Standards. In 1958, they coined the term photolithography as a part of their paper presentation in Washington DC. Later that year, the invention of integrated circuits (ICs) was credited to Jack Kilby and Robert Noyce.

In 1979, Stanford's Stephen Terry initiated the actual work that drove the development of microfluidics. Terry designed a miniaturized gas chromatography (GC) unit and registered it as the first "laboratory-on-a-chip" (LOC) device. During the 1980s, attention was focused on performing the photolithography process on substrates that included polymers and glass (Terry et al. 1979). With roots from Terry's work, Andreas Manz, an analytical chemist, presented the idea of using microchannels to be utilized in a "total analysis system" (μ-TAS) for chemical sensing. This μ-TAS was advantageous in terms of efficiency, fast separation, and reduced transport length. As a result, there was a reduction in reagent consumption compared to conventional chemical sensing methods (Manz et al. 1990, Harrison et al. 1993, Dittrich and Manz 2006).

The year 1994 marked the significant research outcomes of two important microfluidics applications. Firstly, there was the introduction of microfluidics technology for cell biology, biochemistry, biosensors based on cells, and protein/cell sorting or patterning (Hulme et al. 2007, McDonald et al. 1999, Folch and Toner 1998). Secondly, the US Defense Advanced Research Projects Agency (DARPA), worked on developing microelectromechanical systems (MEMs) and portable LOCs for the detection of biological and chemical warfare weapons.

Harvard University professor George Whitesides utilized siloxanes or organic polymers for the first time in 1998 and completely revolutionized the field of microfluidics. In his work, he proposed a material and named it poly(dimethylsiloxane), commonly known as PDMS nowadays, for rapid/cost-effective fabrication of microfluidics devices (Xia and Whitesides 1998).

The year 2000 was noted for the introduction by Richard B. Fair of digital microfluidics for precise control and manipulation of droplets inside the system. Another remarkable development that year was the debut of pneumatic microvalves by Stephen Quake, which were termed Quake microvalves. These microvalves utilized a slight deflection of a PDMS membrane to disrupt the fluid flow.

In the 21st century, tremendous work has been extensively carried out to mimic body organs to understand their functionalities, this was generally called "organ-on-a-chip." In 2007, George Whitesides introduced paper-based microfluidics devices for rapid prototypic and diagnostics (Hulme et al. 2007). In subsequent years, 3D printing was introduced for the fabrication of microfluidics reactors, and this opened doors for the rapid printing of organs-on-a-chip and studying their functionalities. Since then, microfluidics technology has been explored extensively for a large variety of biological/biomedical, diagnostic, energy, sensing, separation applications, etc. (Sinton 2014, Modestino et al. 2016, Capel et al. 2018, Hou et al. 2017, Dong et al. 2019, Zhong et al. 2019, Gong and Sinton 2017, Weng et al. 2019, Ramachandran et al. 2020, Yamada et al. 2017, Yakoh et al. 2019, Zhao et al. 2020, Chibh et al. 2021, Prakash et al. 2020, Katoch et al. 2021, Singh et al. 2018).

1.1.2 Importance of Microfluidics

- Efficient and reliable: The role of microfluidics for point-of-care (POC) devices is crucial, owing to the possibility of miniaturization that enforces a low Reynolds number, thus laminar flow. Consequently, there is a reduction in the required reagent volume and sample size. There is also a possibility of carrying out multiple analyses simultaneously because of the ease of containing various microchannels within a single device. The effectiveness of detection is achieved by tuning the channel design, architecture, and geometry. As a result, the processing time compared to conventional diagnostic methods reduces to few minutes from several hours.

- Portability and ease of handling: Often being smaller than palm size, microfluidics devices carry some inherent advantages, namely accessibility, portability, and ease of handling/usage. The simplicity of such devices also eradicates the requirement to hire trained professionals.
- Cost effectiveness: Besides being efficient and convenient, the cost of fabrication of microfluidics-based diagnostic devices is much lower compared to conventional diagnostic techniques. At an industrial production scale, the cost of the final product directly correlates to the raw materials and the process adopted for production. In conventional methods, the total cost of the above-mentioned aspects adds up to a huge amount. While in this context, polymeric materials like PDMS, polymethylmethacrylate (PMMA), polystyrene (PS), and polycarbonate (PC) are utilized for the production of devices that are not only cost effective but can also be easily processed. Paper is also one such cost-effective material that is being used for the fabrication of such microfluidics devices which is lightweight, disposable, and biocompatible.

1.1.3 Applications of Microfluidics

- Flow chemistry and reactions: Microfluidics is linked to the synthesis of materials, where the reactions governing the synthesis process are carried out inside a microchannel. This platform of synthesis of opens up opportunities for the industrial-scale production of materials.
- Testing and environment applications: The broad classification of testing applications of microfluidics-based devices is the inspection of water and air quality and to identify the presence of impurities present in them. Miniaturized and robust chemical sensing microfluidics devices have also been employed to monitor food quality.
- Analytical devices: These types of microfluidics devices are used to mimic bulky columns for chromatography and mass spectroscopy. Such microfluidics devices utilize relatively low sample concentrations and volumes and provide analytical results faster.
- Drug delivery: Microfluidics-based devices are also used in invasive drug delivery application in which microneedles, micropumps, and inhalers are used to precisely deliver small volumes of a drug at specific target sites. With advancements in microfluidics technology such drug delivery units are also being employed.
- Pharmaceutical research: In pharmaceutics microfluidics devices are used for discovery and screening of new drugs. Cell analysis is also performed within microchannels with high accuracy. Also, microfluidics devices used in synthesis and to study the structure of proteins and genes.
- Point-of-care diagnostic devices: Such microfluidics devices are used for the purpose of diagnostics at places away from the laboratory or locations where complex systems cannot be installed. POC diagnostic devices are a viable tool for testing at home or in remote areas without trained personnel.

1.2 SCALING LAWS AND GOVERNING EQUATIONS

In general, scaling means a reduction in the size of a system in an isomorphic way from all directions. Scaling or miniaturization of products has intensified to achieve the utmost level of sensitivity, robustness, and multi-functionality but lowering the overall cost at the same time. In general, these scaling laws are the relationship between dependent and independent variables. All the physical quantities are expressed in terms of x^a, where "a" is any real number.

In microfluidics devices, if L is the linear dimension, volume ($V = L^3$) and surface ($SA = L^2$), thus following the scaling law of system dimensions:

$$\frac{SA}{V} \propto L^{-1} \tag{1.2}$$

Here, S/V from Equation (1.2) is inversely proportional to the linear dimension, so the reduction in length increases the surface-to-volume ratio in a similar fashion. Hence, surface forces dominate over the body forces in a microfluidic system.

Also, while discussing scaling in fluid mechanics, two quantities, volumetric flow (Q) and pressure drop (ΔP) are considered to be important.

From Equation (1.1) and Figure 1.3, Q, the volumetric flow rate, is directly proportional to the fourth power of radius r.

$$Q \propto r^4 \tag{1.3}$$

This implies, if the radius is reduced by ten times, there is a reduction of volumetric flow by 10^4 times.

Again, from Equation (1.1)

$$\frac{\Delta P}{L} \propto r^{-3} \tag{1.4}$$

With the reduction of the radius by ten times, there is an increase in the pressure drop by a value of 10^3 times per unit length.

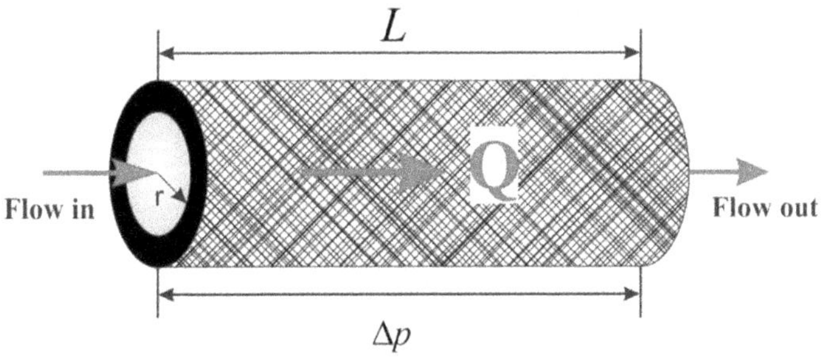

FIGURE 1.3 Fluid flow inside a cylindrical channel with a diameter "r" and length "*L*".

1.2.1 Correlation of Physical Quantities with Length Scale in Microfluidics

Scaling laws are related to the study of change in physical quantities with the size or dimension (*L*) of the system in consideration. Table 1.1 illustrates the relationship of characteristic dimension (*L*) with different physical quantities.

Similar to the scaling of physical quantities, the scaling of length (*L*) to dimensionless numbers in fluid flow is also studied. The Hagen–Poiseuille equation can also be written in velocity form as below:

$$\Delta P = u\frac{32\,\mu L}{D^2} \propto L^{-1} \tag{1.5}$$

$$u \propto \frac{\Delta P}{32\mu} L \propto L \tag{1.6}$$

Here, u is the velocity of the fluid, and D is the diameter of the channel.

1.2.2 Scaling of Dimensionless Numbers in Microfluidics with Length Scale (*L*)

With the understanding of dimensionless numbers in microfluidics, there is a reduction in the variables used to describe a given system, thus there is a decrease in the experimental data that is required in order to make correlations between the physical phenomena to those of the scalable systems.

1.2.2.1 Reynolds Number

A Reynolds number is a dimensionless quantity that helps in predicting the type of flow of a fluid. This concept was introduced in 1851 by George Stokes and popularized in 1883 by Osborne Reynolds, a British engineer, and physicist from the

TABLE 1.1
Relationship of Physical Quantities with Length Scale in Microfluidics

Quantity	Scaling with L
Electrostatic force	L^2
Pressure force	L^2
Capillary force	L^1
Flow speed	L
Diffusion time	L^2
Volume flow rate (with constant velocity)	L^2
Mass flow rate (with constant velocity)	L^2
Inertial force (with constant velocity)	L^2
Viscous force (with constant velocity)	L^1

University of Manchester. A Reynolds number is the ratio of internal forces to viscous forces when a fluid is subjected to relative internal momentum owing to the differences in fluid velocities of individual layers. The detailed derivation of a Reynolds number is deduced in the latter half of the chapter.

$$R_e = \frac{\text{Interial Force}}{\text{Viscous Force}} = \frac{\rho v D}{\mu} = \frac{uD}{v} \tag{1.7}$$

Here, ρ and v are the density and speed of the fluid, D is the tube diameter, μ and v are the dynamic and kinematic viscosities of the fluids.

Scaling a Reynolds number w.r.t. L is given as follows:

- When velocity (u) is constant

$$R_e = \frac{\rho v D}{\mu} \propto L \tag{1.8}$$

- When pressure difference (ΔP) is constant

$$R_e = \frac{\rho v D}{\mu} \propto L^2 \tag{1.9}$$

1.2.2.2 Knudsen Number

A Knudsen number is the ratio of mean free path (λ) and length (L) scale given by Danish physicist Martin Knudsen.

$$K_n = \frac{\text{Mean Free Path}}{\text{Length}} = \frac{\lambda}{L} \tag{1.10}$$

Scaling of a Knudsen number w.r.t. L is given as follows

- When velocity (u) is constant

$$K_n = \frac{\lambda}{L} \propto L^{-1} \tag{1.11}$$

- When pressure difference (ΔP) is constant

$$K_n = \frac{\lambda}{L} \propto L^{-1} \tag{1.12}$$

1.2.2.3 Weber Number

A Weber number was coined by the German naval mechanics professor, Moritz Weber. It is employed to understand the fluid flow behavior in a multiphase flow system.

$$W_e = \frac{\text{Interial Force}}{\text{Capillary Force}} = \frac{\rho u^2 l}{\sigma} \tag{1.13}$$

Here, ρ and u are the density and speed of the fluid respectively, l is the length, and σ is the surface tension.

Scaling of a Weber number w.r.t. L is given as follows:

- When velocity (u) is constant

$$W_e = \frac{\rho u^2 l}{\sigma} \propto L \tag{1.14}$$

- When pressure difference (ΔP) is constant

$$W_e = \frac{\rho u^2 l}{\sigma} \propto L^3 \tag{1.15}$$

1.2.2.4 Froude Number

A Froude number is the dimensionless number given by English engineer William Froude, and defines the ratio of inertial forces to the gravitational force (g).

$$F_r = \frac{\text{Interial force}}{\text{Gravitational Force}} = \frac{u}{gl} \tag{1.16}$$

Scaling of a Froude number w.r.t. L is given as follows:

- When velocity (u) is constant

$$F_r = \frac{u}{gl} \propto L^{-1} \tag{1.17}$$

- When pressure difference (ΔP) is constant

$$F_r = \frac{u}{gl} \propto L^{1} \tag{1.18}$$

1.2.2.5 Capillary Number

A capillary number is the ratio of viscous forces to the surface tension acting on the interface between the two phases.

$$C_a = \frac{\text{Viscous Force}}{\text{Capillary Force}} = \frac{\mu u}{\sigma} \tag{1.19}$$

Scaling of capillary number w.r.t. L is given as follows:

- When velocity (u) is constant

$$C_a = \frac{\mu u}{\sigma} \propto L^0 \tag{1.20}$$

- When pressure difference (ΔP) is constant

$$C_a = \frac{\mu u}{\sigma} \propto L^1 \tag{1.21}$$

1.2.2.6 Péclet Number

The Péclet number was named after French physicist Jean Claude Eugène Péclet and it is used to study the process of transport phenomena. The Péclet number is defined as the ratio of advection to the diffusive transport rates.

$$P_e = \frac{\text{Avdection Velocity}}{\text{Diffusive Velocity}} = \frac{u}{D/l} \tag{1.22}$$

Here, D is the mass diffusivity coefficient.

Scaling of the Péclet number w.r.t. L is given as follows:

- When velocity (u) is constant

$$P_e = \frac{u}{D/l} \propto L^1 \tag{1.23}$$

- When pressure difference (ΔP) is constant

$$P_e = \frac{u}{D/l} \propto L^2 \tag{1.24}$$

1.3 TYPES OF FLUID

Fluids can be categorized as ideal fluids, real fluids, Newtonian fluids, non-Newtonian fluids, and ideal plastic fluids.

- Ideal fluids: If a fluid cannot be compressed or there is no change in the fluid viscosity, it is termed as an ideal fluid. This type of fluid is an imaginary fluid and does not exist in reality.
- Real fluids: This is a broad name given to almost all the fluids that exist so far. In short, these are the fluids that possess some viscosity.
- Newtonian fluids: Fluids obeying Newton's law of viscosity are termed as Newtonian fluids. Here, Newton's law states that stress is directly proportional to strain as given in equation (1.25) and depicted in Figure 1.4. Air, water, a large variety of alcohols, etc. are a few examples of Newtonian fluids.

$$\tau = \mu \frac{\partial u}{\partial x} \tag{1.25}$$

- Non-Newtonian fluids: Fluids that do not obey Newton's law of viscosity are termed as non-Newtonian fluids. Dilatant and pseudoplastic are the two types of non-Newtonian fluids in which the relationship between shear rate

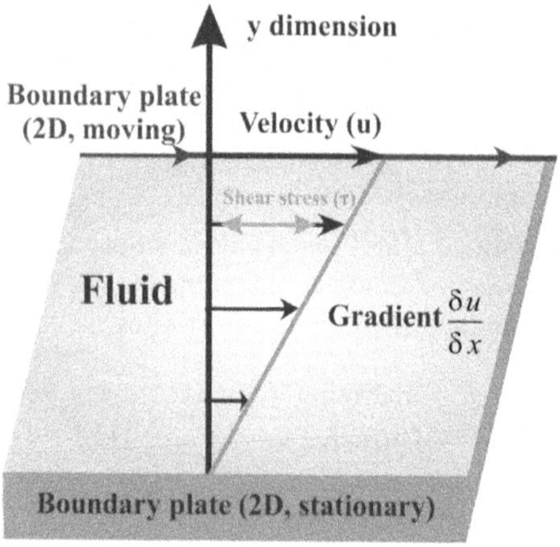

FIGURE 1.4 Velocity profile of the fluid between two parallel plates.

and viscosity is time independent. An example of dilatants is a mixture of sand and water whereas an example of pseudoplastic is blood. The other class of non-Newtonian fluid is the Herschel–Bulkley fluid. Such fluids are materials with both shear thickening and shear thinning properties. The practical examples of such materials are greases, colloidal suspensions, starch pastes, etc.

- Ideal plastic fluids: Ideal plastic fluids or Bingham plastics are fluids that act as a rigid body under low stress conditions and viscous fluids under high stress conditions. One very popular example is toothpaste, in which the pressure has to be applied in order to extrude it from the tube.

The relationship of viscosity with the shear rate for Newtonian and non-Newtonian fluids is given in Figure 1.5.

1.4 TYPES OF FLUID FLOW

While talking about the types of fluid flow, the concept of Reynolds number (R_e) is vital for a clear understanding. Figure 1.6 signifies the velocity profile adopted by the individual layers flowing one over another.

$$\text{Reynolds Number}(R_e) = \frac{\text{Inertial Forces}}{\text{Viscous forces}} \tag{1.26}$$

$$(R_e) = \frac{\rho V \frac{dV}{dx}}{\mu \frac{d^2V}{dx^2}} \tag{1.27}$$

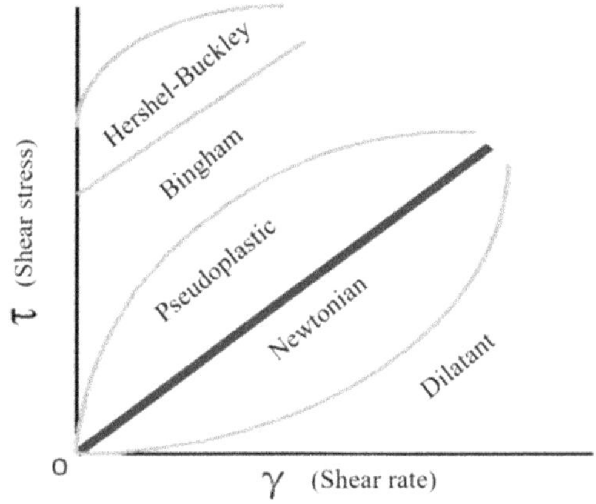

FIGURE 1.5 Relationship of viscosity with the shear rate for Newtonian and non-Newtonian fluids (Wang et al. 2018).

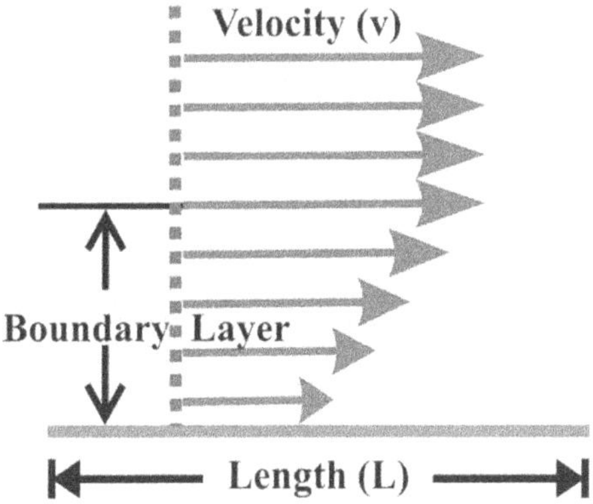

FIGURE 1.6 Velocity of boundary layers in the flow regime on the a plate.

$$(R_e) = \frac{\rho V \frac{V}{dL}}{\mu \frac{V}{L^2}} \tag{1.28}$$

$$(R_e) = \frac{\rho V L}{\mu} \tag{1.29}$$

$$\text{Kinematic Viscocity}(v) = \frac{\mu}{\rho} \tag{1.30}$$

$$(R_e) = \frac{V L}{v} \tag{1.31}$$

Here, ρ and v are the density and velocity of the fluid respectively, μ is the dynamic viscosity of the fluid.

Thus, from the definition of R_e Table 1.2 depicts the significance of R_e with the nature of the fluid flow.

1.5 ROLE OF MECHANICAL PARAMETERS IN THE FLUID FLOW

1.5.1 Shear

Shear forces are termed as forces acting tangentially to the surface of the solid that cause deformation. Solids can resist the deformation, whereas liquids tend to deform and flow by the action of the force. So, when fluid is in a state of motion, shear stress develops owing to the relative motion of fluid layers. Shear force is generated because of the velocity difference at the center and walls of the channel, as depicted in Figure 1.8.

TABLE 1.2
Relationship of R_e Values with the Nature of the Fluid Flow

R_e value	Type of Flow	Observations
$R_e < 2300$	Laminar flow (Figure 1.7(a)) (Slow and predictable mixing)	1. Viscous forces dominate 2. Observed at low flow rates
$2300 < R_e > 4000$	Transitional/transient Flow (Figure 1.7(b))	Generally turbulent at the center of the channel and laminar near the surface of the channel
$R_e > 4000$	Turbulent flow (Figure 1.7(c)) (Rapid or unpredictable mixing)	1. Inertial forces dominate 2. Observed at high flow rates

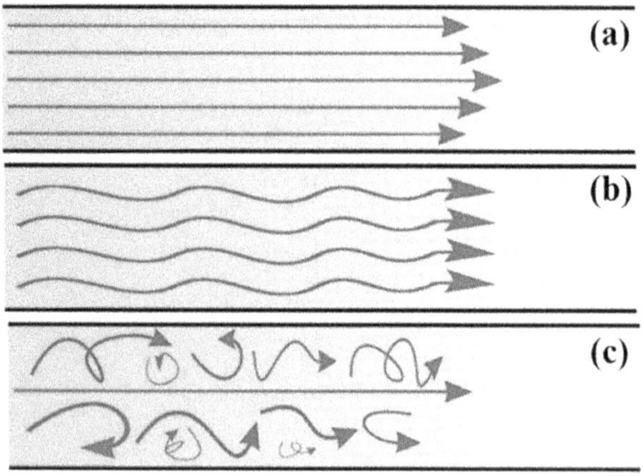

FIGURE 1.7 Flow profile in (a) laminar, (b) transient, and (c) turbulent flow regime.

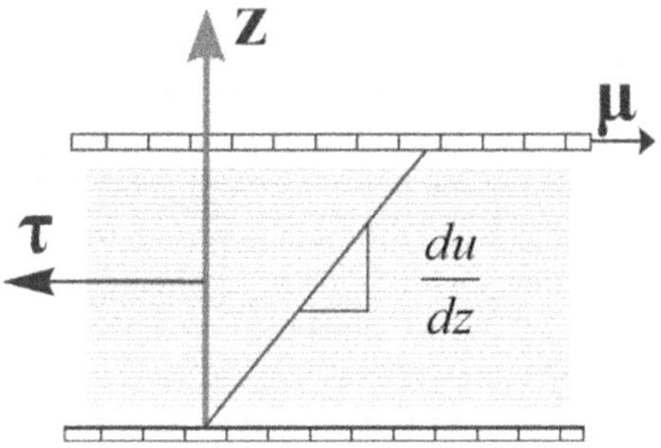

FIGURE 1.8 Relationship of shear stress between the individual layers of fluid and velocity gradient.

$$\tau = \mu \frac{du}{dz} \tag{1.32}$$

Here, τ and μ are the shear stress and viscosity of the fluid respectively and *du*/*dz* is the velocity gradient between the layers normal to the direction of flow.

1.5.2 Viscosity

Viscosity in terms of fluid flow is the resistance that the fluid exerts against shear or deformation. In simpler terms, it is the friction exerted by the fluid to resist its flow. In other words, viscous forces are proportional to the rate of change of fluid velocity within a constant area.

1.5.2.1 Absolute Viscosity

It is the measure of the internal resistance. This form of viscosity is the tangential force acting per unit area, which is required to move a horizontal plane w.r.t. another plane.

By rearranging the above Equation (1.32), dynamic viscosity μ is expressed as

$$\mu = \tau \frac{dz}{du} \tag{1.33}$$

1.5.2.2 Kinematic Viscosity

It is the ratio of dynamic viscosity to the density of the fluid. It is calculated by dividing the absolute viscosity values of a liquid by the density of the fluid.

$$\gamma = \frac{\mu}{\rho} \tag{1.34}$$

Here, γ and μ are the kinematic viscosity and the density of the fluid.

1.5.3 Surface Tension

Surface tension is defined as the tendency of any liquid surface to shrink into a minimum possible surface area. The force of cohesion between the neighboring liquid molecules is responsible for the surface tension phenomenon to occur. The liquid molecules on the surface behave in a different way than that of the molecules underneath because they are in contact with different neighboring sites like the air or other phases, as shown in Figure 1.9. When this force acts at the interface of two different liquids or gases, it is termed as interfacial tension. A water droplet is one excellent example of surface tension acting in three-dimensional space.

1.6 INTERFACE, SURFACE TENSION, AND CAPILLARY ACTION

At the liquid–air interface, the process of surface tension comes into play in order to minimize the surface energy, thus forming a droplet. There is an occurrence of capillary action when the surface adhesive forces of capillary and liquid are greater than the forces of cohesion between liquid molecules. Thus, the height attained by the capillary action is strongly influenced by factors such as gravity and surface tension.

1.6.1 Laplace's Law

This law was put forward by a French mathematician called Pierre-Simon Laplace and his co-workers describing the relationship between internal pressure and the radius of the vessel through which the fluid is flowing. In other words, the pressure change has a relationship with the curvature of the surface, as illustrated in Figure 1.10. This law has been widely used to understand the physiological attributes of the heart and alveoli in medical science.

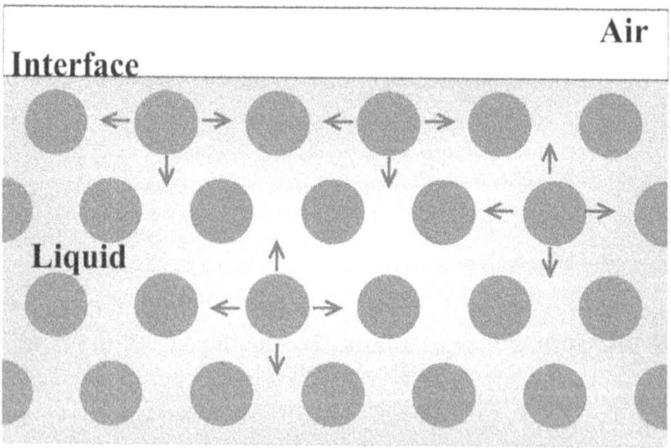

FIGURE 1.9 Force exerted by the liquid molecules within a liquid and at the liquid-air interface.

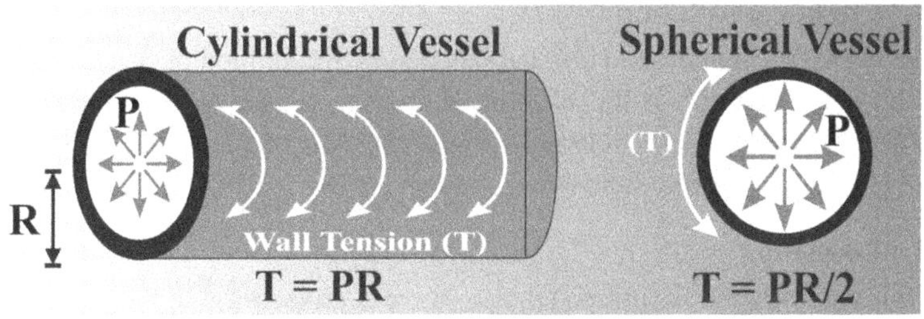

FIGURE 1.10 Relationship of wall tension (T) with the radius of the vessel and pressure exerted by the liquid on the wall of the container.

1.6.2 Measurement of Surface Tension

The measurement of surface tension effects is usually performed at the interface or liquid–air, liquid–solid, or liquid–liquid phases. Taking into account the surface tension acting on the water in a container, it comprises of two different forces. One is the surface interactions between the air and another with the walls of the container. There are several methods to measure the different forms of forces of interaction between the surfaces.

The traditional methods include the du Noüy ring method, which is the measure of the force applied to lift a platinum ring from the surface of a liquid. Firstly, the ring is submerged in the liquid and then raised. The lifting force is then the measure of the surface tension of the liquid. This method of measurement of surface tension in liquids only came to light in 1925 by the efforts of Pierre Lecomte du Noüy, a French biophysicist.

1.6.3 Parameters Affecting Surface Tension

The surface tension properties of any material usually remain stable. Still, factors such as the variation in temperature, chemical modification of bonds, presence of impurities, and extent of oxidation are the parameters that bring about the difference in surface tension.

1.6.3.1 Temperature

Molecules tend to thermally vibrate with an increase in temperature. As a result, there is a decrease in surface tension. The value of surface tension becomes zero at the critical temperature of the liquid. The Eötvös rule is an empirical equation that signifies the relationship of temperature and surface tension, given as following:

$$\gamma V^{\frac{2}{3}} = K\left(T_C - T\right) \tag{1.35}$$

Here, V signifies the molar volume, critical temperature (T_c), Eötvös constant (K) ($K = 2.1 \times 10^{-7}$ J K^{-1} $mol^{2/3}$).

1.6.3.2 Chemical Addition

A change in surface tension can also happen due to the addition of certain chemicals. One such typical example is the addition of soap to water that allows easy removal of dirt from the substrate when treated. Here soap acts as a surfactant which brings about a reduction in surface tension.

1.6.3.3 Oxidation

Firstly, with an increase in surface tension, there is an increase in the intermolecular forces. Oxygen in the atmosphere reacts with the surface impurities and brings about a reduction in surface tension. Thus, if the impurities are highly soluble in water, then this would not bring about much change in the surface tension.

1.6.4 Contact Angle, Drop Thickness, and Wettability

In the presence of an interface between a solid and a liquid, there is an angle between the liquid and solid surface which is termed as the contact angle (θ), as shown in Figure 1.11. This also determines the balance of forces of interaction and repulsion at the point of interface. The forces of interaction are generally cohesive or adhesive forces which are categorized broadly under intermolecular forces. The forces between similar molecules are cohesive in nature, such as hydrogen bonding or van der Waals forces. While the balance between the dissimilar molecules is because of the adhesive nature of forces such as mechanical or electrostatic forces. These forces, either attractive or repulsive, play a crucial role in the determination of the contact angle as depicted in (Table 1.3).

1.6.4.1 Thermodynamics and Force Balance

Thomas Young, a British polymath, proposed that there is a mechanical equilibrium when a drop of liquid is resting on a solid surface. The surface tension will

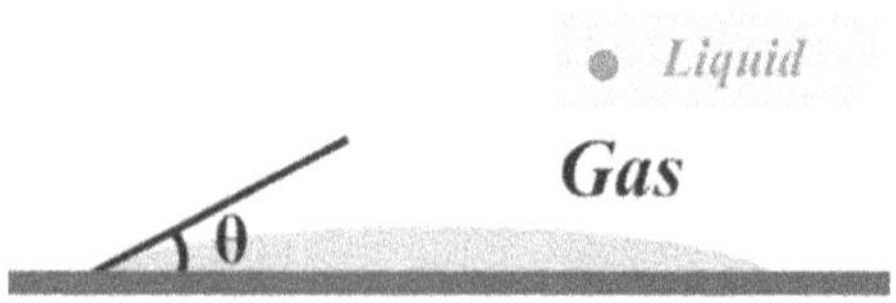

FIGURE 1.11 Contact angle traced between solid–liquid interfaces.

TABLE 1.3
Dependence of Contact Angle with the Forces of Attraction and Repulsion

For smaller contact angle	Adhesive forces stronger than cohesive forces	The interaction of liquid-solid molecules is more significant than liquid–liquid molecules
For larger contact angle	Cohesive forces stronger than adhesive forces	The interaction of liquid–liquid molecules is more significant than liquid–solid molecules

primarily have three components acting on the interface, as depicted by the arrows in Figure 1.12:

1. The surface tension at the liquid–vapor interface (γ_{lv})
2. The surface tension at the solid–liquid interface (γ_{sl})
3. The surface tension at the solid–vapor interface (γ_{sv})

To bring about equilibrium within the droplet, there is an outward pull of each force from the point of equilibrium, defined by Young's Equation (1.36) as follows:

$$\gamma_{sv} = \gamma_{ls} + \gamma_{lv} \cos\theta \tag{1.36}$$

Here, θ is the contact angle.

Rewriting the above equation

$$\cos\theta = \frac{\gamma_{sv} - \gamma_{ls}}{\gamma_{lv}} \tag{1.37}$$

From Table 1.4, the higher the surface energy, the stronger are the attractive forces that pull the droplet down and cause the spread of the liquid. This concept is known as wettability. In other words, wettability is the ability of the liquid to maintain a balance between the intermolecular forces of attraction and repulsion, thus being in contact with the surface of the solid.

The results of surface energy dominating the surface tension within the liquid molecule forces the liquid to attain a shape of a droplet. Since the surface tension of the liquid is a constant quantity, if the value of surface energy dominates the surface tension value, the liquid will spread onto the surface of the solid. Thus, the higher the values of surface energy and interfacial tension, the lower the contact angle. This implies that the surface will be wetting.

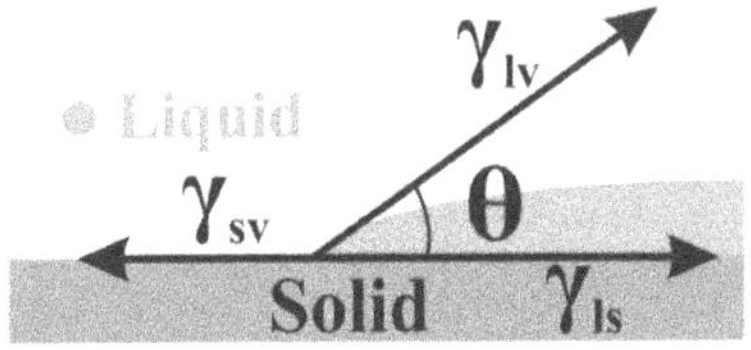

FIGURE 1.12 Contact angle forces witnessed at the interface of solid, liquid, and gas.

TABLE 1.4
Role of Interfacial Forces on the Wettability of the Surface

$\gamma_{sv} < \gamma_{sl}$	$\theta < 0$ (cos θ is negative)	Droplet wets the surface
$\gamma_{sv} > \gamma_{sl}$	$\theta > 0$ (cos θ is positive)	Droplet does not wet the surface

However, for a low wetting surface, the values of surface energy are smaller than the surface tension of the liquids, thus the liquid will be in droplet form rather than spreading. This also correlates that there is a low interfacial surface tension between the solid surface and the liquid droplet. Thus, the higher the contact angle values, the lower the surface energy and surface tension.

1.6.5 Nature-Inspired Phenomenon

The close relationship of the concepts of contact angle and wettability are nature-inspired superhydrophilic surfaces, as illustrated in Figure 1.13. A few examples derived from nature are the lotus leaf, the surface of nepenthes pitcher plants, butterfly wings, the legs of water striders (commonly known as Jesus bugs), etc. All these surfaces have a self-cleaning ability in which the surfaces are cleaned by the rolling action of the water droplets on the surface. The self-cleaning ability of such surfaces is possible due to the phenomenon of superhydrophobicity. Surfaces with a self-cleaning ability have a contact angle greater than 150°. This means that there is no wetting on the solid surface.

Surface roughness affects the contact angle, thus bringing about a change in the wettability of the surface. Therefore, the following models were developed to understand the wettability of the surfaces.

1.6.5.1 Young's Model

Young and co-workers proposed a model that describes the relationship between a perfectly flat solid surface and a water droplet. This model typically correlates the contact angle that the water droplets make with the flat surface, and the surface energies of the solid, liquid, and gas are given in Equation (1.37) and Figure 1.14(a).

1.6.5.2 Wenzel's Model

This model comes into play to describe a water droplet placed on a rough surface. Since the water droplet is in contact with a surface with some topological features,

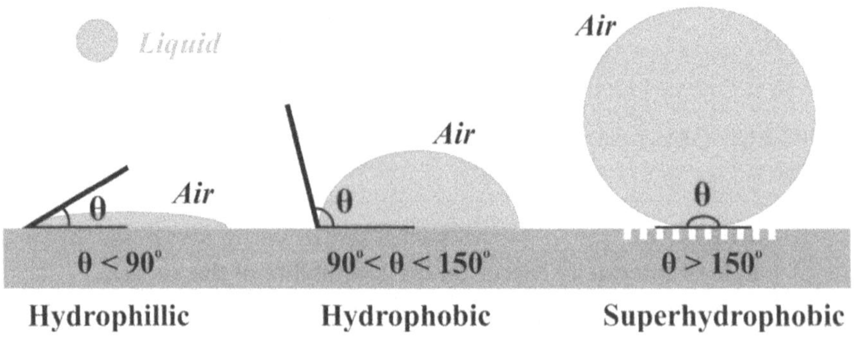

FIGURE 1.13 Extent of surface wettability as a function of contact angle.

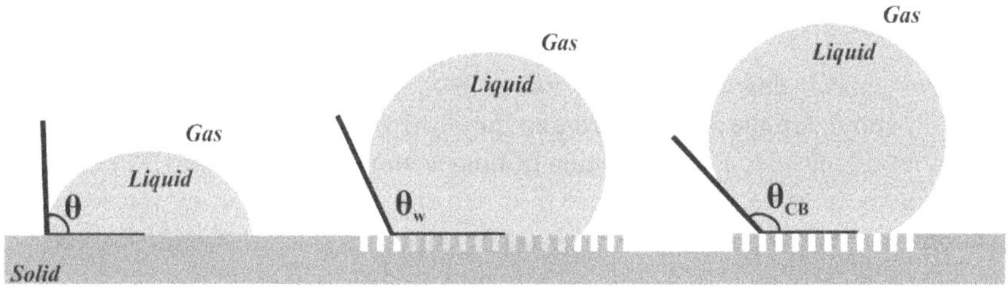

FIGURE 1.14 Scheme representing the (a) Young, (b) Wenzel, and (c) Cassie–Baxter models for the wettability of a surface.

the surface area curved by the droplet is more significant compared to the water droplet placed onto the flat surface, as represented in Figure 1.14(b).

$$\cos\theta = R_f\left(\theta_w\right) \tag{1.38}$$

Here, θ_w is the contact angle deduced by the Wenzel model, θ is the contact angle deduced from Young's model. R_f denotes the factor of surface roughness.

1.6.5.3 Cassie–Baxter model

This model is considered when there is a creation of air pockets by water droplets placed on a solid surface. Thus, the water droplet is in contact with both air pockets and the solid surface as represented in Figure 1.14(c). The equation that describes this model is as follows:

$$\cos\theta_{CB} = R_f\cos\theta_o - f_{LA}\left\{\left(R_f\cos\theta_o\right)+1\right\} \tag{1.39}$$

Again, θ_{CB} is the contact angle deduced by the Cassie–Baxter model, and f_{LA} is the fraction of liquid and air in contact with each other.

1.7 NEWTON'S SECOND LAW VS THE NAVIER–STOKES EQUATION

The Navier–Stokes equation in fluid mechanics is named after a French mechanical engineer, Claude-Louis Navier, and Irish mathematician George Gabriel Stokes, describing the motion of an incompressible fluid in the partial differential equation form.

In a simplified way, the sum of forces in X, Y, and Z directions are equal to the product of mass and acceleration given by the following equations:

$$F_x = Ma_x \tag{1.40}$$

$$F_y = Ma_y \tag{1.41}$$

$$F_z = Ma_z \tag{1.42}$$

Here, F_x, F_y, and F_z denote the sum of forces in X, Y, and Z directions. M is the mass and a_x, a_y, and a_z are the acceleration and the X, Y, and Z directions.

Now, let us consider a small portion of fluid as depicted in Figure 1.15 below.

The forces in three directions act in a per unit volume given by the following equation:

$$F_g = \frac{mg}{V} = \rho g \tag{1.43}$$

F_g is the force due to gravity, m, g, and V are the mass, acceleration due to gravity, and volume, respectively.

Similarly, from Equations (1.40), (1.41), and (1.42),

$$\frac{F}{V} = \frac{Ma}{V} = \rho a \tag{1.44}$$

Let us now apply mathematics for X direction, and we can extend it to Y and Z directions in a similar way.

The X component of velocity and acceleration is denoted by u and a respectively. Here u is the function of $u(x, y, z, \text{and } t)$ and a_x will be given by

$$a_x = \frac{du}{dt} = \frac{\partial u}{\partial t} + \frac{\partial u}{\partial x}\frac{\partial x}{\partial t} + \frac{\partial u}{\partial y}\frac{\partial x}{\partial t} + \frac{\partial u}{\partial z}\frac{\partial z}{\partial t} \tag{1.45}$$

Simplifying Equation (1.45)

$$a_x = \frac{du}{dt} = \frac{\partial u}{\partial t} + u\frac{\partial u}{\partial x} + v\frac{\partial u}{\partial y} + w\frac{\partial u}{\partial z} \tag{1.46}$$

Here $\left(\frac{\partial u}{\partial t}\right)$ is called the local acceleration and $\left(u\frac{\partial u}{\partial x} + v\frac{\partial u}{\partial y} + w\frac{\partial u}{\partial z}\right)$ is called the convective acceleration.

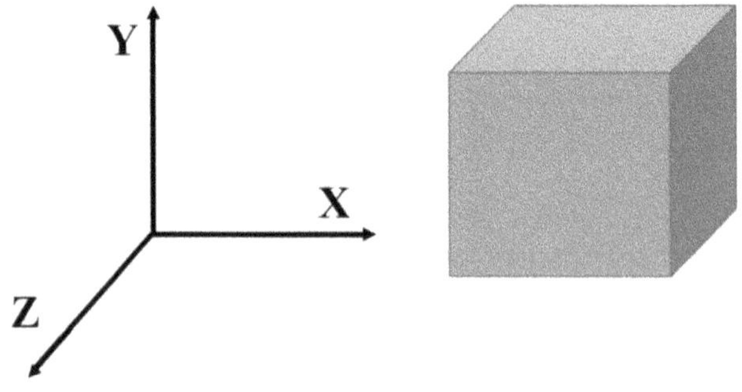

FIGURE 1.15 Small portion of the fluid in 3D space.

Here, *dx*, *dy*, and *dz* are the differential elements along X, Y, and Z directions.

Here, σ denotes stress on the left and right of the cube, T indicates shear stress on the top/bottom or front and back of the cube, as shown in Figure 1.16.

$$\rho g_x dxdydx + \sigma_{xx}(x+dx)dydz - \sigma_{xx}(x)\,dydz + \tau_{yx}(y+dy)dxdz - \tau_{yx}(y)dxdz + \tau_{zx}(z+dz)\,dxdy - \tau_{zx}(z)\,dxdy = \rho a_x \tag{1.47}$$

On dividing Equation (1.47) by volume *dxdydz* and simplifying

$$\rho g_x + \frac{\sigma_{xx}(x+dx) - \sigma_{xx}(x)}{dx} + \frac{\tau_{yx}(y+dy) - \tau_{yx}(y)}{dy} + \frac{\tau_{zx}(z+dz) - \tau_{zx}(z)}{dz} = \rho a_x \tag{1.48}$$

On further simplifying Equation (1.48)

$$\rho g_x + \frac{\partial \sigma_{xx}}{\partial x} + \frac{\partial \tau_{yx}}{\partial y} + \frac{\partial \tau_{zx}}{\partial z} = \rho a_x \tag{1.49}$$

Combining equation across the equality

$$\rho g_x + \frac{\partial \sigma_{xx}}{\partial x} + \frac{\partial \tau_{yx}}{\partial y} + \frac{\partial \tau_{zx}}{\partial z} = \rho\left(\frac{\partial u}{\partial t} + u\frac{\partial u}{\partial x} + v\frac{\partial u}{\partial y} + w\frac{\partial u}{\partial z}\right) \tag{1.50}$$

Similarly on the Y and Z directions Equation (1.50) can be written as

$$\rho g_y + \frac{\partial \tau_{xy}}{\partial x} + \frac{\partial \sigma_{yy}}{\partial y} + \frac{\partial \tau_{zy}}{\partial z} = \rho\left(\frac{\partial v}{\partial t} + u\frac{\partial v}{\partial x} + v\frac{\partial v}{\partial y} + w\frac{\partial v}{\partial z}\right) \tag{1.51}$$

$$\rho g_y + \frac{\partial \tau_{xz}}{\partial x} + \frac{\partial \tau_{yz}}{\partial y} + \frac{\partial \sigma_{zz}}{\partial z} = \rho\left(\frac{\partial w}{\partial t} + u\frac{\partial w}{\partial x} + v\frac{\partial w}{\partial y} + w\frac{\partial w}{\partial z}\right) \tag{1.52}$$

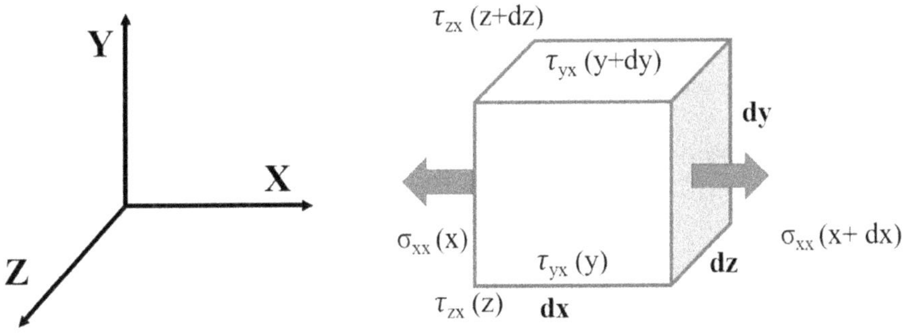

FIGURE 1.16 Stress acting on a small portion of fluid in all directions.

Here, both normal and shear stress are related to the viscosity and velocity profiles.

$$\sigma_{xx} = -p + 2\mu \frac{\partial u}{\partial x}$$

$$\sigma_{yy} = -p + 2\mu \frac{\partial v}{\partial y}$$

$$\sigma_{zz} = -p + 2\mu \frac{\partial w}{\partial z}$$

$$\tau_{xy} = \tau_{yx} = \mu \left(\frac{\partial u}{\partial y} + \frac{\partial v}{\partial x} \right)$$

$$\tau_{yz} = \tau_{zy} = \mu \left(\frac{\partial v}{\partial z} + \frac{\partial w}{\partial y} \right)$$

$$\tau_{zx} = \tau_{xz} = \mu \left(\frac{\partial w}{\partial x} + \frac{\partial u}{\partial z} \right)$$

On substituting the values of σ_{xx}, σ_{yy}, σ_{zz}, τ_{xy}, τ_{yx}, τ_{yz}, τ_{zy}, τ_{zx}, and τ_{xz} in Equations (1.50), (1.51), and (1.52) and solving

$$\rho g_x - \frac{\partial P}{\partial x} + \mu \left(\frac{\partial^2 u}{\partial x^2} + \frac{\partial^2 u}{\partial y^2} + \frac{\partial^2 u}{\partial z^2} \right) + u \frac{\partial}{\partial x} \left(\frac{\partial u}{\partial x} + \frac{\partial v}{\partial y} + \frac{\partial w}{\partial z} \right) = \rho a_x \tag{1.53}$$

From Equation (1.53) the term $\left(\frac{\partial u}{\partial x} + \frac{\partial v}{\partial y} + \frac{\partial w}{\partial z} \right) = 0$, as per the continuity equation for an incompressible fluid, therefore Equation (1.53) becomes

$$\rho g_x - \frac{\partial P}{\partial x} + \mu \left(\frac{\partial^2 u}{\partial x^2} + \frac{\partial^2 u}{\partial y^2} + \frac{\partial^2 u}{\partial z^2} \right) = \rho \left(\frac{\partial u}{\partial t} + u \frac{\partial u}{\partial x} + v \frac{\partial u}{\partial y} + w \frac{\partial u}{\partial z} \right) \tag{1.54}$$

Equation (1.54) is the Navier–Stokes equation applied in X direction

Similarly for Y and Z directions:

$$\rho g_y - \frac{\partial P}{\partial y} + \mu \left(\frac{\partial^2 v}{\partial x^2} + \frac{\partial^2 v}{\partial y^2} + \frac{\partial^2 v}{\partial z^2} \right) = \rho \left(\frac{\partial v}{\partial t} + u \frac{\partial v}{\partial x} + v \frac{\partial v}{\partial y} + w \frac{\partial v}{\partial z} \right) \tag{1.55}$$

$$g_z - \frac{\partial P}{\partial z} + \mu \left(\frac{\partial^2 w}{\partial x^2} + \frac{\partial^2 w}{\partial y^2} + \frac{\partial^2 w}{\partial z^2} \right) = \rho \left(\frac{\partial w}{\partial t} + u \frac{\partial w}{\partial x} + v \frac{\partial w}{\partial y} + w \frac{\partial w}{\partial z} \right) \tag{1.56}$$

∴Equations (1.54), (1.55), and (1.56) give the expression for Navier–Stokes equations.

1.8 MIXING INSIDE A MICROCHANNEL

Mixing in microscale involves miniaturized devices fabricated by precision engineering and micro–nano fabrication technology that promote the mixing of at least two or more different phases that can either be liquid or gas. The channel dimensions are in the sub-millimeter order, where the channel width is in the range of 100–500 μm, length could be in the order of few millimeters or even more, and the channel height is less than or comparable to that of the channel width. In volumetric terms, it implies that the volume of phases interacting with each other is in the order of microliters (μL) to milliliters (mL).

1.8.1 Mechanism of Mixing in Macroscale and Microscale

Mixing at these scales can be categorized into following three different types:

1.8.1.1 Macromixing

This type of mixing takes place in the largest possible dimensional scale for fluid in motion. One typical example of this can be mixing taking place equivalent to that of the diameter of a chemical reactor or batch reactor tanks. The blending or mixing time is a crucial parameter in macromixing phenomena.

1.8.1.2 Mesomixing

This mixing is profound when the dimensional scale is smaller than the batch reactors, but dimensions are higher than the one used for mixing in microscale. This type of mixing is prevalent in feed pipes connecting various batch reactors.

1.8.1.3 Micromixing

Owing to mixing in the smallest scale of motion, termed as the Kolmogorov scale, viscous and molecular diffusions play a vital role in the mixing process. Since reaction takes place in micron scale or molecular level, there is an increase in the available interfacial area for diffusion. Thus, this type of mixing brings about faster reaction kinetics and the overall reaction rates are accelerated compared with the other two mixing types.

1.8.2 Types of Mixing: Passive and Active Mixing

Passive mixing is a type of mixing that takes place without the utilization of external energy and does not require any actuators or moving parts. The concept of passive mixing is entirely based on chaotic advection and molecular diffusion. Chaotic advection is favorable in large molecules with small diffusion coefficients, whereas the molecular diffusion can be subcategorized, further taking into account the arrangement of mixed phases. These arrangements are serial segmentation, injection, parallel, and serial lamination.

Active mixing is a type of mixing that utilizes external energy to initiate and improve mixing. Active mixing can be subcategorized based on a physical phenomenon to bring about disturbance to start mixing. Electrohydrodynamics and electrokinetics, magnetohydrodynamics, pressure-driven flow, acoustics, and application of heat are a few examples that assist in the mixing. Such external energy sources often render complexity, challenge, and expense while designing and fabricating an active mixing system.

From a broader perspective, because of the absence of any external actuators or energy sources, the passive mixers or mixing techniques are generally stable to operate, robust, less expensive, and can be easily integrated with the system.

1.8.3 Brownian Motion, Taylor Dispersion, and Chaotic Advection

The process of diffusive transport can be explained by the help of Brownian motion, Taylor dispersion, and Chaotic Advection as discussed in the sections below.

1.8.3.1 Brownian Motion or Diffusive Transport

The process of diffusive transport takes place because of the random motion of molecules, which is often called as Brownian motion. The relationship of diffusion flux, the concentration of species, and the diffusion coefficient are called Fick's first law given by the Equation (1.57). Depending on the nature of material (solid, liquid or gas) there is variation in diffusion coefficients as illustrated in Figure 1.17.

$$j_{diff} = -D\frac{dc}{dx} \tag{1.57}$$

Here j_{diff} is the diffusion flux, c signifies the concentration of species (kg m^{-3}), and D means diffusion coefficient (m^2 s^{-1}) ()

1.8.3.2 Taylor Dispersion

The process of efficient mixing is carried out by the flow under the action of pressure inside the microchannel, giving rise to velocity profile. The coupling of molecular diffusion and the velocity profile in the transverse direction gives rise to an axial effect. Because of the axial convective transport process, the solvent obeys a parabolic velocity profile in which it stretches at the channel center compared to the channel walls (Nguyen 2012).

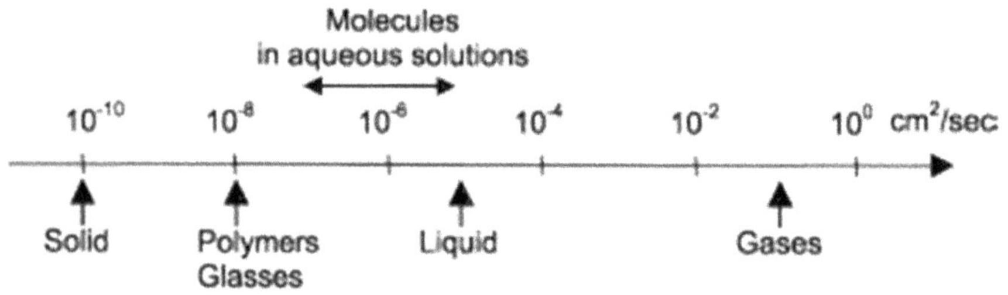

FIGURE 1.17 Range of diffusion coefficients in solids, liquids, and gases (Nguyen 2007).

Thus, owing to the concertation gradient developed between the adjacent layers, the rate of diffusion is much faster in the axial direction as compared to molecular diffusion, as depicted in Figure 1.18. Moreover, there is a relationship between the axial dispersion coefficient of different channel geometries with the Péclet number, diameter, and aspect ratio of channel as described in Table 1.5.

1.8.3.3 Chaotic Advection

When the transverse flow brings about the process of folding, breaking up, and stretching, it is termed chaotic advection. The term advection signifies the transport of species by the flow. Moreover, in a laminar flow regime, even the flow can be chaotic as a result of advection cycles in which the fluid flows through repeated structures. Repeating these advection cycles leads to complete mixing in the system.

1.8.4 Diffusion: Molecular Diffusion, Eddy Diffusion, and Bulk Diffusion

After the completion of the diffusion process there is a concentration equilibrium because of random migration of the elements in a given system. In a gas phase, there

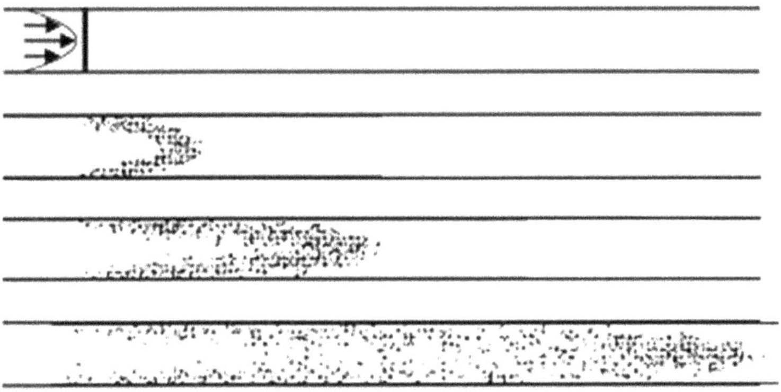

FIGURE 1.18 Particle in a parabolic velocity profile distribution inside a microchannel (Nguyen 2012).

TABLE 1.5
Axial Dispersion Coefficients for Cylindrical Capillary and Rectangular Channel

Cylindrical capillary	$D^* = D\left[1+\frac{1}{48}P_e\left(d_c\right)^2\right]$	$P_e\ (d_c)$ is the Péclet number related to the diameter (d_c)
Rectangular channel	$D^* = D\left[1+\frac{1}{210}P_e\left(W\right)^2 f\left(\frac{H}{W}\right)\right]$	$P_e\ (W)$ is the Péclet number related to channel width $f(H/W)$. Here, H/W is the aspect ratio of channel height and width

D and D^* are the molecular diffusion coefficient and axial dispersion coefficient is a temperature dependency with the mean velocity of molecules and motion of molecules is chaotic in nature. The molecules even collide with one another and tend to change the direction of motion. In the liquids, the process of molecular diffusion occurs in which the molecules jump from one position the another when external energy is applied. This energy should be enough to break the bonds with adjacent molecules allowing the molecule to move. Primarily, there are three types of diffusion, namely molecular diffusion, eddy diffusion, and bulk diffusion; these are discussed in this section.

1.8.4.1 Molecular Diffusion

The thermal motion of liquid or gas particles above absolute zero temperatures is called molecular diffusion. The extent to which the diffusion takes place correlates to the viscosity of the medium, temperatures, and the particle size. The process of self-diffusion directs the process of molecular diffusion as a result of the random motion of the molecules. The diffusion process results in uniform mixing of the medium with equal distribution of units throughout.

1.8.4.2 Eddy Diffusion

An eddy or swirling motion causes the process of diffusion, called eddy diffusion. There is a loss of kinetic energy in the form of heat which results in a decrease in the size of the eddies. Because of the onset of turbulent flow, eddy currents are witnessed giving rise to eddy diffusion.

1.8.4.3 Bulk Diffusion

The diffusion that takes place within a crystalline lattice at an atomic level is called bulk diffusion. Within a crystal lattice, the process of diffusion is governed by either interstitial or substitutional processes. Interstitial diffusion refers to the process of diffusion in which a diffusant diffuses within the crystal lattice of crystalline material. Whereas substitutional diffusion refers to the process in which there is a replacement of the host atom by the diffusant.

1.8.5 Role of Channel Architecture and Physical Forces

The challenge in the application of microfluidic devices in carrying out efficient mixing is because of the low value of the Reynolds number. This imparts the flow to be laminar, and the viscous forces dominate the inertial forces. The dominant effect of viscous forces results in reduction of irregularities in the stream that promotes the mixing of the fluid. Thus, diffusive mixing is one parameter that governs mixing in microfluidics devices compared to the turbulent mixing that is prevalent in bulk or macroscale mixing.

1.8.5.1 Split and Recombine

The first step in this process is bringing together the two streams in a single microchannel. As a result of diffusion at the molecular level, diffusion takes place at the interface of fluids. The process of molecular diffusion is assisted by Brownian motion. In general, when diffusion takes place at the interface of the two liquids, the rate of diffusion is much higher across the interface. Many research studies suggest a spilt-and-recombine method to bring about efficient mixing (Ebnereza et al. 2020, Pascal et al. 2018, Husain et al. 2018, Shakhawat et al. 2017). You. et al. studied the process of mixing in Y-shaped microreactors with channel dimensions of 150 μm. At a flow rate of 2 mL/min and $R_e = 221$, there is a formation of an interface of Janus Green B dye and water instead of mixing. At a flow rate of 10 mL/min and R_e approximately equal to 1106, mixing was observed as a function of channel length. Mixing at the flow rates of 40 mL/min was instantaneous at the Y junction itself. At low values of the Péclet number of the fluid flowing inside the microchannel, the rate of diffusion slows down as shown in Figure 1.19 below and often requires large channel lengths to complete the process of mixing (You et al. 2015).

Mixing in such streams can be enhanced by the addition of a split-and-combine process post initial mixing of the two streams. Figure 1.20 shows the channel

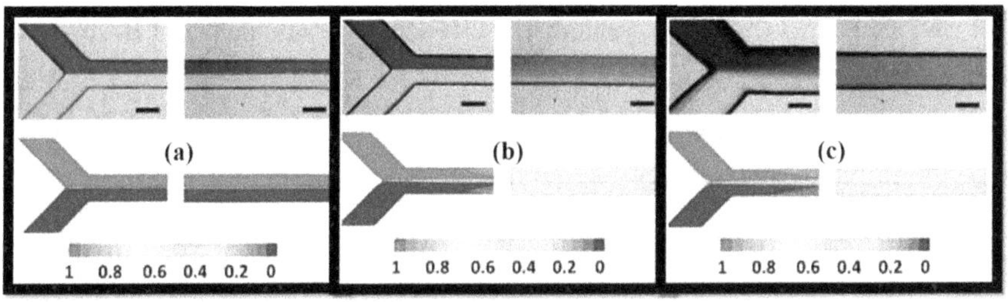

FIGURE 1.19 Simulated concentration depicting the mixing of two fluids at a flow rate of (a) 2 mL/min, (b) 10 mL/min, and (c) 40 mL/min (You et al. 2015).

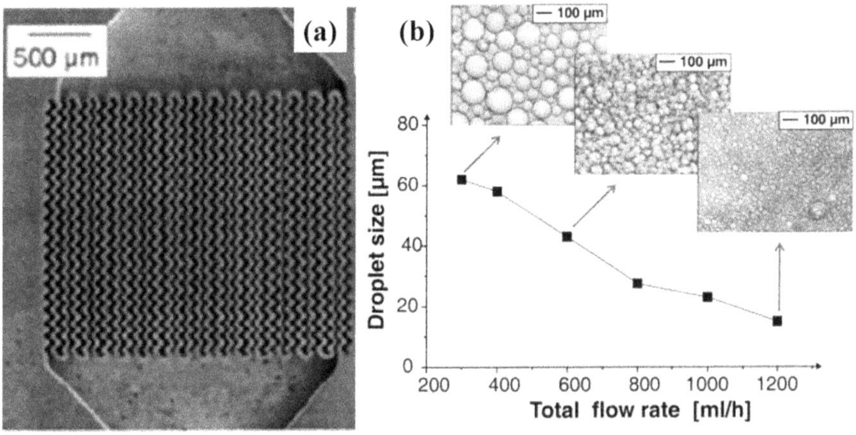

FIGURE 1.20 (a) SEM image of the mixing element comprising of regular arrangement of corrugated walls. (b) Droplet size distribution as a function of flow rate (Haverkamp et al. 1999).

designed to produce 3D flow for mixing of silicon oil and water (dye) at different flow rates (Haverkamp et al. 1999).

In another work by Kim and co-workers, an F-shaped mixer component in a serpentine that relies on the concept of splitting and recombination, is shown in Figure 1.21. Although the split-and-recombine process assists in efficient mixing, apart from molecular diffusion such types of microchannel still require longer channel lengths (Kim et al. 2005).

1.8.5.2 Ridges, Grooves, or Slanted walls

Fabrication techniques for microreactors are vital in designing the channels with slanted walls, micropillars, etc. for efficient mixing (Pawinanto et al. 2020, Zhou et al. 2015, Yu et al. 2016). The mixing can be greatly enhanced by increasing the overall channel length. Figure 1.22 is the work of Johnson and group in which they created slants using an excimer laser within the microchannel made of polycarbonate. Owing to the slanted channel architecture and the voltage applied, there is a surge in the mixing process compared to the non-slanted channel. Also, experiments were performed by varying the number of wells and orientation for higher flow rates (Johnson et al. 2002).

1.8.5.3 Multiphase Mixing

Since the previous studies focus on mixing associated with miscible fluids, there are other devices that allow mixing in multiphase flows. One typical example is the generation of droplets in a carrier fluid such as oil. Since oil acts as an immiscible fluid, the liquids are not able to penetrate to their surroundings and are confined inside a droplet. The process of localization of droplets along with chaotic advection assists in mixing inside a droplet (Pishbin et al. 2020, Serra et al. 2017, Kerr et al. 2019, Geng et al. 2020). Gunther and co-workers worked on the generation of a liquid droplet in a gas as a carrier, as shown in Figure 1.23. The first step was the generation of liquid droplets that assist in the enhanced mixing of two liquids, and later the separation of gas and fluid takes place. Thus, the final phase is a single liquid phase (Gunther et al. 2005).

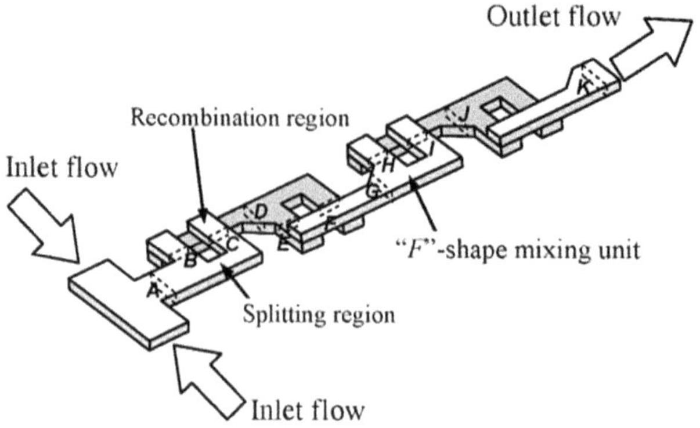

FIGURE 1.21 Representation of serpentine-shaped laminating micromixer fitted with splitting, recombination, and mixing units (Kim et al. 2005).

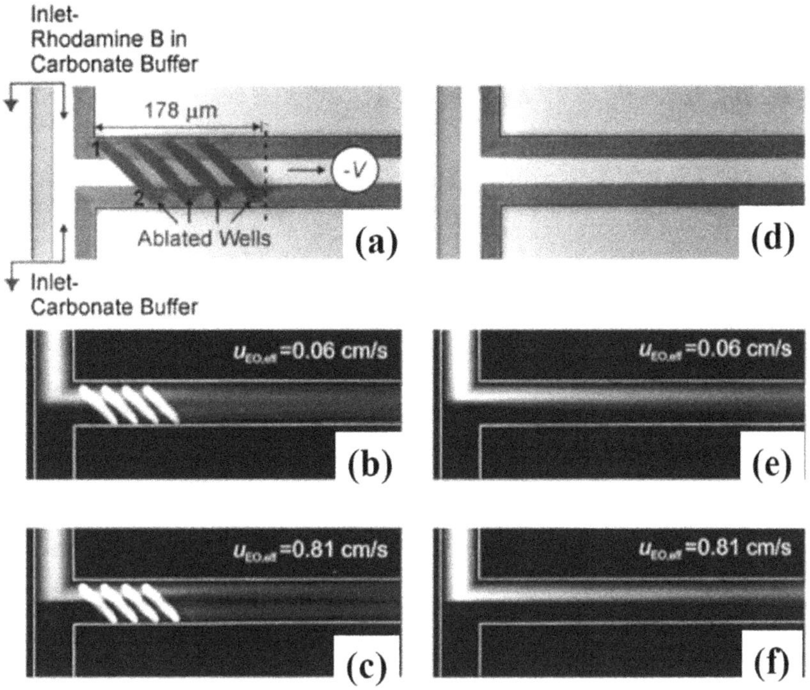

FIGURE 1.22 (a) White-light microscopy image of T-shape microreactors with well architecture designed using an excimer laser. Mixing at the flow rate of (b) 0.06 cm/s and, (c) 0.81 cm/s. (d) White light microscopy image of T-shape microreactors without well architecture. Mixing at the flow rate of (e) 0.06 cm/s and (f) 0.81 cm/s (Johnson et al. 2002).

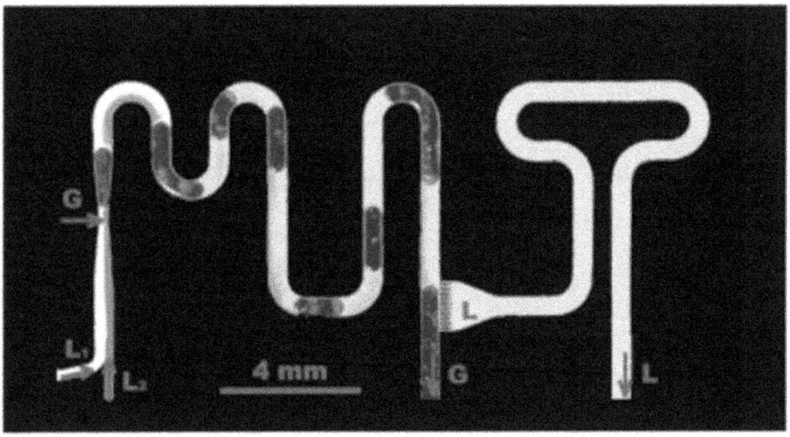

FIGURE 1.23 Microreactors with channel dimensions of 400 μm in width and 150 μm in depth used in the mixing and separation of liquid and gas. Here flow rate of gas stream (G) and each liquid streams (L_1 and L_2) are 30 μL/min and 5 μL/min, respectively (Gunther et al. 2005).

1.8.5.4 Microstirrers

Similar to that of mixing by magnetic stirring at macroscale level, this type of mixing uses a magnetic field to bring about the rotation of microbars in a fluid system. In work by Lee and co-workers, they performed experiments using ferromagnetic particles to understand the extent of mixing by applying a magnetic field, as illustrated in Figure 1.24. Using such particles that align or aggregate under low and high flow conditions, there was an enhancement in the mixing process (Lee et al. 2009).

1.8.5.5 Acoustic Mixing

A unique way to enhance the mixing process is by the usage of surface acoustic waves that are a form of elastic energy, as shown in Figure 1.25. This form of energy induces an acoustic steaming effect in the fluid and thus enhances the mixing even at a low Reynolds number (Frommelt et al. 2008, Song et al. 2018, Pothuri et al. 2019).

1.9 SUMMARY OF MATERIALS AND FABRICATION TECHNIQUES FOR MICROFLUIDICS DEVICES

To fabricate a microchannel for a given application, the factors such as materials, design, and fabrication processes are selected accordingly. A large variety of materials is used for the fabrication of microfluidics reactors, as illustrated in Figure 1.26. Silicon was the first material to be worked upon for the manufacturing of microfluidics devices followed by glass and other materials. However, with the advancements in science and technology, materials such as polymers, composites, and paper have gained much popularity for the fabrication of microreactors. In translation from the

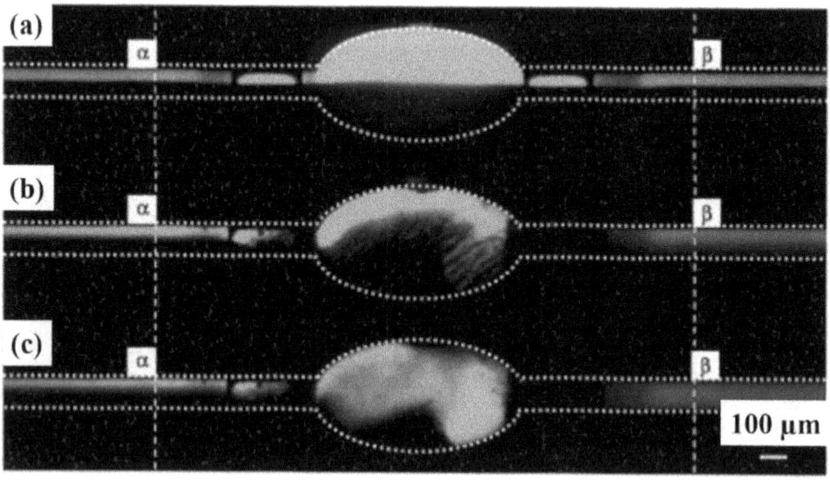

FIGURE 1.24 Fluorescence micrograph image of the mixing chamber (a) in the absence of magnetic particles, (b) in the presence of nonrotating magnetic particles, and (c) in the presence of rotating magnetic particles (Lee et al. 2009).

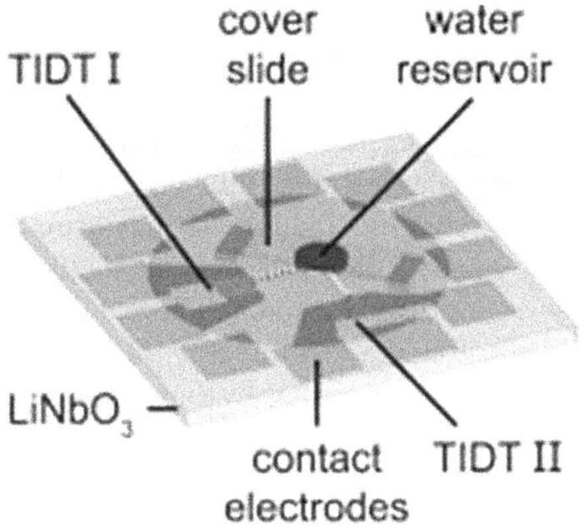

FIGURE 1.25 A 3D representation of a mixer based on a surface acoustic waves (SAW) microreactor. Tapered interdigital transducers (TIDTs)are special tapered interdigital transducers employed for the generation of acoustic waves (Frommelt et al. 2008).

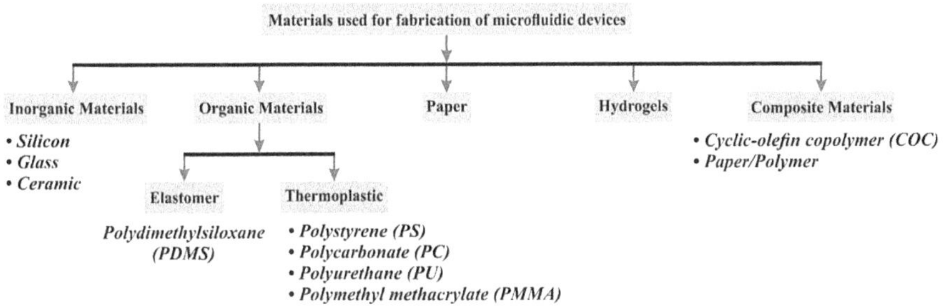

FIGURE 1.26 Flowchart depicting materials used for microreactor fabrication for a large variety of applications.

laboratory to industrial scale, priority is given to the cost of production, performance, and reliability. Thus, a crucial aspect is to bridge the balance between the materials and fabrication techniques in order to bring down the final cost of the product.

Established silicon wafer processing techniques, the resistance of silicon to organic solvents, and the excellent physical properties of silicon have made a significant contribution to the speedy growth of microfluidics technology. The methodology involved in the fabrication of silicon-based microfluidics devices generally begins with the substrate cleaning process. Techniques such as lithography, etching, and lithography, electroplating and moulding (LIGA) are employed for the manufacturing of microfluidics devices. Post channel fabrication processes such as anodic bonding and fusion bonding are of the utmost importance in closing the open channels. However, apart from being expensive, the issues of non-flexibility in silicon wafers, toxic chemicals used in the fabrication process, and limited opacity are directly correlated to the downside of usage of silicon wafers for fabrication techniques.

Thus, glass and polymers are suitable candidates to overcome the shortcomings of conventional silicon-based microfluidics devices. Properties such as optical transparency, electrical insulation, chemical inertness, and the low cost of processing make them a suitable candidate for device fabrication. Glass capillaries are also a promising option used in the fabrication of capillary-based microreactors. Still, the assembly of such capillaries requires handling skills and cumbersome operations. A microchannel on glass or quartz is fabricated via photolithography and wet or dry etching. Although glass is considered biocompatible and has widespread application in biological science, it is brittle and requires extensive care during the fabrication and handling of such devices. Also, similar to those of silicon-based microfluidics devices, the channels made in glass are also open channels and need a cleanroom environment at the time of bonding. Thus, during manufacturing, this adds up to the overall cost of glass-based microfluidics devices.

The underlying disadvantages of both silicon- and glass-based microfluidics devices have triggered extensive research in the field of finding alternate materials for device fabrication. The class of polymers used in the fabrication process is polydimethylsiloxane (PDMS), polystyrene (PS), polycarbonate (PC), and polymethylmethacrylate (PMMA). Polymeric microfluidics devices gained popularity as they are low cost, optically transparent, biocompatible, disposable, and offer design flexibility. Fabrication techniques such as soft micromachining that includes laser ablation and micromachining, computer numerical control (CNC) micromachining, optical/photo/X-ray lithography, hot embossing, soft lithography, and injection molding are the techniques employed for the fabrication of polymeric microfluidics devices.

Cyclic olefin copolymers (COC) fall under the class of cyclic olefin polymers (COP) that are based on a cyclic olefin monomer unit and ethene. COC has very low water absorptivity, is electrically insulating, operatically transparent, and highly rigid. It is inert to the attack of acids, alkalines, polar solvents, and acetone. COC-based microfluidics devices are created using techniques such as micromilling or CNC machining, hot embossing, and injection molding. Recently 3D printing is also being used to fabricate COC-based microfluidics devices.

Apart from inorganic and polymeric materials, paper is one such material that has become a promising option as a substrate for microfluidics devices. Apart from being flexible, biocompatible, and most cost effective, paper-based devices can be disposed of either by burning or the process of natural degradation. By the process of surface/chemical modifications, capillary action on the surface of paper can be controlled. Paper is an excellent option in microfluidics technology, being used for biomedical analysis and forensic diagnostics. The detection or investigation of analytes using paper-based microfluidics devices is carried out through either colorimetric, electrochemical, chemiluminescence, or by electrochemiluminescence.

Moreover, the white background of paper imparts excellent contrast for colorimetric detection. Owing to the porous nature of paper, paper-based devices have extended applications in the field of filtration and separation. Physical and chemical techniques including inkjet printing, wax patterning, lithography, plasma treatment, and laser treatment are the techniques used in the fabrication of paper-based microfluidics devices. Paper origami and stacking are also techniques that are used to produce 3D paper-based microfluidics devices.

1.10 CONCLUSION

This chapter gives an insight into the fundamental laws and equations governing the fluid flow in a microfluidics device. These fluid flow equations are vital to understand before diving into any application related to this field. While characterizing and scaling down the processes from macroscale to microscale, it is important to question and understand the relationship of fluid flow, molecular interactions, and surface forces because of the change in dimensional scale. Within microchannels the flow is usually laminar and thus hinders the mixing process. To overcome this, use of techniques such as split and recombine, ridges, grooves, slanted walls, or even acoustic mixing have shown to significantly improve the mixing process inside microchannels. Understanding fluid flow has perhaps an edge in utilizing the concept of microfluidics for low-cost and rapid POC devices that can be operated without hiring trained professionals or without setting up hefty laboratory infrastructure. Along with advances in microfluidics technology, apart from sensing and diagnostics, the field has opened up plenty of interdisciplinary research opportunities in the biomedical and pharmaceutical industries. However, the list of microfluidics-related applications is endless and ever-expanding.

REFERENCES

Capel, A. J., R. P. Rimington, M. P. Lewis, and S. D. R. Christie. 2018. "3D printing for chemical, pharmaceutical and biological applications." *Nature Reviews Chemistry* no. 2 (12):422–436. doi: 10.1038/s41570-018-0058-y.

Chibh, S., V. Katoch, A. Kour, F. Khanam, A. S. Yadav, M. Singh, G. C. Kundu, B. Prakash, and J. J. Panda. 2021. "Continuous flow fabrication of Fmoc-cysteine based nanobowl infused core–shell like microstructures for pH switchable on-demand anti-cancer drug delivery." *Biomaterials Science* no. 9: 942–959. doi: 10.1039/D0BM01386B.

Dittrich, P. S., and A. Manz. 2006. "Lab-on-a-chip: microfluidics in drug discovery." *Nature Reviews Drug Discovery* no. 5 (3):210–218. doi: 10.1038/nrd1985.

Dong, R., Y. Liu, L. Mou, J. Deng, and X. Jiang. 2019. "Microfluidics-based biomaterials and biodevices." *Advanced Materials* no. 31 (45):1805033–1805051. doi: 10.1002/adma.201805033.

Duffy, D. C. , J. C. McDonald, O. A. J. Schueller, and G. M. Whitesides. 1998. "Rapid prototyping of microfluidic systems in poly(dimethylsiloxane)." *Analytical Chemistry* no. 70 (23):4974–4984. doi: 10.1021/ac980656z.

Ebnereza, E., K. Hassani, M. Seraj, and K. G. Moghaddam. 2020. "Shape optimization of a split-and-recombine micromixer by the local energy dissipation rate." *Proceedings of the Institution of Mechanical Engineers, Part E: Journal of Process Mechanical Engineering* no. 234 (3):243–251. doi: 10.1177/0954408920910588.

Folch, A., and M. Toner. 1998. "Cellular micropatterns on biocompatible materials." *Biotechnology Progress* no. 14 (3):388–392. doi: 10.1021/bp980037b.

Frommelt, T., M. Kostur, M. Wenzel-Schafer, P. Talkner, P. Hanggi, and A. Wixforth. 2008. "Microfluidic mixing via acoustically driven chaotic advection." *Physical Review Letters* no. 100 (3):034502–034506. doi: 10.1103/PhysRevLett.100.034502.

Geng, Y., S. Ling, J. Huang, and J. Xu. 2020. "Multiphase microfluidics: fundamentals, fabrication, and functions." *Small* no. 16 (6):1906357–1906377. doi: 10.1002/smll.201906357.

Gong, M. M., and D. Sinton. 2017. "Turning the page: advancing paper-based microfluidics for broad diagnostic application." *Chemical Reviews* no. 117 (12):8447–8480. doi: 10.1021/acs.chemrev.7b00024.

Gunther, A., M. Jhunjhunwala, M. Thalmann, M. A. Schmidt, and K. F. Jensen. 2005. "Micromixing of miscible liquids in segmented gas-liquid flow." *Langmuir* no. 21 (4):1547–1555. doi: 10.1021/la0482406.

Harrison, D. J. , K. Fluri, K. Seiler, Z. Fan, C. S. Effenhauser, and A. Manz. 1993. "Micromachining a miniaturized capillary electrophoresis-based chemical analysis system on a chip." *Science* no. 261 (5123):895–897. doi: 10.1126/science.261.5123.895.

Haverkamp, V., W. Ehrfeld, K. Gebauer, V. Hessel, H. Löwe, T. Richter, and C. Wille. 1999. "The potential of micromixers for contacting of disperse liquid phases." *Fresenius' Journal of Analytical Chemistry* no. 364 (7):617–624. doi: 10.1007/s002160051397.

Hou, X., Y. S. Zhang, G. Trujillo-de Santiago, M. M. Alvarez, J. Ribas, S. J. Jonas, P. S. Weiss, A. M. Andrews, J. Aizenberg, and A. Khademhosseini. 2017. "Interplay between materials and microfluidics." *Nature Reviews Materials* no. 2 (5):17016–17031. doi: 10.1038/natrevmats.2017.16.

Hulme, S. E., S. S. Shevkoplyas, J. Apfeld, W. Fontana, and G. M. Whitesides. 2007. "A microfabricated array of clamps for immobilizing and imaging C. elegans." *Lab on a Chip* no. 7 (11):1515–23. doi: 10.1039/b707861g.

Husain, A., F. A. Khan, N. Huda, and M. A. Ansari. 2018. "Mixing performance of split-and-recombine micromixer with offset inlets." *Microsystem Technologies* no. 24 (3):1511–1523. doi: 10.1007/s00542-017-3516-4.

Hyde, J. F. 1965. "Chemical background of silicones: the siloxane linkage as a structure-building device gives variety and versatility to the silicones." *Science* no. 147 (3660):829–836. doi: 10.1126/science.147.3660.829.

Johnson, T. J., D. Ross, and L. E. Locascio. 2002. "Rapid microfluidic mixing." *Analytical Chemistry* no. 74 (1):45–51. doi: 10.1021/ac010895d.

Katoch, V., N. Sharma, M. Sharma, M. Baghoria, J. J. Panda, M. Singh, and B. Prakash. 2021. "Microflow synthesis and enhanced photocatalytic dye degradation performance of antibacterial Bi_2O_3 nanoparticles." *Environmental Science and Pollution Research* no. 28:19155–19165. doi: 10.1007/s11356-020-11711-1.

Kerr, C. B., R. W. Epps, and M. Abolhasani. 2019. "A low-cost, non-invasive phase velocity and length meter and controller for multiphase lab-in-a-tube devices." *Lab on a Chip* no. 19 (12):2107–2113. doi: 10.1039/C9LC00296K.

Kim, D. S., S. H. Lee, T. H. Kwon, and C. H. Ahn. 2005. "A serpentine laminating micromixer combining splitting/recombination and advection." *Lab Chip* no. 5 (7):739–747. doi: 10.1039/b418314b.

Lee, S. H., D. van Noort, J. Y. Lee, B. T. Zhang, and T. H. Park. 2009. "Effective mixing in a microfluidic chip using magnetic particles." *Lab Chip* no. 9 (3):479–482. doi: 10.1039/b814371d.

Manz, A. , N. Graber, and H. M. Widmer. 1990. "Miniaturized total chemical analysis systems: a novel concept for chemical sensing." *Sensors and Actuators B: Chemical* no. 1 (1–6):244–248. doi: 10.1016/0925-4005(90)80209-I.

McDonald, J. C., D. C. Duffy, J. R. Anderson, D. T. Chiu, H. Wu, O. J. A. Schueller, and G. M. Whitesides. 1999. "Fabrication of microfluidic systems in poly(dimethylsiloxane)." *Electrophoresis* no. 21 (1):27–40. doi: 10.1002/(SICI)1522-2683(20000101)21:1<27::AID-ELPS27>3.0.CO;2-C.

Modestino, M. A., D. Fernandez Rivas, S. M. H. Hashemi, J. G. E. Gardeniers, and D. Psaltis. 2016. "The potential for microfluidics in electrochemical energy systems." *Energy & Environmental Science* no. 9 (11):3381–3391. doi: 10.1039/C6EE01884J.

Nguyen, Nam-Trung. 2007. *Mixing in Microscale.* Boston, MA: Springer.

Nguyen, Nam-Trung. 2012. *Micromixers. Fundamentals, Design and Fabrication (Micro and Nano Technologies)* 2 ed. William Andrew, Waltham, MA.

Pascal, H., T. Jens, H. Marko, S. Michael, H. Christian, L. Patrick, and Z. Dirk. 2018. "Optimization of a split and recombine micromixer by improved exploitation of secondary flows." *Chemical Engineering Journal* no. 334:1996–2003. doi: 10.1016/j.cej.2017.11.131.

Pawinanto, R. E., J. Yunas, and A. M. Hashim. 2020. "Micropillar based active microfluidic mixer for the detection of glucose concentration." *Microelectronic Engineering* no. 234:111452. doi: 10.1016/j.mee.2020.111452.

Pishbin, E., A. Kazemzadeh, M. Chimerad, S. Asiaei, M. Navidbakhsh, and A. Russom. 2020. "Frequency dependent multiphase flows on centrifugal microfluidics." *Lab on a Chip* no. 20 (3):514–524. doi: 10.1039/C9LC00924H.

Pothuri, C., M. Azharudeen, and K. Subramani. 2019. "Rapid mixing in microchannel using standing bulk acoustic waves." *Physics of Fluids* no. 31 (12):122001–122012. doi: 10.1063/1.5126259.

Prakash, B., V. Katoch, A. Shah, M. Sharma, M. M. Devi, J. J. Panda, J. Sharma, and A. K. Ganguli. 2020. "Continuous flow reactor for the controlled synthesis and inline photocatalysis of antibacterial Ag_2S nanoparticles." *Photochemistry and Photobiology* no. 96 (6):1273–1282. doi: 10.1111/php.13297.

Ramachandran, A., D. A. Huyke, E. Sharma, M. K. Sahoo, C. Huang, N. Banaei, B. A. Pinsky, and J. G. Santiago. 2020. "Electric field-driven microfluidics for rapid CRISPR-based diagnostics and its application to detection of SARS-CoV-2." *Proceedings of the National Academy of Sciences* no. 117 (47):29518–29525. doi: 10.1073/pnas.2010254117.

Serra, M., D. Ferraro, I. Pereiro, J. L. Viovy, and S. Descroix. 2017. "The power of solid supports in multiphase and droplet-based microfluidics: towards clinical applications." *Lab on a Chip* no. 17 (23):3979–3999. doi: 10.1039/C7LC00582B.

Shakhawat, H., L. Insu, K. Sun Min, and K. Kwang-Yong. 2017. "A micromixer with two-layer serpentine crossing channels having excellent mixing performance at low Reynolds numbers." *Chemical Engineering Journal* no. 327:268–277. doi: 10.1016/j.cej.2017.06.106.

Singh, A., A. Baruah, v. Katoch, K. Vaghasiya, B. Prakash, and A. K. Ganguli. 2018. "Continuous flow synthesis of Ag_3PO_4 nanoparticles with greater photostability and photocatalytic dye degradation efficiency." *Journal of Photochemistry and Photobiology A: Chemistry* no. 364:382–389. doi: 10.1016/j.jphotochem.2018.05.017.

Sinton, D. 2014. "Energy: the microfluidic frontier." *Lab on a Chip* no. 14 (17):3127–3134. doi: 10.1039/C4LC00267A.

Song, C., T. Jin, R. Yan, W. Qi, T. Huang, H. Ding, S. H. Tan, Nam-Trung Nguyen, and L. Xi. 2018. "Opto-acousto-fluidic microscopy for three-dimensional label-free detection of droplets and cells in microchannels." *Lab on a Chip* no. 18 (9):1292–1297. doi: 10.1039/C8LC00106E.

Sutera, S. P. 1993. "The history of poiseuille's law." *Annual Review of Fluid Mechanics* no. 25:19. doi: 10.1146/annurev.fl.25.010193.000245.

Terry, S. C. , J. H. Jerman, and J. B. Angell. 1979 "A gas chromatographic air analyzer fabricated on a silicon wafer." *IEEE Transactions on Electron Devices* no. 26 (12):1880–1886. doi: 10.1109/T-ED.1979.19791.

Wang, X., X. Ban, R. He, D. Wu, X. Liu, and Y. Xu. 2018. "Fluid-solid boundary handling using pairwise interaction model for non-newtonian fluid." *Symmetry* no. 10 (4):94–110. doi: 10.3390/sym10040094.

Weng, X., Y. Kang, Q. Guo, B. Peng, and H. Jiang. 2019. "Recent advances in thread-based microfluidics for diagnostic applications." *Biosensors and Bioelectronics* no. 132:171–185. doi: 10.1016/j.bios.2019.03.009.

Xia, Y., and G. M. Whitesides. 1998. "Soft lithography." *Angew. Chem. Int. Ed.* no. 28:26. doi: 10.1146/annurev.matsci.28.1.153.

Yakoh, A., S. Chaiyo, W. Siangproh, and O. Chailapakul. 2019. "3D capillary-driven paper-based sequential microfluidic device for electrochemical sensing applications." *ACS Sensors* no. 4 (5):1211–1221. doi: 10.1021/acssensors.8b01574.

Yamada, K., H. Shibata, K. Suzuki, and D. Citterio. 2017. "Toward practical application of paper-based microfluidics for medical diagnostics: state-of-the-art and challenges." *Lab on a Chip* no. 17 (7):1206–1249. doi: 10.1039/C6LC01577H.

You, J. B., K. Kang, T. T. Tran, H. Park, W. R. Hwang, J. M. Kim, and S. G. Im. 2015. "PDMS-based turbulent microfluidic mixer." *Lab on a Chip* no. 15 (7):1727–1735. doi: 10.1039/c5lc00070j.

Yu, H., Thien-Binh Nguyen, S. H. Ng, and T. Tran. 2016. "Mixing control by frequency variable magnetic micropillar." *RSC Advances* no. 6 (14):11822–11828. doi: 10.1039/C5RA24996A.

Zhao, Y., X. Hu, S. Hu, and Y. Peng. 2020. "Applications of fiber-optic biochemical sensor in microfluidic chips: a review." *Biosensors and Bioelectronics* no. 166:112447–112470. doi: 10.1016/j.bios.2020.112447.

Zhong, Q., H. Ding, B. Gao, Z. He, and Z. Gu. 2019. "Advances of microfluidics in biomedical engineering." *Advanced Materials Technologies* no. 4 (6):1800663–1800690. doi: 10.1002/admt.201800663.

Zhou, B., W. Xu, A. A. Syed, Y. Chau, L. Chen, B. Chew, O. Yassine, X. Wu, Y. Gao, J. Zhang, X. Xiao, J. Kosel, X. Zhang, Z. Yao, and W. Wen. 2015. "Design and fabrication of magnetically functionalized flexible micropillar arrays for rapid and controllable microfluidic mixing." *Lab on a Chip* no. 15 (9):2125–2132. doi: 10.1039/C5LC00173K.

2 Role of Microfluidics-Based Point-of-Care Testing (POCT) for Clinical Applications

Arpana Parihar, Dipesh Singh Parihar, Pushpesh Ranjan, and Raju Khan

CONTENTS

DOI: 10.1201/9781003033479-2

2.1 INTRODUCTION

Point-of-care testing (POCT) entails diagnostics based on biochemical, hematological, coagulation, or molecular tests at or near a patient. The POCT has been widely used in healthcare monitoring for the last two decades. Owing to the advantage of near-patient testing it can be implemented in various settings, from self-testing to outpatient clinics and to the intensive care unit (ICU), its demand is increasing day by day. POCT offers quick test results in low sample volumes with high sensitivity and specificity and can be helpful in improving patient outcomes through regular patient monitoring [1]. The major advantageous features of POCT are that it is simple to operate, can be used by unskilled personnel, is cost effective, has easy bulk fabrication, and has rapid turn-around time [2]. As per quality assessment and satisfaction surveys carried out on medical staff and physicians, the use of POCT devices has been shown to reduce mortality, morbidity, and improve quality of life [3]. Moreover, small volumes of bodily fluids, e.g., blood, saliva, and urine are required for POCT diagnostics. However, due to the contagious nature of certain biological samples, disposable POC devices can be used to protect the end-users from exposure to biohazardous or infectious agents. In order to make cost-effective POC diagnostic devices, one can use (i) minimal expensive reagents, (ii) inexpensive manufacturing for mass production, and (iii) miniaturization processes [4].

Major issues such as the handling of complex assays by non-technical personnel and the unavoidable resistance to changes required in healthcare delivery can be better dealt with by microfluidics-based devices. In the 1980s, POCT devices were relatively crude and gave qualitative results, but gradually, due to technology advancement and supportive developments related to liquid handling, signal detection, miniaturization, and integration helped to address and overcome these issues very well [3–4]. Based on the way they function, microfluidics-based POCT devices can be divided into two categories: Lateral flow assay-based and nucleic acid amplification-based. Figure 2.1 shows a schematic representation of types of POCT devices. The lateral flow assay-based techniques utilize either antibody-based detection or a nucleic acid platform. The techniques which utilize antibody-functionalized surfaces face certain issues of non-specific binding due to the reproducible manufacturing of functionalized surfaces and disease progression. Further, the antibodies are expensive and an excess amount is used in order to fully functionalize the microchannel surfaces of devices. which further increases the cost of POCT. Nevertheless, the nucleic acid-based method displays higher sensitivity and specificity when compared

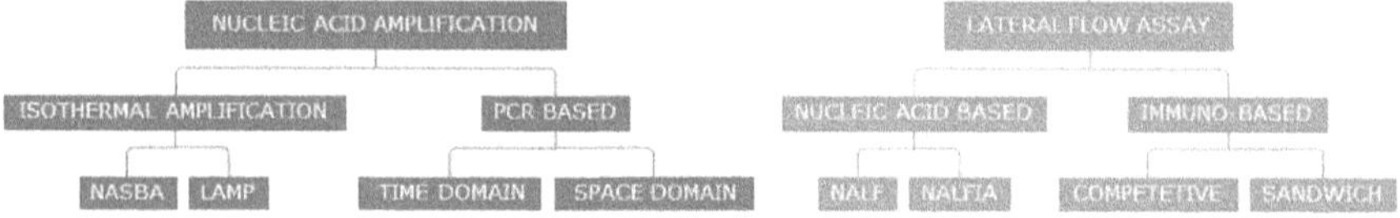

FIGURE 2.1 A schematic of types of POC diagnostics technologies.

to antibody-based lateral flow assays. However, the nucleic acid isolation and amplification process takes a longer diagnostic time compared to antibody assays. Apart from this, precise temperature control for efficient nucleic acid amplification is needed, which enhances the manufacturing complexity of the POCT chip [4–6]. It is expected that the global population will reach 9 billion by 2050 and 90% of the world disease burden comes from developing countries which spend only 12% on health systems. The frequency of communicable and noncommunicable diseases such as diabetes, obesity, cardiovascular, cancer, etc., is much higher in developing/underdeveloped countries compared to the developed countries. Quality community screening for diseases can demonstrably improve public health in developing countries. The accurate diagnosis of disease along with its incidence rate would not only help to figure out the root cause of diseases at both the individual and population levels but also lower the world disease burden and improve the quality of human life. The limited-resource setting along with a weak infrastructure-based healthcare setup needs cost-effective, easy-to-handle diagnostic devices, and most of the commercially available diagnostic tools are unable to meet the present demand. Therefore, the development of POCT for developing countries with limited resources is needed for an improvement in disease management [7–9].

This chapter deals with the role of microfluidics-based POCT devices for clinical applications, such as the diagnosis of various communicable and noncommunicable diseases, development of laboratory assays for research, and their involvement in drug design and development. As well as , there is a brief outline about its impact on developing countries and the material selection for device fabrication; recent advances in POCT devices and future prospects are also discussed.

2.2 IMPACT OF MICROFLUIDIC-BASED POCT IN RESOURCE-LIMITED SETTINGS

Conventional POCT needed a centralized lab facility, sophisticated instruments, and skilled personnel. The emergence of a rapid test platform in a decentralized testing for detection of pregnancy in the 1970s was the crucial, driving, and key development in the field of POCT devices. After this, rapid tests in a dipstick format for communicable infectious diseases, such as tuberculosis, hepatitis B, and human immunodeficiency virus (HIV-1 and HIV-2) were introduced. Later in the 1980s, the invention of the first immunochromatographic strip (ICS)-based lateral flow assay for disease diagnosis was considered as a breakthrough in the disease diagnostic field. Subsequently, POCT technology has continuously evolved owing to its wider applications for the diagnosis of various diseases all over the world. Figure 2.2 represents a schematic which shows the POCT-based device market segment for various communicable and noncommunicable diseases in various regions of the world. It also gives information about users of POCT devices. Owing to the advantageous features of microfluidics-based devices, such as precise performance, interpretation by minimally-trained users, stability for more than a year, capacity of multiplexing, and ease of shipping and storage, this made them a boon for resource-limited countries [10]. A schematic of a platform for POC diagnostics in developing

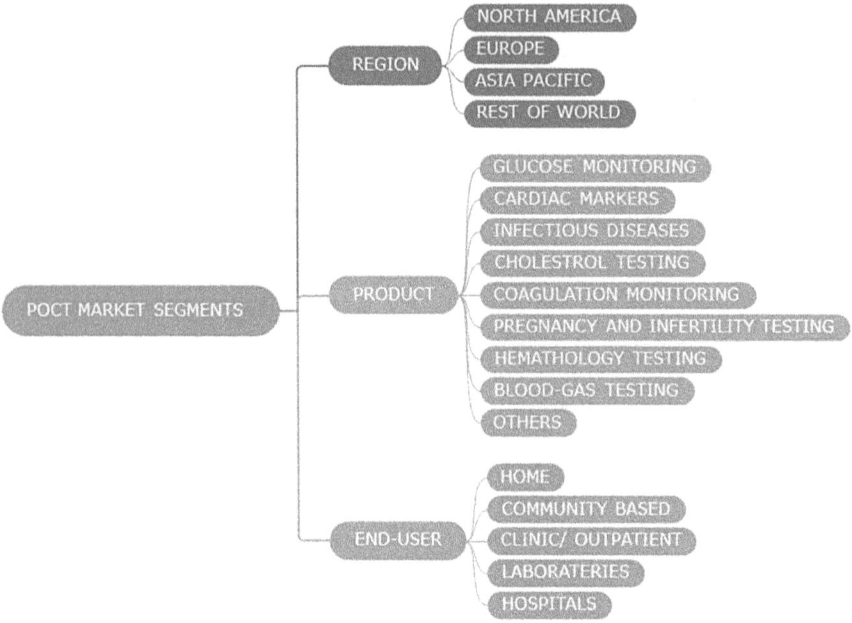

FIGURE 2.2 A schematic of a platform for POC diagnostics in developing and developed countries.

countries showing various steps of diagnosis, data collection, storage, and interpretation followed by disease prediction is shown in Figure 2.3. To date, more than 100 companies all over the world are producing a wide range of lateral flow assay-based POCT devices for infectious diseases, cancer, and cardiac diseases. As per world economic data, the US and European market alone covers 50% (US$1680 million) and 40% (US$1344 million) respectively, compared to the rest of the world which comprises only 10% (US$336 million) of the rapid POCT test market [11].

The market-leading company, Alere, covered approximately 35% of the global market of lateral flow immune assay (LFIA)tests for infectious diseases such as HIV, respiratory viruses, strep A/B, influenza, legionella, filariasis, malaria, meningitis, tuberculosis, measles, mumps, chlamydia, rubella, and gonorrhea [12]. The LFIA for HIV testing has expanded rapidly worldwide and more than 29 million people were tested for HIV infection in developing countries in 2010 [13, 14]. The quantitative test for rapid HIV monitoring based on lateral flow technology called the OraQuick ADVANCE Rapid HIV-1/2 Antibody Test can detect low concentrations of HIV antigens in oral fluid. Another LFIA-based test for early diagnostic assay for pulmonary tuberculosis has shown reduced mortality and transmission in underdeveloped and low- resource settings countries. Furthermore, a LFIA-based test for Cryptococcal antigen and *Plasmodium falciparum* malaria displays promising sensitivity for detection of disease and in developing countries and use of this diagnostic platform reduced the mortality rate among the infected population [15, 16].

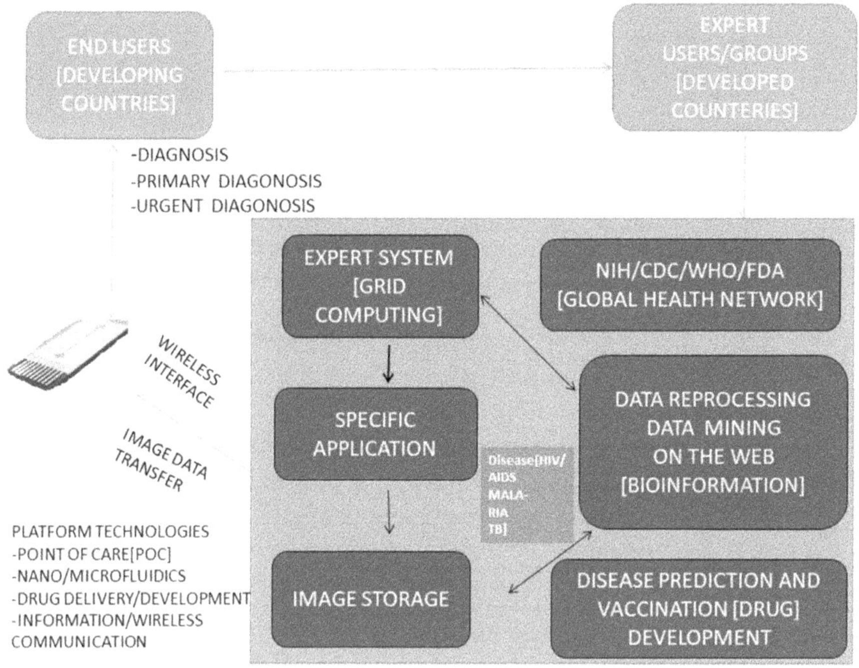

FIGURE 2.3 A schematic of a platform for POC diagnostics in developing countries showing various steps of diagnosis data collection, storage, and interpretation followed by disease prediction.

Figure 2.4 shows a schematic representation of FDA-approved microfluidics-based POC diagnostics in developing and developed countries.

2.3 CLINICAL APPLICATIONS USING MICROFLUIDICS-BASED DEVICES FOR POCT

The widely used POCTs for clinical application are glucose monitoring, blood gas, and electrolyte measurement, urine dipsticks, cardiac markers, pregnancy, and infertility testing devices, etc. While drug efficacy and toxicity testing, lactate, magnesium, lipids, hemoglobin A1C, HIV, influenza, *Helicobacter pylori*, tuberculosis-detecting devices, although present commercially they are used less often. In the following sections we discuss widely used POCT devices in clinical settings.

2.3.1 Glucose Monitoring for Diagnosis of Diabetes

Diabetes is mentioned as a global public health epidemic due to the disturbance in glucose metabolism related with insufficient insulin production. Yao et al. measure insulin levels in diabetic patients using a pneumatic microfluidic device fabricated by 3D-based extrusion printer which exploits the principle of impedance spectroscopy

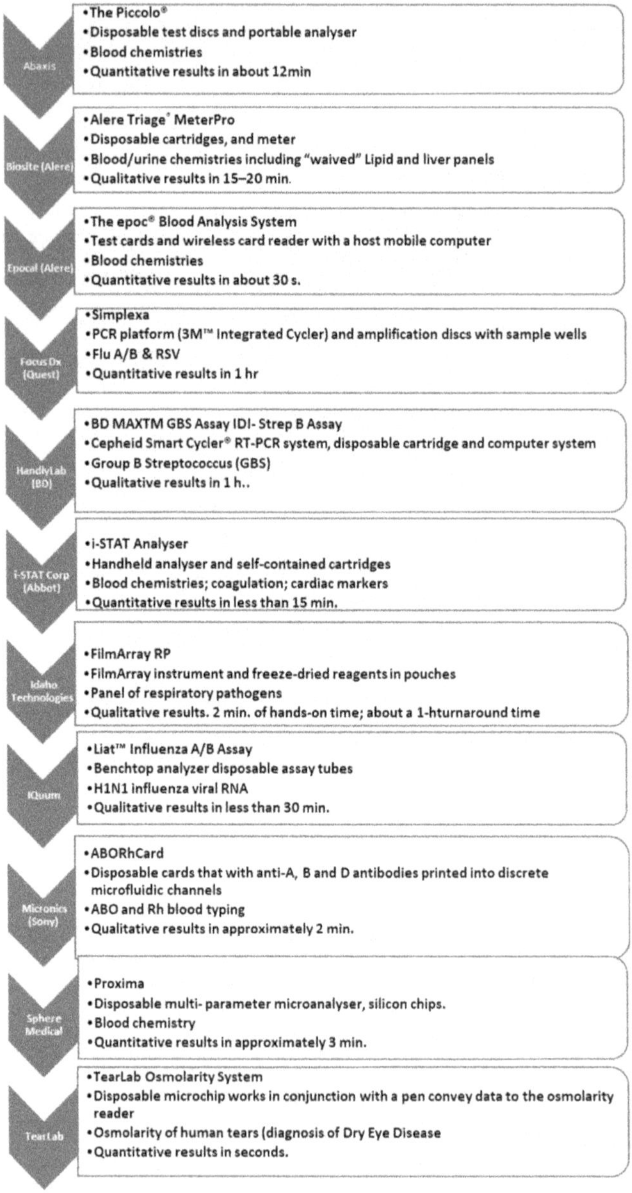

FIGURE 2.4 A schematic representation of FDA-approved microfluidics-based POC diagnostics in developing and developed countries. Adopted and modified.

measurement [17]. This microfluidic device is cost effective, having inbuilt efficient reagent mixing channels and shows a microscale range of detection. Samper et al. have fabricated a 3D-printed microfluidic device which can be easily integrated with the electrochemical biosensor settings. This device records short lactate concentration pulses and can simultaneously detect glucose, lactate, and glutamate [18]. The high incidence of diabetes in China and India has increased the adoption of testing in these countries, hence POCT devices for glucose monitoring have gained considerable interest and replaced conventional laboratory testing [19, 20] in these countries. POCT device-based glucose testing for diabetes is cost effective and user friendly, therefore it can even be done by the patient. Further, regular glucose monitoring is helpful to adjust or change drug doses for patients as per requirement. Moreover, a blood glucose monitoring device based on the electrochemical method can measure glucose in fresh whole blood. In this device the glucose strips are labeled with modified glucose dehydrogenase pyrroloquinoline quinone (GDH-PQQ) enzyme which converts glucose to gluconolactone. This reaction generates DC current which is detected by an analyzer and afterwards the signal is converted into a blood glucose value (Accu-Chek Aviva: Roche). Similarly, a cartridge-based analyzer was introduced to monitor diabetes and the albumin/creatinine ratio in patients. Herein, the automated cartridges contain a sample collecting slot and all the necessary reagents, which offer rapid quantitative results using a biological sample such as urine or whole blood. Another device known as Afinion HbA1c Dx test is Chemiluminescent Immunoassay (CLIA) waived, automated boronate affinity test which quantifies glycated hemoglobin in blood samples. This analyzer is comprised of two diodes, a digital diode and a light-emitting diode (LED), for reflection and transmission measurement respectively. Oh et al. developed a multiplexed viscosity measurements device fabricated with a 3D-printed technique for blood viscosity analysis [19]. This device efficiently measures the rate of viscosity of both Newtonian fluids, having varying viscosities, and non-Newtonian fluids at different shear rate conditions [19]. Another device involved in the clotting time measurement for patients is necessary for the prescription of anti-coagulant therapy. [20]. The clotting rate, also known as INR, can be self-tested by the patient themself using POCT devices and drug doses can be adjusted in conjunction with clinical recommendations.

2.3.2 Cardiac Disease-Associated Marker Detection

Millions of lives are lost due to cardiac disorders every year and early detection along with effective treatment can save the lives of patients and improve their health. A portable, easy-to-use, cost-effective blood analyzer system introduced to the market by Abbott is helpful in cardiac disease monitoring. This system utilizes disposable cartridge technology which gives rapid results within a few minutes. The cartridge comprises a silicon wafer with electrical sensors, waste chamber, sample well, and a pouch with calibrant solutions. These cartridges can detect a range of analytes from lactate to cardiac markers of endocrinology. The wireless version of this device, known as iSTAT1 Wireless, is available for monitoring disease in remote areas [21–23].

2.3.3 Infectious Diseases

Infectious diseases are a leading cause of death and around 95% of these deaths occur due to lack of appropriate diagnosis and timely treatment in developing countries [3]. The World Health Organization (WHO) has established a set of guidelines to develop diagnostic tools for infectious diseases at resource-limited settings – (i) affordable, (ii) sensitive, (iii) specific, (iv) user friendly, (v) rapid and robust, (vi) equipment-free, and (vii) delivery to those who need it, leading to the acronym "ASSURED" [4]. These guidelines can be helpful to fabricate and manufacture more suitable diagnostic devices for underdeveloped countries in low-resource settings [4, 20].

2.3.4 COVID-19 POC Diagnostics

The COVID-19 disease caused by SARS-CoV2, which emerged in late December 2019 in Wuhan, China, has become the worst pandemic for a century. Early diagnosis and isolation of patients is effective in curbing infection. More than 200 commercial manufacturers and laboratories received FDA (Food and Drug Administration) and EUA (Emergency Use Authorization) clearance and approval until August 8, 2020, for diagnostic kits for the detection of the infectious agent SARS-CoV2. The rapid POCT developed by Roche Diagnostics received emergency approval and clearance by the FDA for emergency usage. Similarly, the cobas SARS-CoV-2 is a RT-PCR-based Nucleic Acid Amplification Test (NAAT), which can simultaneously test 96 samples at a single run, has been approved. Another, electrochemiluminescence immunoassay-based test called the Elecsys Anti-SARS-CoV-2 has received approval. The Applied Biosystems division of Thermo Fisher Scientific has introduced the TaqPath COVID-19 Combo kit, to detect COVID-19. This kit detects various SARS-CoV2 targets, ORF1ab, spike protein, and nucleocapsid protein genes and can analyze 96 samples per assay [24–26]. Similarly, Abbott has launched three tests in response to COVID-19, among them the Abbott Real Time SARS-CoV-2 assay was the first launched by Abbott Molecular to detect molecular targets RdRp and N-genes of the novel coronavirus. Exploiting the GeneXpert technology, Cepheid introduced the Xpert Xpress SARS-CoV-2 molecular test which was the first POC test to be approved by the FDA for emergency use. This test can provide results within 30 min [27]. Cellex Inc. has also manufactured a rapid test based on lateral flow technology, named the qSARS-CoV-2 Rapid Test, which detects IgM and IgG antibodies present in serum or whole blood samples and provides results in 15–20 min [28].

2.3.5 Tuberculosis (TB) POC Diagnostics

TB is one of the deadly infectious diseases increasing the death rate and drug resistance of immune-suppressed patients. It became more lethal to HIV-infected patients [29, 30]. Conventionally, microscopic examination of a sputum smear has been used for TB diagnosis which has relatively lower sensitivity in underdeveloped countries

[31, 32]. Sheehan et al. designed a microfluidic device for DNA amplification using polymerase chain reaction (PCR) for diagnosis of *Mycobacterium tuberculosis* [33]. They demonstrated a noncontact thermo-cycling approach for DNA amplification of the pathogen with the help of a low-power halogen lamp. The advantageous feature of this device is the low sample and reagent consumption and usage of lower thermal mass which decreases the required time for PCR from hours to minutes. The user-friendly layout of the microfluidic device, low-cost heat source, and low-power consumption of the system present a possibility to develop a portable field device. More recently, Lee et al. have developed a portable microfluidic nuclear magnetic resonance (NMR) biosensor for rapid, quantitative, and multiplexed detection of biological targets, such as bacteria, cancer biomarkers, and TB [34]. The biosensor containing PMMA-based microfluidic mixers, microcoil arrays, printed circuit board, and a permanent dipole magnet was fabricated by photolithography and electroplating techniques. This sensor is based on a self-amplifying proximity assay and magnetic nanoparticles. In this system, magnetic nanoparticles served as a proximity sensor that bound to target biomolecules and subsequently formed soluble nanoscale clusters, which led to NMR signal changes. The data can be electronically obtained without bulky and expensive optical components. They also demonstrated the use of this system for the detection and characterization of infectious agents, such as bacteria, viruses, fungi, and parasites.

2.3.6 Human Immunodeficiency Virus (HIV) POC Diagnostics

The WHO has stressed the importance of developing POC devices for HIV/AIDS diagnosis and monitoring in resource-constrained areas [35]. These POC devices have the potential to be designed as an accurate, low-cost, and disposable platform for counting CD4+ T-lymphocytes and HIV viral load [36–38]. The POC device must detect and count fewer than 200 CD4+ cells per liter of blood and 400 copies of HIV per liter of blood for clinical applications [39]. Antiretroviral Therapy (ART) aims to suppress virus replication, slowing the progression of HIV infection to disease (AIDS) and increasing the number of CD4+ T-lymphocytes [35, 40]. As a result, HIV is monitored using CD4+ T-lymphocyte counting and viral load quantification. Traditional techniques, such as costly flow cytometry and quantitative PCR, have many drawbacks, including a long diagnostic time and the requirement for a well-trained technician [41].

To monitor the disease, it is necessary to state that the number of CD4+ T-lymphocytes in microliter of sample volumes taken from HIV-infected blood has been counted and used for prescribing ART treatment [40]. HIV binds to the CD4+ T-lymphocytes of host cells [42] through the HIV gp120 envelope glycoprotein which allows the virus entry to the cells and spreads infection which harms cells. While in developed countries, the CD4+ T-lymphocyte count is done four times a year using conventional flow cytometers, it is done only twice a year in developing countries [43].

A number of studies have been conducted to miniaturize flow cytometers as a portable POC device for global health applications [44]. In resource-constrained environments, however, a number of problems, such as cost and complexity, must be tackled. As a result, basic and mass-producible microfluidic devices could be useful for HIV/AIDS POC diagnostics. For example, using a microfluidic immunoassay called "POCKET" (portable and cost effective), a simple and reliable method for quantifying anti-HIV-1 antibodies in the sera of HIV-1 infected patients, was recently developed [45, 46].

For POC diagnostics, Lee and colleagues designed a disposable polymer-based RT-PCR chip with pinched microvalves [47]. This device has been used to detect the HIV p24 and gp120 proteins, which are the major capsid and envelope proteins encoded by the HIV gag and envelope genes, for early HIV infection diagnosis. This technique, however, requires optimization for HIV-infected whole blood processing and simplified POC operation in resource-limited environments for realistic POC applications.

2.3.7 Malaria POC Diagnostics

In order to prevent the spread of malaria, early detection and treatment are critical. Furthermore, incorrect diagnosis and treatment can lead to drug abuse or unanticipated side effects [47]. To reduce the misuse of anti-malarial drugs or antibiotics, simple yet highly accurate diagnostic devices are needed. In resource-constrained environments, lateral-flow Rapid Diagnostic Test (RDTs) are one method for diagnosing malaria infection [48, 49]. Malaria infection in human blood was first observed in the 1880s [50], and microscopy has long been considered the gold standard for malaria diagnosis.

This method of detecting malaria infection and parasites in resource-poor environments necessitates the use of highly qualified researchers [51]. Nano/microfluidic technologies can be used to develop POC devices for rapid and precise malaria diagnosis, reducing the need for well-trained experts and equipment [52]. The RDT strip chip, for example, can detect proteins derived from the blood of malaria parasites in a microfluidic format and can be commercialized. When antibodies are accumulated at the test line, this chip allows for the generation of a series of visible lines to indicate the presence of specific antigens in blood [53].

Rathod et al. created microfluidic channels to investigate the complex interactions between host cell ligands and parasitized erythrocytes that occur during malaria pathogenesis [54]. They were able to observe host–parasite interaction and malaria-infected red blood cells in a capillary environment because the microfluidic channels successfully mimicked the sizes and shapes of small capillary blood vessels.

2.3.8 Sepsis POC Diagnostics

Sepsis is a severe infection-related complication that can be fatal. It is characterized by a systemic inflammatory response syndrome (SIRS) that occurs as a result of chemicals secreted to fight infection [55]. The blood cleansing capacity of organs and the protective functions of immune cells can be overwhelmed by inflammation [56]. Several studies on sepsis, its causes, diagnosis, and prognosis have been published in the literature, and readers who are interested in learning more about sepsis, its causes, diagnosis, and prognosis should read these articles [56, 57].

Multiple organ failure, shock, and death can all result from sepsis (mortality increases by 7.6% for every hour spent waiting for antibiotics in septic shock, for example) [58]. As a result, early, precise, and timely diagnosis and treatment of sepsis are critical. Current diagnostic and therapeutic approaches, such as physical examination, blood and urine analysis, and molecular diagnostic methods, can be suboptimal if not integrated. Furthermore, due to disadvantages such as the high sample volume required and the time and labor-intensive nature of these methods, they do not meet the ASSURED requirements [59].

Su et al. looked at how particular molecular approaches can be used for sepsis/infection management and reviewed the state-of-the-art for bacteria detection [60]. In this review, innovations that aid sepsis management are divided into two categories – (i) diagnosis (i.e., pathogen detection and monitoring biochemical composition of patients' bodily fluids) and (ii) treatment.

2.3.9 Other Infectious Diseases (SARS, Dengue, Tuberculosis) POC Diagnostics

In 2003, SARS, which is caused by a new coronavirus (CoV), began to spread around the world. SARS was finally brought under control by employing management strategies such as isolating suspects and their close contacts. SARS claimed 744 lives in 2003, mostly in the Asia-Pacific region, due to the lack of an effective cure [61].

Nonetheless, antibody levels begin to rise about ten days after infection, while the SARS virus cell culture is laborious and time-consuming. Zhou et al. developed a SARS-CoV diagnostic platform using multiplex real-time PCR and capillary electrophoresis on a microfluidic chip. The system produced positive results in 17 of the 18 clinically diagnosed SARS patients, whereas traditional RT-PCR with agarose gel electrophoresis yielded only 12 positive results [62].

Huh et al. also developed a microfluidic platform with a magnetic force-activated micromixer and functionalized surfaces for cell lysis, intracellular protein purification, and SARS-CoV detection [63]. Similarly, Ramalingan and colleagues used an integrated microfluidic PCR chip to detect SARS virus DNA using an isothermal helicase-dependent amplification method [64].

The mosquito vector spreads the dengue virus, with the majority of infections occurring in Latin America and Asia [65]. Dengue virus has four serotypes that can cause anything from mild dengue fever to severe dengue shock syndrome in humans [66]. The main diagnostic procedures are still nucleic acid or protein/serological based due to the small sizes of dengue viruses.

Clinical testing for dengue, such as hemagglutination inhibition and neutralization, are often non-specific and costly. Su and colleagues used a piezotransducer-like arrangement to coat two surfaces of an immunochip with glycoprotein-E and non-structural protein 1 (NS1) and analyzed signals to diagnose dengue fever [67]. Because dengue virus deposition increased the resonance angle, Kumbhat et al. used surface plasmon resonance to diagnose it [68].

For diagnosis, Zaytseva and her colleagues used fluorescent liposomes conjugated with reporter probes that bind to dengue virus RNA [69]. Zhang and colleagues recently created a multi-stack paper immunoassay that removes proteinaceous material from saliva before detecting target antigens with adsorbed antibodies [70]. This device eliminates the need for centrifugation and improves the detection of dengue-specific immunoglobins (IgG) in serum, which can help distinguish primary from secondary infection.

Apart from the aforementioned issues with the antibody-functionalized surface method for diagnosis, one major limitation in all of these studies is that microfluidic platforms are unable to distinguish between different types and stages of dengue infection. There are four different dengue viruses that can cause dengue fever. Furthermore, disease severity is increased by coinfection and reinfection with another virus. As a result, the current challenge is to develop a microfluidic device that can distinguish virus strains and disease stages for clinical dengue diagnosis.

According to the WHO, TB kills about 1.3 million people each year (WHO, 2013). Despite the fact that this is a disease that is largely treatable, the large number of undiagnosed cases (global detection rate of 63%) prevents timely therapeutic interventions [71]. Traditional TB diagnostic methods suffer from the same drawbacks as other infectious disease diagnostic methods, such as low specificity, high costs, and the need for a large sample volume [72]. To that end, microfabricated devices may be able to fill some of the gaps in TB diagnostic efforts. For example, Wojcik and colleagues developed a colorimetric DNA test based on the conjugation of gold nanoparticles covalently bound to a *Mycobacterium tuberculosis* gene fragment [73].

2.3.10 Cholesterol Monitoring

Cholesterol secreted by cells is usually measured using cholesterol efflux assays with a radioactive or fluorescent label [74] or enzymatic assays on the collected culture medium [75]. Biosensors that use electrophoresis, colorimetry, and amperometry [75–78] have been used to measure cholesterol in microfluidics, but most of the time the measurement was not done in-line for cellular analysis [78].

In-line measurement of cholesterol secreted by human hepatocytes cultured in a microfluidic tissue-chip under perfusion was demonstrated by Karnik et al. This assay device has the advantage of allowing in-line real-time monitoring of biomolecules that use similar assay chemistry to cholesterol, such as glycerol, glucose, and fatty acids [79].

2.3.11 Pregnancy and Infertility Testing

Microfluidic systems have been used in the field of assisted reproductive technologies (ART) to help with sperm sorting [80], oocyte manipulation [81], insemination [82], embryo culturing [83], and sperm and embryo quality assessment by relying on these benefits [84]. Microfluidics is expected to increase sample preparation effectiveness, allow for consistency in cell/embryo culturing and operating environments, and minimize human error.

Despite proof-of-concept studies demonstrating the capabilities of microfluidic systems in a variety of ART areas, the majority of clinical applications of microfluidic systems has been in sperm purification or sorting. A key sample preparation process in the treatment of male-factor infertility is the selection of qualitatively and quantitatively sufficient sperm. Sperm-sorting processes based on microfluidic systems can increase motile sperm recovery from semen, sperm recovery from highly heterogeneous mixtures, and potentially reduce clinician ability requirements for the sperm purification process [85–88].

2.3.12 Hematological and Blood Gas Testing

Electrochemical or optical sensors are commonly used in blood gas analyzers (BGAs) for pH, pO_2, and pCO_2. The IRMA TruPoint by ITC Medical (Edison, NJ), the Critical Care Xpress series of instruments from Nova Biomedical (Waltham, MA), the cobas b 221 from Roche Diagnostics (USA), and the RAPID Point 400/405 from Siemens Medical Solutions (Erlangen, Germany) are some of the commercially available BGAs. The addition of a CO-oximetry unit to a BGA is an optional configuration [89].

The CO-oximetry unit is a multiwavelength spectrophotometer that measures the typical absorption spectra of various hemoglobin (Hb) species in order to differentiate O_2–Hb from other Hb species and calculate the O_2–Hb saturation number [89]. The combined analysis of plasma clotting, thrombocyte function, and fibrinolysis is known as viscoelastic coagulation testing [90]. The ROTEM from TEM International (Munich, Germany) and the Sonoclot series from Sienco Inc. (Boulder, CO) are two commercially available instruments. Platelet function in terms of in vitro bleeding time can be assessed using optical aggregometry [91].

2.3.13 Other POCT Devices

The PFA-100 System from Siemens Healthcare Diagnostics (Erlangen, Germany) and the VerifyNow System from Accumetrics (San Diego, CA) are two commercially available analyzers. The importance of plasma lipid levels in determining cardiovascular risk is well understood. Accutrend Plus (Roche Diagnostics, USA), CardioChek PA (PTS Diagnostics, Indianapolis, IN), and MultiCare-in (BSI, Arezzo, Italy) are some of the small POCT cholesterol systems. The color intensity of a chromogen reaction is proportional to the concentration of cholesterol or triglycerides in blood in the MultiCare systems [92], which are pocket-sized reflectance photometers.

To produce results for cholesterol and triglyceride assays, a drop of capillary blood is applied to the test strip. There are also desktop POCT cholesterol analyzers available, such as the Abaxis Piccolo Xpress and the Alere Cholestech LDX, [93]. Immunoassay test panels for cardiac markers are available on the Biosite Triage MeterPlus [94] (Biosite, San Diego, CA).

Breast cancer is the most commonly diagnosed cancer in women, and it is the second leading cause of cancer death after lung cancer. A fast POC breath test for detecting breast cancer breath biomarkers has been tested. In about six minutes, the BreathLink system [95] (Menssana Research, Newark, NJ) collects, concentrates, and analyzes breath volatile organic compounds (VOCs).

2.4 LIMITATIONS OF CONVENTIONAL POC DIAGNOSTIC

As previously stated, LFIA methods have enabled rapid and easy POC diagnosis of a variety of diseases in a variety of settings. Their simplicity, however, limits their performance, and more complex devices are often required for accurate diagnosis. Several transferable diseases, such as HIV, TB, and sexually transmitted diseases (STDs), have been successfully diagnosed using LFIA or immunochromatographic (ICS)-based strips in low-resource environments. Capillary forces are used to control fluid flow in ICS strips.

Because of the nature of sample addition to such devices, there may be some differences between the amount of sample added originally and the amount actually used in the test, resulting in poor accuracy. Furthermore, pretreatment of samples is required in many test formats when the sample matrix is not a fluid or when major interferents are present. Furthermore, limitations in these platforms' limit of detection may limit their application to the detection of analytes that are abundant in the sample tested, such as lymphocytes in HIV tests or pathogenic microorganisms in malaria tests.

As a result, even when validated by a reader, the results may only be qualitative or semi-quantitative. An LFIA test's high coefficient of variation (CV) limits the test's reproducibility and, as a result, its ability to confirm the presence of the disease [12]. Variable test performance as a result of environmental factors such as temperature, humidity, heat, air, and sunshine add to the complexity [96].

2.5 CURRENT TRENDS, FUTURE PROSPECTS, AND CONCLUDING REMARKS

Microfluidic devices arose from the integrated circuits (IC) industry and have been on the rise since the early 1990s, with early microfluidic devices primarily consisting of silicon-based analyzers with channels for entity separation in a fluid carrier. Recent advances in microfabrication and device design have brought innovative components and platform integration to the forefront, with applications in biology, pharmaceutical sciences, medicine, food, and environmental monitoring.

Higher sensitivities in shorter time spans can be achieved when analyzing entire chemical analyses in miniaturized volumes thanks to the implementation of "micro total analysis systems" (microTAS) or "labs-on-a-chip" (LOC) technology. Due to newer materials, fabrication methods, and device designs, microfluidic devices are becoming more integrated, providing new capabilities at the microscale, and enabling the commercialization of these platforms in life science laboratories as well as non-laboratory environments. MicroTAS in the field of microfluidics has revolutionized the way "fluids" are handled in almost any field, including biotechnology and biomedicine, chemical synthesis, environmental monitoring, cosmetics, and advanced-materials green technologies.

This holds great promise for truly portable and compact systems that can bring biochemistry LOC to the masses in a much more user-friendly manner, enabling the delivery of diagnostic systems for healthcare and greatly facilitating global health efforts. It will aid in the development of standalone, autonomous, and distributed monitors for constant and frequent public health check-ups and biothreat surveillance well ahead of time. Furthermore, genetic information obtained from patients through microfluidics-based home test machines can assist not only in the prescription and administration of drugs, but also in the recommendation of specific personalized medicine to healthcare professionals, significantly increasing the effectiveness and safety of future medicine.

Microfluidics can help with the ubiquitous monitoring of infectious disease agents, biologically engineered pathogens, and toxic substances in the environment, which is important given recent environmental and national security threats. The Artificial Intelligence-flips, which are used to gather biochemical data, will be distributed in public places such as buildings, aircraft, subways, parks, stadiums, schools, and malls, among other places, to gather biochemical data and identify the onset of major infectious diseases or the early signs of a biological or chemical terrorist attack, among other things. This optimization protocol will be used by future cancer researchers to create true microfluidic models of the tumor-vascular interface.

The use of these well-validated models will allow for more in-depth research into the in vivo tumor microenvironment, tumor growth and development, and tumor interactions with nearby blood vessels and extra cellular matrix (ECM) components. For a wider impact and application, our model requires more research. Drug testing for anti-angiogenic therapies, for example, is still needed to see if the model will support the development of more advanced personalized solid tumor cancer therapeutics.

In conclusion, this chapter has focused on recent developments in nano/microfluidic POC devices and their clinical applications in resource-constrained environments. The majority of them showed great promise in terms of meeting the clinical and technical needs of global health care. However, one major challenge remains: How to effectively network these nano/microfluidic POC diagnostics between developed and developing countries, so that information can be shared with interactive input for improved global health. As previously stated, existing mobile phone networks may be useful in creating a network-based platform that can continuously enhance nano/microfluidic POC devices.

Another significant challenge may be establishing a proper regulation and standard for evaluating nano/microfluidic POC diagnostics for clinical use [4]. Several procedures, particularly for developing diagnostic tests, may be included in the evaluation criteria, such as the identification of the diagnostic target, the optimization of test reagents, and the development of a prototype. Nano/microfluidic technologies, as shown in the literature, hold great promise for establishing a standard of detection sensitivity level for POC devices through quantitative, proof-of-principle studies conducted in a rapid, controlled, and high-throughput manner.

Furthermore, in order to obtain FDA or NIH approval for legal clinical use of POC devices, nano/microfluidic technologies must be capable of meeting critical evaluation criteria, such as test characteristics and factors like (i) test performance (sensitivity and specificity), (ii) ease of use, (iii) conditions of use and storage, and (iv) shelf life [4]. For commercialization, efforts in this field should focus on non- and minimally instrumented nano/microfluidic POC diagnostics, as well as platforms that can operate without any peripherals.

ACKNOWLEDGMENTS

The authors thank Dr Avanish Kumar Srivastava, Director, Council of Scientific and Industrial Research–Advanced Materials and Process Research Bhopal, India, for his interest and encouragement in this work. The fellowship provided to Arpana Parihar under the DST-WoS-B scheme is duly acknowledged. Pushpesh Ranjan is thankful to the Council of Scientific and Industrial Research (CSIR), India for an award by the Senior Research Fellow. Raju Khan would like to acknowledge Science and Engineering Research Board for providing funds in the form of the IPA/2020/000130 project.

REFERENCES

1. Luppa, P. B., Bietenbeck, A., Beaudoin, C., & Giannetti, A. 2016. Clinically relevant analytical techniques, organizational concepts for application and future perspectives of point-of-care testing. *Biotechnology Advances*, 34(3), 139–160.
2. Sachdeva, S., Davis, R. W., & Saha, A. K. 2021. Microfluidic point-of-care testing: commercial landscape and future directions. *Frontiers in Bioengineering and Biotechnology*, 8, 602659.
3. Kankaanpää, M., Holma-Eriksson, M., Kapanen, S., Heitto, M., Bergström, S., Muukkonen, L., & Harjola, V. P. 2018. Comparison of the use of comprehensive point-of-care test panel to conventional laboratory process in emergency department. *BMC Emergency Medicine*, 18(1), 43.
4. Lee, W. G., Kim, Y. G., Chung, B. G., Demirci, U., & Khademhosseini, A. 2010. Nano/Microfluidics for diagnosis of infectious diseases in developing countries. *Advanced Drug Delivery Reviews*, 62(4–5), 449–457.
5. Huckle D. 2015. The impact of new trends in POCTs for companion diagnostics, non-invasive testing and molecular diagnostics. *Expert Review of Molecular Diagnostics*, 15(6), 815–827.
6. Srinivasan, B., & Tung, S. 2015. Development and applications of portable biosensors. *Journal of Laboratory Automation*, 20(4), 365–389.
7. Laxminarayan, R., Mills, A. J., Breman, J. G., Measham, A. R., Alleyne, G., Claeson, M., Jha, P., Musgrove, P., Chow, J., Shahid-Salles, S., & Jamison, D. T. 2006. Advancement of global health: key messages from the disease control priorities project. *Lancet*, 367(9517), 1193–1208.
8. Bloom, D. E., Cafiero, E. T., Jané-Llopis, E., Abrahams-Gessel, S. Bloom, L. R., Fathima, S., Feigl, A. B., Gaziano, T., Mowafi, M., Pandya, A., et al. 2011. *The Global Economic Burden of Noncommunicable Diseases*. Geneva, Switzerland: World Economic Forum, pp. 2–25.

9. The Emerging Crisis: Noncommunicable Diseases. Available online: http://www.cfr.org/diseases-noncommunicable/NCDs-interactive/p33802?cid=otr-marketing_use-NCDs_interactive/#!/. Date accessed on 21/03/2021.
10. Learmonth, K. M., McPhee, D. A., Jardine, D. K., Walker, S. K., Aye, T. T., & Dax, E. M. 2008. Assessing proficiency of interpretation of rapid human immunodeficiency virus assays in nonlaboratory settings: ensuring quality of testing. *Journal of Clinical Microbiology*, 46(5), 1692–1697.
11. O'Farrell, B. 2013. Lateral flow immunoassay systems: evolution from the current state of the art to the next generation of highly sensitive, quantitative rapid assays. In *The Immunoassay Handbook: Theory and Applications of Ligand Binding, ELISA and Related Techniques*, 4th ed., Wild, D., Ed. Boston, MA: Newnes, pp. 89–107.
12. Posthuma-Trumpie, G. A., Korf, J., & van Amerongen, A. 2009. Lateral flow (immuno) assay: its strengths, weaknesses, opportunities and threats. A literature survey. *Analytical and Bioanalytical Chemistry*, 393(2), 569–582.
13. McPartlin, D. A., & O'Kennedy, R. J. 2014. Point-of-care diagnostics, a major opportunity for change in traditional diagnostic approaches: potential and limitations. *Expert Review of Molecular Diagnostics*, 14(8), 979–998.
14. UNAIDS World AIDS Day Report 2012. Available online: http://www.unaids.org/sites/default/files/media_asset/JC2434_WorldAIDSday_results_en_1.pdf. Date accessed on 21/03/2021.
15. Kozel, T. R., & Bauman, S. K. 2012. CrAg lateral flow assay for cryptococcosis. *Expert Opinion on Medical Diagnostics*, 6(3), 245–251.
16. Tangpukdee, N., Duangdee, C., Wilairatana, P., & Krudsood, S. 2009. Malaria diagnosis: a brief review. *The Korean Journal of Parasitology*, 47(2), 93–102.
17. Yao, P., Xu, T. & Tung, S. 2018. Pneumatic microfluidic device by 3D printing technology for insulin determination. In IEEE 12th International Conference on Nano/Molecular Medicine and Engineering (NANOMED) 2018, Hawaii, USA, p. 8641565.
18. Samper, I. C., Gowers, S., Rogers, M. L., Murray, D., Jewell, S. L., Pahl, C., Strong, A. J., & Boutelle, M. G. 2019. 3D printed microfluidic device for online detection of neurochemical changes with high temporal resolution in human brain microdialysate. *Lab on a Chip*, 19(11), 2038–2048.
19. Oh, S., & Choi, S. 2018. 3D-printed capillary circuits for calibration-free viscosity measurement of Newtonian and Non-Newtonian fluids. *Micromachines*, 9(7), 314.
20. Pashchenko, O., Shelby, T., Banerjee, T., & Santra, S. 2018. A Comparison of Optical, Electrochemical, Magnetic, and Colorimetric Point-of-Care Biosensors for Infectious Disease Diagnosis. *ACS Infectious Diseases*, 4(8), 1162–1178.
21. Binx IO. Available online at: https://mybinxhealth.com/point-of-care/. Date accessed 21/03/2021.
22. iSTAT Corporation. Available online at: https://www.sec.gov/Archives/edgar/data/882365/000091205702038057/a2090815zex-99_1.htm#toc_ka8573_6
23. iSTAT Cartridges. Available online at: https://www.pointofcare.abbott/us/en/offerings/istat/istat-test-cartridges. Date accessed 21/03/2021.
24. Testing for COVID-19 | CDC. Available online at: https://www.cdc.gov/coronavirus/2019-ncov/symptoms-testing/testing.html. Date accessed 21/03/2021.
25. COVID-19, Antigen Test. Available online at: https://www.fda.gov/newsevents/press-announcements/coronavirus-covid-19-update-fda-authorizes-first-antigen-test-help-rapid-detection-virus-causes. Date accessed 21/03/2021.
26. Roche Diagnostics. Available online at: https://diagnostics.roche.com/global/en/c/covid-19-pandemic.html. Date accessed 21/03/2021.

27. Opollo, V. S., Nikuze, A., Ben-Farhat, J., Anyango, E., Humwa, F., Oyaro, B., Wanjala, S., Omwoyo, W., Majiwa, M., Akelo, V., Zeh, C., & Maman, D. 2018. Field evaluation of near point of care Cepheid GeneXpert HIV-1 Qual for early infant diagnosis. *PloS One*, 13(12), e0209778.
28. Etherington, D. 2020. Mesa biotech gains emergency FDA approval for rapid, point-of-care COVID-19 test. *TechCrunch*. Available online at: https://techcrunch.com/2020/03/24/mesa-biotech-gains-emergency-fda-approval-for-rapid-point-of-care-covid-19-test/. Date accessed 21/03/2021.
29. Padayatchi, N., & Friedland, G. 2008. Decentralised management of drug-resistant tuberculosis (MDR- and XDR-TB) in South Africa: an alternative model of care. *The International Journal of Tuberculosis and Lung Disease*, 12(8), 978–980.
30. Singh, J. A., Upshur, R., & Padayatchi, N. 2007. XDR-TB in South Africa: no time for denial or complacency. *PLoS Medicine*, 4(1), e50.
31. Ridderhof, J. C., van Deun, A., Kam, K. M., Narayanan, P. R., & Aziz, M. A. 2007. Roles of laboratories and laboratory systems in effective tuberculosis programmes. *Bulletin of the World Health Organization*, 85(5), 354–359.
32. Steingart, K. R., Ng, V., Henry, M., Hopewell, P. C., Ramsay, A., Cunningham, J., Urbanczik, R., Perkins, M. D., Aziz, M. A., & Pai, M. 2006. Sputum processing methods to improve the sensitivity of smear microscopy for tuberculosis: a systematic review. *The Lancet. Infectious Diseases*, 6(10), 664–674.
33. Ke, C., Berney, H., Mathewson, A. & Sheehan, M. M. 2004. Rapid amplification for the detection of Mycobacterium tuberculosis using a non-contact heating method in a silicon microreactor based thermal cycler, *Sensors and Actuators B*, 102 (2004), 308–314.
34. Lee, H., Sun, E., Ham, D., & Weissleder, R. 2008. Chip-NMR biosensor for detection and molecular analysis of cells. *Nature Medicine*, 14(8), 869–874.
35. WHO, Patient Monitoring Guidelines for HIV Care and ART, 2005. http://www. who.int/hiv/pub/guidelines/patientmonitoring.pdf. Date accessed 21/03/2021.
36. Willyard C. 2007. Simpler tests for immune cells could transform AIDS care in Africa. *Nature Medicine*, 13(10), 1131.
37. Cohen J. 2004. Monitoring treatment: at what cost?. *Science*, 304(5679), 1936.
38. Linder, V., Sia, S. K., & Whitesides, G. M. 2005. Reagent-loaded cartridges for valveless and automated fluid delivery in microfluidic devices. *Analytical Chemistry*, 77(1), 64–71.
39. Butte, M. J., Wong, A. P., A. H. Sharpe & Whitesides, G. M. 2005. Microfluidic device for low cost screening of newborns for severe combined immune deficiency, *Clinical and Experimental Immunology* 116 (2005) 282.
40. Simon, V., & Ho, D. D. (2003). HIV-1 dynamics in vivo: implications for therapy. *Nature Reviews. Microbiology*, 1(3), 181–190.
41. Hammer, S. M., Eron, J. J., Jr, Reiss, P., Schooley, R. T., Thompson, M. A. Walmsley, S., Cahn, P., Fischl, M. A., Gatell, J. M., Hirsch, M. S., Jacobsen, D. M., Montaner, J. S., Richman, D. D., Yeni, P. G., Volberding, P. A., & International AIDS Society-USA. 2008. Antiretroviral treatment of adult HIV infection: 2008 recommendations of the International AIDS Society-USA panel. *JAMA*, 300(5), 555–570.
42. Fiscus, S. A., Cheng, B., Crowe, S. M., Demeter, L., Jennings, C., Miller, V., Respess, R., & Stevens, W. 2006. Forum for collaborative HIV research alternative viral load assay working group HIV-1 viral load assays for resource-limited settings. *PLoS Medicine*, 3(10), e417.
43. Dalgleish, A. G., Beverley, P. C., Clapham, P. R., Crawford, D. H., Greaves, M. F., & Weiss, R. A. 1984. The CD4 (T4) antigen is an essential component of the receptor for the AIDS retrovirus. *Nature*, 312(5996), 763–767.

44. WHO, Patient Monitoring Guidelines for HIV Care and Antiretroviral Therapy. 2008. http://www.who.int/hiv/. Date accessed 21/03/2021.
45. Sia, S. K., Linder, V., Parviz, B. A., Siegel, A., & Whitesides, G. M. 2004. An integrated approach to a portable and low-cost immunoassay for resource-poor settings. *Angewandte Chemie*, 43(4), 498–502.
46. Spacek, L. A., Shihab, H. M., Lutwama, F., Summerton, J., Mayanja, H., Kamya, M., Ronald, A., Margolick, J. B., Nilles, T. L., & Quinn, T. C. (2006). Evaluation of a low-cost method, the Guava EasyCD4 assay, to enumerate CD4-positive lymphocyte counts in HIV-infected patients in the United States and Uganda. *Journal of Acquired Immune Deficiency Syndromes* (1999), 41(5), 607–610.
47. Lee, S. H., Kim, S. W., Kang, J. Y., & Ahn, C. H. 2008. A polymer lab-on-a-chip for reverse transcription (RT)-PCR based point-of-care clinical diagnostics. *Lab on a Chip*, 8(12), 2121–2127.
48. Evans, J. A., Adusei, A., Timmann, C., May, J., Mack, D., Agbenyega, T., Horstmann, R. D., & Frimpong, E. 2004. High mortality of infant bacteraemia clinically indistinguishable from severe malaria. *QJM*, 97(9), 591–597.
49. Moody A. 2002. Rapid diagnostic tests for malaria parasites. *Clinical Microbiology Reviews*, 15(1), 66–78.
50. Rafael, M. E., Taylor, T., Magill, A., Lim, Y. W., Girosi, F., & Allan, R. 2006. Reducing the burden of childhood malaria in Africa: the role of improved. *Nature*, 444(Suppl 1), 39–48.
51. Laveran C. L. 1982. Classics in infectious diseases: a newly discovered parasite in the blood of patients suffering from malaria. Parasitic etiology of attacks of malaria: Charles Louis Alphonse Laveran (1845–1922). *Reviews of Infectious Diseases*, 4(4), 908–911.
52. Jorgensen, P., Chanthap, L., Rebueno, A., Tsuyuoka, R., & Bell, D. 2006. Malaria rapid diagnostic tests in tropical climates: the need for a cool chain. *The American Journal of Tropical Medicine and Hygiene*, 74(5), 750–754.
53. Houpt, E. R., & Guerrant, R. L. 2008. Technology in global health: the need for essential diagnostics. *Lancet*, 372(9642), 873–874.
54. Bell, D., Wongsrichanalai, C., & Barnwell, J. W. 2006. Ensuring quality and access for malaria diagnosis: how can it be achieved?. *Nature Reviews. Microbiology*, 4(9), 682–695.
55. Levy, M. M., Fink, M. P., Marshall, J. C., Abraham, E., Angus, D., Cook, D., Cohen, J., Opal, S. M., Vincent, J. L., Ramsay, G., & SCCM/ESICM/ACCP/ATS/SIS 2003. 2001 SCCM/ESICM/ACCP/ATS/SIS International Sepsis Definitions Conference. *Critical Care Medicine*, 31(4), 1250–1256.
56. Hotchkiss, R. S., & Karl, I. E. 2003. The pathophysiology and treatment of sepsis. *The New England Journal of Medicine*, 348(2), 138–150.
57. Lever, A., & Mackenzie, I. 2007. Sepsis: definition, epidemiology, and diagnosis. *BMJ*, 335(7625), 879–883.
58. Kumar, A., Roberts, D., Wood, K. E., Light, B., Parrillo, J. E., Sharma, S., Suppes, R., Feinstein, D., Zanotti, S., Taiberg, L., Gurka, D., Kumar, A., & Cheang, M. 2006. Duration of hypotension before initiation of effective antimicrobial therapy is the critical determinant of survival in human septic shock. *Critical Care Medicine*, 34(6), 1589–1596.
59. Ritzi-Lehnert M. 2012. Development of chip-compatible sample preparation for diagnosis of infectious diseases. *Expert Review of Molecular Diagnostics*, 12(2), 189–206.
60. Su, W., Gao, X., Jiang, L., & Qin, J. 2015. Microfluidic platform towards point-of-care diagnostics in infectious diseases. *Journal of Chromatography A*, 1377, 13–26.

61. Hui, D. S., Chan, M. C., Wu, A. K., & Ng, P. C. 2004. Severe acute respiratory syndrome (SARS): epidemiology and clinical features. *Postgraduate Medical Journal*, 80(945), 373–381.
62. Zhou, X., Liu, D., Zhong, R., Dai, Z., Wu, D., Wang, H., Du, Y., Xia, Z., Zhang, L., Mei, X., & Lin, B. 2004. Determination of SARS-coronavirus by a microfluidic chip system. *Electrophoresis*, 25(17), 3032–3039.
63. Choi, J. W., Oh, K. W., Thomas, J. H., Heineman, W. R., Halsall, H. B., Nevin, J. H., Helmicki, A. J., Henderson, H. T., & Ahn, C. H. 2002. An integrated microfluidic biochemical detection system for protein analysis with magnetic bead-based sampling capabilities. *Lab on a Chip*, 2(1), 27–30.
64. Ramalingam, N., San, T. C., Kai, T. J., Mak, M., & Gong, H. Q. 2009. Microfluidic devices harboring unsealed reactors for real-time isothermal helicase-dependent amplification. *Microfluidics and Nanofluidics*, 7(3), 325.
65. Rigau-Pérez, J. G., Clark, G. G., Gubler, D. J., Reiter, P., Sanders, E. J., & Vorndam, A. V. 1998. Dengue and dengue haemorrhagic fever. *Lancet*, 352(9132), 971–977.
66. Gubler D. J. 2002. Epidemic dengue/dengue hemorrhagic fever as a public health, social and economic problem in the 21st century. *Trends in Microbiology*, 10(2), 100–103.
67. Su, C. C., Wu, T. Z., Chen, L. K., Yang, H. H., & Tai, D. F. 2003. Development of immunochips for the detection of dengue viral antigens. *Analytica Chimica Acta* 479(2), 117–123.
68. Kumbhat, S., Sharma, K., Gehlot, R., Solanki, A., & Joshi, V. 2010. Surface plasmon resonance based immunosensor for serological diagnosis of dengue virus infection. *Journal of Pharmaceutical and Biomedical Analysis*, 52(2), 255–259.
69. Zaytseva, N. V., Montagna, R. A., & Baeumner, A. J. 2005. Microfluidic biosensor for the serotype-specific detection of dengue virus RNA. *Analytical Chemistry*, 77(23), 7520–7527.
70. Zhang, Y., Bai, J., & Ying, J. Y. 2015. A stacking flow immunoassay for the detection of dengue-specific immunoglobulins in salivary fluid. *Lab on a Chip*, 15(6), 1465–1471.
71. McNerney, R., & Daley, P. 2011. Towards a point-of-care test for active tuberculosis: obstacles and opportunities. *Nature Reviews. Microbiology*, 9(3), 204–213.
72. Mani, V., Wang, S., Inci, F., De Libero, G., Singhal, A., & Demirci, U. 2014. Emerging technologies for monitoring drug-resistant tuberculosis at the point-of-care. *Advanced Drug Delivery Reviews*, 78, 105–117.
73. Dheda, K., Ruhwald, M., Theron, G., Peter, J., & Yam, W. C. 2013. Point-of-care diagnosis of tuberculosis: past, present and future. *Respirology*, 18(2), 217–232.
74. Tsai, A. G., Williamson, D. F., & Glick, H. A. 2011. Direct medical cost of overweight and obesity in the USA: a quantitative systematic review. *Obesity Reviews*, 12(1), 50–61.
75. Krueger, W. H., Tanasijevic, B., Barber, V., Flamier, A., Gu, X., Manautou, J., & Rasmussen, T. P. 2013. Cholesterol-secreting and statin-responsive hepatocytes from human ES and iPS cells to model hepatic involvement in cardiovascular health. *PloS One*, 8(7), e67296.
76. Ruecha, N., Siangproh, W., & Chailapakul, O. 2011. A fast and highly sensitive detection of cholesterol using polymer microfluidic devices and amperometric system. *Talanta*, 84(5), 1323–1328.
77. Wisitsoraat, A., Sritongkham, P., Karuwan, C., Phokharatkul, D., Maturos, T., & Tuantranont, A. 2010. Fast cholesterol detection using flow injection microfluidic device with functionalized carbon nanotubes based electrochemical sensor. *Biosensors & Bioelectronics*, 26(4), 1514–1520.

78. Ali, M. A., Srivastava, S., Solanki, P. R., Reddy, V., Agrawal, V. V., Kim, C., John, R., & Malhotra, B. D. 2013. Highly efficient bienzyme functionalized nanocomposite-based microfluidics biosensor platform for biomedical application. *Scientific Reports*, 3, 2661.
79. Karnik, S., Lee, C., Cancino, A., & Bhushan, A. 2018. Real-time measurement of cholesterol secreted by human hepatocytes using a novel microfluidic assay. *Technology*, 6(3–4), 135–141.
80. Nosrati, R., Graham, P. J., Zhang, B., Riordon, J., Lagunov, A., Hannam, T. G., Escobedo, C., Jarvi, K., & Sinton, D. 2017. Microfluidics for sperm analysis and selection. *Nature Reviews. Urology*, 14(12), 707–730.
81. Sadani, Z., Wacogne, B., & Pieralli, C. 2005. Microsystems and microfluidic device for single oocyte transportation and trapping: toward the automation of in vitro fertilising. *Sensors and Actuators A*, 121, 364–372.
82. Matsuura, K., Uozumi, T., Furuichi, T., Sugimoto, I., Kodama, M., & Funahashi, H. 2013. A microfluidic device to reduce treatment time of intracytoplasmic sperm injection. *Fertility and Sterility*, 99(2), 400–407.
83. Krisher, R. L., & Wheeler, M. B. 2010. Towards the use of microfluidics for individual embryo culture. *Reproduction, Fertility, and Development*, 22(1), 32–39.
84. Yanez, L. Z., & Camarillo, D. B. 2017. Microfluidic analysis of oocyte and embryo biomechanical properties to improve outcomes in assisted reproductive technologies. *Molecular Human Reproduction*, 23(4), 235–247.
85. Samuel, R., Badamjav, O., Murphy, K. E., Patel, D. P., Son, J., Gale, B. K., Carrell, D. T., & Hotaling, J. M. 2016. Microfluidics: the future of microdissection TESE? *Systems Biology in Reproductive Medicine*, 62(3), 161–170.
86. Son, J., Murphy, K., & Samuel, R. 2015. Non-motile sperm cell separation using a spiral channel. *Analytical Methods*, 7, 8041–8047.
87. Son, J., Samuel, R., Gale, B. K., Carrell, D. T., & Hotaling, J. M. 2017. Separation of sperm cells from samples containing high concentrations of white blood cells using a spiral channel. *Biomicrofluidics*, 11(5), 054106.
88. Jenkins T., Samuel, R., & Jafek, A., 2011. Rapid microfluidic sperm isolation from microtesesamples in men with non-obstructive azoospermia. *Fertility and Sterility*, 1(108), e244.
89. Luppa, P. B., Müller, C., Schlichtiger, A., & Schlebusch, H. 2011. Point-of-care testing (POCT): current techniques and future perspectives. *Trends in Analytical Chemistry*, 30(6), 887–898.
90. Ganter, M. T., & Hofer, C. K. 2008. Coagulation monitoring: current techniques and clinical use of viscoelastic point-of-care coagulation devices. *Anesthesia and Analgesia*, 106(5), 1366–1375.
91. Dyszkiewicz-Korpanty, A., Olteanu, H., Frenkel, E. P., & Sarode, R. 2007. Clopidogrel anti-platelet effect: an evaluation by optical aggregometry, impedance aggregometry, and the platelet function analyzer (PFA-100). *Platelets*, 18(7), 491–496.
92. Rapi, S., Bazzini, C., Tozzetti, C., Sbolci, V., & Modesti, P. A. 2009. Point-of-care testing of cholesterol and triglycerides for epidemiologic studies: evaluation of the multicare-in system. *Translational Research: The Journal of Laboratory and Clinical Medicine*, 153(2), 71–76.
93. Shephard, M. D., Mazzachi, B. C., & Shephard, A. K. 2007. Comparative performance of two point-of-care analysers for lipid testing. *Clinical Laboratory*, 53(9–12), 561–566.
94. Sykes, E., Karcher, R. E., Eisenstadt, J., Tushman, D. A., Balasubramaniam, M., Gusway, J., & Perason, V. J. 2005. Analytical relationships among Biosite, Bayer, and Roche methods for BNP and NT-proBNP. *American Journal of Clinical Pathology*, 123(4), 584–590.

95. Phillips, M., Beatty, J. D., Cataneo, R. N., Huston, J., Kaplan, P. D., Lalisang, R. I., Lambin, P., Lobbes, M. B., Mundada, M., Pappas, N., & Patel, U. 2014. Rapid point-of-care breath test for biomarkers of breast cancer and abnormal mammograms. *PloS One*, 9(3), e90226.
96. Novak, M. T., Kotanen, C. N., Carrara, S., Guiseppi-Elie, A., Moussy, F. G. 2013 Diagnostic tools and technologies for infectious and non-communicable diseases in low-and-middle-income countries. *Health Technology*, 3, 271–281.

3 Microfluidic Paper-Based Analytical Devices for Glucose Detection

Shristi Handa, Vibhav Katoch, and Bhanu Prakash

CONTENTS

DOI: 10.1201/9781003033479-3

LIST OF ABBREVIATIONS

TBHBA	2,4,6-tribromo-3-hydroxybenzoic acid
TMB	3,3′,5,5′-tetramethylbenzidine
4-APP	4-aminoantipyrine
AKD	alkyl ketene dimer
CdS	cadmium sulfide
CdTe	cadmium telluride
CPP	casting polypropylene
CL	chemiluminescence
CIJ	continuous inkjet printing
CuO–ZnO	copper oxide–zinc oxide
DSA	drop shape analysis
DOD	drop-on-demand inkjet printing
ECL	electrochemical detection
ECD	Electrochemical detection
ePADs	electrochemical paper-based analytical devices
EDX	energy dispersive X-ray
FLASH	fast lithographic activation of sheets
FITC	fluorescein iso-thiocyanate
FTO	fluorine-doped tin oxide
GOx	glucose oxidase
GO	graphene oxide
HRP	horseradish peroxidase
(H_2O_2)	hydrogen peroxide
IgG	immunoglobulin G
ITO	indium tin oxide
CIE	International Commission on Illumination
MMA/DVB	methyl methacrylate/1,2-divinylbenzene
μPADs	microfluidic paper-based analytical devices
NRs	nanorods
pl	picoliter
Pt	platinum
POCTs	point-of-care tests
PEDOT PSS	poly(3,4-ethylenedioxythiophene) polystyrene sulfonate

PVDF	poly(vinylidene fluoride)
PEMs	polyelectrolyte multilayers
PEO	polyethylene oxide
PCB	printed circuit boards
PB-SPEs	Prussian blue modified screen-printed electrode
microRNA	ribonucleic acid
SEM	scanning electron microscopy
Ag/AgCl	silver/silver chloride
NaOH	sodium hydroxide
SI-ATRP	surface-initiated atom transfer radical polymerization
TiO_2 NPs	titanium dioxide nanoparticles
TTIP	titanium isopropoxide
UV	ultraviolet
UV–Vis	ultraviolet–visible
W/O/O	water-in-oil-in-oil
W/O/W	water-in-oil-in-water
WHO	World Health Organization
ZnO	zinc oxide

3.1 PAPER-BASED MICROFLUIDIC DEVICES: AN INTRODUCTION

Of all the essential metabolic intermediates, glucose is a major energy source for cellular activity in the living body. It is crucial to maintain a proper glucose concentration in the blood. However, many people in the world suffer from metabolic disorders such as those affecting glucose processing, which result in harmful consequences like diabetes mellitus. Diabetes mellitus is one of the most widespread chronic diseases, leading to multiple complications such as blindness (Cui et al. 2017), cardiovascular disorders (Avogaro 2016, Appleton et al. 2013), and kidney failure (Wanner et al. 2016). Because of the severe medical ramifications and associated complications, there is a critical need to monitor the blood glucose level in the human body (Nathan et al. 1993, Nathan 2014). Therefore, according to the American Diabetes Association, frequent monitoring of blood glucose concentration is crucial for medical diagnosis and prevention of complications originating because of diabetes. Thus, the first step in curing a disease is detecting the disease effectively (Kaefer et al. 2010, Chen et al. 2012).

To estimate and regulate the concentration of blood glucose levels, different glucose sensors have been developed. Among various methods for glucose measurements, optical and electrochemical analysis have been widely investigated. Optical methods work on the principle of the color pigmentation that occurs during an enzymatic reaction in the presence of an indicator to reflect the amount of glucose concentration through the color change. Although this method provides patients with a spontaneous way to check their blood glucose level, it is not an effective method for quantifying glucose levels, even for measuring low glucose levels. This is because of the requirement for bulky instrumentation such as a spectrophotometer, which makes it less feasible for commercial use. Electrochemical analysis involves a quantitative mode of operation; therefore, it can be used over a broad detection range. Electrochemical signals are measured and directly converted to the corresponding concentration of glucose. Extensive research has been performed to establish more sensitive measurements and selective modes of operation, improving the structure of the electrodes, surface functionalization techniques, and electrochemical analysis methods. Several glucose monitoring methods have been developed for non-invasive glucose monitoring and direct blood glucose monitoring, making patient-friendly diabetes management possible. Thus, microfluidic technology plays a vital role in the evolution of biomedical research. From the vast pool of microfluidic device technologies for medical diagnostics, paper-based microfluidic devices construct a framework that leaves the complication level and cost low while keeping efficiency relatively high.

The term "ASSURED" signifying "affordable, sensitive, specific, user friendly, rapid and robust, equipment-free, and delivered to those in need," was coined by the World Health Organization (WHO) as the necessary guidelines to be taken into account during the design of diagnostic point-of-care tests (POCTs). Cellulose is one of the most abundant biopolymers available on earth, mainly used to produce paper, being composed of a network of hydrophilic cellulose fibers (Martinez et al. 2010). Paper, due to its natural porous microstructure, makes it responsive to lateral flow via capillary action, thus using it for on-site analysis that requires no external forces assisted by microfluidic pumps, etc. (Martinez et al. 2007, Yager et al. 2006). A promising and powerful platform consisting of microfluidic paper-based analytical devices (μPADs) has shown potential in the development of POCTs.

In 2007, the Whitesides group was the first to use the concept of paper-based microfluidic devices (Martinez et al. 2007). Patterned paper for microfluidic devices was employed, allowing liquid transport through capillary force without using external equipment. Due to the advantages of simplicity, portability, nominal reagent consumption, and economic affordability, μPADs are a popular tool in applications, such as clinical diagnostics (Martinez et al. 2007, Aragay et al. 2011), point-of-care multiplexed assays of nucleic acids using microcapillary-based loop-mediated isothermal amplification (Zhang et al. 2014), and food safety (Aid et al. 2015).

The surface of the paper is hydrophilic by nature, therefore, to fabricate μPADs, hydrophobic barriers are created to direct the flow of the fluid either through desired trails or to restrict the flow within the desired location. Several methods such as photolithography, plasma treating (Li et al. 2008), wax printing (Yu and Shi 2015, Lu et al. 2009), flexography (Sameenoi et al. 2014), laser treating (Chitnis et al. 2011), and screen-printing, have been developed to manufacture hydrophobic barriers. With the advancements in research related to μPADs, the detection techniques, namely colorimetric detection (Wei et al. 2016), electrochemical detection (ECL) (Zhao et al. 2013), chemiluminescence (CL) (Chen et al. 2014), and fluorescence, have been applied to paper-based devices for the fabrication of rapid diagnosis and are discussed in detail in the later sections of the chapter.

3.2 FABRICATION METHODS

For the fabrication of μPADs or the creation of hydrophobic barriers, some techniques, such as photolithography, wax printing, screen printing, plasma treating, laser treating, etc., have been established. In the photolithography process, photoresists are used in the fabrication of μPADs, which are costly, and the process requires sophisticated and expensive photolithography equipment, thus making it a less popular choice for device fabrication. The patterning of paper with wax-printing technology is considered an easy process for fabricating μPADs which provides high resolution. On the contrary, the high running cost of commercial wax printers and low melting points of wax restrict their use during batch production. The screen-printing method has shown higher resolutions than wax printing but has limitations in terms of the prefabrication of screens. Fabricating channels by plasma treating produces channels that do not affect the surface topography, but this cannot be used for mass production. Since each fabrication method has its own merits and demerits, the principal issue is still associated with the economic benefit of μPADs for mass production, especially the fabrication of microfluidic devices in glucose detection. Thus, a balance must be made between cost, performance, and research on developing new materials and processing techniques.

3.2.1 Lithography

The word lithography, coined from two Greek words, "Lithos" – signifying stone, and "graphene," meaning to write, was invented by German author Alois Senefelderin in 1796 in search of a practical way to publish his plays. The history of lithography occurs mainly in four significant steps: – (i) the invention and first use of the process, (ii) the introduction of photography to the process, (iii) the addition of the offset press to the process, and (iv) the revolution of lithography.

With the advances in research, photolithography's rebirth took place the early 19th century, which is heralded as a game-changer in the study and industry. A brief timeline of microfluidics technology development came into the picture in early 1947 with the first application of printed circuit boards (PCB). Lithography can be further subdivided into photolithography, electron beam lithography, X-ray and extreme ultraviolet (UV) lithography, focused ion beam and neutral atomic beam lithography, soft lithography, colloidal lithography, nanoimprint lithography, scanning probe lithography, atomic force microscope nanolithography, and others.

3.2.1.1 Basic Principle

The term photolithography refers to the use of light to alter the thin film's solubility. Using a mask or stencil, the photosensitive materials' selected parts are usually protected while the uncovered ones change their properties during exposure. After immersion into a chemical developer, the parts that become more soluble are dissolved, leaving the substrates' desired pattern. Here the photosensitive materials play a crucial role in the success of the process. Such materials are organic polymers called a photoresist, which can go through a series of photochemical reactions when exposed to light.

Photolithography primarily consists of three process steps, as illustrated in Figure 3.1:

- Positioning process: Lateral positioning of the mask and the substrate, which is coated with a resist.

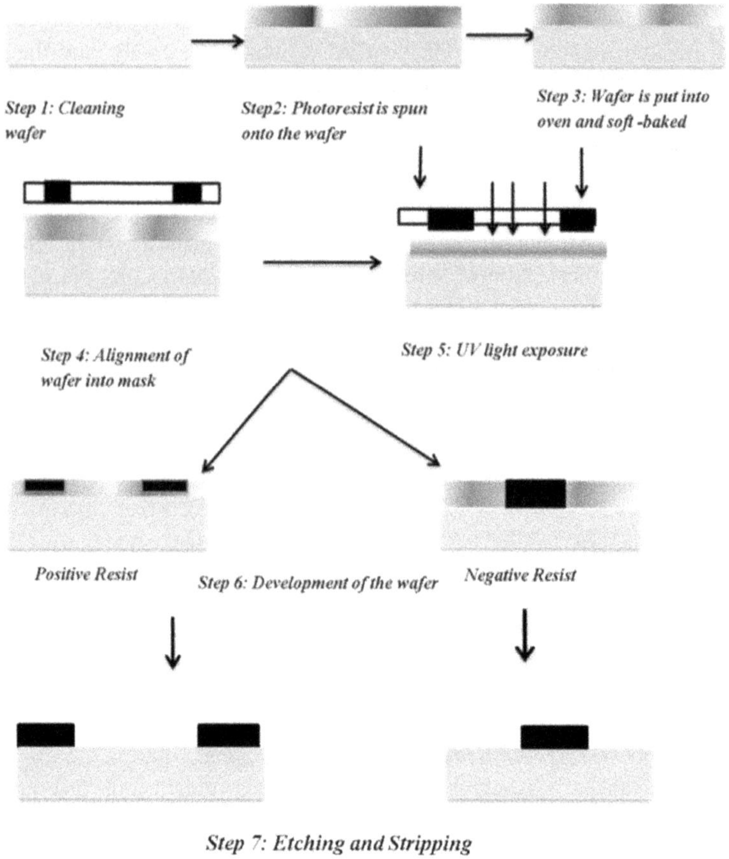

FIGURE 3.1 Steps involved in the photolithography process.

- Exposure process: Optical or X-ray exposure of the resist layer, transferring patterns to the photoresist layer by changing properties of the exposed area.
- Development process: Dissolution (for negative resist) or etching (for positive resist) of the resist pattern in a developer solution.

In 2007, Whitesides and coworkers reported the first microfluidic device fabricated using the photolithography technique (Martinez et al. 2007). To create a rapid, convenient, and cost-effective process, Whitesides and coworkers further tried to simplify their process with the development of another method called fast lithographic activation of sheets (FLASH) (Martinez et al. 2008). The FLASH method replaced the expensive photolithography equipment and cleanroom environment with a simple ultraviolet lamp and a hotplate. Another method used a hydrophobic surface created by thermally depositing titanium dioxide nanoparticles (TiO_2 NPs). A laser-based fabrication method was investigated by Sones et al. Light-sensitive polymer-soaked paper was patterned directly using a UV laser, achieving a significantly high resolution of 80 μm in the fluidic channel for paper-based microfluidic devices (Sones et al. 2014).

Currently, photolithography is the most widespread technique for the fabrication and patterning of integrated circuits. It offers reproducibility and reliability for the fabrication of electronic devices with high resolution and accuracy. Its primary disadvantage is the limitation in the topography, as it can only be implemented on planar substrates.

3.2.2 Wax Printing

The use of wax technology to fabricate μPADs began in 2009. The first step in this process is to create a mask with computer-aided software, which is then printed using a solid wax printer on filter paper, followed by heating to melt the wax (Liu et al. 2016). Due to the porous structure of the paper, the wax penetrates the paper to form well-defined micro-channels. The whole process is simple and requires only a wax printer and a heating source. Nevertheless, the only disadvantage lies in the resolution of the μPADs fabricated by this method which is limited to millimeters. In 2012, Whitesides developed a fully printing wax and toner for fabricating the μPAD (He et al. 2015a). Further improvement in the wax printing method came with the introduction of the wax dipping method introduced by Songjaroen et al. used to fabricate μPADs (Songjaroen et al. 2011).

The wax printing method for the fabrication of μPADs has emerged as one of the most promising approaches owing to its cost effectiveness, robustness, rapid prototyping, and ease of handling. It holds great potential to play a part in the fabrication of future diagnostic devices (Bihar et al. 2018). Wax-printed diagnostic devices can be a solution for early diagnosis and treatment without wasting time and money on clinical and laboratory analyses, such as in remote or rural areas. The wax can be patterned using either a pen or a wax printer, which is then heated, allowing the wax to penetrate through the paper and create a hydrophobic barrier (Altundemir et al. 2017). Later, Kevin et al. developed an enclosing method to avoid contaminating the open wax-printed channels (Sher et al. 2017). They printed both the top and bottom sides of the wax-patterned paper using a printing toner, which permitted the device to be easily handled, and additionally, prohibited the evaporation of the liquid reagents, which saved the solution and maintained the level of concentration.

3.2.2.1 Basic Principle

The emergence of the patterning of microstructures with wax is a novel alternative to any other methods like photolithography. The wax was the first material to be tested within the paper, which gained attention in order to produce low-cost bioassay paper-based sensors on a large scale for point-of-care testing. This is one of the easiest techniques that can be used to pattern the microchannels on paper using either a pen or a wax printer (Lu et al. 2009). After patterning, it is then heated in an oven or on a hotplate, which allows the wax to penetrate through the paper and create a hydrophobic barrier, as depicted in Figure 3.2.

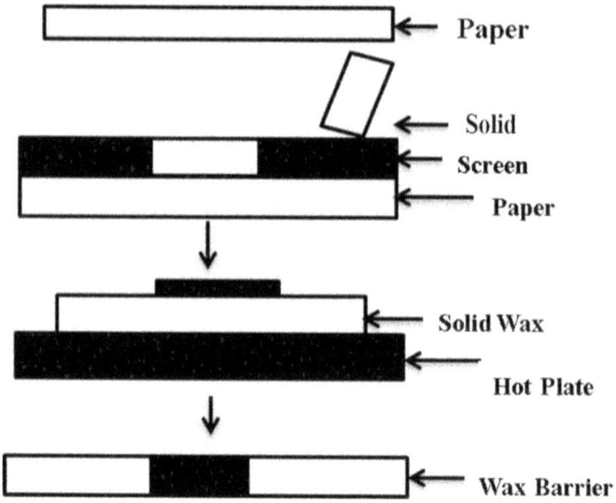

FIGURE 3.2 Schematic diagram showing steps involved in the fabrication of paper-based microfluidic devices using wax printing.

TABLE 3.1
Comparisons between Wax-Based Technology and Photoresist-Based Methods

Wax-Based Technology	Photoresist Technology
Wax-based technology includes printing and heating steps	Photoresist technology includes cumbersome steps such as exposure, developing, and curing
Less time-consuming	Time-consuming
Fast production speed	Slow production speed
Cheaper	Costly

This emerging technology is helpful for many applications in the sciences and is promising as a blood separator device. A micro pad formed using this technology can be widely used in biological analysis, such as quantifying male fertility by detecting live and motile sperm on a wax-printed device. Moreover, a μPAD fabricated by the wax-printing method is also used to measure the concentration of nitrite in saliva, which helps detect oral diseases like periodontitis, which might result in infection around the teeth tissues (Table 3.1).

On the other hand, wax dipping is the less expensive and faster technique compared to photolithography for fabricating microfluidic channels on paper. This is a single-step process that consists only of wax dipping, and channels can be fabricated in less than a minute via successive dipping and standard heating arrangements (Songjaroen et al. 2011). An iron mold is used to protect the hydrophilic channels, and melted wax is coated on the hydrophobic region. The iron mold is placed over the paper using the magnetic field, which forms an assembly. When this assembly is dipped into molten wax, the wax gets absorbed, as shown in Figure 3.3, while the iron mold prevents its penetration. The actual width of the fabricated microfluidic channel is determined by the width of the iron mold utilized (Songjaroen et al. 2014). The wax-dipping method was reported by Zhang et al. to fabricate μPADs using printed-circuit technology (He et al. 2015a). This method involved designing the pattern and printing the channels, then transferring to a copper sheet using a thermal transfer printer. Then, etching the design of the copper sheets is done by dipping into a ferric chloride solution, which is further coated with a film of paraffin and then a piece of filter paper. Finally, a standard electric iron is used to heat the reverse side of the copper sheet, and the melted paraffin penetrates the full thickness of the filter paper and forms a hydrophobic wall. In wax dipping, preparation of the mask mold is a time-consuming process that limits its flexibility. They printed both the top and bottom sides of the wax-patterned paper using a printing toner, which permitted the device to be easily handled and prohibited the evaporation of the liquid reagents; this saved the solution and maintained the level of concentration (Schilling et al. 2012).

3.2.3 Inkjet Printing

The physics behind the inkjet was described in 1878 by Lord Rayleigh, whereas Elmqvist filed the first patent in 1951 at the Siemens-Elema Company. By the start of the 1960s, Richard Sweet developed a model at Stanford University to electrostatically control the trajectory of ejected droplets under the action of the external electric field and this phenomenon was termed “continuous inkjet.” In late 1970, John Vaught from Hewlett Packard and Ichiro Endo from Canon developed drop-on-demand (DOD) systems in which a heating element was used to form and eject the

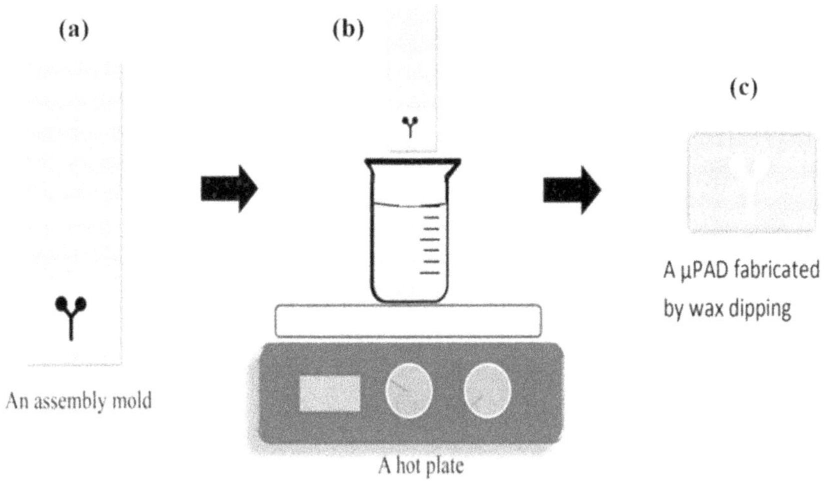

FIGURE 3.3 Fabrication process utilizing wax-dipping method (Songjaroen et al. 2014).

ink and called it thermal inkjet printing technology (Sweet 1965, Ben-Tzvi and Rone 2009). In 1990 with the standardization of personal computers, the inkjet became a technology for daily use because of the commercialization of inexpensive printers.

Inkjet printing is a contactless digital printing technique for precise and rapid deposition of material to fabricate microfluidic devices (Derby 2010). The method allows the deposition of small volume droplets, with a minimum volume of approximately one picoliter (pl), on an enormous range of compatible surfaces such as glass, plastic, or metal with higher precision and reproducibility. Also, contactless transfer of ink from the print head to substrate reduces the chance of contamination and allows the placement of materials on contact-sensitive substrates. The advantages of the inkjet printing method make it compatible with the mass production of smaller and cost-efficient sensors. Inkjet printing can be categorized into two forms distinguished by their drop generation mechanisms, i.e., continuous inkjet printing (CIJ) and drop-on-demand (DOD) inkjet printing (Martin et al. 2008). The difference between the two drop generation mechanisms is that in CIJ, a liquid (the ink) is passed through a small diameter orifice resulting in a fluid jet that breaks up into a train of identical, regularly spaced droplets through Rayleigh instability. After generation, the droplets are electrically charged, enabling them to be steered by an electrical field. In DOD printing, droplets are generated by transmitting a pressure pulse in a fluid-filled chamber.

3.2.3.1 Basic Principle

The formation and ejection of fluid droplets released from a chamber under the variation of internal pressure in the print head cavity and through nozzles form the primary basis of the mechanism of inkjet printing. The fluid ejection depends on the rheological parameters of the ink and the chamber's pressure as illustrated in (Figure 3.4) (Li et al. 2015).

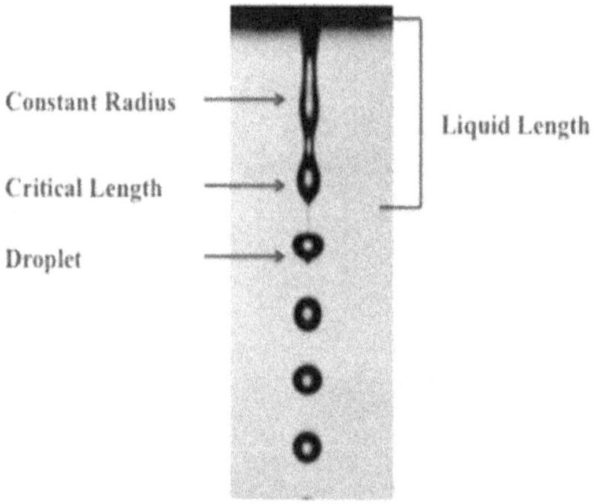

FIGURE 3.4 Idea proposed by Rayleigh-Plateau showing the instability of jets (Li et al. 2015).

3.2.3.2 Continuous Inkjet (CIJ) Printing

The generation and ejection of the continuous drops are the main features of CIJ technology. Following the process of droplets passing through an electrostatic charging electrode, which when deflected by a high-voltage deflector leads to deviation toward a recycling recipient (Li et al. 2015) as shown in Figure 3.5. There are two different categories of CIJ technology that have been developed, namely binary and multiple deflection systems. In a binary deflection system, an electrostatic charge regulates the two states. Firstly, the drops that get charged during the printing step go directly to the substrate, and on the other hand the deviation of non-charged drops takes place which are later collected in a recycle bin, whereas in the other system of deflection, the charged drops are deviated and deposited onto the substrate. The non-charged drops move to the reprocess unit.

3.2.3.3 Drop-On-Demand (DOD) Inkjet Printing

The printing method in which the ink is rapidly heated to a high temperature to vaporize, which creates a bubble at the surface of a heater, causing a pressure pulse that exudes ink droplets through the nozzle, is called drop-on-demand (DOD) inkjet printing. The vapor bubble collapses as the ink droplets are ejected, thereby generating a force to refill the ink, as depicted in Figure 3.6. The DOD technique is divided into piezoelectric, thermal, electrostatic, and acoustic inkjet printing techniques (Tan et al. 2016). The most successful technology adopted in laboratories and industries is thermal heating, which is widely used for graphic printing. In a thermal inkjet, the heating element is integrated within the print head chamber. An increase

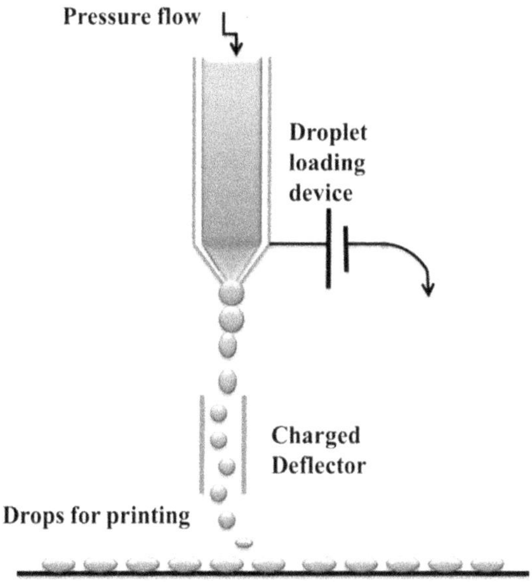

FIGURE 3.5 Schematic showing principal of continuous inkjet printing (CIJ) (Konta et al. 2017).

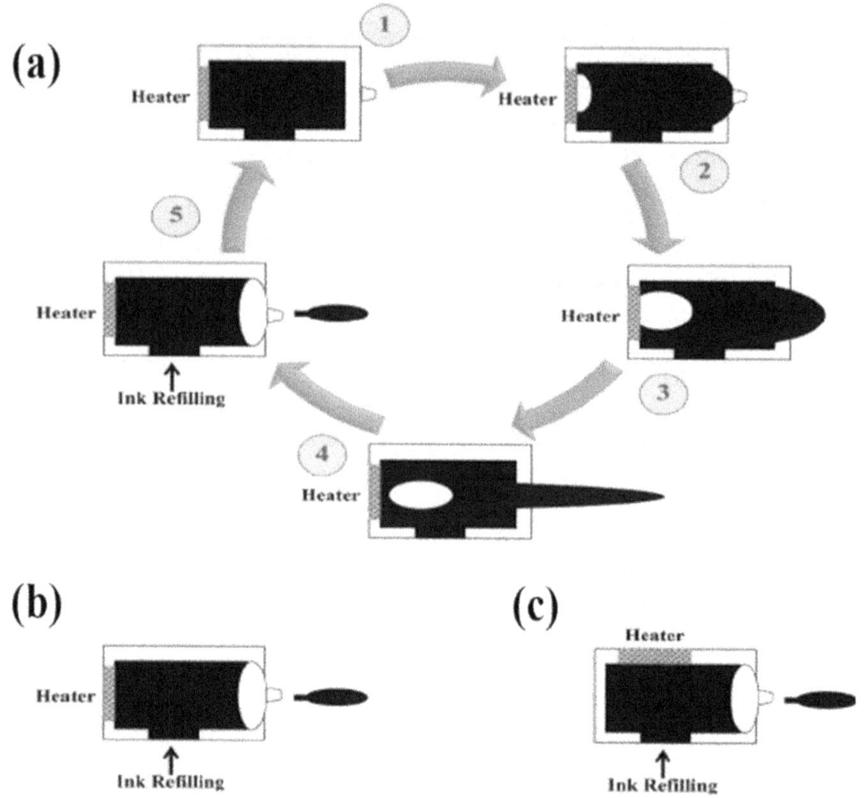

FIGURE 3.6 Mechanistic insight into the drop-on-demand thermal inkjet printer (Li et al. 2015).

in the temperature of the ink is generally achieved up to 300°C, which results in the generation of an air bubble within the ejection chamber. However, heating requirements before the ejection process make this technology limited in terms of the inks that can be used. Setti et al. reported making an amperometric sensor for glucose detection using this technique (Setti et al. 2007).

Shen et al. used the inkjet printing method to fabricate μPADs. The inkjet printing was done on the filter paper using a reconstructed digital inkjet printer employing an alkyl ketene dimer-heptane solution followed by heating the printed paper in an oven at 100ºC for 8 min to cure alkyl ketene dimer (AKD) onto the cellulose fibers (Li et al. 2010). After the paper was dried, the μPAD fabrication was completed. A similar method of inkjet printing was also used by Maejima et al. to fabricate μPADs. Here, the authors used a hydrophobic UV curable acrylate composition composed of nonvolatile and nonflammable compounds to the AKD, the paper was cured under UV light for 60 s after printing the special ink on the paper. Thus, the formation of hydrophobic barriers was achieved (Maejima et al. 2013).

Since most glucose monitoring tools depend on piercing the skin to draw blood, the pain and discomfort linked with pricking the finger need to develop non-invasive and portable glucose assays. Inkjet printing allows the depositing of organic molecules on ecofriendly, low-cost, and recyclable paper substrates. Using this method, an easy-to-use sensor can be designed to regularly monitor glucose in body fluids such as saliva. Developing the μPAD sensor using inkjet printing involves printing all the components of the sensors in a layer-by-layer arrangement. The conducting polymer poly(3,4-ethylenedioxythiophene) polystyrene sulfonate (PEDOT: PSS) is used as an electronic component, whereas biological film comprising of the enzyme/mediator, dielectric, and encapsulation layer are the components of the microfluidic reactor, as shown in Figure 3.7 (Setti et al. 2005).

3.2.4 Role of Semiconductor Oxides for the Fabrication of Microchannels

Nanomaterials and nanohybrids present a new class of functional materials to be used for the fabrication of paper-based devices that have led to advances in the area of biosensing technology. The fabrication of μPADs with nanomaterials brings high selectivity and sensitivity, thereby enabling the integration of a nanoscale phenomenon in biomedical devices (Raza and Ahmad 2018). Nanostructured metal oxides have received significant attention for biosensing applications owing to several characteristics such as ease of fabrication, controllable size/shape, biocompatibility, catalytic and optical properties, and chemical stability, creating new avenues for therapeutic disease management and, most importantly, diagnosis (Hahn et al. 2012). Of all the nanostructures, semiconductor nanomaterials have characteristic properties like a wider bandgap, large exciton binding energy, and enhanced electron mobility. They have gained attention for a broader range of applications in the field of biomedical and clinical sciences. It has been found that nanostructured zinc oxide (ZnO) possesses advantages such as high crystallinity with minor structural defects, low-temperature synthesis, and good electrical conductivity, making it highly suitable for developing rapid, stable, and reliable sensor devices (Foo et al. 2014,

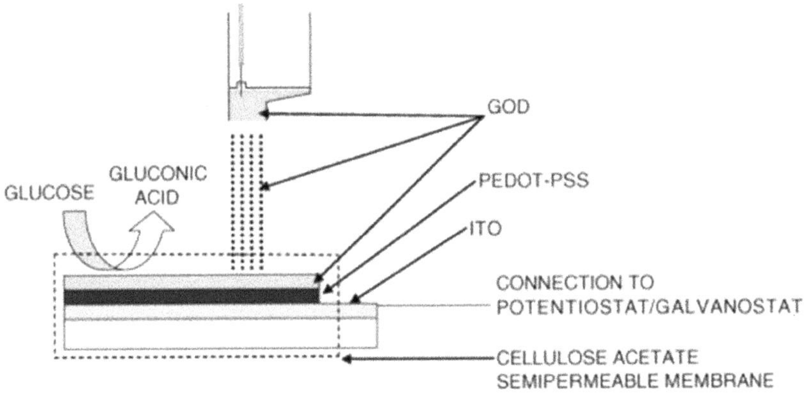

FIGURE 3.7 Diagram illustrating the prototype of a glucose oxidase (GOx) inkjet printer electrode (Setti et al. 2005).

Tak et al. 2014). Nanotextured surfaces, combined with metal oxides, result in the creation of nanohybrid materials. A functionalized form of graphene–graphene oxide (GO), with hydroxyl groups and epoxy on the basal plane and carboxyl groups at its edge, has also been explored in paper-based sensing because of its electron transfer capability, biomolecule immobilization, and ease of functionalization with other molecules on its surface. Yujun et al. demonstrated that carboxyl-modified GO has a peroxidase-like activity that has the ability to catalyze the reaction of peroxidase substrate 3,3′,5,5′-tetramethylbenzidine (TMB) in the presence of hydrogen peroxide (H_2O_2) to produce a color change (Song et al. 2010). In addition to the existing studies, Song et al., through experimentation, showed the improvement of the homogeneity of color distribution in colorimetric assays using graphene oxide by increasing the linkage of aromatic molecules through π–π stacking or van der Waals forces and absorbing biomolecules via electrostatic interactions or hydrogen bonding (Song et al. 2010, Figueredo et al. 2016). Therefore, the properties of GO present a simple, universally applicable approach that features GO-coated μPADs coupled with smartphone-based colorimetric detection for the direct quantification of glucose in saliva. The combination of ZnO and metal oxide nanomaterials provides a new avenue for the development of highly sensitive biosensors. Modification of ZnO with metal oxide nanomaterials provides improved properties for ZnO for the sensing of biomolecules. Wu and Yin used the electrospinning technique in order to deposit nanostructured composite on platinum (Pt) electrodes (Wu and Yin 2013). Nanostructures with a vertical orientation have received attention due to their larger surface area. Soejima et al. synthesized a nanoarray composite of copper oxide–zinc oxide (CuO–ZnO) using an easy one-step and low-temperature route on brass plates (Soejima et al. 2013). The nano arrays synthesized were of ZnO nanorods (NRs) and CuO nanoflowers, found to be electro-catalytically active during glucose oxidation, resulting in high sensitivity, a low limit of detection, and fast response. Also, SoYoon et al. synthesized ZnO NRs and CuO nanoleaf architectures attached to the Cu substrate to fabricate a nonenzymatic glucose sensor (SoYoon et al. 2014). The nanohybrid composite was found to hold excellent electro-catalytic activity during glucose oxidation in the presence of sodium hydroxide (NaOH). Ahmad et al. used vertically grown ZnO NRs on fluorine-doped tin oxide (FTO) to fabricate a nonenzymatic glucose biosensor by employing the FTO electrode after modifying the ZnO surface with CuO NPs as shown in Figure 3.8 (Ahmad et al. 2017).

In another work, Tripathy et. al., demonstrated that vertically grown ZnO NRs synthesized by a low-temperature solution route offered outstanding surface-binding sites for CuO NP loading (Tripathy and Kim 2018). A schematic of the detection mechanism is illustrated in Figure 3.9.

3.2.5 Other Methodologies

3.2.5.1 Plasma Treatment

In this process, the paper is dipped in a solution AKD–heptane to make the surface of the paper hydrophobic, followed by immediately placing it in a fume hood to facilitate evaporation of the heptane (Li et al. 2008). Finally, the paper-coated AKD

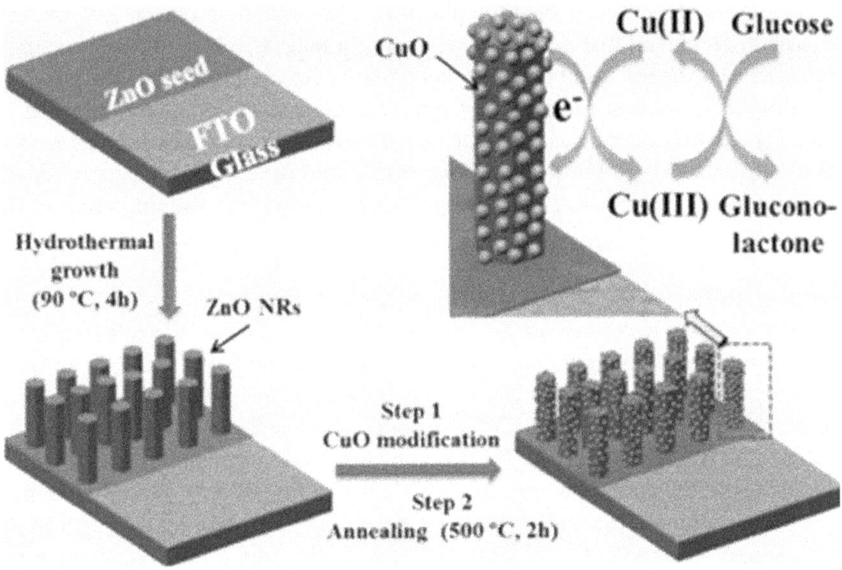

FIGURE 3.8 Representation of a nonenzymatic glucose sensor electrode (Ahmad et al. 2017).

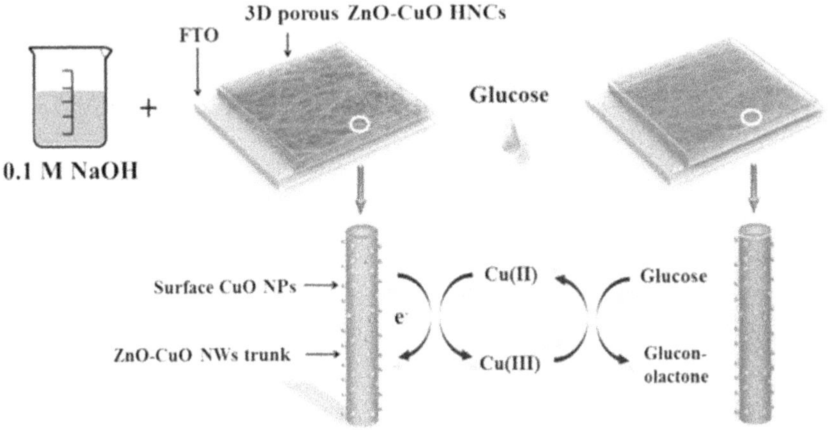

FIGURE 3.9 Schematic illustration of fabricated electrode and glucose detection mechanism over 3D ZnO-CuO HNC's surface (Tripathy and Kim 2018).

reactors are heated at about 100°C for 45 min to cure the AKD. This treatment process has the advantage of maintaining the flexibility and surface topography of paper, thus making it possible to print different patterns and functional components such as control switches, microreactors, and microfilters using this technique. However, this process requires customized masks, vacuum plasma reactors, and a hotplate that restrict this technique's broad applicability. Fluorocarbon plasma polymerization is another method that is alternatively used for making the hydrophobic channel boundaries. The positive and negative metal masks are used to confine the liquid by the hydrophobic walls surrounding it from three sides instead of two (Yetisen et al. 2013). The advantages of this treatment are the simplicity of fabrication as well as the cost effectiveness of the reagents. However, metal masking is expensive, requiring vacuums and costly instrumentation, thus circumventing the limitation of the process.

3.2.5.2 Spray Drying

Spray drying is the technique for the fabrication of hydrophobic channels by directly spraying the paper's hydrophobic material. A mask is employed to create a hydrophobic barrier (Liu et al. 2017). Cardoso et al. fabricated hydrophobic channels using this technique for the first time in 2017, by spraying a scholar glue commonly known as white glue (Cardoso et al. 2017). The steps involved in the fabrication of microchannels using the spray glue process are shown in Figure 3.10. Initially, the process involves aligning the mask with filter paper, and then aligning the paper to the acrylic support magnets. The next step is spraying the glue using a magnetic mask and exposing the paper to ultraviolet-visible (UV–Vis) light using a halogen light source that promotes the cross-linking to form hydrophobic barriers. This is an easy-to-use and equipment-free method for the fabrication of μPADs; however, it demonstrates low resolution and uniformity compared to other methods (Xing et al. 2013).

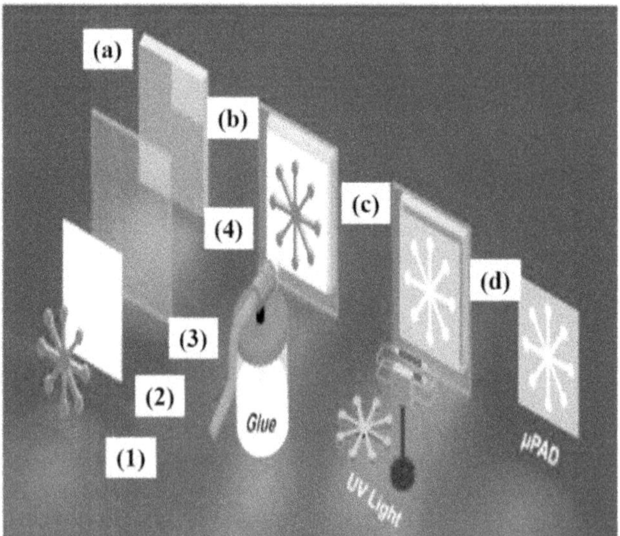

FIGURE 3.10 Schematic illustration spraying of scholar glue for the fabrication of paper-based devices. Magnetic masks are used to mark the microfluidic regime (Cardoso et al. 2017).

3.2.5.3 3D Printing

3D printed μPADs are significantly inexpensive and accessible for mass production (He et al. 2016). Fu et al. initially printed a substrate for the microchannel coated with polydimethylsiloxane (PDMS) to seal the microgap caused by 3D printing. Then, the cellulose powder with deionized water was dispensed into the microchannel and dried in an oven. He et al. used a desktop stereolithography 3D printer and a dynamic mask to fabricate μPADs (He et al. 2015b). In their method, the paper was first immersed in a UV resin followed by UV light exposure through the dynamic mask. This was followed by curing to make hydrophobic barriers. The uncured regions were washed with anhydrous alcohol. The overall process drastically reduced the fabrication time to only 2 min (Lim et al. 2019).

3.3 SURFACE MODIFICATION AND CHARACTERIZATION

The microfluidics industry comprises a billion-dollar market, with the number of publications on microfluidics doubling each year. The interest in this field is growing and is driven mainly by bio microfluidics devices such as point-of-care devices, drug manufacturing micro-reactors, organs-on-chips, and microneedles/pumps for drug delivery. However, choosing suitable materials is critical for biomedical research purposes. Wetting behavior is essential, especially when two different phases, such as water and air, or water and oil, are involved in the system. There are two ways to control the wettability of solid surfaces, i.e., either by enhancing or altering the chemical composition or introducing changes in the geometrical structures of the surface. Since glass and oxidized silicon have been widely used materials for microfluidic devices, polymer-based microfluidic devices have recently gained popularity because of their outstanding properties, such as their low-cost and easy-to-fabricate channels. However, most polymeric materials have pristine surfaces with low surface energies and are hydrophobic rather than hydrophilic. This leads to problems such as high flow resistance, therefore surface modification is the method that produces chemically reactive groups that bind small molecules, polymers, biomolecules, or nanoparticles covalently to a surface.

Further, the method can be categorized into chemical and physical methods. Chemical modification involves the immobilization of the functional molecules that create the desired surface properties to activate the surfaces with attached chemicals. However, on the other hand, physical modification is brought about by changing surface roughness, grain size, and grain boundaries with laser, plasma, heat, or polishing exposure. This section covers the different methods of the surface modification of microfluidic device substrates.

3.3.1 Surface Functionalization

Elastomers such as PDMS offer a range of favorable properties for bio microfluidics applications, such as (i) simple fabrication, (ii) good mechanical properties, (iii) biocompatibility, (iv) non-toxicity, (v) high gas permeability, and (vi) excellent optical transparency ranging from 240 to 1100 nm. Despite the plethora of merits, hydrophobicity is one such issue that limits the applicability of PDMS, especially in biological samples. The hydrophobicity of the PDMS surface results in undesired and nonspecific adsorption of proteins. As a result, there is an effect in analyte transport and reduction in separation efficiency and detection sensitivity. Another challenge due to hydrophobicity is the use of aqueous solutions or mixtures of aqueous and organic solutions (Gokaltun et al. 2019, Gokaltun et al. 2017). Since microfluidics work relies on using polar liquids, this causes a significant obstacle in many applications. This led to developing approaches to render the PDMS surface hydrophilic. The strategies included introduction of alkoxysilanes or chloro-silanes for surface functionalization, using charged surfactants coating PDMS surfaces with polar functionalities. Other conventional techniques include polyelectrolyte multilayers (PEMs), chemical vapor deposition, phospholipid bilayers, and attaching hydrophilic polymer brushes to the surface PDMS via grafting-from and grafting-to approaches. High-energy treatments like oxygen (O_2) plasma and UV/ozone treatments are also utilized to render the surface's hydrophilic nature. While these interventions had proven success in improving surface hydrophilicity, their broader use was often limited by chemical stability, the need for special equipment or hazardous routes, or the length and complexity of their process for fabrication that is restrictive for large-scale implementation. Methods or techniques adopted for the surface functionalization include wet-chemical methods, adsorption of surfactants, polymers, polyelectrolytes, sol–gel coatings, modification with grafting polymers, etc.

3.3.1.1 Sol–Gel Coatings Method

Sol–gel technology is based on wet-chemical processes that allow the production of materials such as fibers, monoliths, or coatings from solutions of low-molecular precursors. Usually, an ester (orthosilicic acid ($Si(OH)_4$)) with alcohol (tetraethyl orthosilicate ($(C_2H_5O)_4Si$)) is mixed, followed by the addition of water to initiate the process of hydrolysis. Poly-condensation via reactive Si–OH groups forms a silicon dioxide (SiO_2) network of Si–O–Si linkages. Thus, the precursor solution turns into a sol, i.e., a colloidal dispersion of solid oxide nanoparticles. With time, due to condensation reactions between the nanosized polycondensates, there is a generation of three-dimensional, sub-micrometer-sized entities. Polymer chain lengths formed beyond the "μm" range are coined as "gels," as shown in Figure 3.11.

Went et al. describe the sol–gel-derived phase's modification for isolation and purification of nucleic acids in microfluidic devices (Wen et al. 2008). Sol–gel techniques have also been used to immobilize fluorophores or chemical sensing indicators (Eichler et al. 2016). Habouti et al. used a particular sol–gel coating variant to deposit zinc oxide/polyvinyl alcohol (ZnO/PVA) nanocomposite layers with optical transparency and a very low water contact angle in a PDMS-based microfluidic chip for optical biosensing (Habouti et al. 2014). In situ sol–gel reactions were used to coat super hydrophilic titanium dioxide (TiO_2) in a silicon wafer microchannel before covering it with a glass lid. The deposition was achieved at 25°C by adding titanium isopropoxide (TTIP) in ethanol to a stirred mixture of ethanol and water in which the structured wafer was immersed (Ma and Zhang 2014).

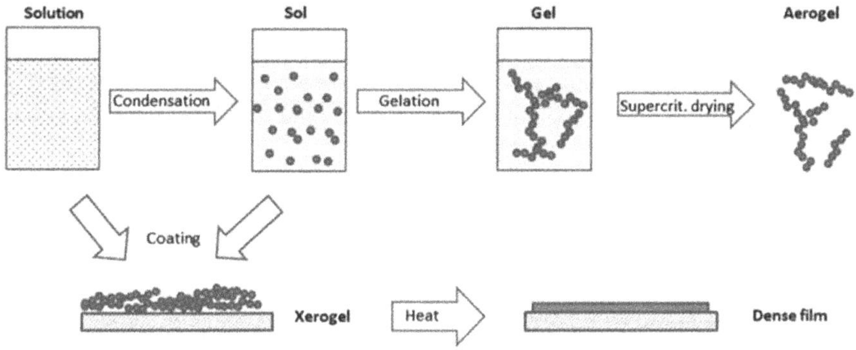

FIGURE 3.11 Figure depicting steps involved in the sol–gel process (Eichler et al. 2016).

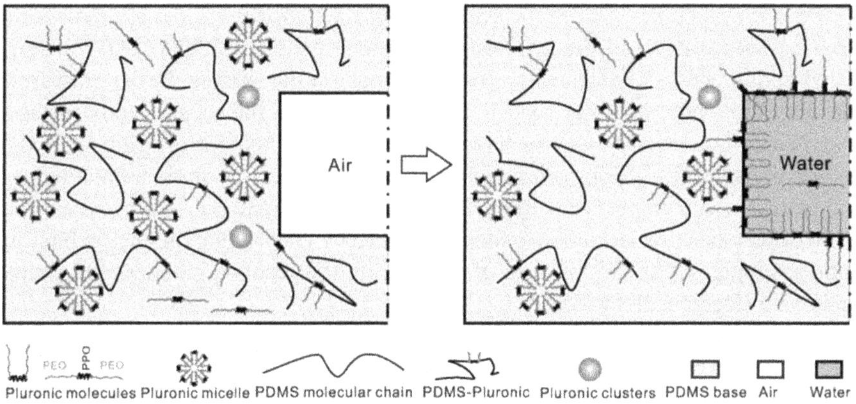

FIGURE 3.12 Schematics illustrating the surface modification of PDMS using Pluronic F127 (Wu and Hjort 2009).

3.3.1.2 Modification Using Surfactant

Surfactants are used to pretreat microchannels which are often added to the running buffer of analytical microfluidic separations to render the surface hydrophilic. Various types of surfactants are used for dynamic surface modification. For instance, Wu and Hjort worked on the addition of a nonionic surfactant, Pluronic F127 ((polyethylene oxide)$_{100}$(polypropylene oxide)$_{65}$(polyethylene oxide)$_{100}$ ((PEO)$_{100}$(PPO)$_{65}$(PEO)$_{100}$)), into a PDMS pre-polymer before curing. The surfactant's addition showed that the surfactant molecules embedded in the PDMS migrated toward the water/PDMS interface that minimized the surface energy of the water-filled microchannel system, as illustrated in Figure 3.12 (Wu and Hjort 2009). This resulted in the hydrophobic interaction between the PPO segments and PDMS, causing the PEO to extend outward from the surface. The modified PDMS surface changed the water contact angle from 99 to 63° after immersing the sample in water for 24 h, when compared with a water contact angle of native PDMS that was 104° (Eichler et al. 2016). Furthermore, the surfactant-modified PDMS showed significant results in suppressing the nonspecific adsorption of a fluorescein iso-thiocyanate (FITC)-labeled immunoglobulin G (IgG) antibody due to the improved hydrophilicity, compared to native PDMS.

3.3.1.3 Grafting Polymer Method

Due to the limitation of most of the coating methods, the graft polymer coating is used to tailor the wettability and surface chemistry of microfluidic devices or incorporate surface functional moieties for further modification. The technique of graft polymer surface modification is divided into two main categories, as depicted in Figure 3.13, as follows:

- Grafting-onto
- Grafting-from

Grafting-onto consists of pre-synthesized polymer chains, which are end-functionalized with a chemical anchoring group adsorbed on the surface. In the graft-from approach, the dense polymer layer is grown in situ from the surface via a surface-adsorbed initiation group. One such technique that falls under the grafting method is surface photo-grafting polymerization which offers versatility in manipulating and tuning the surface properties without damaging the bulk material. It is a process that involves surface polymerization, which is induced by UV irradiation, thus exhibiting some advantages, like the low cost of processing, fast reaction kinetics, and simple equipment (Deng et al. 2009).

In a photoinitiator or photosensitizer, UV light is used to execute the surface graft polymerization. Norrish type I and photo-Norrish type II photoinitiators were more frequently used because of higher grafting efficiency, higher polymerization yield,

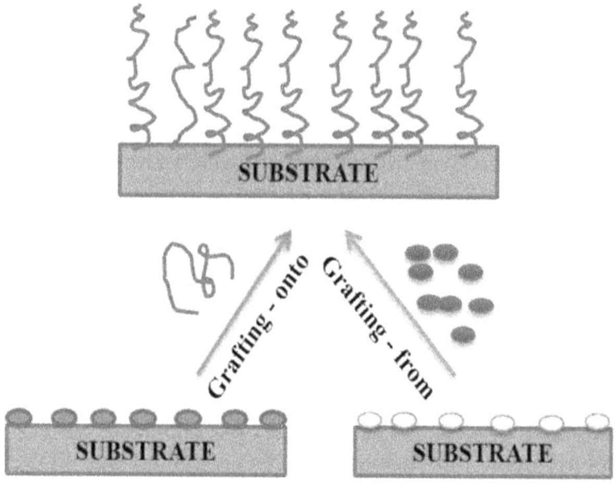

FIGURE 3.13 Sketch of surface polymer grafting approaches – graft-onto (left) and graft-from (right).

and higher polymerization rates. Benzophenone (BP) and its derivatives among the existing Norrish type II photoinitiators are also most widely used, which have been showing effective initiating or co-initiating of several radical-induced surface photo-grafting polymerizations (Deng and Yang 2000, He 2008). In principle, when UV light is irradiated, the BP-based molecules are excited from the singlet state and then jump to a triplet state by intersystem crossing. It has been demonstrated that BP and its derivatives in a triplet state undergo hydrogen-abstracting reactions from substrates, consequently providing surface radicals (R•) capable of initiating surface graft polymerization. The resulting hydrogel-modified PDMS microchannel could be employed to generate both water-in-oil-in-oil (W/O/O) and water-in-oil-in-water (W/O/W) emulsions. Wang and Yang developed an alternative strategy to attain surface photo-grafting polymerization and 3D construction on the surface, firstly, coated with methyl methacrylate/1,2-divinylbenzene (MMA/DVB) microemulsion on casting polypropylene (CPP) films, and then conducted photo-grafting with BP as a photoinitiator (Wang and Yang 2004).

3.3.1.4 Surface-Initiated Atom Transfer Radical Polymerization (SI-ATRP)

Another popular choice for the grafting form is surface-initiated atom transfer radical polymerization (SI-ATRP). ATRP typically offers the possibility to select the most convenient and appropriate initiator for the monomer. However, grafting using the ATRP method limits the generation of the initiating sites at the original surface.

3.3.2 Characterization

Since the fabrication of a microreactor is a vital step, careful examination of the characteristic properties of paper-based microfluidic devices with techniques such as drop shape analysis (DSA), scanning electron microscopy (SEM), and energy dispersive X-ray (EDX) is therefore carried out. A detailed discussion of these techniques with examples appears in the section below.

3.3.2.1 Drop Shape Analysis (DSA)

DSA is an image analysis method for determining the contact angle from the shadow image of a sessile drop and the surface tension or interfacial tension from the shadow image of a pendant drop. An image of the drop is recorded with a camera and transferred to the drop shape analysis software. The two methods of measuring the contact angle are as follows:

- Static: In static contact angle measurement, a droplet of a liquid is placed on top of a substrate, and the resulting contact angle is measured using a contact angle microscope. In this method, the size of the drop does not alter during the measurement.
- Dynamic: The term "dynamic contact angle measurements" refers to techniques that measure the contact angle during movement. This is usually accomplished by adding liquid to a static droplet on a surface and thus pushing the front of the fluid across the unwetted surface.

Mani et al. fabricated paper-based devices using a cost-effective and single-step method to fabricate devices without requiring any expensive instrumentation, simply by positioning correction pens (Mani et al. 2019). Further, they investigated the wetting behavior of the uncoated and coated filter paper by measuring contact angle, from which they inferred that since there is a higher contact angle of coated paper, this results in the formation of a hydrophobic surface, whereas lower contact angle values signify that the uncoated paper surface is hydrophilic in nature, as shown in Figure 3.14.

3.3.2.2 Scanning Electron Microscopy (SEM)

SEM is the method for imaging the microstructure and morphology of the materials. This technique uses a low-energy electron beam, which is radiated to the material to scan the surface of the sample. SEM exists in different ways to characterize materials (including biomaterials) such as X-ray mapping, secondary electrons imaging, backscattered electrons imaging, electron channeling, etc.

Some SEM micrographs for the characterization of devices that are fabricated using the various method are given below. The devices are reported by varying the type of paper used for the fabrication of paper-based glucose sensors. Shown are the SEM images of the PDMS pattern on paper. There are four types of papers used, namely, PES-80, PES-22, printing paper, and filter paper, as shown in Figure 3.15.

3.3.2.3 Energy Dispersive X-Ray (EDX) Microanalysis

EDX is a technique for elemental analysis associated with the electron, which depends on the generation of characteristic X-rays that reveal the specimens' elements. The microanalysis of EDX has usage in different biomedical fields by many researchers and clinicians, containing both semi-qualitative and semi-quantitative information.

Figure 3.16 illustrates some EDX micrographs depicting the change in the elemental concentration of the paper-substrate when fabricated and functionalized for

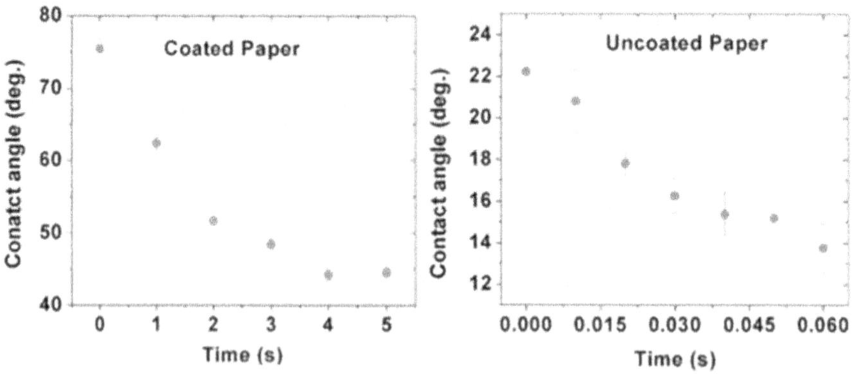

FIGURE 3.14 Contact angle traced between water droplet with the coated and uncoated paper surface (Mani et al. 2019).

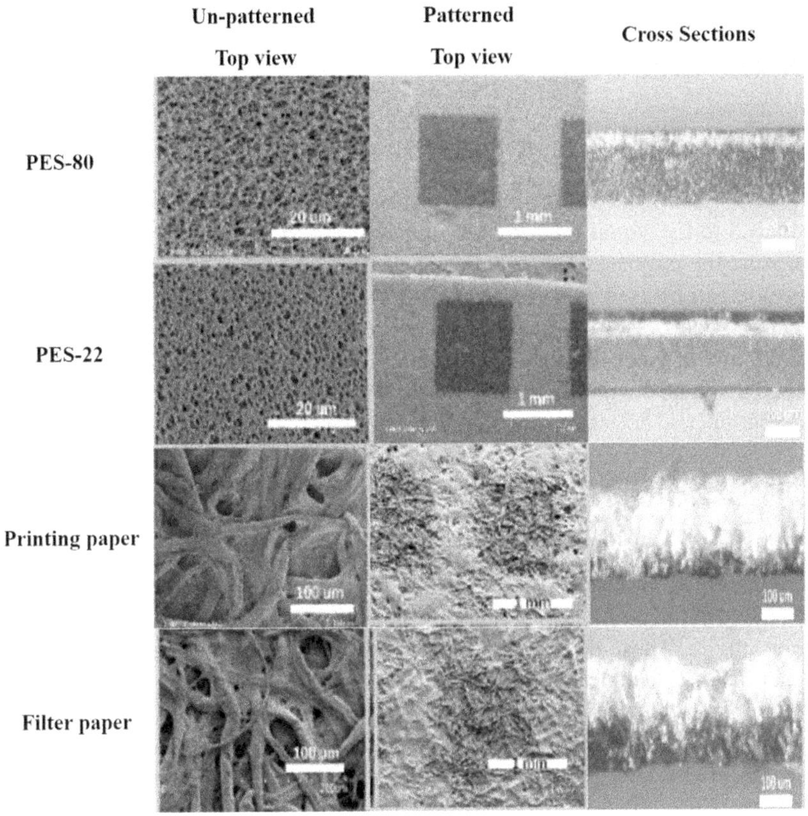

FIGURE 3.15 SEM images of the PES-80, PES-22, printing paper, and filter paper (Shangguan et al. 2016).

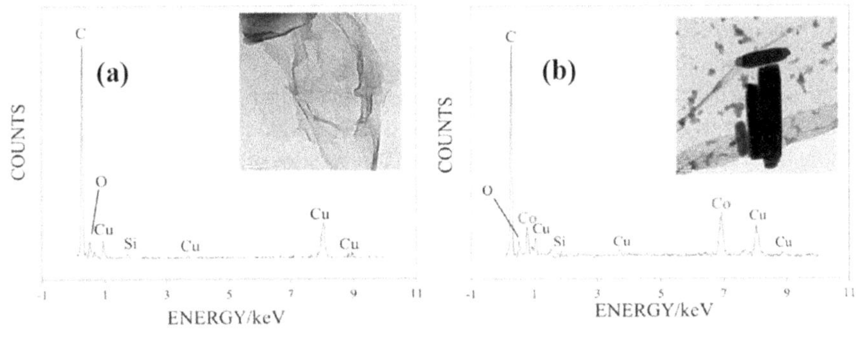

FIGURE 3.16 EDX of (a) CoPc and (b) EDX of CoPc/G/IL composite (Chaiyo et al. 2018).

analysis of the nonenzymatic electrochemical paper-based sensor developed using the composite of a cobalt (II) phthalocyanine/ionic liquid/graphene composite (CoPc/IL/G) for the determination of glucose levels.

3.4 METHODS FOR THE DETECTION OF GLUCOSE

Glucose sensors work on the principle that the recognition of the element that is introduced to the signal transducer correlates with the concentration of the analyte to measure the response. An enzyme such as glucose oxidase (GOx) or glucose dehydrogenase can be used as a glucose recognition element for the generation of redox mediator species resulting in electrochemical signals for detection. GOx catalyzes the oxidation of glucose, which generates hydrogen peroxide, and the level of H_2O_2 released associated with this reaction is then measured by using a suitable electrode. By amperometric detection, the output current generated in the glucose oxidation is directly proportional to glucose concentration.

3.4.1 Electrochemical Method

Electrochemical detection (ECD) is a well-developed detection method for paper-based microfluidics due to its small size, portability, low cost, high sensitivity, and selectivity. These advantages have led to the future use of the ECD method, i.e., hand-held glucometers for monitoring diabetes using screen-printed carbon electrodes.

Coupling of the μPADs with electrochemical detection or electrochemical paper-based analytical devices (ePADs) enables them to detect the analyte at low levels selectively. Currently, there are two techniques reported for fabricating ePADs: (i) direct printing of electrodes onto paper and (ii) by placing a PAD onto a screen-printed electrode. Zang et al. described a 3D μPAD that contains eight screen-printed working electrodes and were able to perform multiplexed electrochemical immunoassays as illustrated in Figure 3.17 (Zang et al. 2012). Kubota et al. published a new approach for the determination of glucose on paper using graphite pencil electrodes (Santhiago and Kubota 2013).

To determine the glucose from the blood sample using ePADs, integrated plasma isolation has been developed. This ePAD is fabricated utilizing the wax-dipping method and is in a dumbbell shape, consisting of two blood separation zones and a middle detection zone as depicted in Figure 3.18. The design of the device assists in the separation of plasma and the generation of a uniform flow to the central detection zone of the ePAD. The glucose detection is done using glucose oxidase immobilized in the middle of the paper device. The hydrogen peroxide is generated from the reaction with glucose. The enzyme is then allowed to pass through to a Prussian blue modified screen-printed electrode (PB-SPEs), where the current is generated. The current generated is measured using chronoamperometry at the optimal detection potential for H_2O_2 using silver/silver chloride (Ag/AgCl) as a reference electrode. It was found that there is a relationship between the current generated and glucose concentrations in the whole blood (Noiphung et al. 2013).

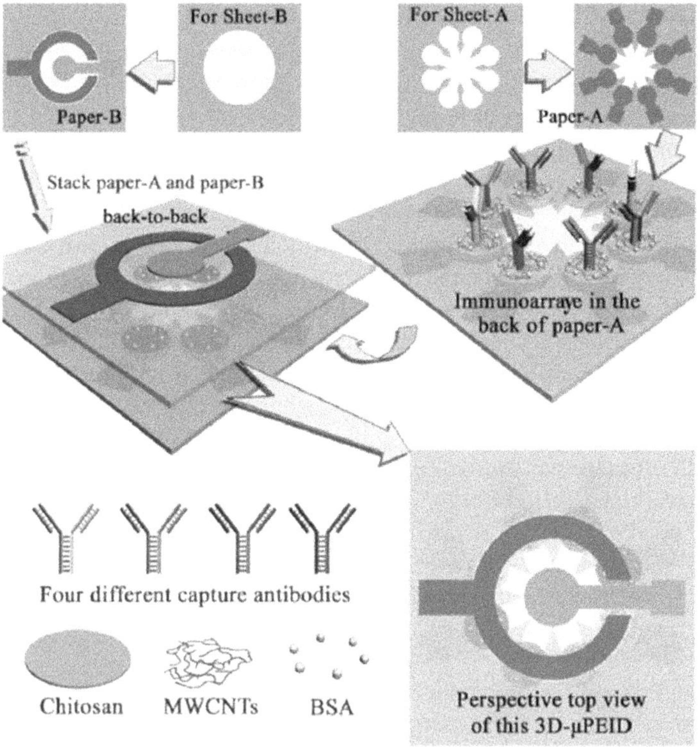

FIGURE 3.17 Figure illustrating the steps involved in the fabrication of 3D PAD (Zang et al. 2012).

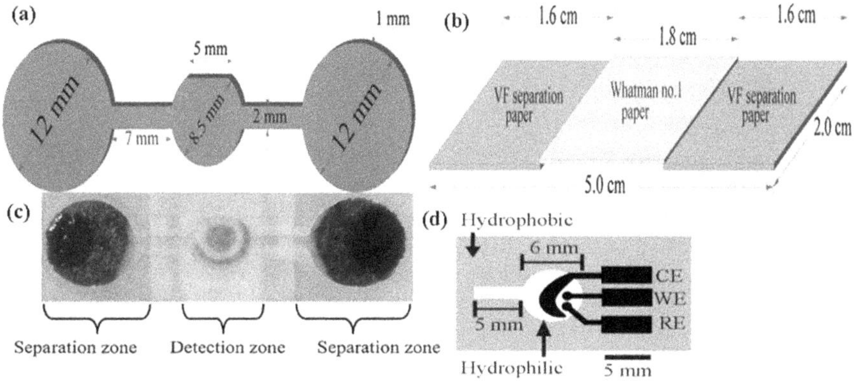

FIGURE 3.18 (a) Picture representing dimensions of an iron mold, (b) arrangement of the filter paper (Whatman No. 1) and papers used for blood separation, (c) digital image depicting the well-defined separation and detection zones, and (d) schematic diagram depicting counter electrode (CE), working electrode (WE) and a reference electrode (RE) used for detection (Noiphung et al. 2013).

3.4.1.1 Electro-Chemiluminescence (ECL) Detection

A combination of chemical luminescence and electrochemical techniques forms an electro-chemiluminescence (ECL) detection system. This technique results in the generation of light and has been integrated with paper-based microfluidic devices. ECL has numerous advantages, such as better sensitivity and an increased dynamic concentration-response range. It also has some prominent features, such as the requirement of smaller sample volumes, lack of a light source, and simple instrumentation. ECL is most widely used in clinical diagnosis. More than 150 different immunoassays are available on the market for detecting tumor markers and treating thyroid disease and various infectious diseases.

Zang et al. developed a multiplexed ECL immunoassay detection device by integrating eight screen-printed carbon working electrodes in a 3D μPAD (Ge et al. 2012). To enhance the reproducibility of ECL emissions of H_2O_2 solutions, Shi et. al. immobilized and stabilized cadmium sulfide (CdS) quantum dots on a double-sided carbon adhesive tape which were supported by an indium tin oxide (ITO) glass that acted as a working electrode for ECL (Shi et al. 2012). Delaney et al. performed ECL detection in μPADs using tris(2,2′bipyridyl)ruthenium(II)($Ru(bpy)_3^{2+}$) and inkjet-printed μPADs laminated with screen-printed electrodes for ECL detection as shown in Figure 3.19 (Delaney and Hogan 2015).

3.4.1.2 Chemiluminescence (CL) Detection

The phenomenon of the generation of light due to a chemical reaction, i.e., the process of conversion of chemical energy into light energy by the movement of the electrons from excited-state levels to lower energy levels is called chemiluminescence (CL). Several compounds react with hydrogen peroxide or oxygen, which emit light due to the decomposition of the compounds. For instance, when organic compounds such as luminol (5-amino-2, 3-dihydro-1, 4-phthalazinedione, or 3-aminophthalhydrazide) react with hydrogen peroxide, they produce light. CL detection has various

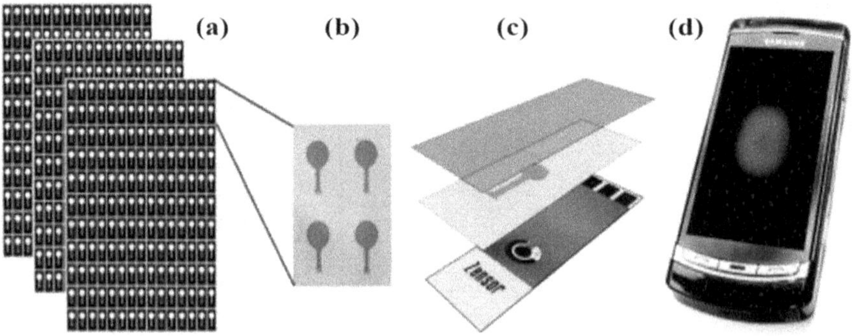

FIGURE 3.19 The μPADs used for ECL detection – (a) several microfluidic devices embedded on a sheet of paper, (b) $Ru(bpy)_3\ ^{2+}$ used to modify the paper-based devices, (c) arrangement of transparent film, paper microfluidic devices and screen-printed electrode, and (d) mobile phone used to capture and analyze via ECL detection (Delaney and Hogan 2015).

advantages, such as less expensive instrumentation, a wide dynamic range, and lower detection limits.

CL detection depends on the capacity to measure light emitted during chemical reactions. Yu et al. used the CL method for the measurement of hydrogen peroxide (produced) and a rhodamine derivative for the development of a multiplexed μPAD that could simultaneously determine the presence of glucose and uric acid by their respective reactions with glucose oxidase and uricase (urate oxidase) (Yu et al. 2011).

3.4.2 Enzymatic Determination (Colorimetric Method)

Colorimetric sensing of the glucose is an enzymatic reaction that allows the binding of the substrates loaded on the fabricated channel using paper as a substrate. The binding of the substrate to the activated site of the enzyme is done using a "lock and key" mechanism to carry out the reaction. In this system, when a glucose solution is dropped onto the loading zone, the solution is transferred to the detection zone modified with GOx, horseradish peroxidase (HRP), and chromogenic compounds. The clinical screening of serum glucose is based on two enzyme-catalyzed reactions.

- The first reaction involves the oxidation of the glucose-by-glucose oxidase to produce hydrogen peroxide.
- Measurement of the peroxides formed by catalyzing the reaction, i.e., causing its chromogenic (indicators) substrate to generate a colorful solution.

Measurement of blood glucose is based on an enzymatic reaction using enzyme glucose oxidase, which is commonly released in the growth medium by *Penicillium notatum*. This enzyme causes catalytic oxidation of β-D-glucopyranose into D-glucono-1, 5-lactone, and formation of hydrogen peroxide. D-glucono-1,5-lactone is then further catabolized to D-gluconic acid. Therefore, glucose solution, which contains peroxidases, utilizes H_2O_2 to oxidize o-toluidine, resulting in the formation of a blue-colored product. The chemical reactions involved are given by equations (3.1), (3.2), and (3.3). This reaction is highly specific in nature and a change in color can be read in less than 8 min. The intensity of the color formed is proportional to the glucose concentration in the sample.

$$\alpha\text{-D-glucose} \rightarrow \beta\text{-D-glucose} \tag{3.1}$$

$$\text{D-glucose} + O_2 + H_2O \rightarrow H_2O_2 + \text{D-gluconic acid} \tag{3.2}$$

$$H_2O_2 + \text{O-tuluidine} \rightarrow \text{oxidized tuluidine}\left(\text{blue colour product}\right) \tag{3.3}$$

In this system, when a glucose solution is dropped onto the loading zone, the solution is transferred to the detection zone, which is modified with GOx, HRP, and chromogenic compounds through the connected fluidic channel. In the presence of Gox-generated H_2O_2, HRP converts chromogenic compounds into the final product exhibiting a blue color. Thereby H_2O_2 plays a significant role in the signal generation and acts as a redox signaling molecule. The byproduct of general oxidation reactions from the catalytic process of glucose oxidase, amino-acid oxidase, and urinase is the formation of H_2O_2 as illustrated in Figure 3.20(b). Levels of H_2O_2 are a relatively significant indicator of several biological processes such as the immune system, etc. Therefore, accurate detection of H_2O_2 or fabricating a unique sensing method based on the medium of H_2O_2 is of practical importance for clinical needs. As it can be seen in Figure 3.20(c), the gradual increase in the color intensity signifies the presence of a higher concentration of glucose, thus sharing a direct proportional relationship. Therefore, it can be said that visual colorimetric detection can be helpful in allowing rapid access to the glucose concentration levels, thus contributing mainly to diagnosis or at the point-of-care (Gabriel et al. 2017).

3.4.2.1 Alternative Color Indicators for Glucose μPADs

Due to the weak color signaling produced by the indicator, such as potassium iodide, some organic compounds and nanoparticles are used as color indicators in glucose μPADs. According to Chen et al., 2,4,6-tribromo-3-hydroxybenzoic acid (TBHBA) and 4-aminoantipyrine (4-APP) are used as substrates catalyzed by HRP for the generation of the color signal for glucose detection due to excellent water solubility of TBHBA and TBHBA/4-APP. The positive charge can be attached firmly onto the paper substrate to the negative charged TBHBA with N-ethyl-N (3-sulfopropyl)-3-methyl-aniline sodium salt (TOPS). These are then used for glucose detection of TOPS/4-APP in the μPAD and showed the limit of detection (LOD) of 38.1 μM. The substrates GOx and colorless 3,3′-diaminobenzidine (DAB) were immobilized in a hydrophilic channel on the μPADs, and the sample solution traveled along the channel by capillary action. The H_2O_2 was generated by the action of the GOx and further reacted with DAB to form a visible brown, insoluble product (poly (DAB)) in the presence of peroxidase, as shown in Figure 3.21.

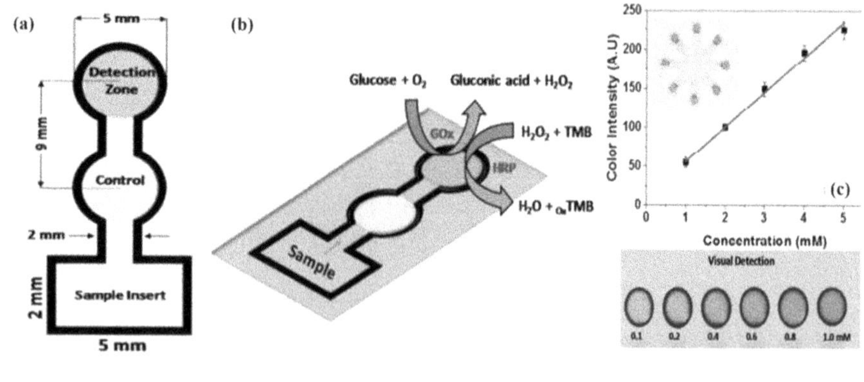

FIGURE 3.20 (a) Outline of μPAD employed for detection of glucose using colorimetric assay, (b) reaction of enzyme and chromogenic reagent for the sensing of glucose, and (c) the color scale bar utilized for glucose detection and different concentrations (Gabriel et al. 2017).

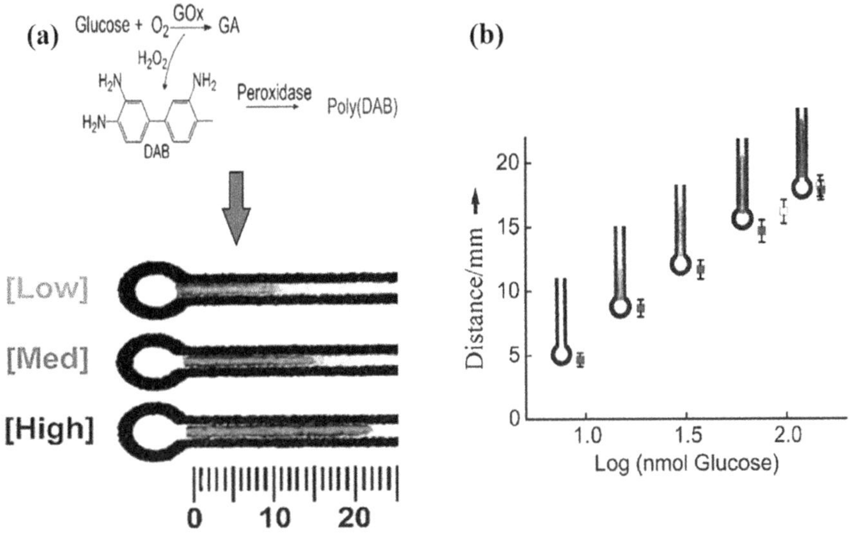

FIGURE 3.21 (a) Colorimetric signals generated corresponding to the variable concentration of glucose and (b) calibration curve for a standard glucose solution (Cate et al. 2013).

3.4.2.2 Fluorescence

The fluorescence method of sensing is based on signal detection, which occurs during the interaction of target molecules and fluorescent dyes known as fluorophores. It was first used in paper microzone plates, and the fluorescence process consists of three stages:

- Excitation
- Excited-state lifetime
- Fluorescence emission

Also, the process of sensing consists of the following steps

- A source of light at a particular wavelength induces luminescence in a fluorophore.
- The light is filtered, and emission photons are isolated from excitation photons.
- Emission photons are detected, producing an electrical signal as an indicator.

Paper can also show fluorescence by the addition of several fluorescent agents. However, these materials may enhance the auto-fluorescence and result in false positives. DNA has been successfully detected using strips of paper immobilized with synthetic DNA oligonucleotides. Yuan et al. fabricated and functionalized the paper by encapsulating cadmium telluride (CdTe) quantum dots and enzymes within a film of poly(diallyldimethylammonium chloride). They were able to detect glucose and catechol in a μPAD via fluorescence detection (Zhu et al. 2014). Several methods were discovered that involved the pretreatment of paper with diagnostic analytes to target areas for specific assays, which could be predesigned into paper-based diagnostic tools. One such research, demonstrated by Yu et al., was that the modification of cellulose paper with divinyl sulfone resulted in the covalent immobilization of carbohydrates, proteins, and DNA for colorimetric and fluorometric bioassays (Yu et al. 2012). In another work, by Yildiz et al. a non-paper-focused approach was employed to fabricated a poly(vinylidene fluoride) (PVDF) porous membrane, which was further modified with poly(3-alkoxy-4-methylthiophene) to detect a ribonucleic acid (microRNA) sequence associated with lung cancer fluorometrically (Yildiz et al. 2013).

3.5 COLOR CALIBRATION TECHNIQUES, TOOLS, AND METHODS ADOPTED

Color calibration is the method to measure the degree of discoloration of the sample and analyze glucose concentration quantitatively. This technique can detect the optical characteristics of the sample when the reaction gives color pigmentation. The discoloration characteristics depend on the change of concentration of glucose observed by using the indicator.

The International Commission on Illumination (CIE) 1931 color space is one of color spaces defined mathematically, based on research on human color recognition. Human eyes recognize colors by combining three colors (red, green, and blue (RGB)), and therefore, the distribution map of all visible lights takes a 3D shape.

Figure 3.22 is a 2D image showing the chromaticity distribution table of the CIE 1931 color space. The outer curve-shaped boundary corresponds to monochromatic light and wavelength (nm). All the colors can be expressed in the visible light as positive X, Y, and Z values.

The *R*, *G*, and *B* values can be converted to *X*, *Y*, and *Z* values by an equation. Then, chromaticity coordinates *x* and *y* can be determined from the *X*, *Y*, and *Z* values depicted in equations (3.4, 3.5, and 3.6). The *Y* value indicates the degree of the luminance or brightness of a particular color in CIE coordinates. The chromaticity of color can be expressed by *x* and *y* parameters dependent on tri-stimulus values *X*, *Y*, and *Z* (Kim et al. 2017).

$$X = (0.412 * R) + (0.358 * G) + (0.180 * B) \tag{3.4}$$

$$Y = (0.213 * R) + (0.715 * G) + (0.072 * B) \tag{3.5}$$

$$Z = (0.019 * R) + (0.119 * G) + (0.950 * B) \tag{3.6}$$

$$x = \frac{X}{(X + Y + Z)} \tag{3.7}$$

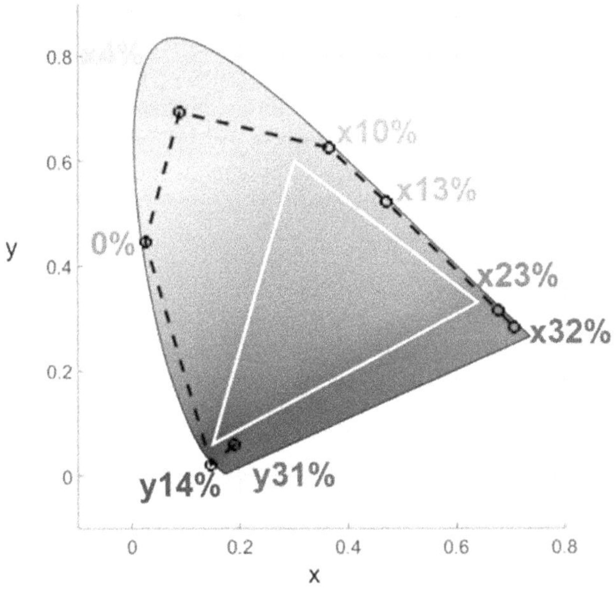

FIGURE 3.22 International Commission on Illumination (CIE) 1931 diagram depicting the color gamut (Tseng et al. 2017).

$$y = \frac{Y}{(X+Y+Z)} \tag{3.8}$$

3.6 CONCLUSION

Since glucose is an essential indicator of the amount of metabolic activity, a rapid and convenient test for glucose has become essential in underdeveloped as well as in developing countries. Thus, over a few years, the emergence of microfluidics advanced the quest for miniaturization of laboratory diagnostic functions, unlocking the possibilities for an immediate response, bringing the concept of portability and low cost in the field of point-of-care diagnostics. Paper-based analytic devices have the edge over other microfluidic devices in terms of cost effectiveness, disposability, ease of manufacture, etc. Thus, they hold greater potential in various sciences such as cell biology, clinical diagnosis, and drug screening. Various techniques, such as photolithography, wax printing, screen printing, spray drying, plasma treating, and laser treatment which are used for the fabrication of hydrophobic barriers have been discussed. Though photolithography provides higher reproducibility, the entire process is often time-consuming, tedious, and requires sophisticated equipment. Therefore, the adoption of wax-printing technology over the photolithography method offers advantages such as high resolution, easy processing, and cost effectiveness, thus gaining more popularity. Numerous methods have been developed offering multiple detection applied to glucose diagnostics. Still, colorimetric and electrochemical detection are undoubtedly essential techniques. With the development of point-of-care diagnostics (POCT), the devices tend towards miniaturization and integrating smartphone spectrometric functions or electronic measurements. Besides this, the consideration of biocompatibility, portability of papers, etc., gives potential for developing invasive or non-invasive μPADs for real-time glucose detection. Improvements in the stability, sensitivity, and accuracy of glucose detection could make the devices commercially available in the future and even operated by everyone without leaving their homes.

REFERENCES

Ahmad, R., N. Tripathy, M. S. Ahn, K. S. Bhat, T. Mahmoudi, Y. Wang, J. Y. Yoo, D. W. Kwon, H. Y. Yang, and Y. B. Hahn. 2017. "Highly efficient non-enzymatic glucose sensor based on CuO modified vertically-grown ZnO nanorods on electrode." *Sci Rep* no. 7 (1):5715–5724. doi: 10.1038/s41598-017-06064-8.

Aid, T., M. Kaljurand, and M. Vaher. 2015. "Colorimetric determination of total phenolic contents in ionic liquid extracts by paper microzones and digital camera." *Anal Methods* no. 7 (7):3193–3199. doi: 10.1039/c5ay00194c.

Altundemir, S., A. K. Uguz, and K. Ulgen. 2017. "A review on wax printed microfluidic paper-based devices for international health." *Biomicrofluidics* no. 11 (4):041501. doi: 10.1063/1.4991504.

Appleton, S. L., C. J. Seaborn, R. Visvanathan, C. L. Hill, T. K. Gill, A. W. Taylor, and R. J. Adams. 2013. "Diabetes and cardiovascular disease outcomes in the metabolically healthy obese phenotype: a cohort study." *Diabetes Care* no. 36 (8):2388–2394. doi: 10.2337/dc12-1971.

Aragay, G., J. Pons, and A. Merkoci. 2011. "Recent trends in macro-, micro-, and nanomaterial-based tools and strategies for heavy-metal detection." *Chem Rev* no. 111 (5):3433–3458. doi: 10.1021/cr100383r.

Avogaro, A. 2016. "Cardiovascular disease prevention in adults with type 2 diabetes mellitus according to the recent statement from the American Heart Association/American Diabetes Association." *G Ital Cardiol* no. 17 (3 Suppl 2):5s–11. doi: 10.1714/2206.23816.

Ben-Tzvi, P., and W. Rone. 2009. "Microdroplet generation in gaseous and liquid environments." *Microsystem Technologies* no. 16 (3):333–356. doi: 10.1007/s00542-009-0962-7.

Bihar, Eloïse, Shofarul Wustoni, Anna Maria Pappa, Khaled N. Salama, Derya Baran, and Sahika Inal. 2018. "A fully inkjet-printed disposable glucose sensor on paper." *NPJ Flex Electron* no. 2 (1):30–8. doi: 10.1038/s41528-018-0044-y.

Cardoso, T. M. G., F. R. de Souza, P. T. Garcia, D. Rabelo, C. S. Henry, and W. K. T. Coltro. 2017. "Versatile fabrication of paper-based microfluidic devices with high chemical resistance using scholar glue and magnetic masks." *Anal Chim Acta* no. 974:63–68. doi: 10.1016/j.aca.2017.03.043.

Cate, D. M., W. Dungchai, J. C. Cunningham, J. Volckens, and C. S. Henry. 2013. "Simple, distance-based measurement for paper analytical devices." *Lab Chip* no. 13 (12):2397–2404. doi: 10.1039/c3lc50072a.

Chaiyo, S., E. Mehmeti, W. Siangproh, T. L. Hoang, H. P. Nguyen, O. Chailapakul, and K. Kalcher. 2018. "Non-enzymatic electrochemical detection of glucose with a disposable paper-based sensor using a cobalt phthalocyanine-ionic liquid-graphene composite." *Biosens Bioelectron* no. 102:113–120. doi: 10.1016/j.bios.2017.11.015.

Chen, X., J. Chen, F. Wang, X. Xiang, M. Luo, X. Ji, and Z. He. 2012. "Determination of glucose and uric acid with bienzyme colorimetry on microfluidic paper-based analysis devices." *Biosens Bioelectron* no. 35 (1):363–368. doi: 10.1016/j.bios.2012.03.018.

Chen, Xu, Yong Luo, Bo Shi, Xianming Liu, Zhigang Gao, Yuguang Du, Weijie Zhao, and Bingcheng Lin. 2014. "Chemiluminescence diminishment on a paper-based analytical device: high throughput determination of β-agonists in swine hair." *Anal Methods* no. 6 (24):9684–9690. doi: 10.1039/c4ay02116a.

Chitnis, G., Z. Ding, C. L. Chang, C. A. Savran, and B. Ziaie. 2011. "Laser-treated hydrophobic paper: an inexpensive microfluidic platform." *Lab Chip* no. 11 (6):1161–1165. doi: 10.1039/c0lc00512f.

Cui, Y., L. Zhang, M. Zhang, X. Yang, L. Zhang, J. Kuang, G. Zhang, Q. Liu, H. Guo, and Q. Meng. 2017. "Prevalence and causes of low vision and blindness in a Chinese population with type 2 diabetes: the Dongguan Eye Study." *Sci Rep* no. 7 (1):11195. doi: 10.1038/s41598-017-11365-z.

Delaney, J. L., and C. F. Hogan. 2015. "Mobile phone based electrochemiluminescence detection in paper-based microfluidic sensors." *Methods Mol Biol* no. 1256:277–289. doi: 10.1007/978-1-4939-2172-0_19.

Deng, J., L. Wang, L. Liu, and W. Yang. 2009. "Developments and new applications of UV-induced surface graft polymerizations." *Prog Polym Sci.* no. 34 (2):156–193. doi: 10.1016/j.progpolymsci.2008.06.002.

Deng, Jianping, and Wan Yang. 2000. "Surface photografting polymerization of vinyl acetate (VAc), maleic anhydride, and their charge transfer complex. II. VAc(2)." *J Appl Polym Sci* no. 77:1522–1531. doi: 10.1002/1097-4628(20000815)77:73.0.CO;2-Y.

Derby, Brian. 2010. "Inkjet printing of functional and structural materials: fluid property requirements, feature stability, and resolution." *Annu Rev Mater Res* no. 40 (1):395–414. doi: 10.1146/annurev-matsci-070909-104502.

Eichler, M., C. P. Klages, and K. Lachmann. 2016. "Surface functionalization of microfluidic devices." In: Dietzel A. (eds) *Microsystems for Pharmatechnology.* Springer, Cham, 59–97. doi: 10.1007/978-3-319-26920-7_3.

Figueredo, Federico, Paulo T. Garcia, Eduardo Cortón, and Wendell K. T. Coltro. 2016. "Enhanced analytical performance of paper microfluidic devices by using Fe3O4 nanoparticles, MWCNT, and graphene oxide." *ACS Appl Mater Interfaces* no. 8 (1):11–15. doi: 10.1021/acsami.5b10027.

Foo, Kai Loong, Uda Hashim, Kashif Muhammad, and Chun Hong Voon. 2014. "Sol–gel synthesized zinc oxide nanorods and their structural and optical investigation for optoelectronic application." *Nanoscale Res. Lett* no. 9:10.

Gabriel, E., P. Garcia, F. Lopes, and W. Coltro. 2017. "Paper-based colorimetric biosensor for tear glucose measurements." *Micromachines* no. 8 (4):104. doi: 10.3390/mi8040104.

Ge, L., J. Yan, X. Song, M. Yan, S. Ge, and J. Yu. 2012. "Three-dimensional paper-based electrochemiluminescence immunodevice for multiplexed measurement of biomarkers and point-of-care testing." *Biomaterials* no. 33 (4):1024–31. doi: 10.1016/j.biomaterials.2011.10.065.

Gokaltun, A., M. L. Yarmush, A. Asatekin, and O. B. Usta. 2017. "Recent advances in nonbiofouling PDMS surface modification strategies applicable to microfluidic technology." *Technology* no. 5 (1):1–12. doi: 10.1142/S2339547817300013.

Gokaltun, A., Y. B. A. Kang, M. L. Yarmush, O. B. Usta, and A. Asatekin. 2019. "Simple surface modification of poly(dimethylsiloxane) via surface segregating smart polymers for biomicrofluidics." *Sci Rep* no. 9 (1):7377. doi: 10.1038/s41598-019-43625-5.

Habouti, S., C. Kunstmann-Olsen, J. D. Hoyland, Horst-Günter Rubahn, and M. Es-Souni. 2014. "In situ ZnO–PVA nanocomposite coated microfluidic chips for biosensing." *Appl. Phys. A* no. 115 (2):645–649. doi: 10.1007/s00339-014-8397-0.

Hahn, Y. B., R. Ahmad, and N. Tripathy. 2012. "Chemical and biological sensors based on metal oxide nanostructures." *Chem Commun* no. 48 (84):10369–10385. doi: 10.1039/c2cc34706g.

He, Y., Y. Wu, Jian-Zhong Fu, and Wen-Bin Wu. 2015a. "Fabrication of paper-based microfluidic analysis devices: a review." *RSC Adv.* no. 5 (95):78109–78127. doi: 10.1039/c5ra09188h.

He, Yong, Wen-bin Wu, and Jian-zhong Fu. 2015b. "Rapid fabrication of paper-based microfluidic analytical devices with desktop stereolithography 3D printer." *RSC Adv.* no. 5 (4):2694–2701. doi: 10.1039/c4ra12165a.

Kaefer, M., S. J. Piva, J. A. De Carvalho, D. B. Da Silva, A. M. Becker, A. C. Coelho, M. M. Duarte, and R. N. Moresco. 2010. "Association between ischemia modified albumin, inflammation and hyperglycemia in type 2 diabetes mellitus." *Clin Biochem* no. 43 (4–5):450–454. doi: 10.1016/j.clinbiochem.2009.11.018.

Kim, J. S., Han-Byeol Oh, A. H. Kim, J. S. Kim, Eun-Suk Lee, Jin-Young Baek, Ki Sung Lee, Soon-Cheol Chung, and Jae-Hoon Jun. 2017. "A study on detection of glucose concentration using changes in color coordinates." *Bioengineered* no. 8 (1):99–104. doi: 10.1080/21655979.2016.1227629.

Konta, A. A., M. Garcia-Pina, and D. R. Serrano. 2017. "Personalised 3D printed medicines: which techniques and polymers are more successful?" *Bioengineering* no. 4 (4):79–95. doi: 10.3390/bioengineering4040079.

Li, J., F. Rossignol, and J. Macdonald. 2015. "Inkjet printing for biosensor fabrication: combining chemistry and technology for advanced manufacturing." *Lab Chip* no. 15 (12):2538–2558. doi: 10.1039/c5lc00235d.

Li, X., J. Tian, T. Nguyen, and W. Shen. 2008. "Paper-based microfluidic devices by plasma treatment." *Anal Chem* no. 80 (23):9131–9134. doi: 10.1021/ac801729t.

Li, X., J. Tian, G. Garnier, and W. Shen. 2010. "Fabrication of paper-based microfluidic sensors by printing." *Colloids Surf B* no. 76 (2):564–70. doi: 10.1016/j.colsurfb.2009.12.023.

Lim, H., A. T. Jafry, and J. Lee. 2019. "Fabrication, flow control, and applications of microfluidic paper-based analytical devices." *Molecules* no. 24 (16):2869–2901. doi: 10.3390/molecules24162869.

Liu, Ning, Jing Xu, Hong-Jie An, Dinh-Tuan Phan, Michinao Hashimoto, and Wen Siang Lew. 2017. "Direct spraying method for fabrication of paper-based microfluidic devices." *J. Micromech. Microeng.* no. 27 (10):104001. doi: 10.1088/1361-6439/aa82ce.

Liu, S., W. Su, and X. Ding. 2016. "A review on microfluidic paper-based analytical devices for glucose detection." *Sensors* no. 16 (12):2086–2103. doi: 10.3390/s16122086.

Lu, Y., W. Shi, L. Jiang, J. Qin, and B. Lin. 2009. "Rapid prototyping of paper-based microfluidics with wax for low-cost, portable bioassay." *Electrophoresis* no. 30 (9):1497–500. doi: 10.1002/elps.200800563.

Ma, H., and M. Zhang. 2014. "Superhydrophilic titania wall coating in microchannels by in situ sol–gel modification." *J Mater Sci* no. 49 (23):8123–8126. doi: 10.1007/s10853-014-8521-8.

Maejima, K., S. Tomikawa, K. Suzuki, and D. Citterio. 2013. "Inkjet printing: an integrated and green chemical approach to microfluidic paper-based analytical devices." *RSC Adv.* no. 3 (24):9258–9263. doi: 10.1039/C3RA40828K.

Mani, N. K., A. Prabhu, S. K. Biswas, and S. Chakraborty. 2019. "Fabricating paper based devices using correction pens." *Sci Rep* no. 9 (1):1752. doi: 10.1038/s41598-018-38308-6.

Martin, G. D., S. D. Hoath, and I. M. Hutchings. 2008. "Inkjet printing: the physics of manipulating liquid jets and drops." *J Phys Conf Ser* no. 105:012001. doi: 10.1088/1742-6596/105/1/012001.

Martinez, A. W., S. T. Phillips, M. J. Butte, and G. M. Whitesides. 2007. "Patterned paper as a platform for inexpensive, low-volume, portable bioassays." *Angew Chem Int Ed Engl* no. 46 (8):1318–1320. doi: 10.1002/anie.200603817.

Martinez, A. W., S. T. Phillips, B. J. Wiley, M. Gupta, and G. M. Whitesides. 2008. "FLASH: a rapid method for prototyping paper-based microfluidic devices." *Lab Chip* no. 8 (12):2146–2150. doi: 10.1039/b811135a.

Martinez, A. W., S. T. Phillips, G. M. Whitesides, and E. Carrilho. 2010. "Diagnostics for the developing world: microfluidic paper-based analytical devices." *Anal Chem* no. 82 (1):3–10. doi: 10.1021/ac9013989.

Nathan, D. M. 2014. "The diabetes control and complications trial/epidemiology of diabetes interventions and complications study at 30 years: overview." *Diabetes Care* no. 37 (1):9–16. doi: 10.2337/dc13-2112.

Nathan, D. M., S. Genuth, J. Lachin, P. Cleary, O. Crofford, M. Davis, L. Rand, and C. Siebert. 1993. "The effect of intensive treatment of diabetes on the development and progression of long-term complications in insulin-dependent diabetes mellitus." *N Engl J Med* no. 329 (14):977–986. doi: 10.1056/nejm199309303291401.

Noiphung, J., T. Songjaroen, W. Dungchai, C. S. Henry, O. Chailapakul, and W. Laiwattanapaisal. 2013. "Electrochemical detection of glucose from whole blood using paper-based microfluidic devices." *Anal Chim Acta* no. 788:39–45. doi: 10.1016/j.aca.2013.06.021.

Raza, W., and K. Ahmad. 2018. "A highly selective Fe@ZnO modified disposable screen printed electrode based non-enzymatic glucose sensor (SPE/Fe@ZnO)." *Mater. Lett.* no. 212:231–234. doi: 10.1016/j.matlet.2017.10.100.

Sameenoi, Y., P. N. Nongkai, S. Nouanthavong, C. S. Henry, and D. Nacapricha. 2014. "One-step polymer screen-printing for microfluidic paper-based analytical device (muPAD) fabrication." *Analyst* no. 139 (24):6580–8. doi: 10.1039/c4an01624f.

Santhiago, M., and L. T. Kubota. 2013. "A new approach for paper-based analytical devices with electrochemical detection based on graphite pencil electrodes." *Sens Actuators B* no. 177:224–230. doi: 10.1016/j.snb.2012.11.002.

Schilling, K. M., A. L. Lepore, J. A. Kurian, and A. W. Martinez. 2012. "Fully enclosed microfluidic paper-based analytical devices." *Anal Chem* no. 84 (3):1579–1585. doi: 10.1021/ac202837s.

Setti, L., A. Fraleoni-Morgera, B. Ballarin, A. Filippini, D. Frascaro, and C. Piana. 2005. "An amperometric glucose biosensor prototype fabricated by thermal inkjet printing." *Biosens Bioelectron* no. 20 (10):2019–26. doi: 10.1016/j.bios.2004.09.022.

Setti, L., A. Fraleonimorgera, I. Mencarelli, A. Filippini, B. Ballarin, and M. Dibiase. 2007. "An HRP-based amperometric biosensor fabricated by thermal inkjet printing." *Sens Actuators B* no. 126 (1):252–257. doi: 10.1016/j.snb.2006.12.015.

Shangguan, J. W., Y. Liu, J. B. Pan, B. Y. Xu, J. J. Xu, and H. Y. Chen. 2016. "Microfluidic PDMS on paper (POP) devices." *Lab Chip* no. 17 (1):120–127. doi: 10.1039/c6lc01250g.

Sher, M., R. Zhuang, U. Demirci, and W. Asghar. 2017. "Paper-based analytical devices for clinical diagnosis: recent advances in the fabrication techniques and sensing mechanisms." *Expert Rev Mol Diagn* no. 17 (4):351–366. doi: 10.1080/14737159.2017.1285228.

Shi, Chuan-Guo, Xia Shan, Zhong-Qin Pan, Jing-Juan Xu, Chang Lu, Ning Bao, and Hai-Ying Gu. 2012. "Quantum dot (QD)-modified carbon tape electrodes for reproducible electrochemiluminescence (ECL) emission on a paper-based platform." *Anal. Chem.* no. 84 (6):3033–3038. doi: 10.1021/ac2033968.

Soejima, T., K. Takada, and S. Ito. 2013. "Alkaline vapor oxidation synthesis and electrocatalytic activity toward glucose oxidation of CuO/ZnO composite nanoarrays." *Appl Surf Sci* no. 277:192–200. doi: 10.1016/j.apsusc.2013.04.024.

Sones, C. L., I. N. Katis, P. J. He, B. Mills, M. F. Namiq, P. Shardlow, M. Ibsen, and R. W. Eason. 2014. "Laser-induced photo-polymerisation for creation of paper-based fluidic devices." *Lab Chip* no. 14 (23):4567–4574. doi: 10.1039/c4lc00850b.

Song, Y., K. Qu, C. Zhao, J. Ren, and X. Qu. 2010. "Graphene oxide: intrinsic peroxidase catalytic activity and its application to glucose detection." *Adv. Mater. Lett.* no. 22 (19):2206–2210. doi: 10.1002/adma.200903783.

Songjaroen, T., W. Dungchai, O. Chailapakul, and W. Laiwattanapaisal. 2011. "Novel, simple and low-cost alternative method for fabrication of paper-based microfluidics by wax dipping." *Talanta* no. 85 (5):2587–93. doi: 10.1016/j.talanta.2011.08.024.

Songjaroen, Temsiri, J. Noiphung, I. Hongwarittorrn, K. Talalak, and Wanida Laiwattanapaisal. 2014. "Assay time reduction and thermal stability improvement of a lowcost, Wax-dipping paper-based microfluidic device." *J Chem Pharm* no. 6:2895–2903.

SoYoon, S., A. Ramadoss, B. Saravanakumar, and S. J. Kim. 2014. "Novel Cu/CuO/ZnO hybrid hierarchical nanostructures for non-enzymatic glucose sensor application." *J Electroanal Chem* no. 717–718:90–95. doi: 10.1016/j.jelechem.2014.01.012.

Sweet, R. G. 1965. "High frequency recording with electrostatically deflected ink jets." *Rev Sci Instrum* no. 36 (2):131–136. doi: 10.1063/1.1719502.

Tak, M., V. Gupta, and M. Tomar. 2014. "Flower-like ZnO nanostructure based electrochemical DNA biosensor for bacterial meningitis detection." *Biosens Bioelectron* no. 59:200–207. doi: 10.1016/j.bios.2014.03.036.

Tan, H. W., T. Tran, and C. K. Chua. 2016. "A review of printed passive electronic components through fully additive manufacturing methods." *Virtual Phys Prototyp* no. 11 (4):271–288. doi: 10.1080/17452759.2016.1217586.

Tripathy, N., and D. H. Kim. 2018. "Metal oxide modified ZnO nanomaterials for biosensor applications." *Nano Converg* no. 5 (1):27. doi: 10.1186/s40580-018-0159-9.

Tseng, M. L., J. Yang, M. Semmlinger, C. Zhang, P. Nordlander, and N. J. Halas. 2017. "Two-dimensional active tuning of an aluminum plasmonic array for full-spectrum response." *Nano Lett* no. 17 (10):6034–6039. doi: 10.1021/acs.nanolett.7b02350.

Wang, Y., and W. Yang. 2004. "MMA/DVB emulsion surface graft polymerization initiated by UV light." *Langmuir* no. 20 (15):6225–6231. doi: 10.1021/la0493924.

Wanner, Ch, S. E. Inzucchi, and B. Zinman. 2016. "Empagliflozin and progression of kidney disease in type 2 diabetes." *N Engl J Med* no. 375 (18):1801–1802. doi: 10.1056/NEJMc1611290.

Wei, X., T. Tian, S. Jia, Z. Zhu, Y. Ma, J. Sun, Z. Lin, and C. J. Yang. 2016. "Microfluidic distance readout sweet hydrogel integrated paper-based analytical device (muDiSH-PAD) for visual quantitative point-of-care testing." *Anal Chem* no. 88 (4):2345–52. doi: 10.1021/acs.analchem.5b04294.

Wen, J., L. A. Legendre, J. M. Bienvenue, and J. P. Landers. 2008. "Purification of nucleic acids in microfluidic devices." *Anal. Chem.* no. 80 (17):6472–6479. doi: 10.1021/ac8014998.

Wu, J., and F. Yin. 2013. "Easy fabrication of a sensitive non-enzymatic glucose sensor based on electrospinning CuO-ZnO nanocomposites." *Integr Ferroelectr* no. 147 (1):47–58. doi: 10.1080/10584587.2013.790695.

Wu, Z., and K. Hjort. 2009. "Surface modification of PDMS by gradient-induced migration of embedded Pluronic." *Lab Chip* no. 9 (11):1500–1503. doi: 10.1039/b901651a.

Xing, S., J. Jiang, and T. Pan. 2013. "Interfacial microfluidic transport on micropatterned superhydrophobic textile." *Lab Chip* no. 13 (10):1937–1947. doi: 10.1039/c3lc41255e.

Yager, P., T. Edwards, E. Fu, K. Helton, K. Nelson, M. R. Tam, and B. H. Weigl. 2006. "Microfluidic diagnostic technologies for global public health." *Nature* no. 442 (7101):412–418. doi: 10.1038/nature05064.

Yetisen, A. K., M. S. Akram, and C. R. Lowe. 2013. "Paper-based microfluidic point-of-care diagnostic devices." *Lab Chip* no. 13 (12):2210–2251. doi: 10.1039/c3lc50169h.

Yildiz, U. H., P. Alagappan, and B. Liedberg. 2013. "Naked eye detection of lung cancer associated miRNA by paper based biosensing platform." *Anal Chem* no. 85 (2):820–824. doi: 10.1021/ac3034008.

Yu, A., J. Shang, F. Cheng, B. A. Paik, J. M. Kaplan, R. B. Andrade, and D. M. Ratner. 2012. "Biofunctional paper via the covalent modification of cellulose." *Langmuir* no. 28 (30):11265–11273. doi: 10.1021/la301661x.

Yu, J., L. Ge, J. Huang, S. Wang, and S. Ge. 2011. "Microfluidic paper-based chemiluminescence biosensor for simultaneous determination of glucose and uric acid." *Lab Chip* no. 11 (7):1286–1291. doi: 10.1039/c0lc00524j.

Yu, L., and Z. Z. Shi. 2015. "Microfluidic paper-based analytical devices fabricated by low-cost photolithography and embossing of Parafilm(R)." *Lab Chip* no. 15 (7):1642–1645. doi: 10.1039/c5lc00044k.

Zang, D., L. Ge, M. Yan, X. Song, and J. Yu. 2012. "Electrochemical immunoassay on a 3D microfluidic paper-based device." *ChemComm* no. 48 (39):4683–4685. doi: 10.1039/C2CC16958D.

Zhang, Y., L. Zhang, J. Sun, Y. Liu, X. Ma, S. Cui, L. Ma, J. J. Xi, and X. Jiang. 2014. "Point-of-care multiplexed assays of nucleic acids using microcapillary-based loop-mediated isothermal amplification." *Anal Chem* no. 86 (14):7057–7062. doi: 10.1021/ac5014332.

Zhao, C., M. M. Thuo, and X. Liu. 2013. "A microfluidic paper-based electrochemical biosensor array for multiplexed detection of metabolic biomarkers." *Sci Technol Adv Mater* no. 14 (5):054402. doi: 10.1088/1468-6996/14/5/054402.

Zhu, Y., X. Xu, N. D. Brault, A. J. Keefe, X. Han, Y. Deng, J. Xu, Q. Yu, and S. Jiang. 2014. "Cellulose paper sensors modified with zwitterionic poly(carboxybetaine) for sensing and detection in complex media." *Anal Chem* no. 86 (6):2871–2875. doi: 10.1021/ac500467c.

4 Microfluidics-Based Point-of-Care Diagnostic Devices

Ashis K. Sen, Amal Nath, Aremanda Sudeepthi, Sachin K. Jain, and Utsab Banerjee

CONTENTS

4.1 INTRODUCTION

The World Health Organization (WHO) recently reported that infectious diseases such as tuberculosis (TB), malaria, and acquired immunodeficiency syndrome (AIDS) are the second leading cause of mortality throughout the world after cardiovascular disease [1]. One of the primary reasons behind this fact is the lack of accurate, user-friendly, cost-effective testing facilities in developing countries [2, 3]. Precise and early clinical diagnosis is of paramount importance for the interpretation of symptoms, prevention, and in determining the potency of treatments. Microfluidic devices employ analytical components and techniques at microscopic scales (1 to 500 μm) which involve microfilters, microchannels, microarrays, micropumps, microvalves, and bioelectronic chips for various clinical diagnostic applications for point-of-care testing (POCT).

POCT is the realm of healthcare in which clinical testing and assaying are executed at the patient bedside, either by medical professionals or by the patient. For

DOI: 10.1201/9781003033479-4

example, POC systems can enable early diagnosis of cancer by detecting a few cancer cells, like circulating tumor cells (CTCs), out of a million healthy ones, which eventually helps medical professionals to plan treatments in advance. POCT offers several significant advantages over conventional techniques – (i) providing results in a short time (<15 min), (ii) user-friendly usability (sample preparation, automated sample processing), (iii) small sample volumes (<30 μL), and (iv) integration and functionality of multiple segments for various tests. The global POC technology market is expected to grow from US$23.16 billion in 2016 to US$36.96 billion in 2021 at the compound annual growth rate (CAGR) of 9.8% from 2016 to 2021 [4]. POCT is expected to increase considerably in India and China, which have enormous populations with an increased prevalence of chronic and infectious diseases. In recent years, POC devices incorporating both disposable microelectronic and microfluidic components have been developed (examples include products from companies like Abbott, Metrika, Biosite, iSTAT, and Unipath). Glucose testing is the most successful and widespread test performed at the point of care and comprises the majority of the POCT market segment. The various POC devices can be polymer based, which require both complex fabrication and external instruments indicating a limitation in the case of POCT. On the other hand, paper-based POC devices, being affordable, sensitive, specific, user friendly, rapid, and robust offer several advantages over polymer devices. The paper-based devices can be utilized for protein-based assays, cell-based assays, glucose level detection, etc., as discussed in the chapter.

This chapter is organized as follows. In Section 4.2, we discuss the current developments in POC technology, including a discussion on existing devices and their applications. The different components to be integrated to realize a full "sample-in result-out" POC device are detailed in Section 4.3. In Section 4.4, we look at a particular class of POC devices, paper-based devices, that has seen considerable research over recent years due to its unique features. Different commercialized POC devices are looked at in Section 4.5, and we conclude with a discussion on current challenges in POC technology and what lies ahead.

4.2 POINT-OF-CARE: THE CURRENT SCENARIO

The outbreak of the recent global pandemic, COVID-19, due to a highly infectious strand of coronavirus, sudden acute respiratory syndrome coronavirus-2 (SARS-CoV-2), has put the entire world into a stall. At the time of writing, around 15 million cases of COVID-19 have been detected worldwide and are estimated to increase more than tenfold. The meteoric surge in the number of cases is widely attributed to lapses in quick detection of the disease due to use of traditional diagnosis techniques and has underlined the need for rapid POCT where early detection can help in faster isolation of the patient and prevention of the spread of the disease. As opposed to traditional benchtop medical diagnostics which require bulky equipment and invasive tests, POC diagnostics require that devices be extremely easy to use, small, portable, low cost, and lightweight, require very small sample volume, have a low assay time, and do not require experienced personnel to operate.

POC devices have already been developed and are in extensive use for a range of applications like blood chemistry and gas tests (determining levels of blood glucose, blood lactate, electrolytes, hemoglobin). These devices require very low sample volumes of only ~100 μL and have processing and result times of ~1–2 min. The use of dipsticks for pregnancy testing and lateral flow tests for the detection of cardiac disease or prostate cancer are some of the other routine devices based on POC technology. In addition, commercialized devices also exist for the detection of a whole range of pathogens like viruses, parasites (like malaria), and bacteria (like syphilis and tuberculosis). For example, a viral diagnosis has been achieved by PCR-based devices by detecting viral RNA and DNA for a whole range of viruses, including the Ebola virus, ZIKA, HIV, HPV, and HCV [5]. Related efforts, in light of the recent pandemic, are still underway to develop devices for the rapid detection of the SARS-CoV-2 coronavirus. The current standard protocol to detect COVID-19 requires reverse real-time polymerase chain reaction (RT-PCR) assay where the clinical specimen from the patient can include bronchoalveolar lavage fluid, fibrobronchoscope brush biopsies, sputum, nasal swabs, pharyngeal swabs, feces, or blood [6, 7]. Although RT-PCR tests for the detection of COVID-19 were distributed by the WHO as early as January 2020, the complexity and cost associated with the initial testing protocol, the requirement of well-established clinical laboratories, in addition to the long test result duration of 4–6 h, proved testing rates to be ineffective, leading to a large surge in cases. Development of POC devices in the months following the initial outbreak have now made the test results available in a few minutes. For example, the Abbott ID NOW COVID-19 test developed by Abbott Laboratories is an FDA-approved PCR-based test which delivers a positive result in five minutes and a negative result in 13 min [8]. Commercial COVID-19 test kits relying on lateral flow immunoassays to detect viral antigens or patient antibodies have also been developed by Mammoth Biosciences, Luminex, Pharmact, and other companies.

The present trend of POC diagnostics has skewed towards optimizing the cost of the device to make it more accessible to the developing world. In this light, significant research has been carried out to develop the use of low-cost paper-based and polymer-based platforms. One of the advantages of using paper-based devices is that in addition to being low cost and affordable, no additional pumping is required, and the fluid to be analyzed flows due to capillary effect. A large number of recent studies has aimed at developing POC devices using paper-based substrates, and there are ongoing studies to standardize the platform and improve device sensitivity and reproducibility of results [9]. Accordingly, paper-based POC devices are revisited in more detail in this chapter in Section 4.4.

The need to make healthcare more personalized and mobile with minimum intervention has also led to considerable research in developing cellphone-based smart devices. Equipped with components like powerful processors, cameras, microphones, and a variety of sensors and features like real-time location tracking with Global Positioning System (GPS), high data storage and management capacity, and wireless connectivity, the current generation of cellphones serves as a digital platform for the development of bioanalytical POC devices. To be used in POC detection and analysis, cellphones are attached with sample mounting systems and auxiliary optical components like customized lenses to improve the field of view and are turned into effective, lightweight microscopes which can operate in bright-field, fluorescence, dark-field, or bright-field microscopy imaging modes [10]. They can then be used as test readers to analyze results from colorimetric, fluorescent, or chemiluminescent data. Device designs that act as readers for urine, pH test strips, biochemical and immunoassays for clinical imaging of sickle red blood cells (RBCs), visual detection and evaluation of hemoglobin from blood [11] (see Figure 4.1(a)), detecting biomarker for malaria [12] (see Figure 4.1(b)), tuberculosis and HIV [13] and other viruses have already been developed (see Figure 4.1(c)). Personalized monitoring of different health parameters like weight, heart activity, pulse rate, blood pressure,

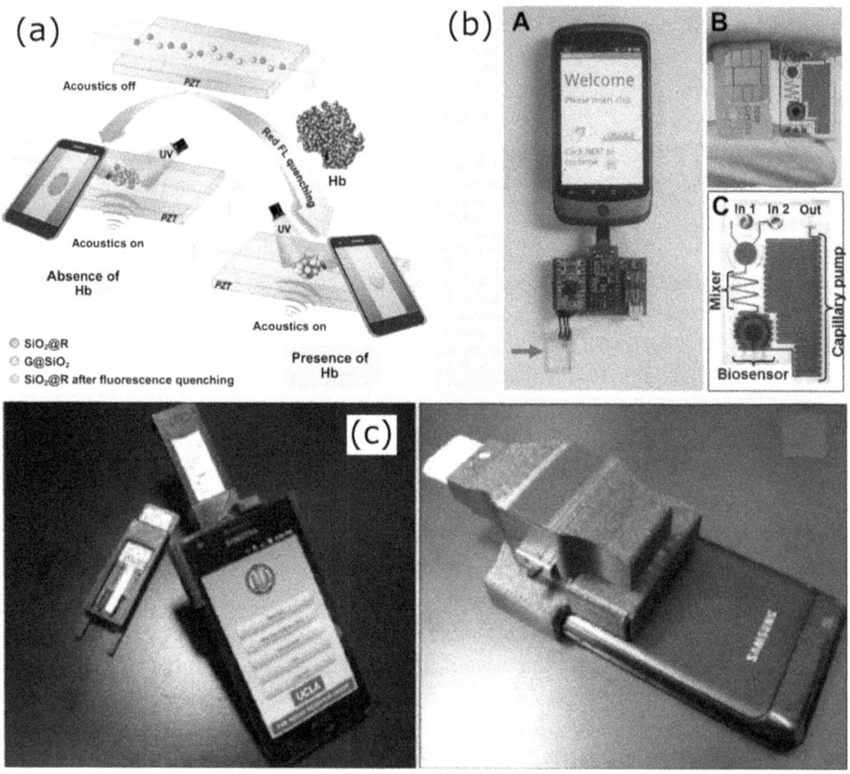

FIGURE 4.1 (a) Colorimetric detection of hemoglobin from human blood using a cellphone-based acoustofluidic platform where the emitted signal shifts to green in presence of hemoglobin, reprinted from [11] Copyright (2020) Royal Society of Chemistry. (b) A – cellphone-based biomolecular detection of a malaria biomarker using a disposable capillary flow-based microchip, B – size of the chip as compared to a SIM card, and C – an enhanced view of the fluidic network of the chip is shown using a dye, reprinted from [12] Copyright (2013) Royal Society of Chemistry. (c) Rapid Diagnostic Test (RDT) reader attachment powered by the cellphone battery installed on a Samsung Galaxy S II phone for tuberculosis and HIV, reprinted from [13] Copyright (2012) Royal Society of Chemistry.

blood oxygen saturation, physical activity, and sleep, has also been achieved by the use of different commercialized devices [4, 14].

4.3 COMPONENTS OF A GENERALIZED MICROFLUIDIC SYSTEM FOR POC APPLICATIONS

In this section, a general discussion is provided on the different components of an integrated POC device. An integrated POC device to facilitate a complete assay consists of three main modules: Flow pumping and control module, sample preparation and processing module, and target detection and analysis module [15]. A large number of microfluidic techniques has been developed to achieve each of these specific functions. We discuss some of them in detail. Further references are provided under each section for the interested reader.

4.3.1 Flow Pumping and Control

Micropumps and microvalves are integrated onto a POC device to enable precise flow control of the sample, buffer, and reagent flow and delivery. The most routine pumping technique used in laboratories involves utilizing a stepper motor to move the plunger in a syringe pump at a constant speed, which defines the sample flow rate. Syringe pumps can be used over a wide range of flow rates and sample volumes. Another class of pump, peristaltic pumps, is fabricated using an elastomeric sheet where the fluid-filled straight channel made of the flexible wall is subjected to a sequence of compression and expansion cycles by an external force to facilitate fluid transport [16] as shown in Figure 4.2(a). Electro-osmotic pumping is another well-known technique where the charged surface creates an ion distribution close to the wall, and applying a tangential electric field leads to Coulomb forces on mobile counter ions. The resulting ion drag establishes a pressure gradient and flow of liquid [17, 18], as shown in Figure 4.2(b). Other active pumping techniques include digital manipulation and targeted transport of an array of reagents or droplets on a substrate achieved through the use of electric [19], acoustic [20] as shown in Figure 4.2(c), optical [21], or magnetic forces [22, 23].

Flow pumping can also be performed by passive forces. Typical driving forces for passive pumping include chemical gradients on surfaces, osmotic pressure, permeation in poly dimethyl siloxane (PDMS), or capillary effects [24]. The most common pumping and control technique used in lateral flow assays is the capillary-driven flow, which makes use of the wetting properties of the substrate materials. Capillary-driven flow is advantageous over active techniques like electro-osmotic pumping, electro-wetting, centrifugal force-driven flows (lab-on-a-CD approach), etc. due to simplicity in design, disposability and low cost, absence of moving parts, and the lack of need for external power [25]. Capillary wetting can be controlled by using arrays of asymmetrical micropillars, and the evaporation rate of the liquid in capillary pumps can be tuned to control the flow rates [26] as shown in Figure 4.2(d). The applicability of capillary-based pumping is, however, limited by changes in the

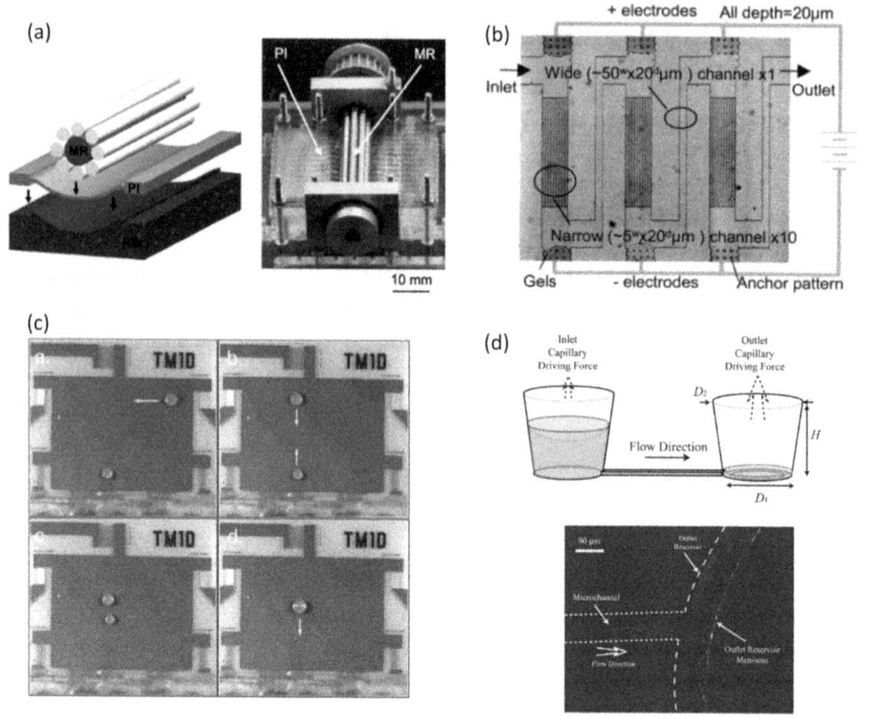

FIGURE 4.2 (a) PDMS-based peristaltic micropump where the multi-roller (MR) is connected to a stepper motor and the pumping inlay (PI) secured by PMMA brackets, reprinted from [16], Copyright (2009) Royal Society of Chemistry. (b) Cascade-type electro-osmosis pump, reprinted from [18], Copyright (2003) Wiley-VCH Verlag GmbH. (c) Droplet transport on a hydrophobic silanized surface by using six different interdigitated transducers, reprinted from [20], Copyright (2005) Royal Society of Chemistry. (d) Schematic illustrating the principle of a passive capillary-drive pump with coupled evaporation effect, transport of 0.5 µm fluorescent tracer beads towards the outlet reservoir is also shown, reprinted from [26], Copyright (2009) Royal Society of Chemistry.

surface properties of the device over time, and the lack of standardization due to poor reproducibility of the material, for example, in paper-based devices.

Flow control is enabled by the use of microvalves, which ideally switch between two channel states – open or closed. While active valves rely on the use of an external agency, passive valves exploit energy potential within the device. Active valves generally involve actuating a flexible membrane (for example, made of PDMS, PEG, poly(N-isopropylacrylamide, etc.) by an external force or by heating. The well-known PDMS-valve (or Quake valve) uses an elastomeric PDMS membrane, which is pressed down by externally pumped compressed air to close the flow channel [27]. Other active valves include pneumatic or torque-actuated screw valves (TWIST valves), electrokinetic valves [28], ferrofluid-based valves [29], and solenoid valves [30, 31]. Passive valves can be constructed within the channel geometry as part of the design, and some of them include capillary-based, pH-controlled valves [32], use of hydrophobic barrier, etc. Capillary valves, for example, use an abruptly changing geometry of the flow path where the curvature of the filling front changes, flattening the meniscus and stopping the flow. They can be used to handle liquids of volumes from nanoliters to picoliters, stopping flow from a few seconds to several minutes [33]. Different microvalves, their fabrication, and actuation are discussed in more detail in a later chapter in this book.

4.3.2 Sample Preparation and Processing

In POC devices, sample processing involves isolation, sorting, separation of a target analyte from an untreated sample of blood, saliva, or urine, etc., buffering of the sample containing a large number of biomarkers and mixing of multiple reagents through active or passive mechanisms. When the concentration of biomarkers in the sample is low, detection by biosensors may prove difficult, and sometimes, a pre-concentration step is necessary to increase the analyte concentration [24]. The sample preparation technique used is sometimes specific to the sample being analyzed, the target analyte in the sample, or the reagent used. In this section, we discuss advances in different microfluidic techniques used in sample processing with an increased focus on target sorting and separation.

Acoustophoresis has been employed either in continuous or batch-wise processing for cell washing, isolation, enrichment, patterning, tweezing, or trapping [34, 35], for the separation of blood plasma [36], viable, and non-viable cells based on the difference in their compressibility [37, 38], CTCs from different leukocyte subpopulations in the blood [39, 40] as shown in Figure 4.3(a), blood samples [41] as shown in Figure 4.3(b), lipids from erythrocytes [42], bacteria from a blood sample [43, 44], and enrichment and separation of rare cells [41, 45, 46]. Although acoustophoresis-based POC devices use silicon or glass substrates (for bulk acoustic wave (BAW)) or lithium niobate substrates (for surface acoustic wave (SAW)) which are relatively expensive, recent research has aimed at integrating existing acoustophoretic applications with low-cost disposable substrates, for example, the development of polystyrene-based microchannel devices [47]. Dielectrophoresis (DEP)-based techniques have exploited the differences in dielectric properties of analytes within the sample to enable separation. Dielectric properties of lymphocytes, monocytes, and granulocytes are sufficiently different, and hence manipulation is made possible [48]. Likewise, the difference in electrical signatures has also been used for selective isolation of living human leukemia cells from dead cells [49], separation of *E. coli* bacteria from the blood [50] as shown in Figure 4.3(c), blood plasma separation [51] and separation of RBCs and WBCs from whole blood in a micro-separator [52].

Sample processing using magnetophoresis involves initial labeling of the analyte with magnetic beads functionalized with antibodies specific to the target. Magnetic force by use of permanent magnets or electromagnets has been used in separation of RBCs and WBCs from whole blood [53, 54], malaria-infected red blood cells (i-RBCs) from blood samples [55], *E. coli* bacteria from whole bovine blood [56] as shown in Figure 4.3(d), separation of CTCs [57, 58]. The disadvantage with

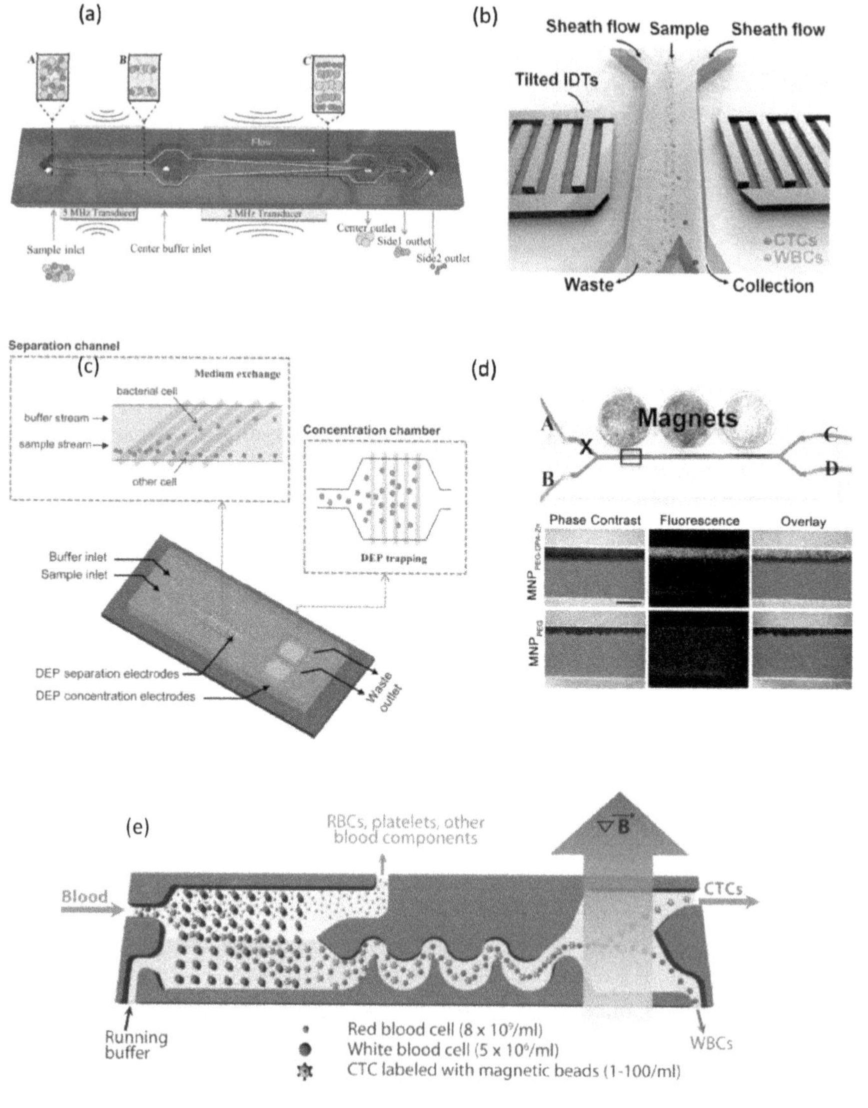

FIGURE 4.3 (a) Acoustofluidic design with a prefocusing zone for label-free separation of leukocyte subpopulations, reprinted from [39] Copyright (2019) Royal Society of Chemistry. (b) Schematic of a chip using tilted angle Standing Surface Acoustic Wave (taSSAW) actuated by two interdigitated transducers for separation of CTCs from blood, reprinted from [41] Copyright (2015) National Academy of Sciences. (c) Separation and concentration of bacteria using DEP, reprinted from [50] Copyright (2011) Royal Society of Chemistry. (d) Separation of *E. coli* bacteria from whole blood containing magnetic nanoparticles (MNPs) using magnetophoresis, reprinted from [56] Copyright (2014) American Chemical Society. (e) CTC-iChip which combines deterministic lateral displacement, inertial focusing, and magnetophoresis to separate a range of CTCs from whole blood, reprinted from [62] Copyright (2013) American Association for the Advancement of Science.

magnetophoretic techniques vis-à-vis other active techniques is the need for additional processing steps due to the introduction of magnetic nanoparticles or beads to the system. While the use of an active force (acoustic, electric, magnetic) requires additional external or integrated components in the chip, passive mechanisms have proven to be more cost effective. Introducing high-flow-rate inertial effects, modification of channel geometries, use of micro-pillars and filter membranes have enabled blood plasma separation [59], separation of leukocytes [60] and RBCs [61], isolation of rare CTCs [62] as shown in Figure 4.3(e), CTC clusters [63], and bacteria from the blood [64]. An elaborate exposition of the different microfluidic approaches in sorting, isolation, and separation can be found in [65]. The sample preparation stage also involves the efficient mixing of reagents in microfluidic platforms, which are achieved through a number of active or passive mixing mechanisms. Different microfluidic mixing techniques are discussed in detail in recent reviews [25, 66, 67], and different micromixers are also elaborated in detail in a later chapter in this book.

4.3.3 Target Detection and Analysis

Target detection and analysis involve converting the biochemical recognition in analytes into electrical signals (for example, in blood gas and chemistry analyzers for electrolyte levels, hematocrit, pH, blood glucose, etc.) or optical signals (in pregnancy test kits, etc.). Electrochemical detection involves transduction of the signal into equivalent electrical current, voltage, or impedance, which is proportional to the target analyte concentration. Label-free electrochemical detection is possible when the target species is electrically active, for example, as demonstrated in [68] where glucose, lactate, and uric acid were detected in biological samples using oxidase enzyme or in the detection of food-borne bacteria using a thermostable reporter enzyme, esterase 2 (EST2) [69]. When the target species is not inherently electroactive, they are tagged by another species or enzyme, which causes an enzyme-enhanced electrochemical reaction, which amplifies the signal [70].

Optical target detection may involve colorimetric (absorbance), fluorescence, or chemiluminescence detection. Colorimetric detection is the most commonly used due to its use in lateral flow assays based on polymer or gold nanoparticles (which exhibit surface plasmon resonance and are easy to synthesize and modify optical properties) [71], while fluorescence-based detection systems are used due to their increased sensitivity and easy availability of different colors of fluorophores, fluorescent nanoparticles, and quantum dots. Optical methods are based on labeling the analyte by attaching a chromophore, fluorophore, or particle to an antibody or nucleic acid strand that confers specific recognition [25].

Colorimetric detection involves the color change of the detection system, which can be observed with the naked eye. The color change occurs due to the chemical or biochemical interaction between the target analytes in the assay sample and a colorimetric molecular probe. They are simple and easy to perceive with yes/no accuracy of the test results. However, colorimetric detection methods do not provide any quantitative analysis of the detected analytes. Due to their simplicity and easy readability, they have been widely employed for the detection of analytes related to infectious diseases, cardiovascular diseases, and various cancers. A well-known application of the colorimetric technique is the detection of glucose in the blood for diabetics [72–74].

The reference standard of many heterogeneous immunoassays is the enzyme-linked immunosorbent assay (ELISA), which uses multiwell plates and can be automated for high-throughput processing. ELISAs generally use colorimetric or chemiluminescence-based detection. The fluorescent signal-based system also finds extensive use, for example, in the detection of cardiac marker C-reactive protein (CRP) [75] or gases like hydrogen sulfide in the blood [76]. Devices incorporating photodiodes, LEDs, and affordable mini microscopes for the detection of fluorescence signals have also seen significant development.

In the next section, we elaborate on a particular class of POC devices, paper-based devices, due to burgeoning recent research into their development. Fabrication of paper-based devices, their applications, and present uses and limitations with current technology are also discussed. Some of the detection schemes explained in this section are also revisited.

4.4 LOW-COST PAPER-BASED DEVICES

The cost of fabrication, as well as the operation of a POC device, depends significantly on the material used as the microfluidic platform. There are many commercialized POC devices that use glass-based or silicon-based platforms for diagnostic purposes. However, for cost optimization and to make the technology more widely available, low-cost alternatives are needed. Paper-based devices have received a lot of attention due to simplicity, availability, and the lack of need for external peripherals like pump mechanisms.

Paper-based POC devices are especially useful and suitable in locations where resources are scarce. While everyday paper is produced almost everywhere around the world, paper required for fabrication of POC devices needs certain features (hydrophilicity, porosity, and adequate pore size) depending on the utility. Paper is generally produced using pressed cellulosic fibers of mainly wood, cotton, grass, bamboo, jute, etc. The mixture of these materials tends to degrade fast, making them undesirable for diagnostics purposes. Paper-based POC devices generally use 100% pure cotton, which comes at a slightly increased production cost. While the pore size determines the largest size of the particles that can pass through the paper-based filter, porosity is the void fraction of the 3D skeleton of the membrane or absorption layer. Porosity determines the maximum capillary flow rate of the analyte. Paper-based devices are also preferred over hydrophobic nitrocellulose-based membranes (NCMs) as the paper has more affinity towards aqueous-based samples and reagents than NCMs. Patterning paper [77] into several hydrophilic and hydrophobic channels offers four fundamental capabilities, all on a single analytical device [78] – (i) enabling multiple assays by the distribution of a sample into several locations on a single device, (ii) actuating the sample movement using capillary action, thus no pumps required, (iii) using small volumes as the sample, and (iv) disposing of hazardous wastes by incineration. Several waterborne diseases remain a global issue due to contaminated water and inadequate sanitation. Thus, the requirement of a low-cost and environmentally safe POC device remains at the top of the list, especially in the developing world. The WHO has set seven essential guidelines for developing a diagnostic tool that can overcome the challenges faced in rural areas. These tests kit must be affordable, sensitive, specific, user friendly, rapid and robust, equipment-free, and delivered to the needy areas which correspond to the acronym "ASSURED" [79].

Thus, the development of any paper-based POC device should ensure the following – (i) it should be able to process low volumes of biological samples (blood, urine, etc.), (ii) accomplish the test within a short period of time, and (iii) it should be comparable to existing nitrocellulose-based assays in specificity and sensitivity.

In the case of medical research, diagnosis of blood [80], urine [81], etc. play a vital role in detecting various diseases. Paper-based devices facilitate the detection of multiple analytes on a single device, by facilitating reaction between the target analyte in the sample with previously immobilized reagents. The reagents can be enzymes, acid-base indicators, or dyes. Colorimetric sensing on the paper-based platform has been exploited using pH, glucose, and protein assays in artificial urine [77]. The samples deposited on the paper-based devices get distributed to various reaction zones, which lead to a change in color and enable visual estimation of analyte levels using a calibration chart. For example, in the case of glucose assay, a color change from clear to brown indicates a positive result. Colorimetric detection of liver function is depicted in Figure 4.4(a) [82]. Electrochemical detection is also widely used in paper-based systems. In general, electrochemical sensors consist of three electrodes – one counter electrode, one working electrode, and one or multiple reference electrodes. On the paper-based platform, reaction zones include multiple-electrode mechanisms, as shown in Figure 4.4(b). Electrochemical

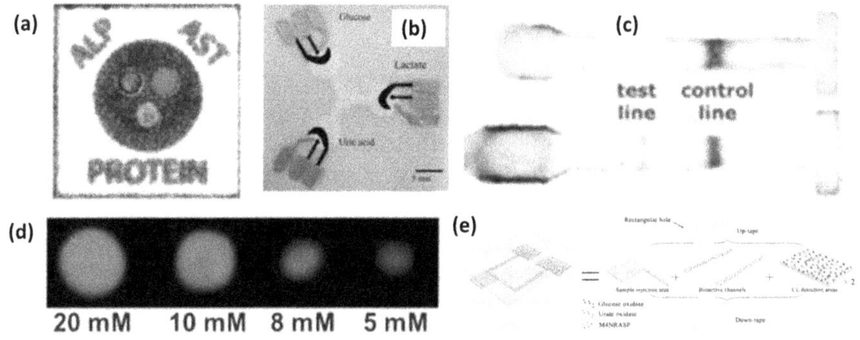

FIGURE 4.4 (a) Colorimetric detection of liver function, reprinted from [82], Copyright (2012) American Chemical Society. (b) Electrochemical sensing of glucose, lactate, and uric acid, reprinted from [68], Copyright (2009) American Chemical Society. (c) Antibody conjugated/gold nanoparticle sensing of immunoglobulin G., reprinted from [83], Copyright (2010) American Chemical Society. (d) ECL emission from a paper-based device at different concentrations, reprinted from [84], Copyright (2011) American Chemical Society. (e) Schematic of the construction of a CL sensor in paper-based microfluidics, reprinted from [87], Copyright (2011) Royal Society of Chemistry.

detection is utilized for sensing several analytes [68] glucose, lactate, uric acid, etc. Nanoparticle-based detection exploits colloidal gold and monodisperse latex as the detector reagents for lateral flow assays [78]. The rationale behind using gold nanoparticles (20–40 nm) lies in its high extinction coefficients as compared to organic dyes. In addition to that, their qualitative interpretation is sufficient and does not require any reader. A systematic study employing nanoparticle-based detection was in multianalyte immunochemical detection on filter paper utilizing inkjet printers [83] is depicted in Figure 4.4(c). Electrochemiluminescence involves the generation of luminescence due to electrochemical reactions. Electrochemically generated intermediates undergo exergonic reactions leading to an electronically excited state. This configuration emits light upon migrating to a lower-level state, thus enabling readouts without the requirement of a photodetector. Several applications include the detection of analytes [84], as shown in Figure 4.4(d), tumor markers [85], ions [86], etc. A chemiluminescence (CL) detection mechanism is based on the emission of light generated due to a chemical reaction. In paper-based microfluidics, CL has been implemented for detecting biological analytes (Figure 4.4(e)) [87] and tumor markers [88].

4.4.1 Blood Diagnostics

Blood is considered as a window through which human health can be analyzed [80]. Blood is primarily responsible for transporting nutrients and wastes from one part of the body to another, maintaining body temperature, proper osmotic pressure, pH of body fluids (~7.4). Thus, blood remains one of the vital fluids which contains a lot of information about the human body. To collect blood, two methods are generally used – finger pricking and venipuncture. A finger prick is mainly used for POC testing and venipuncture for laboratory testing. Paper-based devices facilitate fluid flow without the incorporation of an external force and completely rely on capillary action caused due to the intermolecular force between the fluid and the porous cellulose matrix. There exist several applications involving paper-based devices for blood diagnostics, but here we discuss only a few vital applications in this section – plasma separation, blood grouping, hematology, blood coagulation, and malaria detection.

The noncellular component of whole blood, plasma (accounting for ~55% of whole blood) and is widely analyzed to detect various diseases including Alzheimer's disease, cancer, and sepsis [89]. Paper-based blood plasma separation using an H-filter from whole blood has been achieved (see Figure 4.5(a)). Plasma separation occurs due to the capillary-driven diffusion of blood samples. Images captured during the experiments were processed to estimate the plasma separation [90]. Grouping of blood involves a simple blood test that examines a person's blood type. Figure 4.5(b) shows a device for determining blood typing [91]. It consists of three hydrophilic channels present on a Kleenex paper towel treated with anti-A, anti-B, and anti-D antibodies. The blood samples are introduced and allowed a reaction time of 30 s. Finally, the eluting length in each channel is estimated using a bar code for blood typing. A hematocrit (Hct) test is the estimation of the volume fraction of RBCs in a person's blood. The healthy limit [80] of hematocrit is 40–54% for adult men

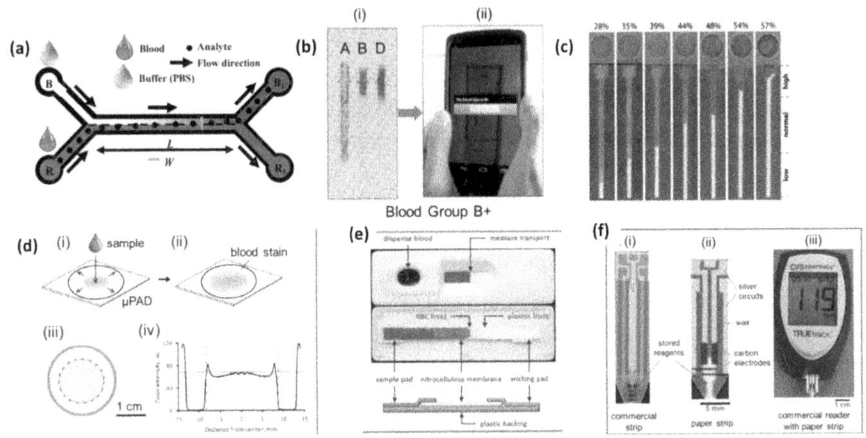

FIGURE 4.5 (a) Schematic representation of the device setup, where the blood and dilution buffer are dispensed in reservoirs R and B, respectively. At the end of the experiments, it is seen that the plasma solution is separated in reservoir B_1, whereas reservoir R_1 contains the separated RBCs, reprinted from [90], Copyright (2015) Royal Society of Chemistry. (b) Barcode-like paper device with designated phone app for blood grouping – (i) photo of paper device when applying B+ blood to blood with matching antibodies traveling a shorter distance and vice versa, and (ii) reading the test result using a smartphone and app, reprinted from [91], Copyright (2014) American Chemical Society. (c) Hematocrit measurement on a paper device; higher hematocrit resulted in shorter blood travel distance and vice versa, reprinted from [92], Copyright (2016) Royal Society of Chemistry. (d) Hemoglobin measurement on a paper device using digital image analysis: (i, ii) illustrations of test procedures, (iii) photo of a paper device after the formation of blood stain, (iv) plot of red color intensity vs distance from the center when digitally analyzing blood stain image from (iii), reprinted from [93], Copyright (2013) American Chemical Society. (e) Photos of the no-reaction lateral flow assay (nrLFA) device (top: plastic cassette and nrLFA test strip) and schematic of nrLFA test strip (bottom: sample pad, analytical membrane, wicking pad, and plastic backing), reprinted from [95], Copyright (2014) Royal Society of Chemistry. (f) Glucose detection using novel wax- and silver-printed paper strip paired with a commercial glucometer – (i) commercial glucose test strip, (ii) novel paper strip with wax-defined fluid channels, silver circuits, and prestored reagents for glucose testing, and (iii) glucose testing using a novel paper strip with a commercial glucometer, reprinted from [97], Copyright (2010) Royal Society of Chemistry.

and 36–48% for adult women. An abnormal Hct level indicates anemia, infection, WBC disorder such as leukemia or lymphoma, vitamin or mineral deficiencies, lung or heart disease, etc. [80]. A low-cost and straightforward paper-based microfluidic for the estimation of Hct is reported [92]. The device constitutes of a top lamination sheet, a sample adding layer (made of chromatography paper) that prevents the trapped WBCs from entering the channel, a cutter-patterned double-sided adhesive, a wax-patterned microfluidic channel which is pre-treated with 4.5 mM EDTA and 5 mM NaCl solutions for blood flow enhancement, and a bottom lamination sheet. Once the blood sample is dispensed, WBCs are trapped in the sample adding layer, while RBCs and plasma travel through it and onto the microfluidic channel, as shown in Figure 4.5(c). Similarly, hemoglobin (Hb) is the protein present in RBCs, and its primary function is to deliver oxygen to cells and tissues. Most paper-based hemoglobin detection is based on colorimetric detection followed by image analysis or spectrophotometric detection using a reader. A simple paper-based test for the measurement of blood hemoglobin concentration using chromatography paper is demonstrated [93] and shown in Figure 4.5(d). The device fabrication involves printing a circular wax pattern on paper followed by heating, in which a 20 μL mixture of whole blood and reagent (for Hb colorimetric analysis) is applied, and the subsequent bloodstain is scanned and digitally analyzed.

When a blood vessel is damaged, there is a possibility of blood loss, and the prevention of such loss is known as hemostasis. This process consists of four mechanisms – (i) constriction of the vascular wall, (ii) formation of a platelet plug, (iii) formation of a blood clot due to coagulation, and (iv) formation of fibrous tissues in the clot to permanently close the vascular wall. Activation of coagulation leads to a gradual increase in blood viscosity and the physical properties of blood change from that of a viscoelastic fluid to a viscoelastic solid after the formation of cross-linked fibrin clot [94]. The study of blood coagulation is carried out in a device [95] which consists of a paper-based test strip incorporating a fiberglass sample pad, a nitrocellulose analytical membrane, and a cellulose wicking pad. The assembled strip is inserted into a plastic housing with openings for sample dispensing and flow observation, as shown in Figure 4.5(e). Constant monitoring of glucose levels is of paramount importance for a diabetes patient under insulin medication [96]. A simple paper-based analytical device capable of rapid and quantitative detection of metabolites, including glucose, cholesterol, lactate, and alcohol has been designed [97] to operate in conjunction with a commercial handheld glucometer (i.e., TRUEtrack from CVS/Pharmacy). The device is a strip consisting of microfluidic channels, carbon electrodes, and silver wires on one sheet of chromatography paper. The device, as shown in Figure 4.5(f), typically requires a volume of ~1 μL of the fluid sample (blood, plasma, or aqueous solution) to wet the channels completely, and the chemical reagents required for the tests are stored in the detection zone of the device. Malaria has been a primary cause of mortality and morbidity in tropical and subtropical counties, resulting in over one million deaths every year around the globe [98]. Current research is also exploring various low-cost paper-based devices to detect malaria in rural areas [99].

4.4.2 The Road Ahead for Paper-Based Diagnostics

Although paper-based devices are simple, cost effective, and provide useful applications in disease detection, analyte estimation, etc. [100], they need further attention relating to improvement in terms of clinical performance. Paper-based devices show varying specificity and sensitivity [101], which may result in false-negatives, false-positive results, or incorrect estimates. For example, home pregnancy tests may sometimes yield false-negative results and lead to contraceptive non-adherence, and a higher rate of sexually transmitted infections and pregnancy [102], while on the other hand, a false-positive pregnancy report can result in significant patient anxiety and unnecessary interventions [103]. Similarly, POC glucose meters also show varying levels of clinical performance and accuracy [104]. Inappropriate glucose meters can provide overestimated glucose levels, which can lead to improper insulin dose adjustment [105]. In general, paper-based devices suffer from lack of standardization due to differences in batch-to-batch variation, relatively heightened sensitivity to temperature and humidity, reduced long term stability of the reagents, and the need to withstand harsh environmental conditions.

4.5 COMMERCIALIZATION OF POC DEVICES

A large number of microfluidic POC devices has been developed and commercialized in the recent past based on their applicability, i.e., the type of detection elements. Typical detection elements used in POC diagnostic assays are blood chemistries, cells, and nucleic acids (DNA and RNA). Some of the commercialized microfluidic POC devices approved by the FDA are listed in Table. 4.1 [117]. For a comprehensive review, interested readers may further refer to [14, 118–120].

Despite the enormous amount of fundamental research carried in microfluidics for the development of POC devices, the transition from laboratory prototypes to end products is still challenged by various factors. Integration is one of the factors that puts a limitation on the commercialization of POC devices. An ideal POC device should directly connect the patient to the result. In general, any laboratory diagnostic assay encompasses the processes of sample collection and processing, necessary chemical or biological reactions, signal generation and detection, analysis, and reporting of the results. A POC device as a successful end product should smoothly integrate all these processes into a single platform in a well-automated

TABLE 4.1
Some FDA-Approved Microfluidic POC Devices with the Manufacturer, Type of Assay, and the Corresponding Detection Element

Company	Test	Detection Element
Abaxis [106]	The Piccolo®	Blood chemistries
Biosite (Alere) [107]	Alere Triage®	Blood/urine chemistries
Epocal (Alere) [108]	The epoc® Blood Analysis System	Blood chemistries
Focus Dx (Quest) [109]	Simplexa	Flu A/B & RSV (viral)
HandlyLab (BD) [110]	BD MAXTM GBS Assay IDIStrep B Assay	Group B Streptococcus (GBS)
i-STAT Corp (Abbott) [111]	i-STAT Analyser	Blood chemistries, coagulation, cardiac marker
BioFire Diagnostics [112]	FilmArray RP	DNA or RNA
IQuum (Roche) [113]	Liat™ Influenza A/B Assay	H1N1 influenza viral RNA
Sphere Medical [114]	Proxima	Blood chemistry
TearLab [115]	TearLab Osmolarity System	Osmolarity of human tears
Achira [116]	ACIX100	Thyroid stimulant hormone

manner such that the device works on the theme of unprocessed sample-in (from the patient directly) and result-out.

However, the vast majority of the microfluidic research on the development of POC devices is focused on developing independent technologies as proofs-of-concept for POC diagnostics. For example, the development of different microfluidic technologies for fluid control systems, detection modules, data analysis, and result interpretation modules independently. Discretization is undoubtedly needed during the development stage of technology for troubleshooting and rectifying problems, but to realize an end product, these technologies must be integrated seamlessly. In addition, it is known that most of these concepts are proven only for laboratory conditions, device designs, and materials specific to their respective experimental conditions, and thus integration with other existing technologies may not be easy.

Universality and integration flexibility with other platforms need to be considered right from the development stage of any technology for commercialization. Besides, licensing issues associated with the original laboratory ideas and the patenting rights of various components and reagents also limit the integration. Another challenge that limits the commercialization of POC devices is the use of bulky peripheral setups such as pumps, power supply sources, optical detection, and data analysis platforms during the demonstration of the technology in laboratory conditions. The miniaturization of all the components is an absolute necessity for the conversion of a lab-on-a-chip (LOC) technology into a POC device. However, it is worth mentioning that considerable research has been initiated with the intention of miniaturizing this peripheral setup to integrate into portable POC devices [15].

Apart from technical aspects, marketing strategies also play a significant role in the commercialization of POC devices. Fundraising remains a challenge for researchers in bringing a novel laboratory technology to the market in the form of POC devices as their field of application, i.e., healthcare technology, in general, incurs considerable capital investment. Furthermore, the device should offer significant benefits over the existing clinical techniques in terms of convenience to the patient, sensitivity, and the cost to sustain it in the market.

Material considerations, manufacturing methods, and mass production aspects also play key roles in the commercialization of POC devices as they are closely related to the cost of the product. The majority of the LOC technologies developed in the laboratory use material like silicon, glass and, PDMS and employed the manufacturing techniques of micromachining, photolithography (cleanroom fabrication methods), and soft lithography. On the other hand, plastic- and paper- or membrane-based microfluidic devices have shown great potential for mass production. These devices are of an affordable price owing to the low cost of their fabrication materials and suitability for high-volume production. However, it should be noted that the choice of POC device materials is again constrained by the application (detection elements) and the compatibility of the material with assay samples (chemical and biological).

4.6 OUTLOOK AND FUTURE PERSPECTIVES

In this chapter, an overview of low-cost microfluidics POC devices was presented discussing the state of the art of the field, components of a general POC device including different microfluidic technologies used, a specific look at low-cost paper-based devices, and finally we concluded with commercialization aspects for POC devices. Research into POC devices remains extremely important with regard to making healthcare affordable in resource-constrained areas [121]. Various low-cost POC devices have already been developed for the rapid diagnosis of diseases in healthcare technology. Importantly, the research into plastics and paper or membrane-based microfluidics devices has shown considerable growth and development over the recent years and now offers a range of multiplexing capabilities and can perform ever more tests whose results are easily readable. In the coming years, research into enhancing existing capabilities, standardization, improving the breadth of applications, and integration of POC with electronic devices like smartphones are expected to generate significant attention. The ongoing COVID-19 global pandemic outbreak has once again shown the limitations of traditional healthcare facilities and outlined the need for simple, cost effective, and quick personal healthcare.

REFERENCES

1. Foudeh AM, Fatanat Didar T, Veres T, et al. Microfluidic designs and techniques using lab-on-a-chip devices for pathogen detection for point-of-care diagnostics. *Lab Chip.* 2012;12:3249–3266.
2. Yager P, Edwards T, Fu E, et al. Microfluidic diagnostic technologies for global public health. *Nature.* 2006;442:412–418.
3. Lee WG, Kim Y-G, Chung BG, et al. Nano/Microfluidics for diagnosis of infectious diseases in developing countries. *Adv Drug Deliv Rev.* 2010;62:449–457.
4. Vashist SK. Point-of-care diagnostics: Recent advances and trends. *Biosensors.* 2017;7:10–13.
5. Zhu H, Fohlerová Z, Pekárek J, et al. Recent advances in lab-on-a-chip technologies for viral diagnosis. *Biosens Bioelectron.* 2020;153:112041.
6. Yang T, Wang YC, Shen CF, et al. Point-of-care RNA-based diagnostic device for Covid-19. *Diagnostics.* 2020;10:9–11.
7. Nguyen T, Bang DD, Wolff A. 2019 Novel coronavirus disease (COVID-19): Paving the road for rapid detection and point-of-care diagnostics. *Micromachines.* 2020;11:1–7.
8. Sheridan C. Fast, portable tests come online to curb coronavirus pandemic. *Nat Biotechnol.* 2020;38:515–518.
9. Tavakoli H, Zhou W, Ma L, et al. Paper and paper hybrid microfluidic devices for point-of-care detection of infectious diseases. *Nanotechnol Microfluid.* 2020;177–209.
10. Vashist SK, Mudanyali O, Schneider EM, et al. Cellphone-based devices for bioanalytical sciences multiplex platforms in diagnostics and bioanalytics. *Anal Bioanal Chem.* 2014;406:3263–3277.
11. Zhang L, Tian Z, Bachman H, et al. A cell-phone-based acoustofluidic platform for quantitative point-of-care testing. *ACS Nano.* 2020;14:3159–3169.
12. Lillehoj PB, Huang MC, Truong N, et al. Rapid electrochemical detection on a mobile phone. *Lab Chip.* 2013;13:2950–2955.
13. Mudanyali O, Dimitrov S, Sikora U, et al. Integrated rapid-diagnostic-test reader platform on a cellphone. *Lab Chip.* 2012;12:2678–2686.
14. Vashist SK, Luppa PB, Yeo LY, et al. Emerging technologies for next-generation point-of-care testing. *Trends Biotechnol.* 2015;33:692–705. Available from: http://dx.doi.org/10.1016/j.tibtech.2015.09.001.
15. Boyd-Moss M, Baratchi S, Di Venere M, et al. Self-contained microfluidic systems: A review. *Lab Chip.* 2016;16:3177–3192.

16. Skafte-Pedersen P, Sabourin D, Dufva M, et al. Multi-channel peristaltic pump for microfluidic applications featuring monolithic PDMS inlay. *Lab Chip.* 2009;9:3003–3006.
17. Wang X, Cheng C, Wang S, et al. Electroosmotic pumps and their applications in microfluidic systems. *Microfluid Nanofluidics.* 2009;6:145–162.
18. Takamura Y, Onoda H, Inokuchi H, et al. Low-voltage electroosmosis pump for standalone microfluidics devices. *Electrophoresis.* 2003;24:185–192.
19. Sista R, Hua Z, Thwar P, et al. Development of a digital microfluidic platform for point of care testing. *Lab Chip.* 2008;8:2091–2104.
20. Guttenberg Z, Müller H, Habermüller H, et al. Planar chip device for PCR and hybridization with surface acoustic wave pump. *Lab Chip.* 2005;5:308–317.
21. Chiou PY, Park SY, Wu MC. Continuous optoelectrowetting for picoliter droplet manipulation. *Appl Phys Lett.* 2008;93:1–4.
22. García AA, Egatz-Gómez A, Lindsay SA, et al. Magnetic movement of biological fluid droplets. *J Magn Magn Mater.* 2007;311:238–243.
23. Wang K, Fatoyinbo HO. Digital microfluidics. *RSC Detect Sci.* 2015;2015-January:84–135.
24. Gervais L, De Rooij N, Delamarche E. Microfluidic chips for point-of-care immunodiagnostics. *Adv Mater.* 2011;23:151–176.
25. Gubala V, Harris LF, Ricco AJ, et al. Point of care diagnostics: Status and future. *Anal Chem.* 2012;84:487–515.
26. Lynn NS, Dandy DS. Passive microfluidic pumping using coupled capillary/evaporation effects. *Lab Chip.* 2009;9:3422–3429.
27. Unger MA, Chou HP, Thorsen T, et al. Monolithic microfabricated valves and pumps by multilayer soft lithography. *Science.* 2000;288:113–116.
28. Jacobson SC, Ermakov S V., Ramsey JM. Minimizing the number of voltage sources and fluid reservoirs for electrokinetic valving in microfluidic devices. *Anal Chem.* 1999;71:3273–3276.
29. Hartshorne H, Backhouse CJ, Lee WE. Ferrofluid-based microchip pump and valve. *Sens Actuators B Chem.* 2004;99:592–600.
30. Elizabeth Hulme S, Shevkoplyas SS, Whitesides GM. Incorporation of prefabricated screw, pneumatic, and solenoid valves into microfluidic devices. *Lab Chip.* 2009;9:79–86.
31. Weibel DB, Kruithof M, Potenta S, et al. Torque-actuated valves for microfluidics. *Anal Chem.* 2005;77:4726–4733.
32. Beebe DJ, Moore JS, Bauer JM, et al. Functional hydrogel structures for autonomous flow control inside microfluidic channels. *Nature.* 2000;404:588–590.
33. Zimmermann M, Hunziker P, Delamarche E. Valves for autonomous capillary systems. *Microfluid Nanofluidics.* 2008;5:395–402.
34. Petersson F, Åberg L, Swärd-Nilsson AM, et al. Free flow acoustophoresis: Microfluidic-based mode of particle and cell separation. *Anal Chem.* 2007;79:5117–5123.
35. Lin S-CS, Mao X, Huang TJ. Surface acoustic wave (SAW) acoustophoresis: Now and beyond. *Lab Chip.* 2012;12:2766.
36. Lenshof A, Ahmad-Tajudin A, Järås K, et al. Acoustic whole blood plasmapheresis chip for prostate specific antigen microarray diagnostics. *Anal Chem.* 2009;81:6030–6037.
37. Yang AHJ, Soh HT. Acoustophoretic sorting of viable mammalian cells in a microfluidic device. *Anal Chem.* 2012;84:10756–10762.
38. Olofsson K, Hammarström B, Wiklund M. Acoustic separation of living and dead cells using high density medium. *Lab Chip.* 2020;15:1350–1359.
39. Urbansky A, Olm F, Scheding S, et al. Label-free separation of leukocyte subpopulations using high throughput multiplex acoustophoresis. *Lab Chip.* 2019;19:1406–1416.
40. Grenvall C, Magnusson C, Lilja H, et al. Concurrent isolation of lymphocytes and granulocytes using prefocused free flow acoustophoresis. *Anal Chem.* 2015;87:5596–5604.

41. Dao M, Suresh S, Huang TJ, et al. Acoustic separation of circulating tumor cells. *Proc Natl Acad Sci U S A*. 2015;112:4970–4975.
42. Petersson F, Nilsson A, Holm C, et al. Separation of lipids from blood utilizing ultrasonic standing waves in microfluidic channels. *Analyst*. 2004;129:938–943.
43. Antfolk M, Muller PB, Augustsson P, et al. Focusing of sub-micrometer particles and bacteria enabled by two-dimensional acoustophoresis. *Lab Chip*. 2014;14:2791–2799.
44. Van Assche D, Reithuber E, Qiu W, et al. Gradient acoustic focusing of sub-micron particles for separation of bacteria from blood lysate. *Sci Rep*. 2020;10:3670.
45. Antfolk M, Laurell T. Continuous flow microfluidic separation and processing of rare cells and bioparticles found in blood: A review. *Anal Chim Acta*. 2017;965:9–35.
46. Wu Z, Jiang H, Zhang L, et al. The acoustofluidic focusing and separation of rare tumor cells using transparent lithium niobate transducers. *Lab Chip*. 2019;19:3922–3930.
47. Dow P, Kotz K, Gruszka S, et al. Acoustic separation in plastic microfluidics for rapid detection of bacteria in blood using engineered bacteriophage. *Lab Chip*. 2018;18:923–932.
48. Yang J, Huang Y, Wang XB, et al. Differential analysis of human leukocytes by dielectrophoretic field- flow-fractionation. *Biophys J*. 2000;78:2680–2689. Available from: http://dx.doi.org/10.1016/S0006-3495(00)76812-3.
49. Shafiee H, Sano MB, Henslee EA, et al. Selective isolation of live/dead cells using contactless dielectrophoresis (cDEP). *Lab Chip*. 2010;10:438–445.
50. Park S, Zhang Y, Wang TH, et al. Continuous dielectrophoretic bacterial separation and concentration from physiological media of high conductivity. *Lab Chip*. 2011;11:2893–2900.
51. Liao SH, Chang CY, Chang HC. A capillary dielectrophoretic chip for real-time blood cell separation from a drop of whole blood. *Biomicrofluidics*. 2013;7.
52. Han KH, Frazier AB. Lateral-driven continuous dielectrophoretic microseparators for blood cells suspended in a highly conductive medium. *Lab Chip*. 2008;8:1079–1086.
53. Jung J, Han KH. Lateral-driven continuous magnetophoretic separation of blood cells. *Appl Phys Lett*. 2008;93:2006–2009.
54. Han KH, Bruno Frazier A. Continuous magnetophoretic separation of blood cells in microdevlce format. *J Appl Phys*. 2004;96:5797–5802.
55. Nam J, Huang H, Lim H, et al. Magnetic separation of malaria-infected red blood cells in various developmental stages. *Anal Chem*. 2013;85:7316–7323.
56. Lee JJ, Jeong KJ, Hashimoto M, et al. Synthetic ligand-coated magnetic nanoparticles for microfluidic bacterial separation from blood. *Nano Lett*. 2014;14:1–5.
57. Kim S, Han SI, Park MJ, et al. Circulating tumor cell microseparator based on lateral magnetophoresis and immunomagnetic nanobeads. *Anal Chem*. 2013;85:2779–2786.
58. Kang H, Kim J, Cho H, et al. Evaluation of positive and negative methods for isolation of circulating tumor cells by lateral magnetophoresis. *Micromachines*. 2019;10:1–10.
59. Nivedita N, Papautsky I. Continuous separation of blood cells in spiral microfluidic devices. *Biomicrofluidics*. 2013;7:1–14.
60. Zhou J, Papautsky I. Size-dependent enrichment of leukocytes from undiluted whole blood using shear-induced diffusion. *Lab Chip*. 2019;15:3962–3979.
61. Zeming KK, Ranjan S, Zhang Y. Rotational separation of non-spherical bioparticles using I-shaped pillar arrays in a microfluidic device. *Nat Commun*. 2013;4:1625–1628.
62. Ozkumur E, Shah AM, Ciciliano JC, et al. Inertial focusing for tumor antigen-dependent and -independent sorting of rare circulating tumor cells. *Sci Transl Med*. 2013;5.
63. Au SH, Edd J, Stoddard AE, et al. Microfluidic isolation of circulating tumor cell clusters by size and asymmetry. *Sci Rep*. 2017;7:1–10.
64. Mach AJ, di Carlo D. Continuous scalable blood filtration device using inertial microfluidics. *Biotechnol Bioeng*. 2010;107:302–311.

65. Dalili A, Samiei E, Hoorfar M. A review of sorting, separation and isolation of cells and microbeads for biomedical applications: Microfluidic approaches. *Analyst.* 2019;144:87–113.
66. Cai G, Xue L, Zhang H, et al. A review on micromixers. *Micromachines.* 2017;8:1–27.
67. Lee CY, Chang CL, Wang YN, et al. Microfluidic mixing: A review. *Int J Mol Sci.* 2011;12:3263–3287.
68. Dungchai W, Chailapakul O, Henry CS. Electrochemical detection for paper-based microfluidics. *Anal Chem.* 2009;81:5821–5826.
69. Pöhlmann C, Wang Y, Humenik M, et al. Rapid, specific and sensitive electrochemical detection of foodborne bacteria. *Biosens Bioelectron.* 2009;24:2766–2771.
70. Cooper J, Yazvenko N, Peyvan K, et al. Targeted deposition of antibodies on a multiplex CMOS microarray and optimization of a sensitive immunoassay using electrochemical detection. *PLoS One.* 2010;5:1–13.
71. Radwan SH, Azzazy HME. Gold nanoparticles for molecular diagnostics. *Expert Rev Mol Diagn.* 2009;9:511–524.
72. Sanjay ST, Fu G, Dou M, et al. Biomarker detection for disease diagnosis using cost-effective microfluidic platforms. *Analyst.* 2015:7062–7081.
73. Maria MS, Rakesh PE, Chandra TS, et al. Capillary flow-driven microfluidic device with wettability gradient and sedimentation effects for blood plasma separation. *Sci Rep.* 2017;7:1–12.
74. Samy RA, Sen AK. Elastocapillary flow driven lab-on-a-membrane device based on differential wetting and sedimentation effect for blood plasma separation. *J Micromech Microeng.* 2019;29:1–10.
75. Gervais L, Delamarche E. Toward one-step point-of-care immunodiagnostics using capillary-driven microfluidics and PDMS substrates. *Lab Chip.* 2009;9:3330–3337.
76. Karunya R, Jayaprakash KS, Gaikwad R, et al. Rapid measurement of hydrogen sulphide in human blood plasma using a microfluidic method. *Sci Rep.* 2019;9:3258. Available from: http://www.nature.com/articles/s41598-019-39389-7.
77. Martinez AW, Phillips ST, Butte MJ, et al. Patterned paper as a platform for inexpensive, low-volume, portable bioassays. *Angew Chemie.* 2007;119:1340–1342.
78. Yetisen AK, Akram MS, Lowe CR. Paper-based microfluidic point-of-care diagnostic devices. *Lab Chip.* 2013;13:2210–2251.
79. Urdea M, Penny LA, Olmsted SS, et al. Requirements for high impact diagnostics in the developing world. *Nature.* 2006;444 Suppl:73–79.
80. Li H, Steckl AJ. Paper microfluidics for point-of-care blood-based analysis and diagnostics. *Anal Chem.* 2019;91:352–371.
81. Lepowsky E, Ghaderinezhad F, Knowlton S, et al. Paper-based assays for urine analysis. *Biomicrofluidics.* 2017;11.
82. Vella SJ, Beattie P, Cademartiri R, et al. Measuring markers of liver function using a micropatterned paper device designed for blood from a fingerstick. *Anal Chem.* 2012;84:2883–2891.
83. Abe K, Suzuki K, Citterio D. Inkjet-printed microfluidic multianalyte chemical sensing paper. *Anal Chem.* 2008;80:6928–6934.
84. Delaney JL, Hogan CF, Tian J, et al. Electrogenerated chemiluminescence detection in paper-based microfluidic sensors. *Anal Chem.* 2011;83:1300–1306.
85. Zang D, Ge L, Yan M, et al. Electrochemical immunoassay on a 3D microfluidic paper-based device. *Chem Commun.* 2012;48:4683–4685.
86. Zhang M, Ge L, Ge S, et al. Three-dimensional paper-based electrochemiluminescence device for simultaneous detection of Pb2+ and Hg2+ based on potential-control technique. *Biosens Bioelectron.* 2013;41:544–550. Available from: http://dx.doi.org/10.1016/j.bios.2012.09.022.

87. Yu J, Ge L, Huang J, et al. Microfluidic paper-based chemiluminescence biosensor for simultaneous determination of glucose and uric acid. *Lab Chip.* 2011;11:1286–1291.
88. Ge L, Wang S, Song X, et al. 3D Origami-based multifunction-integrated immunodevice: Low-cost and multiplexed sandwich chemiluminescence immunoassay on microfluidic paper-based analytical device. *Lab Chip.* 2012;12:3150–3158.
89. Mielczarek WS, Obaje EA, Bachmann TT, et al. Microfluidic blood plasma separation for medical diagnostics: Is it worth it? *Lab Chip.* 2016;16:3441–3448.
90. Kar S, Maiti TK, Chakraborty S. Capillarity-driven blood plasma separation on paper-based devices. *Analyst.* 2015;140:6473–6476. Available from: http://xlink.rsc.org/?DOI=C5AN00849B.
91. Guan L, Tian J, Cao R, et al. Barcode-like paper sensor for smartphone diagnostics: An application of blood typing. *Anal Chem.* 2014;86:11362–11367.
92. Berry SB, Fernandes SC, Rajaratnam A, et al. Measurement of the hematocrit using paper-based microfluidic devices. *Lab Chip.* 2016;16:3689–3694.
93. Yang X, Piety NZ, Vignes SM, et al. Simple paper-based test for measuring blood hemoglobin concentration in resource-limited settings. *Clin Chem.* 2013;59:1506–1513.
94. Merrill W. Rheology of Blood. *Physiological Reviews.* 1969;49:863–888.
95. Li H, Han D, Pauletti GM, et al. Blood coagulation screening using a paper-based microfluidic lateral flow device. *Lab Chip.* 2014;14:4035–4041.
96. Olansky L, Kennedy L. Finger-stick glucose monitoring: Issues of accuracy and specificity. *Diabetes Care.* 2010;33:948–949.
97. Nie Z, Deiss F, Liu X, et al. Integration of paper-based microfluidic devices with commercial electrochemical readers. *Lab Chip.* 2010;10:3163–3169.
98. Tangpukdee N, Duangdee C, Wilairatana P, et al. Malaria diagnosis: A brief review. *Korean J Parasitol.* 2009;47:93–102.
99. Reboud J, Xu G, Garrett A, et al. Paper-based microfluidics for DNA diagnostics of malaria in low resource underserved rural communities. *Proc Natl Acad Sci U S A.* 2019;116:4834–4842.
100. Whitesides GM. Cool, or simple and cheap? Why not both? *Lab Chip.* 2013;13:11–13.
101. Pike J, Godbert S, Johnson S. Comparison of volunteers' experience of using, and accuracy of reading, different types of home pregnancy test formats. *Expert Opin Med Diagn.* 2013;7:435–441.
102. Rahman M, Berenson AB. Pregnancy test taking is a correlate of unsafe sex, contraceptive nonadherence, pregnancy, and sexually transmitted infections in adolescent and young adult women. *J Women's Heal.* 2013;22:339–343.
103. Nakhal RS, Wood D, Woodhouse C, et al. False-positive pregnancy tests following enterocystoplasty. *BJOG An Int J Obstet Gynaecol.* 2012;119:366–368.
104. Watkinson PJ, Barber VS, Amira E, et al. The effects of precision, haematocrit, pH and oxygen tension on point-of-care glucose measurement in critically ill patients: A prospective study. *Ann Clin Biochem.* 2012;49:144–151.
105. Perera NJ, Stewart PM, Williams PF, et al. The danger of using inappropriate point-of-care glucose meters in patients on icodextrin dialysis. *Diabet Med.* 2011;28:1272–1276.
106. Abaxis Global Diagnostics, (n.d.). https://www.abaxis.com/piccolo-xpress
107. Abbott Point of Care, (n.d.). https://www.globalpointofcare.abbott
108. Siemens Healthineers, (n.d.). https://www.siemens-healthineers.com/blood-gas/blood-gas-systems/epoc-blood-analysis-system
109. Focus Diagnostics, (n.d.). https://www.focusdx.com/product-catalog/simplexa
110. Becton Dickinson, (n.d.). https://moleculardiagnostics.bd.com/
111. Abbott Point of Care, (n.d.). https://www.pointofcare.abbott/int/en/home
112. BioFire Diagnostics, (n.d.). https://www.biofiredx.com
113. Roche Diagnsotics, (n.d.). https://diagnostics.roche.com

114. Sphere Medical, (n.d.). http://www.spheremedical.com/
115. TearLab, (n.d.). https://www.tearlab.com/
116. Achira - Medical Diagnostics Tool, (n.d.). https://achiralabs.com/
117. Sharma S, Zapatero-Rodríguez J, Estrela P, et al. Point-of-care diagnostics in low resource settings: Present status and future role of microfluidics. *Biosensors*. 2015;5:577–601.
118. Chin CD, Linder V, Sia SK. Commercialization of microfluidic point-of-care diagnostic devices. *Lab Chip*. 2012;12:2118–2134.
119. Jung W, Han J, Choi JW, et al. Point-of-care testing (POCT) diagnostic systems using microfluidic lab-on-a-chip technologies. *Microelectron Eng*. 2015;132:46–57. Available from: http://dx.doi.org/10.1016/j.mee.2014.09.024.
120. Wang P, Kricka LJ. Current and emerging trends in point-of-care technology and strategies for clinical validation and implementation. *Clin Chem*. 2018;64:1439–1452.
121. Gomez FA. The future of microfluidic point-of-care diagnostic devices. *Bioanalysis*. 2013;5:1–3.

5 Microfluidics Device for Isolation of Circulating Tumor Cells in Blood

Ashis K. Sen, Utsab Banerjee, Sachin K. Jain, Amal Nath, and Aremanda Sudeepthi

CONTENTS

5.1 INTRODUCTION

Cancer is a medical condition in which cells with mutated DNA undergo uncontrolled cell division to produce many abnormal (cancerous) cells. These cancer cells form clusters in tissues causing the formation of malignant tumors in non-hematologic cancers. Circulating tumor cells (CTCs) are the cancer cells that have entered into the vascular system from the malignant tumors. CTCs flow in the circulatory system and possess the potential to invade other distant organs and form secondary tumors, a condition called metastasis. Metastasis causes the majority of cancer-related deaths [1]. According to the World Health Organization (WHO), deaths due to metastasis will rise to more than 20 million per year by 2030 [2]. During metastasis, the cells escape from the tumors and intravasate into the circulatory system (thus becoming CTCs) through the vascular endothelium. These CTCs extravasate into the distant organs and multiply to cause the formation of new tumors [3]. CTCs were first witnessed in the blood of a breast cancer patient by an Australian physician, Thomas Ashworth, in 1869 [4]. Since then, the detection and study of CTCs has attracted a lot of attention due to their high value in cancer diagnosis, prognosis, treatment monitoring, metastasis research, and drug development. CTCs act as potential biomarkers for early-stage cancer diagnosis in liquid biopsy assays, which offer minimal

DOI: 10.1201/9781003033479-5

invasiveness compared to the tissue biopsy assays [5]. The number of CTCs in the peripheral blood is related to the overall and progression-free survival rates of cancer patients [6]. Since the CTCs in the peripheral bloodstream can be from either primary or secondary tumors or both, their enumeration and analysis could provide information about cancer progression and the patient's response to the treatment [7]. Besides their clinical significance, CTC isolation and their downstream analysis would be useful for their genotyping and phenotyping and could provide worthwhile insights about tumorigenesis. CTC isolation and their consequent culturing can help conduct in vitro tests for drug development and choose a drug type for patient-specific treatments (personalized therapy). However, the low number (1–100 CTCs per mL of whole blood among 5 billion of erythrocytes, 4 million leukocytes, and 300 million platelets approximately) and heterogeneity of CTCs impose persisting challenges on their detection and isolation from the blood samples [8].

An ideal CTC isolation system should possess high capture efficiency, purity, and throughput [9]. Broadly, the different isolation methods can be classified into immunoaffinity-based techniques that use the biological properties of CTCs, and label-free methods that rely on the physical properties of the CTCs for detection and isolation [10]. Immunoaffinity-based detection and isolation of CTCs are carried out by utilizing epithelial expression markers like epithelial cell adhesion molecules (EpCAM) and cytokeratins which are also called positive isolation methods. Conversely, there are also negative isolation methods that use the depletion of WBCs using anti-CD45 antibodies to enrich the CTCs. The majority of the immunoaffinity-based CTC isolation techniques rely on EpCAM expression, which is a glycoprotein involved in the intracellular adhesion and signal transduction of the epithelial cells. Due to their epithelial origin, CTCs are presumed to express EpCAM. In general, it is overexpressed in cancer conditions and thus paves the way for detection [8]. However, CTC detection based on epithelial expression markers is challenged by certain limitations. Some of the CTCs may undergo epithelial to mesenchymal transition (EMT) and lower the epithelial expression markers like EpCAM and cytokeratins.

Moreover, the EpCAM is not only specific to malignant tumor cells but also found to be expressed by the cells from benign tumor origins [11]. At the macro-scale, the CellSearch system utilizes immunoaffinity techniques to enrich CTCs. Currently, the CellSearch system remains the only US Food and Drug Administration (FDA) approved CTC isolation technique, and no microfluidic systems have been granted approval. In the CellSearch system, antibodies specific to the EpCAM are coated over magnetic particles to bind them to the epithelial cell adhesion molecules of the CTCs. The antibody-captured CTCs are then identified and counted using immunostaining techniques. Though the CellSearch system is approved by the FDA, it is limited to only breast, prostate, and colorectal cancers and the technique is challenged by its inadequate sensitivity [12]. Moreover, the technique is semi-automatic and does not retain cell viability after isolation for the downstream analysis. Also. it relies on EpCAM expression to isolate CTCs which could be dynamic and patient specific.

Label-free isolation, on the other hand, is performed by utilizing differences in the physical properties of the CTCs such as size, density, compressibility, deformability, and dielectric nature [13]. At the macro scale, CTC isolation based on their physical properties is carried by employing the differences in size and density of CTCs using the methods, isolation by size of epithelial tumor cells (ISET), and OncoQuick respectively [14, 15]. Though these techniques result in high throughput, they suffer from inadequate isolation purity [9].

Microfluidics, a technology that handles fluids in sub-millimeter channels, has shown great potential for the precise and accurate manipulation of microscale objects. It is thus reckoned as a potential alternative to the macroscale CTC isolation methods. Microfluidic chip-based CTC isolation methods render remarkable purity and capture efficiency compared to macroscale techniques due to their high sensitivity and specificity [5, 9]. Besides, they offer automation, better integration capabilities with the other downstream analysis modules (like CTC culture and characterization), and preserve the cells viable and intact after isolation for the downstream analysis. Moreover, the manipulation at cellular scale dimensions facilitates single-cell analysis, which is advantageous when the cell population is heterogeneous, like CTCs. CTC isolation at the microscale reduces the consumption of reagents and hence the expenses incurred. Therefore, many microfluidics techniques have been proposed to perform CTC isolation based on the biological or physical properties of the cells by employing both passive and active techniques. Passive techniques accomplish CTC separation by employing flow hydrodynamics. In contrast, active methods utilize external force fields to achieve the same. This chapter provides an overview of various active, passive, and combined microfluidics techniques employed to detect CTCs in the blood samples with their respective advantages and limitations.

5.2 METASTASIS AND IMPORTANCE OF CTC ISOLATION

The primary cause of death in cancer patients is metastasis. Metastasis consists of several sub-processes. Initially, cancer cells infiltrate into the adjacent tissue, and then tumor cells migrate into the bloodstream; the process is termed intravasation. Next, these cells escape the immune system attack and exit the blood to enter the organ tissues, this process is termed extravasation. Finally, these cells proliferate and develop into a newly formed tumor, a secondary tumor. The whole process of metastasis is depicted in Figure 5.1. The cells that detach from the primary tumor and migrate into the bloodstream are known as circulating tumor cells (CTCs). Detection of CTCs in cancer patients' blood is a promising biomarker for cancer diagnosis, prognosis, stratification, and pharmacodynamics without invasive tissue biopsy [17]. Early detection and diagnosis of cancer cells are of paramount interest to provide clinical guidance on the treatment of cancer. A quantitative estimation of cancer patients' blood shows it to contain approximately 7 million leukocytes, nearly 5 billion RBCs, and less than about 200 CTCs in a sample of 1 mL of whole blood. These data bolster the statement that CTCs are rare cells in the blood and require sensitive and efficient techniques for the detection and isolation of CTCs from the patients' whole blood. Another quantitative estimation shows that 57% of 123 tested patients with metastatic prostate cancer had two or more CTCs per 7.5 mL of blood; however, only 32% of these patients had ≥10 CTCs per mL. Thus,

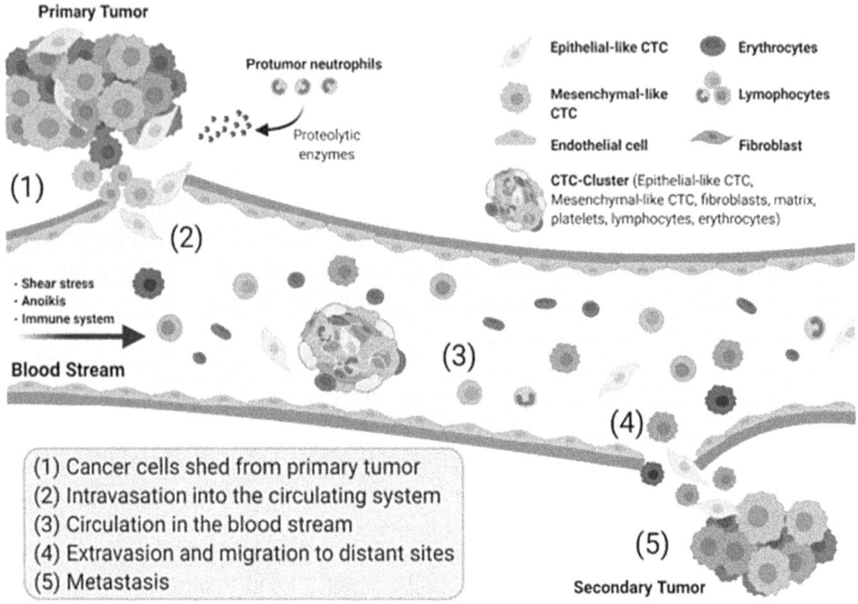

FIGURE 5.1 Pictorial depiction of the metastasis process, leading to the secondary tumor (reprinted from [16] with permission from MDPI).

to capture sufficient numbers of CTCs in a large number of patients for statistically reliable and reproducible diagnostic applications, it is likely that large volumes of blood (≥20 mL) need to be processed [18].

5.3 PRINCIPLES OF CTC ISOLATION FROM BLOOD

The primary goal of any CTC isolation technique is to identify and isolate the cells which are capable of causing metastasis [9]. Thus, any CTC detection and isolation technique should specifically possess these features as – (i) capable of isolating all the CTCs present in the blood, i.e., high capture efficiency, (ii) capable of isolating only CTCs, i.e., no other cells should be present along with the CTCs after isolation, indicating high isolation purity, (iii) isolation should be a quick process, i.e., high throughput. Most of the CTC detection and isolation techniques exploit the characteristics of CTCs, which makes them distinguishable from healthy blood cells. These characteristics include biological properties (surface protein, presence of mutations, expressions of specific genes, viability, etc.) and/or physical properties (size, density, electrical charge, deformability, etc.). Based on the system and external forces, manipulation techniques can also be classified as passive techniques, active techniques, and combined techniques (see Figure 5.2). Passive methods rely on the system forces such as inertial or non-inertial lift, Dean drag, etc. In addition to that, it can also utilize the fluid properties and channel geometry to demonstrate CTC detection and isolation successfully. On the other hand, incorporating one or

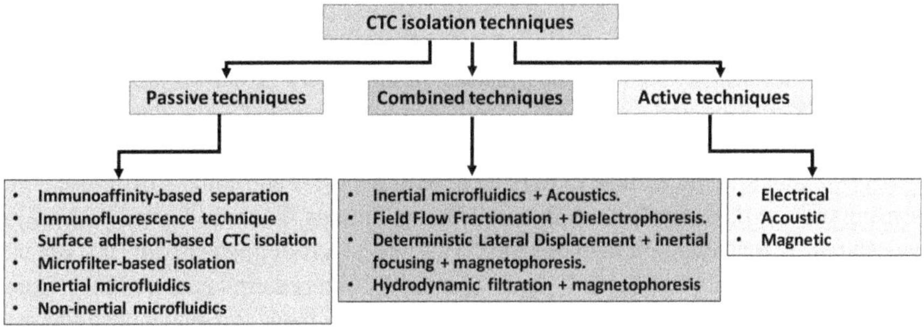

FIGURE 5.2 A summary of the various CTC isolation techniques.

more external agents can also serve as a useful tool for manipulating CTCs on a microfluidic platform. These are termed as active techniques. The external force can be electrical, magnetic, optical, acoustic, etc. Similarly, a combination of passive and active techniques is also found to demonstrate the efficient sorting of CTCs inside the microchannel. All these techniques are discussed in detail, along with their working principles.

5.3.1 Passive Techniques

Passive microfluidic-based techniques exploit cell characteristics such as size, surface antigens, shape, compressibility, and density, without incorporating an external agent, e.g., magnetic, electric, acoustic, optics, etc. Based on the above distinguishable cell characteristics, we discuss the different passive techniques in this section.

Immunoaffinity-Based Separation

The immunoaffinity-based technique is a label-based technique which can facilitate positive and negative isolation. In the case of positive separation, cells of interest are labeled and isolated. On the contrary, in the case of negative sorting, cells that are not of interest are labeled, detected, and isolated. CTCs express various cell surface markers compared to blood cells and, therefore, can be isolated from the circulatory cells. Explicitly, a CTC shows epithelial cell adhesion molecules (EpCAM) and cytokeratin (CK) to a variable extent, but not CD45, which is a differential marker for white blood cells (WBCs). These antibodies can be either immobilized on the walls of the microfluidics devices or can be conjugated to the magnetic nanoparticles for capturing CTCs. Depending on the detection and isolation, immunoaffinity-based technique can be further classified as immunofluorescence technique, surface adhesion-based techniques, or immunomagnetic technique. The first two techniques are discussed in this section, and the last method is covered under active techniques.

Immunofluorescence Technique

In this technique, CTCs are labeled using dyes or nanoparticles [19] by targeting CTCs via antibodies depending on the antigen present on their surface. The cells expressing those biomarkers are tagged to fluorescently labeled antibodies, which emit a light signal of a particular wavelength upon illumination. This signal is detected using optical instruments, which indicates the presence of antibodies on the surface of the cell. Continuous detection and separation of cells can be accomplished by flow cytometry. These flow cytometers are based on fluorescence-activated cell sorting (FACS) technology. A fundamental micro-flow cytometer consists of three sub-units – hydrodynamic focusing unit, optical detection unit, and isolation unit, as shown in Figure 5.3(a). The hydrodynamic focusing unit focusses the CTCs into a well-defined stream. The optical detection unit detects the optical signal emitted by fluorescent-labeled CTCs. Upon detection, the isolation unit comes into play and isolates the CTCs into a desired location. The isolation of CTCs can be accomplished in various ways, such as direct isolation and droplet-based isolation, as shown in Figure 5.3(b). In the case of a direct isolation method, CTCs are separated directly through the stream. Whenever a CTC is detected, the detection unit generates a pulse, and the isolation unit is triggered, which isolates the CTCs into different channel outlets.

Similarly, various forces can be exploited to accomplish CTC isolation as delineated under active techniques. In the case of droplet-based separation, CTCs are encapsulated inside droplets, as shown in Figure 5.3(c), and efforts are made to isolate those droplets using conventional droplet microfluidic techniques. One of the advantages of droplet-based detection and isolation is that droplets preserve the CTCs intact inside it, ruling out the possibility of cell damage.

Surface Adhesion-Based CTC Isolation

The surface adhesion-based technique is a label-free technique independent of the cell size. Interaction between solid surfaces and cells can be exploited to separate CTCs with different kinds of surfaces. The basic principle for surface adhesion relies on differential adhesion potential between the target cells and non-target cells. There can be two types of approaches for adhesion-based isolation: Implementing specific surface patterns and immobilizing immunoaffinity-binding molecules. Due to the laminar nature of the fluid in microchannels, there are limitations in cell and surface interaction; hence various approaches are taken, like modifying the surface or creating passive turbulence inside the channel [20]. This approach can often be combined with other isolation techniques for further enrichment of CTCs. PDMS microfluidic channels are fabricated that contain PUA flat or nanostructured polymer surfaces on the bottom to separate MCF-7 cells from MCF-10A cells [21] as shown in Figure 5.3(c). The adhesion between MCF-10A and MCF-7 cells was evaluated by enhancing the flow rate of Hank's balanced salt solution. Immunoaffinity-binding molecules are even attached to a silicon surface having thousands of micro-posts etched inside the channel for efficient detection and isolation of CTCs [5]. It has been observed that the capture efficiency of CTCs can be further improved by the integration of the microfluidic system with silicon nanowire (SiNW) [22].

Microfilter-Based Isolation

Since cancer cells are larger than the leukocytes and erythrocytes, the micro-filtration technique can be employed for the separation [23]. A new fluid filter structure in

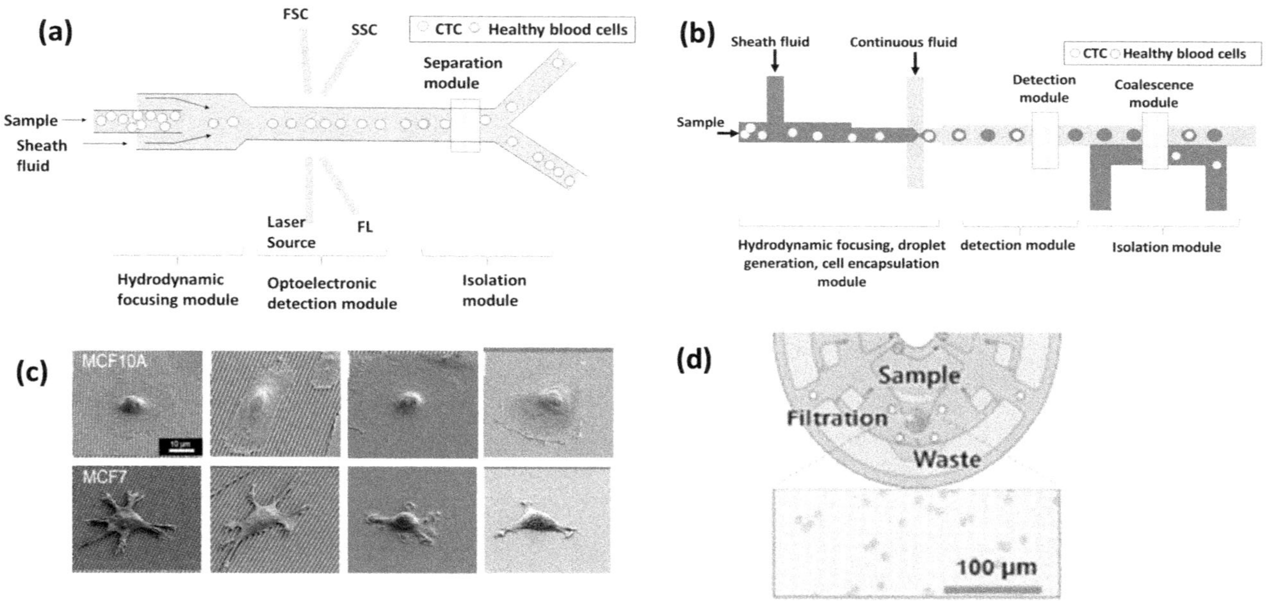

FIGURE 5.3 (a) Schematics of micro-FACS system. (b) Schematics of CTC detection and isolation using droplet coalescence. (c) SEM images of MCF10A and MCF7 cells on three different PUA nanostructures, PUA flat surface, and glass control after 2 h pre-culture (reprinted from [21] with permission from RSC). (d) Image of the standalone lab-on-a-disc system for isolating various CTCs from whole blood (reprinted from [26] with permission from ACS).

silicon fabricated using a self-aligning technique [24] is exploited to demonstrate CTC isolation. The membranes are positioned in such a way that during the filtration process, the flow passes the first membrane, and it finally reaches the hole by overcoming the gap between the two membranes. This configuration ensures a double filtration by recapitulating membrane filters with a microstructure design. Zheng's research group demonstrated a 3D microfilter device to sort MCF-7 cells from diluted blood samples [25]. They employed two layers of parylene membrane (top and bottom) are modified with pores, with the gap in between accurately defined with photolithography. Cancer cells could be trapped by the pores of the top membrane, and the bottom membrane reduced the stress concentration on cell membranes and maintained cell viability. The experimental results showed that the cancer cells were captured in 1 mL of whole blood within 3–5 min with an efficiency of 86.5 ± 5.3%. The filter-based device, however, easily got clogged when dealing with a large volume of samples. Cho's research group presented a standalone lab-on-a-disc system to eliminate this problem [26]. The system used Fluid Assist Separation Technology (FAST) to isolate various CTCs (MDA-MB-231, MDA-MB-436, HCC78, and the AGS gastric cell) from whole blood via a label-free process that required no pretreatment of samples. The FAST disc included three separate filtration units, each consisting of three chambers, as shown in Figure 5.3(d). After the whole blood sample was injected into the loading chamber and pushed into the filtration chamber, the CTCs were captured by the membrane, and the healthy cells could pass through the membrane into the waste chamber. The results demonstrated that the system could isolate the five cancer cell lines with a recovery rate of 95.9 ± 3.1%.

Inertial Microfluidics

Inertial microfluidics has emerged as one of the promising approaches for CTC isolation with better performance. The earliest interpretation of particle focusing was reported by Segre [27] in the 1960s. It was observed that along the flow length, the particles tend to align between the pipe centerline and the wall circumference. This effect is primarily due to the interplay of two forces: The shear-induced lift force and wall-induced lift force. In addition to this, Stokes's drag induced by the secondary flow is also a responsible factor for inertial migration and focusing. The presence of secondary flow in the curved channels is reported as Dean vortices. It arises because of the mismatch in fluid momentum across the channel cross-section due to centrifugal acceleration. The strength of this secondary flow is characterized by the Dean number (De). A microchannel consisting of a contraction–expansion array is exploited to separate cancer cells from whole blood at a low Reynolds number [28], as shown in Figure 5.4(a). The contraction-expansion array (CEA) microfluidic device utilizes a hydrodynamic field effect for cancer cell separation, two kinds of inertial effects: Inertial lift force and Dean flow, which result in label-free size-based separation with high throughput. The separation of cancer cells from whole blood was demonstrated with a cancer cell recovery rate of 99.1%, a blood cell rejection ratio of 88.9%, and a throughput of 1.1×10^8 cells/min. Another study [29] shows the introduction of a high-throughput size-based separation method for processing diluted blood using inertial microfluidics as shown in Figure 5.4(b). The technique

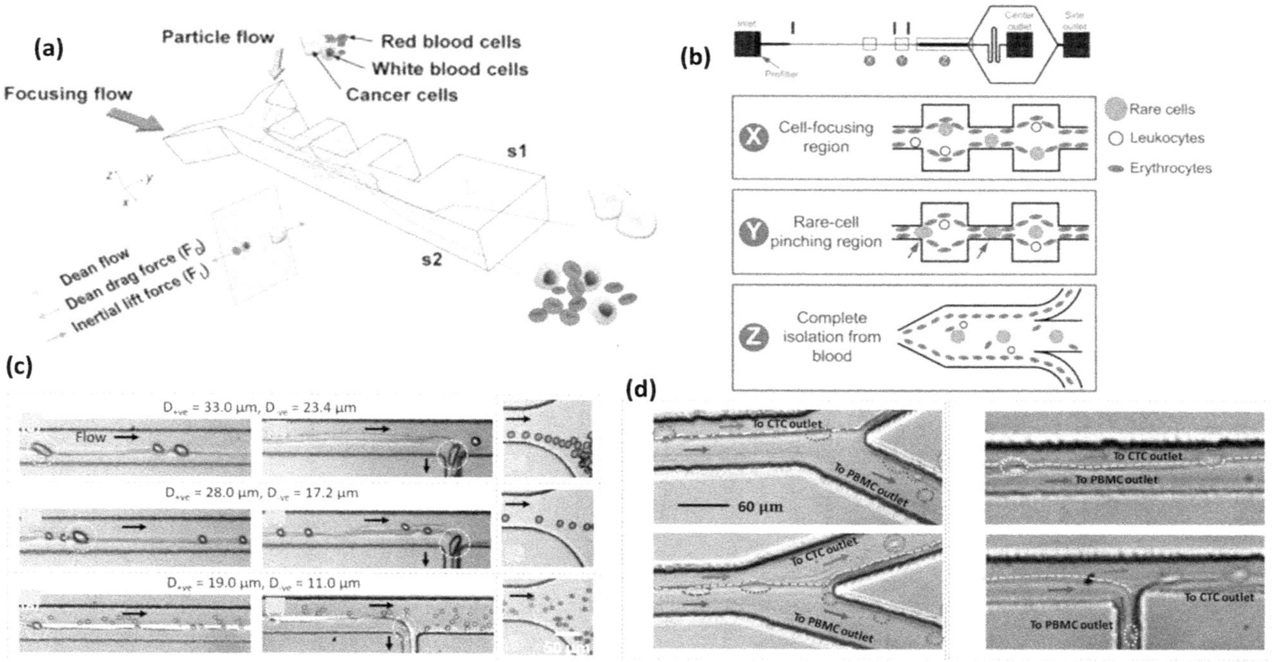

FIGURE 5.4 (a) Schematic of the experimental setup (reprinted from [28] with permission from ACS). (b) Schematic illustration of the microfluidic device (reprinted from [29] with permission from RSC). (c) Sorting of positive droplets encapsulating HeLa cells, MDA-MB 231 and PBMCs across the interface (reprinted from [31], with permission from AIP). (d) Experimental images showing sorting of (a) MCF-7 (dotted blue encircled) and (b) MDA-MB-231 cells (dotted yellow encircled) from PBMCs (reprinted from [32], with permission from RSC).

takes advantage of preferential cell focusing in high aspect-ratio microchannels, coupled with pinched flow dynamics for isolating low abundance cells from the blood. Results from experiments conducted with MCF-7 cells spiked into whole blood indicate >80% cell recovery with an impressive 3.25×10^5-fold enrichment over red blood cells (RBCs) and 1.2×10^4-fold enrichment over peripheral blood leukocytes (PBL). Despite a 20× sample dilution, the fast-operating flow rate allows the processing of ~10^8 cells/min through a single microfluidic device. The device design can be easily customized for isolating other rare cells from blood, including PBLs and fetal nucleated red blood cells, by merely varying the "pinching" width.

Non-Inertial Microfluidics

Apart from the lift forces mentioned in the previous section, droplets and CTCs experience deformability-induced lift forces when subjected to a microchannel flow at a low Reynolds number. This lift force depends on the size of the cell or droplet, shear rate, viscosity of the continuous medium in which cells or droplets are flowing, the lift coefficient, and the location of the cells from the wall [30]. Non-inertial lift-induced interfacial migration phenomena have also been employed to sort cell-encapsulating aqueous droplets (positive droplets) from empty droplets (negative droplets) as shown in Figure 5.4(c). Utilizing the fact that positive droplets were always larger compared to negative droplets, positive droplets were migrated across silicone oil-mineral oil interface, and maximum efficiency of ~95% and purity of ~65% was attained. Similarly, non-droplet-based sorting of CTCs using non-inertial lift as the driving force is also exploited as shown in Figure 5.4(d) and found to manifest a sorting efficiency ~99%.

5.3.2 Active Techniques

Microfluidics-based devices are particularly suitable and advantageous for detection and isolation of CTCs as there have been various techniques developed in the recent past [23] employing acoustics, electricity, magnetism, etc. These active methods offer functionalities that help accelerate the detection process with enhanced accuracy, utilizing specific properties of the CTCs. The principles and some applications-oriented literature regarding these techniques are discussed in this section.

Electrical

Impedance-based detection of CTCs has seen a rise in the recent past as most CTCs have distinct electrical properties compared to the normal healthy cells. Dielectrophoresis (DEP) is one of the most used tools for separating CTCs from healthy cells, as shown in Figure 5.5(a). When an electrically polarizable object immersed in a medium with different polarizability is subjected to a nonuniform electrical field, they experience the dielectrophoretic force [33] which can be expressed as:

$$\boldsymbol{F}_{DEP} = 2\pi\varepsilon_m R_c^3 Re\left(\bar{C}\nabla\bar{\boldsymbol{E}}^2\right); \bar{C} = \frac{\bar{\varepsilon}_c - \bar{\varepsilon}_m}{\bar{\varepsilon}_c + 2\bar{\varepsilon}_m} \tag{5.1}$$

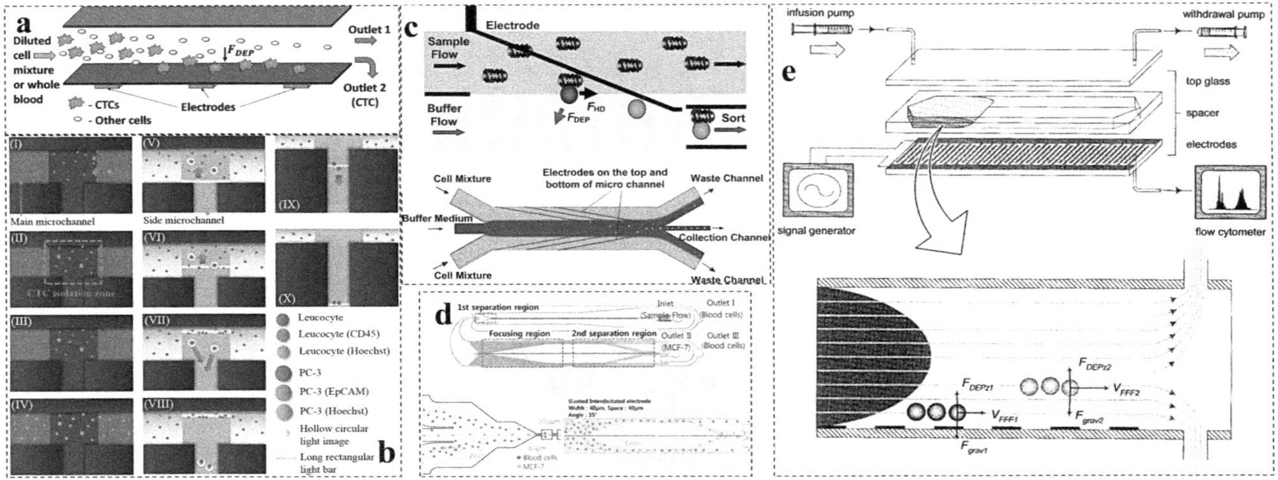

FIGURE 5.5 (a) Schematic of dielectrophoresis (DEP)-based cell separation. (b) Adjustable electrode formation using optically induced DEP (ODEP) (reprinted from [36]). (c) Polarizable particle labeled enhanced and controlled DEP-based cell sorting (reprinted from [37] Copyright (2005) National Academy of Sciences, USA). (d) Schematic of multi-orifice flow fractionation (MOFF) aided DEP for improved throughput. Reprinted from [38] Copyright (2011) Royal Society of Chemistry. (e) DEP field-flow fractionation (DEP-FFF)-based levitation at different height and separation of CTCs (reprinted from [40] Copyright (2000) Elsevier).

Where, $\bar{E}$ is the complex electrical field, a function of space, and frequency (ω). $\bar{C}$ is the Clausius-Mossotti (CM) factor. *Re* represents the real part of the term inside the parenthesis. $\bar{\varepsilon}_i\left(=\epsilon_i+\sigma_i/\left(j\omega\right)\right)$ is the complex permittivity of the *i*th phase (*m* – surrounding media, *c* – cell) with ϵ_i and σ_i being the permittivity and conductivity of the *i*th phase, respectively. The cell is assumed to be a perfect sphere with an outer radius R_c. The above equation is valid for sufficiently small cells when compared to the spatial non-uniformity of the field.

DEP force can either be attractive, i.e., positive DEP (or pDEP) or repulsive, i.e., negative DEP (or nDEP) depending on the permittivity difference of the medium and objects. In early works, Becker et al. [34] used electrorotation for measuring dielectric properties of MDA231 (human metastatic breast cancer cells) and demonstrated the isolation of the tumor cells from whole blood containing erythrocytes and T-lymphocytes. This study was one of the very first successful demonstrations of DEP-based CTCs separation. The same group (Gascoyne et al. [35]) later showed separation of MDA-435, MDA-468, and MDA-231 cells from peripheral blood mononuclear cells (PBMCs) in whole human blood with a recovery rate as high as 92%. The recovery efficiency of the cells dropped as they increased cell loading in the channel and increased the cell concentration in the blood. They also found that both the tumor cells and WBCs have distinct CM factors at different applied frequencies of the alternating current (AC) field, which can be employed for the separation of the CTCs. Optically induced DEP (or ODEP) has been developed [36] for the adjustable and on-demand formation of electrodes made of optically sensitive Si layer on indium tin oxide (ITO) glass substrates as shown in Figure 5.5(b). The photoconductive layer becomes activated and starts functioning as electrodes when it is externally illuminated. Apart from DEP, electrical signal sensing-based CTCs detection has also been investigated. Hu et al. [37] showed that using labeling particles to control cell polarization can have a considerable enhancement in DEP response, as shown in Figure 5.5(c). They indicated an improvement over the cell sorting controllability and a recovery rate of more than 95% of the labeled cells (*E. coli.*) with higher than 200-fold enrichment. This method can be effectively used to improve DEP-based cell sorting of the CTCs. Moon et al. [38] successfully isolated human breast cancer cells (MCF-7) from whole blood by improving the DEP system, as shown in Figure 5.5(d). They added a multi-orifice flow fractionation (MOFF) prior to the DEP module to achieve high throughput and were able to attain 162-fold enrichment of MCF-7. Methods like dielectrophoretic field-flow fractionation (DEP-FFF) [35, 39, 40] have been used for the separation of streamlines for deflecting the trajectories or even trapping of target cells at electrodes when the DEP force overcomes the viscous drag as shown in Figure 5.5(e). The nonuniform field is created by microfabrication of the electrodes or array of electrodes on the microchannel wall. Suzuki et al. [41] very recently used a mechanical low filter of 6 μm constriction-based bridge circuit to detect and separate HeLa, A549, and MDA-MB-231 cells from T-lymphocyte cells. They optimized the constriction size to 6 μm after measuring CTCs and other blood cell residence time, and the corresponding sensor signal at the bridge circuit ends, reaching a separation efficiency of more than 95%.

Acoustic

Acoustophoresis, much like DEP, has also been utilized for CTC detection and separation by exploiting the difference in acoustic impedance of the different cells and the suspending medium as shown in the schematic Figure 5.6(a). The acoustic impedance of a phase primarily depends upon the density and the compressibility. Standing or traveling bulk acoustic waves (SBAW or TBAW) exert primary acoustic radiation force [42] (F_{PAR}) on the particles, driving them towards either pressure node or antinode.

$$F_{PAR} = 4\pi R_c^3 K \phi E_{ac} \sin(2Ky); \tag{5.2}$$

$$\phi = \frac{1-k^*}{3} + \frac{\rho^* - 1}{2\rho^* - 1}; \tag{5.3}$$

$$k^* = \frac{k_c}{k_m}; \quad k_i = \frac{1}{\rho_i c_i^2}; \quad \rho^* = \frac{\rho_c}{\rho_m} \tag{5.4}$$

Where, R_c is the radius of the cell, assuming it to be a sphere. E_{ac} is the acoustic energy density, $K = 2\pi/\lambda$ is the wavenumber, and ϕ is called the acoustic contrast factor. k_i, ρ_i and c_i are the compressibility, density, and speed of sound in the ith phase, respectively. Laurell's group successfully demonstrated [43] separation of DU145 and PC3 from WBCs using BAW-based acoustophoresis as shown in Figure 5.6(b). They were able to achieve recovery rates of 87% and 83% for DU145 and PC3, respectively, at a sampling rate of 70 μl min^{-1}. Their innovative pre-alignment step greatly enhanced the separation efficiency with a ~10^2 to ~10^3-fold increase in enrichment of the tumor cells. The same group (Antfolk et al. [44]) again modified their design to add a concentration module for enrichment of the CTCs. MCF-7 and DU145 were spiked in whole human blood with red (lysed) and white blood cells, respectively. They achieved more than 90% and 82% separation efficiency, respectively, with more than 99% purity for both the tumor cells with a concentration of 23.8 ± 1.3-fold for MCF-7 and 9.6 ± 0.4-fold for DU145. Karthick et al. [42] also showed that HeLa and MDA-MB-231 cells could effectively be isolated from the PBMCs by controlling the impedance of the PBMCs to match it with the suspending medium. This technique helped them in reducing the acoustic streaming that unfavorably affected the separation process. They achieved greater than 50-fold enrichment of the cancer cells with a recovery rate of more than 86% from a sample volume of 2 mL within an hour. Similar to bulk acoustic waves (BAWs), there are also surface acoustic waves (SAWs). Unlike the bulk wave where it is necessary, SAWs need not have half-wavelengths integral in the channel dimensions. BAWs are more tunable than SAWs but costlier both in terms of fabrication and power consumption. Huang's group developed a state-of-the-art device [45] based on tilted-angle standing SAWs (or taSSAWs) and achieved 20 times more efficiency than its predecessors as shown in Figure 5.6(c). They optimized the angle of tilt of the interdigital transducers (IDTs) with respect to the flow direction for different flow rates of the sample. They found

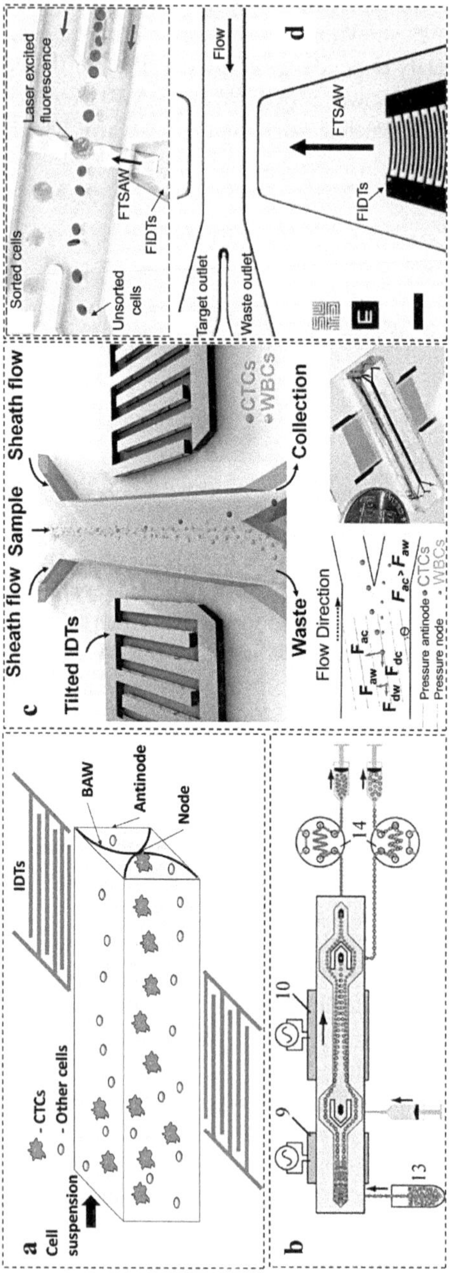

FIGURE 5.6 (a) Schematic of acoustophoretic cell separation. (b) Acoustic (BAW-based) separation of DU145 and PC3 from whole blood using a pre-alignment step (reprinted from [43] Copyright (2012) American Chemical Society). (c) Isolation of CTCs of multiple cell lines using taSSAW (reprinted from [45] Copyright (2015) National Academy of Sciences, USA). (d) MCF-7 separation from diluted whole blood using FTSAW (reprinted from [46] Copyright (2017) Royal Society of Chemistry).

that the tilt-angle required for efficient separation of the tumor cells decreased with a higher flow rate. Their device efficiently isolated multiple lines of cancer cells from WBCs with low concentrations of ~100 cells/mL with a recovery rate of more than 83%. Ma et al. [46] employed fluorescence-activated cell sorting (FACS) along with acoustics for CTCs separation as shown in Figure 5.6(d). They used focused traveling SAW (FTSAW) generated by focused IDTs (FIDT) to isolate fluorescently labeled breast cancer cells (MCF-7) from diluted whole blood samples. They reported a high purity of the separated MCF-7 of more than 86%, and the viability of more than 95% with cell loading of ~2×10^6 per mL of human diluted whole blood. In a novel work, Lu et al. [47] used arrays of acoustic microstreaming traps for label-free and sheath-less parallel isolation of CTCs of sizes larger than 10 µm. They fabricated thousands of micropillars acting as traps for the CTCs. The capture and release of the trapped cancer cells were controlled by switching the ultrasonic (US) wave source on and off, respectively. They isolated three different breast cancer cells, including MCF-7, with isolation efficiencies of more than 90%.

Magnetic

Magnetophoresis (MAP) is also a widely employed method for specific cell sorting either by tagging them with magnetic beads or suspending them in a magnetic medium. The schematic shown in Figure 5.7(a) depicts the overall scheme of CTC manipulation. Similar to DEP, there is positive and negative MAP depending upon the magnetic susceptibilities of the target cells and the suspending medium. The force [48, 49] experienced by the cells can be expressed as:

$$F_{MAP} = \frac{4\pi R_c^3}{\mu_0}\left(\chi_c - \chi_m\right)\left(\boldsymbol{B}\cdot\nabla\right)\boldsymbol{B} \tag{5.5}$$

Where, R_c is the radius of the spherical cell, μ_0 is the vacuum permeability, $\boldsymbol{B}$ is the applied magnetic field density and χ_i is the susceptibility of the ith phase. This equation is valid for cells with the assumption that cells exhibit homogeneous magnetization behavior across its dimension.

Talasaz et al. [50] designed a novel circulating epithelial cells (CEpCs) separator called MagSweeper. The device was able to sweep HLA-A2 positive human PBMCs labeled with anti-HLA-A2 antibody functionalized 4.5 µm magnetic beads from negative PBMCs with a capture efficiency of ≈60%. They observed that with an increase in the number of background cells, capture efficiency first decreased then increased. In contrast, the purity of the sample collected sharply reduced from almost 100% to less than 40%. Kang et al. [51] also showed a similar technique of separating CTCs (mouse 4T1 breast cancer cells) from whole blood with epithelial cell adhesion molecule (EpCAM) antibody-coated superparamagnetic microbeads with sorting efficiencies as high as 90%. Weissenstein et al. [52] combined magnetic microbeads coated with anti-EpCAM and anti-cytokeratin to enhance the sensitivity and specificity of the separation process. They reported an average cell recovery rate of 84% when as low as five HCC1937 cells were spiked in 7.5 mL of whole

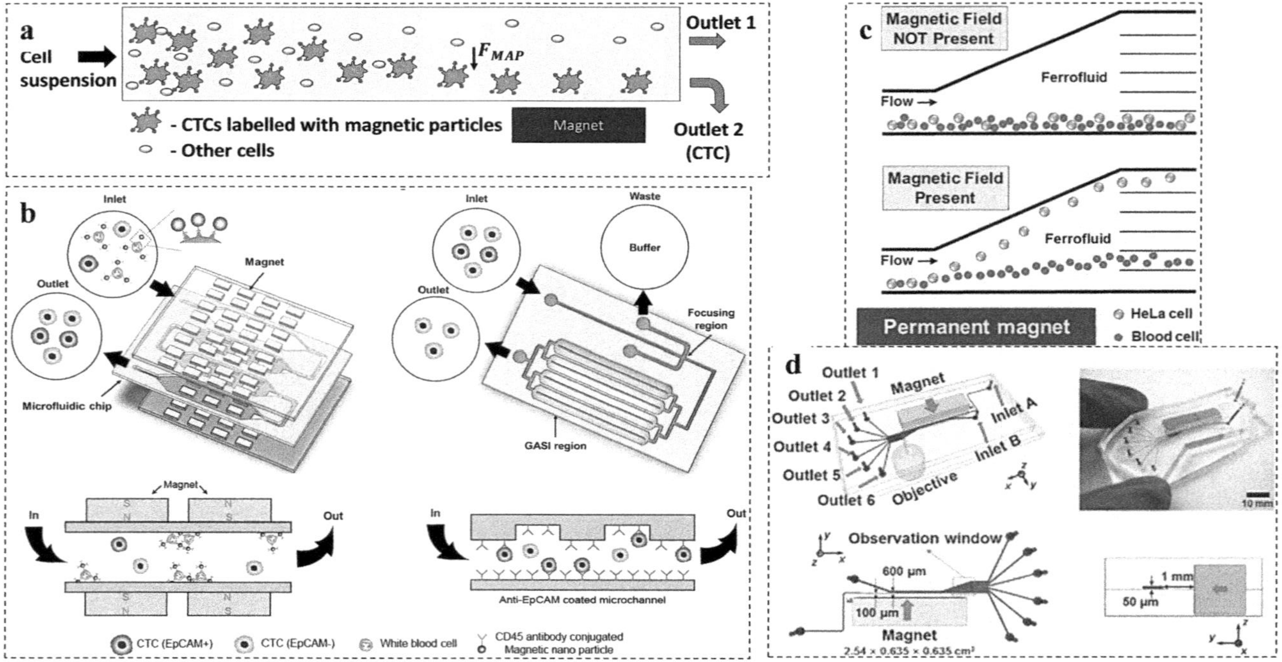

FIGURE 5.7 (a) Schematic of magnetophoretic (MAP) cell separation. b) Two-stage separation process of positive WBC and negative CTC enrichment (reprinted from [53] Copyright (2015) Elsevier). (c, d) Negative MAP-based separation of HeLa cells from blood cells in paramagnetic medium (reprinted from [54] Copyright (2016) John Wiley & Sons).

blood. Hyun et al. [53] employed a two-stage separation method, where they first isolated magnetically tagged WBCs (negative enrichment) as shown in Figure 5.7(b). In the second stage, they captured CTCs (positive enrichment) based on their surface protein expression of EpCAM or HER2 with a magnetically activated channel surface. They achieved 763-fold enrichment of the CTCs at high throughput of around 400 $\mu L\ min^{--1}$, though their separation efficiency was low (<25%). Zhao et al. [54] applied negative MAP to successfully isolate human breast cancer cells (HeLa) from other blood cells (RBCs) as shown in Figures 5.7(c) and 5.7(d). Their in-house-prepared water-based low concentration ferrofluid allowed them to have a higher magnetic response than paramagnetic salt solutions while staying biocompatible for up to two hours. They claimed to have achieved separation efficiency of more than 99% with channel loading of 10^6 cells/h^{-1}.

5.3.3 Combined Techniques

Hybrid methods that use a combination of both active and passive techniques in parallel or in succession have also been studied previously and are discussed in this section.

The most commonly used hybrid methods involve a multi-stage process. In the first stage of the device, a passive technique is employed to facilitate the focusing of all the cells in the sample or a filtering out of the non-target cells. The consequent stages of the process involve differentially separating the CTCs from the remaining cells using a single or a combination of active manipulation techniques (electrophoresis, magnetophoresis, dielectrophoresis, acoustophoresis, etc.). Moon et al. [38] used a combination of hydrodynamic flow fractionation in the first stage and dielectrophoresis in the second stage to accomplish the label-free separation of human breast cancer cells (MCF-7) from a spiked blood cell sample. In the first stage, MOFF is employed, and the larger MCF-7 cells traverse through the center to the DEP channel while the RBCs are separated away. In the DEP module, the frequency and voltage of the field are tuned such that only the larger MCF-7 cells experience sufficient force and are isolated. Ozkumur et al. [55] reported a three-stage device ("CTC-iChip") for isolating magnetically labeled CTCs from whole blood. The first stage of sorting involves lateral displacement by the use of micropillars to remove RBCs, platelets, and other components from whole blood, leaving only WBCs and magnetically labeled CTCs. In the second stage, cells are then inertially focused to a thin band using a serpentine channel before proceeding to the final stage of sorting. In the third stage, magnetically labeled CTCs are deflected due to the magnetophoretic force leading to separation. The iChip technique is high-throughput and is capable of sorting rare CTCs from whole blood at 10^7 cells/s.

In the work by Mizuno et al. [56], a two-stage device is developed. In the first stage, by use of several transverse drain channels, size-based sorting due to hydrodynamic filtration takes the smaller sized cells in a channel (or separation lane) closer to the magnet. In comparison, the larger cells travel into a lane, which is further away from the magnet. Based on the expression of surface marker CD4, the second stage involves magnetophoretic separation. Continuous sorting of JM (human lymphocyte cell line) cells by immunomagnetic labeling of CD4 using anti-CD4 immunomagnetic beads is achieved. Sorting a mixture of two cell types, HeLa cells and JM cells, is also shown. Another technique [57] combining MAP makes use of a negative enrichment process to isolate CTCs. Two cell lines, MCF-7 and NCI-H1975, are spiked in whole blood, and separation is effected through a negative enrichment process. First, immunomagnetic depletion is employed to deplete CD45-positive WBCs. The RBCs are then depleted by filtration using a micro-filter membrane.

More recent studies [58, 59] have demonstrated the separation of fluorescently labeled rare cells using acoustophoretic force as shown in Figure 5.8(a). First, the cells are inertially focused into a small band in a spiral channel before entering the interrogation zone. An acoustic beam is generated from an interdigitated transducer (IDT), which translates the fluorescently labeled cancer cells away from other blood cells. Fluorescence-activated acoustic sorting offers some advantages over other hybrid sorting methods. For example, in the hybrid electrophoresis method, the heating due to the large electric field may damage the cells while in hybrid MAP, extra steps are necessary due to the use of magnetic beads that need to be removed from the isolated, rare cells.

In some studies, instead of a multi-stage process, the passive and active forces are designed to act concurrently on the cells. Gupta et al. [60] demonstrated a combination of dielectrophoresis and hydrodynamic effect by operating at a signal frequency between crossover frequencies of CTCs and PBMCs to effect the separation. At the operating frequency, CTCs experience positive dielectrophoretic force. The PBMCs experience negative DEP force and are levitated and washed away by an eluting buffer flow. Shim et al. [61] use a similar technique by operating between the crossover frequencies of tumor cells and PBMNs; the tumor cells are pulled towards the chamber floor by positive DEP while PBMNs are repelled by negative DEP forces as shown in Figure 5.8(b). The fluid is withdrawn through the bottom opening, and tumor cells within the separating streamline are captured.

Further development of hybrid techniques to separate CTCs can build on some studies from existing literature. In one study, viscoelastic suspending fluid was used to focus the cells into a thinner band (compared to a Newtonian suspending fluid) before the separation of magnetic particles in the second stage [62]. The use of dielectrophoresis with deterministic lateral displacement using a virtual pillar array [63] to effect the separation of RBCs and WBCs was also demonstrated. Kim et al. [64] used a combination of dielectrophoretic and magnetophoretic forces to separate different bacterial cell types. The device features a dielectrophoretic module followed by a magnetic separation module. Based on surface markers, Target A cells were labeled with DEP tags, while Target B cells were labeled with streptavidin-coated magnetic tags. The cells which do not express cell surface markers were left unlabeled. The excellent separation performance of the device was demonstrated.

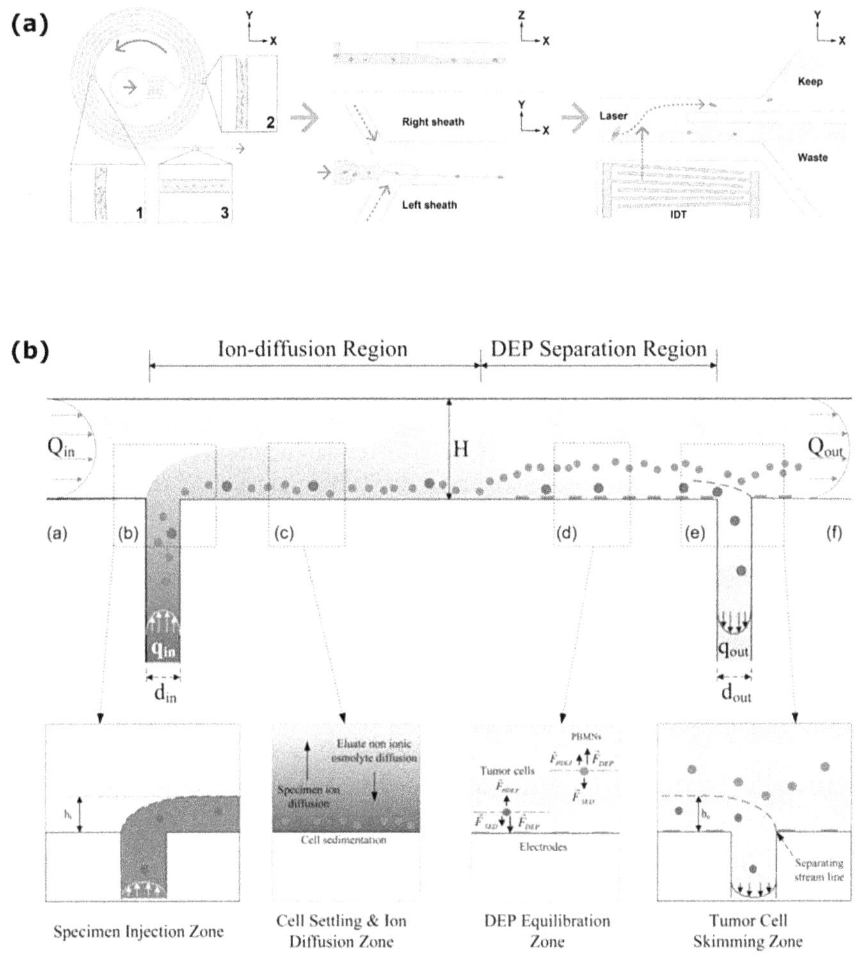

FIGURE 5.8 (a) Inertial focusing of the cell sample is performed in a spiral channel. After alignment with flanking sheath flows, fluorescence-activated sorting is achieved with an acoustic pulse generated by an IDT (reprinted from [59] with permission from RSC). (b) Working principle of the device for isolation of CTCs from PBMNs using DEP-FFF technique is shown (reprinted from [61] with permission from AIP).

5.4 COMMERCIALIZATION OF MICROFLUIDICS DEVICES FOR CTC ISOLATION

Successful commercialization of any CTC detection and isolation technique depends on the smooth transition from laboratory research to the development of the end product. Performance of the CTCs detection microfluidic chip, its universality, and cost effectiveness ratio play a vital role in the commercialization. Furthermore, the commercialized products should be robust, user friendly, and well automated such that they can be operated by non-professionals. Numerous microfluidic chip-based CTC isolation techniques are commercialized such as OncoCEE™48 (Biocept, USA), BioFluidica CTC Detection System (BioFluidica Inc., USA), IsoFlux system (Fluxion Bioscience, USA) that employ immunoaffinity techniques and ClearCell FX (Clearbridge Biomedics, Singapore), Vortex HT device (Vortex Bioscience, USA), and Apostream (ApoCell, USA) that use physical properties of the CTCs for detection and isolation. A detailed review of microfluidic chip-based CTC detection and isolation technologies that have been commercialized can be found elsewhere [8, 13].

Currently, the technical challenges limiting the commercialization of microfluidic chips for CTC isolation techniques are – dealing with large volumes of liquid with high throughput as these devices usually operate at low flow rates [12], epithelial to mesenchymal transition (EMT) of CTCs that affect the capture efficiency in the case of immunoaffinity-based detection methods and lack of universality as the research is carried under experimental conditions specific to that technique. However, these limits may be overcome by employing the well-advanced technologies that utilize multifunctional antibodies and nanomaterials [65]. The universality of the devices can be improved by carefully considering the aspects of commercialization from the point-of-development stage (Table 5.1).

5.5 SUMMARY AND OUTLOOK

In summary, detection and isolation of CTCs are of paramount importance for the early diagnosis and treatment of cancer. This chapter discusses all the detection and isolation techniques exploited on the microfluidics platforms along with their advantages, disadvantages, and performance parameters. The techniques are broadly classified into passive, active, and combined techniques. Passive microfluidics-based techniques exploit the cell characteristics such as size, surface antigens, shape, compressibility, and density, without incorporating an external agent, e.g., magnetic, electric, acoustic, optics, etc. On the contrary, active techniques utilize the external forces, e.g., magnetic, electric, acoustic, optics, etc. to demonstrate CTC isolation. Combined techniques utilize both passive and active means to accomplish CTC detection and isolation on a microfluidic platform. Finally, the chapter discusses the performance parameters, features of various techniques used in CTC detection, and separation. A brief discussion on the commercialization of CTC detection and isolation devices is also presented, which provides an overall idea about the present scenario. It is also seen that CTC isolation and detection have some limitations, such as low abundance, non-specificity, etc. The performance of these techniques is evaluated based on purity, throughput, and biocompatibility. These innovative microfluidics techniques have great potential to change the strategy in prognosis and diagnosis of cancer.

TABLE 5.1
A Comparison of Different CTC Isolation Techniques

Reference	Technique	Target Cell Line	Throughput and Efficiency	Features
[38]	Dielectrophoresis coupled with Field-Flow Fractionation (DEP-FFF)	MCF-7	Sample throughput of 126 μL/min, 162.4-fold MCF-7 enrichment ratio, separation efficiency 75.18% from RBCs and PBMCs	Label-free, high-speed continuous flow-through separation
[55]	Deterministic Lateral Displacement (DLD), inertial focusing, and magnetophoresis	MCF10A, MDA-MB-231, SKBR3	Sorting of rare CTCs from large volumes of whole blood (8 mL/h) at high throughput of 10^7 cells/s, the efficiency of 96.7% for MCF10As, 98.6% for SKBR3, 77.8% for MDA-MB-231	Applicable to all types of cancer cells, lower number of immunomagnetic beads used, can be run in the positive selection or negative depletion modes, CTC-iChip platform
[56]	Combines hydrodynamic size-dependent filtration with magnetophoresis	HeLa and JM cells	Sorting of HeLa cells and JM cells with a throughput of 100 cells/s and purity enriched ~2-fold	Two factors of sorting: Size and immunomagnetic label are combined, the sample is sorted into four outlets, low throughput
[57]	Combines magnetophoresis and micro-slit membrane-based filtration	MCF-7 and NCI- H1975	Throughput of 2 mL/h, ~90% cell recovery from whole blood	Scalable, amenable to standardization
[58]	Inertial focusing and acoustophoresis	MCF-7 in diluted whole blood	2500-fold purity enrichment with the viability of 91%	Fluorescence activated acoustic sorting. Unlike DEP methods, no damage to cells arising from heating, no magnetic labeling involved

(Continued)

TABLE 5.1 (CONTINUED)
A Comparison of Different CTC Isolation Techniques

Reference	Technique	Target Cell Line	Throughput and Efficiency	Features
[59]	Inertial focusing, sheath flow alignment, and acoustophoresis	K562	A sample flow rate of 1.5 mL/h. ~20-fold purity enrichment at 2000 events per second, 91.4% viability	Fluorescence-activated acoustic sorting where a 25 μs TSAW pulse is used to deflect cells
[60]	Continuous isolation of CTCs using DEP-FFF	SKOV3 and MDA-MB-231 spiked in 1 mL samples containing PBMCs	Recovery of 75.4% for SKOV3 and 71.2% for MDA-MB-231 for n = 12 and n = 6 in ~10^6 PBMCs respectively, cell viability of 97.1%	Continuous flow device for independent antibody isolation of CTCs commercialized as ApoStream™
[61]	Continuous isolation of CTCs using DEP-FFF	MDA-MB-231, MDA-MB-435 spiked in samples with PBMNs	The sample flow rate of 20 μL/min, isolation efficiency of 70–80%	Antibody-independent isolation, the continuous flow device

REFERENCES

1. Zhong X, Zhang H, Zhu Y, et al. Circulating tumor cells in cancer patients: Developments and clinical applications for immunotherapy. *Mol Cancer.* 2020;19:1–12.
2. Burinaru TA, Avram M, Avram A, et al. Detection of circulating tumor cells using microfluidics. *ACS Comb Sci.* 2018;20:107–126.
3. Lin Z, Luo G, Du W, et al. Recent advances in microfluidic platforms applied in cancer metastasis: Circulating tumor cells' (CTCs) isolation and tumor-on-a-chip. *Small.* 2020;16:1–21.
4. Ashworth T. A case of cancer in which cells similar to those in the tumours were seen in the blood after death. *Aust Med J.* 1869;14:146–149.
5. Nagrath S, Sequist L V., Maheswaran S, et al. Isolation of rare circulating tumour cells in cancer patients by microchip technology. *Nature.* 2007;450:1235–1239.
6. Ma X, Xiao Z, Li X, et al. Prognostic role of circulating tumor cells and disseminated tumor cells in patients with prostate cancer: A systematic review and meta-analysis. *Tumor Biol.* 2014;35:5551–5560.
7. Gogoi P, Sepehri S, Zhou Y, et al. Development of an automated and sensitive microfluidic device for capturing and characterizing circulating tumor cells (CTCs) from clinical blood samples. *PLoS One.* 2016;11:1–12.
8. Gwak H, Kim J, Kashefi-Kheyrabadi L, et al. Progress in circulating tumor cell research using microfluidic devices. *Micromachines.* 2018;9:1–21.
9. Li P, Stratton ZS, Dao M, et al. Probing circulating tumor cells in microfluidics. *Lab Chip.* 2013;13:602–609.
10. Zhang J, Chen K, Fan ZH. *Circulating Tumor Cell Isolation and Analysis.* 1st ed. Adv. Clin. Chem. Elsevier Inc.; 2016.
11. Gabriel MT, Calleja LR, Chalopin A, et al. Circulating tumor cells: A review of non-EpCAM-based approaches for cell enrichment and isolation. *Clin Chem.* 2016;62:571–581.
12. Shen Z, Wu A, Chen X. Current detection technologies for circulating tumor cells. *Chem Soc Rev.* 2017;46:2038–2056.
13. Cho H, Kim J, Song H, et al. Microfluidic technologies for circulating tumor cell isolation. *Analyst.* 2018;143:2936–2970.
14. Vona G, Sabile A, Louha M, et al. Isolation by size of epithelial tumor cells: A new method for the immunomorphological and molecular characterization of circulating tumor cells. *Am J Pathol.* 2000;156:57–63.
15. Müller V, Stahmann N, Riethdorf S, et al. Circulating tumor cells in breast cancer: Correlation to bone marrow micrometastases, heterogeneous response to systemic therapy and low proliferative activity. *Clin Cancer Res.* 2005;11:3678–3685.
16. Habli Z, Alchamaa W, Saab R, et al. Circulating tumor cell detection technologies and clinical utility : Challenges and opportunities. *Cancers.* 2020;12:1–30.
17. Myung JH, Hong S. Microfluidic devices to enrich and isolate circulating tumor cells. *Lab Chip.* 2015;15:4500–4511.
18. Dong Y, Skelley AM, Merdek KD, et al. Microfluidics and circulating tumor cells. *J Mol Diagnostics.* 2013;15:149–157.
19. Chinen AB, Guan CM, Ferrer JR, et al. Nanoparticle probes for the detection of cancer biomarkers, cells, and tissues by fluorescence. *Chem Rev.* 2015;115:10530–10574.
20. Shi W, Wang S, Maarouf A, et al. Magnetic particles assisted capture and release of rare circulating tumor cells using wavy-herringbone structured microfluidic devices. *Lab Chip.* 2017;17:3291–3299.
21. Kwon KW, Choi SS, Lee SH, et al. Label-free, microfluidic separation and enrichment of human breast cancer cells by adhesion difference. *Lab Chip.* 2007;7:1461–1468.

22. Wang C, Ye M, Cheng L, et al. Simultaneous isolation and detection of circulating tumor cells with a microfluidic silicon-nanowire-array integrated with magnetic upconversion nanoprobes. *Biomaterials* 2015;54:55–62.
23. Liang W, Liu J, Yang X, et al. Microfluidic-based cancer cell separation using active and passive mechanisms. *Microfluid Nanofluidics.* 2020;24:1–19.
24. Stemme G, Kittilsland G. New fluid filter structure in silicon fabricated using a self-aligning technique. *Appl Phys Lett.* 1988;53:1566–1568.
25. Zheng S, Lin HK, Lu B, et al. 3D microfilter device for viable circulating tumor cell (CTC) enrichment from blood. *Biomed Microdevices.* 2011;13:203–213.
26. Kim TH, Lim M, Park J, et al. FAST: Size-selective, clog-free isolation of rare cancer cells from whole blood at a liquid-liquid interface. *Anal Chem.* 2017;89:1155–1162.
27. Segre G, Silberberg A. Radial Particle displacements in Poiseuille flow of suspensions. *Nature.* 1961;189:209–210.
28. Lee MG, Shin JH, Bae CY, et al. Label-free cancer cell separation from human whole blood using inertial microfluidics at low shear stress. *Anal Chem.* 2013;85: 6213–6218.
29. Bhagat AAS, Hou HW, Li LD, et al. Pinched flow coupled shear-modulated inertial microfluidics for high-throughput rare blood cell separation. *Lab Chip.* 2011;11:1870–1878.
30. Jayaprakash KS, Banerjee U, Sen AK. Dynamics of aqueous droplets at the interface of coflowing immiscible oils in a microchannel. *Langmuir.* 2016;32:2136–2143.
31. Jayaprakash KS, Sen AK. Droplet encapsulation of particles in different regimes and sorting of particle-encapsulating-droplets from empty droplets. *Biomicrofluidics.* 2019;13:034108.
32. Hazra S, Jayaprakash KS, Pandian K, et al. Non-inertial lift induced migration for label-free sorting of cells in a co-flowing aqueous two-phase system. *Analyst.* 2019;144:2574–2583.
33. Demircan Y, Özgür E, Külah H. Dielectrophoresis: Applications and future outlook in point of care. *Electrophoresis.* 2013;34:1008–1027.
34. Becker FF, Wang XB, Huang Y, et al. Separation of human breast cancer cells from blood by differential dielectric affinity. *Proc Natl Acad Sci U S A.* 1995;92:860–864.
35. Gascoyne PRC, Noshari J, Anderson TJ, et al. Isolation of rare cells from cell mixtures by dielectrophoresis. *Electrophoresis.* 2009;30:1388–1398.
36. Chiu TK, Chou WP, Huang S Bin, et al. Application of optically-induced-dielectrophoresis in microfluidic system for purification of circulating tumour cells for gene expression analysis-Cancer cell line model. *Sci Rep.* 2016;6:1–14.
37. Hu X, Bessette PH, Qian J, et al. Marker-specific sorting of rare cells using dielectrophoresis. *Proc Natl Acad Sci U S A.* 2005;102:15757–15761.
38. Moon HS, Kwon K, Kim S Il, et al. Continuous separation of breast cancer cells from blood samples using multi-orifice flow fractionation (MOFF) and dielectrophoresis (DEP). *Lab Chip.* 2011;11:1118–1125.
39. Vykoukal J, Vykoukal DM, Freyberg S, et al. Enrichment of putative stem cells from adipose tissue using dielectrophoretic field-flow fractionation. *Lab Chip.* 2008;8:1386–1393.
40. Yang J, Huang Y, Wang XB, et al. Differential analysis of human leukocytes by dielectrophoretic field- flow-fractionation. *Biophys J.* 2000;78:2680–2689.
41. Suzuki T, Kaji N, Yasaki H, et al. Mechanical low-pass filtering of cells for detection of circulating tumor cells in whole blood. *Anal Chem.* 2020;92:2483–2491.
42. Karthick S, Pradeep PN, Kanchana P, et al. Acoustic impedance-based size-independent isolation of circulating tumour cells from blood using acoustophoresis. *Lab Chip.* 2018;18:3802–3813.

43. Augustsson P, Magnusson C, Nordin M, et al. Microfluidic, label-free enrichment of prostate cancer cells in blood based on acoustophoresis. *Anal Chem.* 2012;84:7954–7962.
44. Antfolk M, Magnusson C, Augustsson P, et al. Acoustofluidic, label-free separation and simultaneous concentration of rare tumor cells from white blood cells. *Anal Chem.* 2015;87:9322–9328.
45. Dao M, Suresh S, Huang TJ, et al. Acoustic separation of circulating tumor cells. *Proc Natl Acad Sci U S A.* 2015;112:4970–4975.
46. Ma Z, Zhou Y, Collins DJ, et al. Fluorescence activated cell sorting: Via a focused traveling surface acoustic beam. *Lab Chip.* 2017;17:3176–3185.
47. Lu X, Martin A, Soto F, et al. Parallel label-free isolation of cancer cells using arrays of acoustic microstreaming traps. *Adv Mater Technol.* 2019;4:1–9.
48. Pamme N. Magnetism and microfluidics. *Lab Chip.* 2006;6:24–38.
49. Nguyen NT. Micro-magnetofluidics: Interactions between magnetism and fluid flow on the microscale. *Microfluid Nanofluid.* 2012;12:1–16.
50. Talasaz AH, Powell AA, Huber DE, et al. Isolating highly enriched populations of circulating epithelial cells and other rare cells from blood using a magnetic sweeper device. *Proc Natl Acad Sci U S A.* 2009;106:3970–3975.
51. Kang JH, Krause S, Tobin H, et al. A combined micromagnetic-microfluidic device for rapid capture and culture of rare circulating tumor cells. *Lab Chip.* 2012;12:2175–2181.
52. Weissenstein U, Schumann A, Reif M, et al. Detection of circulating tumor cells in blood of metastatic breast cancer patients using a combination of cytokeratin and EpCAM antibodies. *BMC Cancer.* 2012;12.
53. Hyun KA, Lee TY, Lee SH, et al. Two-stage microfluidic chip for selective isolation of circulating tumor cells (CTCs). *Biosens Bioelectron.* 2015;67:86–92.
54. Zhao W, Zhu T, Cheng R, et al. Label-free and continuous-flow ferrohydrodynamic separation of HeLa cells and blood cells in biocompatible ferrofluids. *Adv Funct Mater.* 2016;26:3990–3998.
55. Ozkumur E, Shah AM, Ciciliano JC, et al. Inertial focusing for tumor antigen-dependent and -independent sorting of rare circulating tumor cells. *Sci Transl Med.* 2013;5:1–20.
56. Mizuno M, Yamada M, Mitamura R, et al. Magnetophoresis-integrated hydrodynamic filtration system for size-and surface marker-based two-dimensional cell sorting. *Anal Chem.* 2013;85:7666–7673.
57. Sajay BNG, Chang CP, Ahmad H, et al. Microfluidic platform for negative enrichment of circulating tumor cells. *Biomed Microdevices.* 2014;16:537–548.
58. Zhou Y, Ma Z, Ai Y. Hybrid microfluidic sorting of rare cells based on high throughput inertial focusing and high accuracy acoustic manipulation. *RSC Adv.* 2019;9:31186–31195.
59. Mutafopulos K, Spink P, Lofstrom CD, et al. Traveling surface acoustic wave (TSAW) microfluidic fluorescence activated cell sorter (μFACS). *Lab Chip.* 2019;19:2435–2443.
60. Gupta V, Jafferji I, Garza M, et al. ApoStream™, a new dielectrophoretic device for antibody independent isolation and recovery of viable cancer cells from blood. *Biomicrofluidics.* 2012;6:1–14.
61. Shim S, Stemke-Hale K, Tsimberidou AM, et al. Antibody-independent isolation of circulating tumor cells by continuous-flow dielectrophoresis. *Biomicrofluidics.* 2013;7:1–12.
62. Del Giudice F, Madadi H, Villone MM, et al. Magnetophoresis "meets" viscoelasticity: Deterministic separation of magnetic particles in a modular microfluidic device. *Lab Chip.* 2015;15:1912–1922.
63. Chang S, Cho YH. Continuous blood cell separation using a dielectrophoretic virtual pillar array. In 3rd IEEE International Conference on Nano/Micro Engineered and Molecular Systems (NEMS), Sanya, China. 2008;974–977.

64. Kim U, Soh HT. Simultaneous sorting of multiple bacterial targets using integrated dielectrophoretic-magnetic activated cell sorter. *Lab Chip.* 2009;9:2313–2318.
65. Wang W, Cui H, Zhang P, et al. Efficient capture of cancer cells by their replicated surfaces reveals multiscale topographic interactions coupled with molecular recognition. *ACS Appl Mater Interfaces.* 2017;9:10537–10543.

6 3D-Printed Microfluidic Device with Integrated Biosensors for Biomedical Applications

Priyanka Prabhakar, Raj Kumar Sen, Neeraj Dwivedi, Raju Khan, Pratima R. Solanki, Satanand Mishra, Avanish Kumar Srivastava, and Chetna Dhand

CONTENTS

DOI: 10.1201/9781003033479-6

6.1 INTRODUCTION

Microfluidics is the science and technology which is used for the manipulation of a small amount of fluid (10^{-9} to 10^{-18} L) in channels with a dimension of 10 to 100 μm. The two prominent characteristics of microfluidics are the small size and less obvious characteristics of fluids in microchannels, such as laminar flow. It provides radically new capabilities in regulating molecular concentration in space and time (Bragheri et al. 2016, Slapar and Poberaj 2008, Tarn et al. 2014).

6.1.1 History of Microfluidics

Whitesides describes the four parents of microfluidics, they are molecular analysis, microelectronics, national security, and molecular biology. According to Whitesides, **molecular analysis** is the oldest (first) parent of microfluidics, this involves methods such as chromatography in the gas phase (GPC) or capillary electrophoresis. The techniques were developed in the 1950s–1960s and used for the separation of chemical compounds or biomolecules by flowing small amounts of samples in narrow tubes or capillaries, reaching high sensitivity and resolution. **Microelectronics** is the most famous parent of microfluidics. In the beginning, researchers tried to directly fit fabrication methods and materials from microelectronics to microfluidics: Photolithography, as well as silicon and glass, were the first players in the microfluidics stages. Microfluidics later separated from microelectronics and semiconductor technology, utilizing more different methods and materials for microfabrication. **Molecular biology** is the fourth parent of microfluidics, in the 1980s molecular biology contributed to the birth of microfluidics. Early the 1980s Kary Mullis developed the polymerase chain reaction technique, which is used to amplify a DNA sequence utilizing heat, in this reaction a very small amount of liquid is needed (nearly 10–200 μL), the precise result was not obtained due to the small amount used, because it was manually operated the reaction was time consuming. The first commercial machine, a simple thermal cycler, was developed in 1987; it rendered the process reliable, and its small dimensions gave the possibility to miniaturize operations as well as to work outside of the lab. In 1994 the Defense Advanced Research Projects Agency of the USA contributed to the growth of microelectromechanical systems (MEMS) and the development of miniaturized and portable "laboratories on a chip" (Castillo-León et al. 2015, Whitesides 2006, Convery et al. 2016).

Microfluidics normally constitutes the basis of integrated devices known as miniaturized total analysis (μTAS) or lab-on-a-chip (LOC) technologies. The attractive properties of these technologies are that large devices are converted into small-size devices and become more portable, have low power consumption, and short assay time. Due to these attractive properties microfluidics devices are used in so many applications in diverse fields, such as chemistry, biology, medicine, and physical science (Lei 2018). Early silicon material is used for fabricating the substrate of microfluidics devices. Microfluidics devices need to fabricate high-aspect-ratio microstructures and bond multi-substrates. The substrate of silicon is relatively cheap, and it is not optically transparent. Due to this property, silicon is used in the limited application. Microfluidics technology became specific in the 2000s when glass and polymer materials were introduced. Compared with a silicon substrate, glass and polymer materials are less expensive and optically transparent. Polymer materials such as polymethylmethacrylate (PMMA), polystyrene (PS), polycarbonate (PC), and polydimethylsiloxane (PDMS) were used to demonstrate the fabrication of microfluidics devices Among these, PDMS is one of the most commonly used materials for fabricating microfluidics devices in current research laboratories. For the fabrication of microfluidics devices, a cleanroom environment is needed. There are different techniques used for the fabrication of microfluidics devices, (Wu et al. 2003, Xia and Whitesides 1998, Klank 2002, Chen et al. 2008, Wabuyele 2001). The techniques are micromachining, soft lithography, embossing, in situ construction infection molding, and laser ablation. They are used for large-scale production and reproduction. But some of these techniques required more space for their equipment, were labor intensive, and there was time wastage to facilitate the change in design and limited biological materials. Quick and easy manufacturing techniques are needed for the fabrication of microfluidics devices, and these are preferred. This drawback can be overcome by the 3D printing technique.

6.2 3D PRINTING

In 1981 Hideo Kodama of Nayoga Municipal Industrial Research Institute first studied and published the manufacture of a printed solid model, the starting point for "additive manufacturing," "rapid prototyping," or "3D-printing technology." The first 3D printer was designed and realized by Charles W. Hull from 3D Systems Corp in 1984 (Pîrjan et al. 2021, Matias et al. 2015, Savini et al. 2015). 3D printing is also referred to as additive manufacturing. This term precisely describes how this technology works to create 3D solid objects. "Additives" refers to the successive addition of a thin layer of between 16 and 180 μm or more to create an object. All 3D-printing technologies are similar, as they build objects layer-by-layer to create complex shapes. 3D printing helps industries to revolutionize and increase production lines. The speed of production increases with this technology while cutting costs. Currently, this technology is widely used all over the world, 3D-printing technology is widely used in the automotive industry, healthcare sector, medical field, aerospace industry, and agricultural sector for the excessive production of various customized designs (https://www.sculpteo.com/en/3d-printing/3d-printing-technologies, https://www.autodesk.in/solutions/3d-printing, https://www.3dhubs.com/guides/3d-printing/).

6.2.1 Working of 3D Printer

Three steps are required to create a 3D object. Firstly, we must create a 3D object file that we want to print. The 3D file can be created by using CAD software or a 3D scanner. Once the file is ready then we can proceed to the next step. In the next step, we process for printing. To print the object, we must choose the materials which are suitable for the specific properties required for our object

(Figure 6.1(A)). There are variety of materials used in 3D printing. They are plastics, ceramics, resins, metal, sand, textiles, biomaterials, glass, food, etc. In the final step, the final object is printed, but it cannot be directly used or delivered until it has been sanded, lacquered, or painted to complete it as intended (Awad et al. 2018, Mpofu et al. 2014).

6.2.2 3D-Printing Techniques

The techniques used for 3D printing are based on the material types. Selective laser melting (SLM), selective laser sintering (SLS), and fused deposition modeling (FDM) are used when the materials are soft or melt to produce the layer. For liquid material stereolithography (SLA), laminated object manufacturing (LOM), digital light processing (DLP), electronic beam melting (EBM) are used. Different types of materials are used, like metals, polymers, ceramics, composites, smart materials, and special materials (food, lumar dust, textile) (Vinod et al. 2017, Bakhtiar et al. 2018). These techniques do not need a cleanroom environment for the fabrication of microfluidics devices. The 3D-printing technology has numerous existing and prospective applications in various domains like the medical, aerospace, fashion, entertainment, automotive, jewelry, and education industries, etc. (Liu et al. 2017, Wang et al. 2018).

6.2.2.1 Role of 3D Printing in the Fabrication of Microfluidics Devices

To streamline microfluidics through a higher number of laboratories, a novel manufacturing method needs to evolve. 3D printing, a prototype technique that has recently emerged as an alternative microfluidics manufacturing method has shown potential to solve many of the PDMS product problems. 3D printing easily fabricates any microfluidics device in a single step, compared to the multistep and labor-intensive soft lithography process. Also, device features can easily be adjusted in each print by changing the design in the CAD software. It can be applied for various manufacturing technologies but printing microfluidics devices are not suitable for all. 3D printing can be categorized according to the application of microfluidics, viz., extrusion-based technology (e.g., FDM), liquid resin-based technology (e.g., SLA, DLP, and two-photon polymerization [2PP]) which also includes inkjet-based 3D printing (e.g., material jetting) due to the similar curing mechanism, powder-based technology (e.g., multi-jet fusion [MJF], SLS, SLM, and EBM), and other less common 3D-printing technologies. 3D-printing methods are versatile because they use polymeric materials such as PVA, PET, nylon, wood-PLA, and thermoplastic elastomer (TPE). 3D printing fabricates microfluidics devices by using the direct or indirect method. Direct 3D printing uses microchannels and other microfluidics components with the ink materials to develop a microfluidic chip, whereas in indirect 3D printing the chip is fabricated by casting PMDS against the mold made by link materials. The final microfluidics chip does not include any ink materials (Chen et al. 2016, Amin et al. 2016, Weisgrab et al. 2019, Zhang 2019).

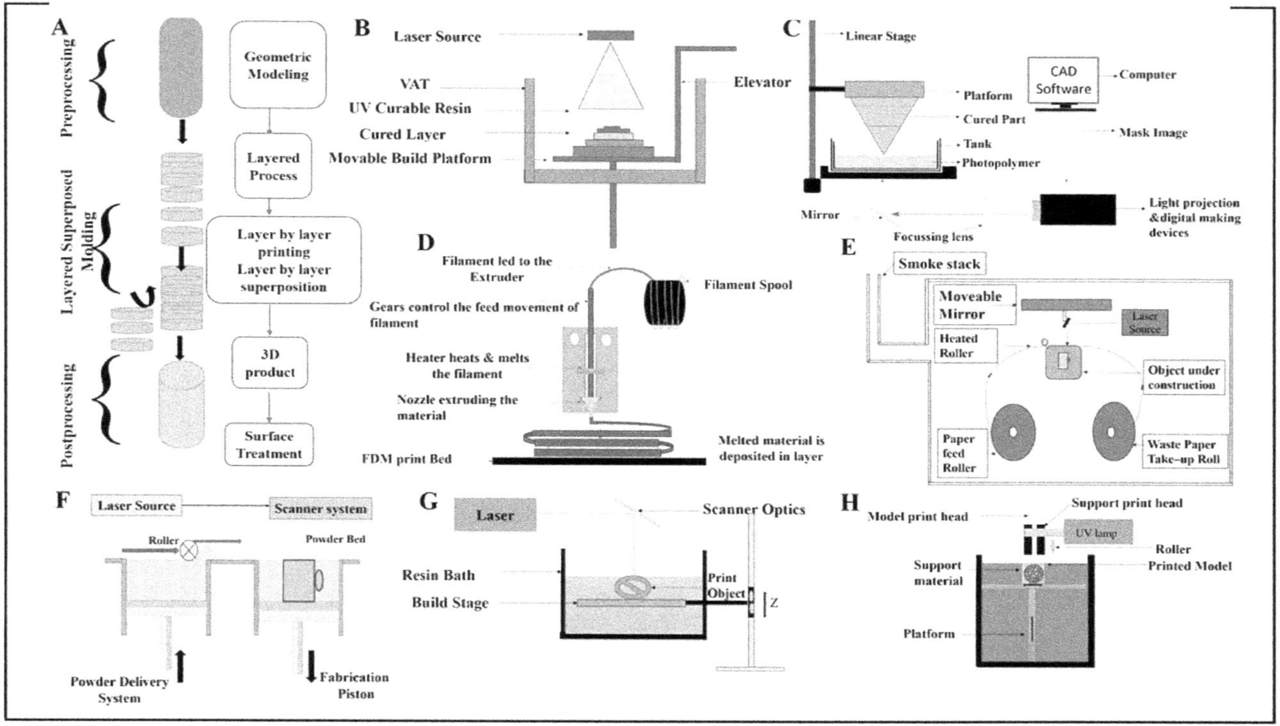

FIGURE 6.1 (A) Block diagram for working of 3D printer. Schematic showing the setup for, (B) stereolithography, and (C) digital light processing technique, (D) fused deposition modeling, (E) laminated object manufacturing, (F) selective laser sintering, (G) direct laser writing, and (H) PolyJet 3D printing.

6.3 3D TECHNOLOGIES

6.3.1 Stereolithography (SLA)

SLA is an early and commonly used 3D-printing technique. In 1986 Chuck Hull first introduced this technique and also the first commercialized printing technology. SLA 3D printers use a UV laser or similar light source to introduce polymerization in a photosensitive resin (Figure 6.1(B)). UV light, generated and patterned using a UV laser design or polymer, is exposed to light where the UV laser draws a cross-section layer by layer. The materials used in SLA are photosensitive thermoset polymers that come in a liquid form. This process is repeated until the final structure has been created. By controlling the position of the UV laser, polymerization of the resin can be controlled to achieve the desired structure and design. The major advantage of the SLA technique is the high precision on the surface resolution. High-resolution products are produced by the SLA system while keeping the cost low due to the relatively low usage of the liquid medium (Melchels et al. 2010).

6.3.2 Digital Light Processing (DLP)

DLP is a type of vat polymerization. These techniques use photopolymer resin and it can cure the resin under the light source. This technique is somewhat similar to the SLA process, except that a safelight (light bulb) is used instead of using a UV laser to cure the photopolymer resin; both are pulled out of the resin, which creates space for the uncured resin at the bottom of the container, thus forming the next layer of the object, down into the tank with the next layer being cured at the top (Figure 6.1(C)). Objects that are printed by DLP have fewer visible layers compared to other 3D-printing techniques such as FDM. Due to the single-layer objects created in one singular digital image, the speed of creating an object is faster compared to the SLA technique. As with SLA, DLP is often used to produce extremely detailed artwork and non-functional prototypes and can be used to create molds for investment casting applications (https://all3dp.com/2/what-is-a-dlp-3d-printer-3d-printing-simply-explain).

6.3.3 Fused Deposition Modeling (FDM)

FDM is the extrusion-based 3D-printing technique. FDM is the most popular printing method among all the printing methods because it is the cheapest, easiest mode of printing and the number of printers available on the market. During this process, the materials are melted and extruded through a 3D nozzle to print a cross-section of each layer of an object at any time, by layer deposition of semi-fused thermoplastic filament on a printing platform (Figure 6.1(D)). The bed lowers for each layer, and this process repeats until the object is finished. The thickness of the layers defines the consistency of the 3D-printing operation. Most FDM 3D printers have two or more print heads available for printing in various colors and enable overhanging areas of complex 3D printing. The most common thermoplastic filaments are polylactic acid (PLA) and acrylonitrile butadiene styrene (ABS) (Hagedorn 2017).

6.3.4 Laminated Object Manufacturing (LOM)

LOM is 3D-printing technology developed by Helisys Inc. (now Cubic Technologies). For this process, a laser cutter is used to render layers of adhesive-coated paper, plastic, or metal laminates successively bonded and cut to the appropriate shape. Objects printed with this technique can be further modified by machining after printing (Figure 6.1(E)). The typical layer resolution for this process is defined by materials feedstock and usually ranges in thickness from one to many sheets of copy paper (https://www.techpats.com/3d-printing-technologies-overview/).

6.3.5 Selective Laser Sintering (SLS)

SLS is the 3D-printing technique in which a laser is used as the power source to sinter powdered materials (mostly metal), aiming the laser at the point in space defined by a 3D model, binding the materials to create a solid structure (https://www.3dhubs.com/knowledge-base/introduction-sls-3d-printing/). A block diagram is shown in Figure 6.1(F).

6.3.6 Selective Laser Melting (SLM)

This technique is the same as the SLS technique but in SLM the materials are fully melted then sintered, allowing different properties (crystal structure, porosity) (http://3dprintingfromscratch.com/common/types-of-3d-printers-or-3d-printing technologies-overview/).

6.3.7 Direct Laser Writing (DLW)

DLW is a form of laser-based 3D printing in which a single focal point of focused laser light is illuminated on a surface or within a volume of photopolymerizable materials; a digitally operated motorized stage and/or mirror galvanometer can then trace this illuminated focal point in 3D space to create a 3D structure (Prabhakar et al. 2021, Hanada et al. 2011, Hwang et al. 2017). A schematic diagram of direct laser writing is shown in Figure 6.1(G).

6.3.8 PolyJet Process

PolyJet printing is similar to inkjet printing. Thousands of photopolymer droplets are jetted onto a built-in substrate and solidified with a UV light source in this 3D-printing technique, as shown in Figure 6.1(H). The printer is made up of several printing heads and movable platforms. Each printing head is filled with a different type of liquid resin (rubber or rigid, transparent or opaque materials). This method has the potential to produce complex multi-material objects with smooth surface texture and high accuracy. Using this technology, high-resolution objects of varying modular strengths can be printed in three dimensions with high dimensional precision. Because of these advantages, the PolyJet process is widely used in a variety of biomedical applications. (Tappa and Jammalamadaka 2018, Matter-Parrat and Liverneaux 2019).

6.3.9 Multi Jet Fusion (MJF)

MJF is a powder-based technology; in this process a fusing agent is applied on a material layer where the particles are destined to fuse. Then a detailing agent is applied to modify fusing and create fine detail and smooth surfaces. To finish, the area is exposed to the energy that will lead to reactions between the agents and the material to create the part. It is the fastest plastic 3D-printing technology, perfect for prototyping or manufacturing. When the printing process is complete, the build box removes the remaining powder thanks to brushes and air blowers. The MJF manufacturing technique is particularly useful to create unique plastic parts with a good surface finish; very complex shapes can be created in a very short amount of time (https://www.techpats.com/3d-printing-technologies-overview/).

6.4 ADVANTAGEOUS FEATURES OF 3D-PRINTED MICROFLUIDICS DEVICES

3D printing is a very efficient method to fabricate structurally robust microfluidics devices. 3D printing enables the fabrication of unique but advantageous features integrated on the microfluidics device. Generally, the integrated component of microfluidics devices has been created with PDMS devices, it is time consuming and costly to integrate multiple functional units on one device. It can produce any shape, complex, and functional 3D-printing microfluidics features based on users' requirements.

1. Provides strong connection ports. A microfluidics device functionality can be improved by integrating chips with other devices/instrumentation with fluidic tubing.
2. Complex flow regulating components. 3D-printed microfluidics devices can flow any complex fluidics with the help of 3D-printed valves and pumps, a common technique for fluid manipulation.
3. Integration of detectors. In a 3D-printed microfluidics device electrodes can be integrated and removed easily by using connection mechanisms such as threads.
4. On-chip cell integration. When an insert cell is contaminated or the cells are not cultured correctly, the insert can be replaced rather than causing a failure of the entire device.
5. Less time consuming and cost effective. 3D-printed microfluidics devices can be carried in the laboratory, which can save a lot of outsourcing processing time and therefore costs associated with using different machines for fabrication (Chen et al. 2016, He et al. 2016).

6.5 BIOSENSOR

A biosensor is an analytical device which is used to convert a biological response into an electrical signal. It has three important functional components; the first part of the biosensor is a biological component that detects analytes and generates a responsible signal. The biological reaction generates a signal which is converted into a detectable response by the second component, called a transducer. The transducer is the most critical component in any biosensing element, the detector is the third part of the biosensor which is used to amplify and process the signal before displaying it on an electronic display. Based on the transducer, the biosensor is classified into three major groups, i.e., electrochemical, optical, and others. When testing a sample the electrochemical biosensor is triggered due to the interaction between the biological element and analyte, the output comes in the form of current, potential, conductivity, or capacitance, Optical properties change due to the interaction between the biocatalyst and the analyte; the biosensors are categorized as bioluminescence/chemiluminescence biosensors, fluorescence biosensors, or colorimetric biosensors (change in UV/visible absorption). Biosensors have a wide range of applications in diverse areas ranging from clinical through environmental and agricultural (Mehrotra 2016, Bhalla et al. 2016).

6.6 HOW 3D-PRINTED MICROFLUIDICS DEVICES INTEGRATE WITH BIOSENSORS

In biosensors, three electrodes are required, i.e., a working electrode, a counter (or auxiliary) electrode, and a reference electrode. For the working of biosensors, the biological component/bioreceptor element is needed which is in direct contact with the electrode to get an analytically useful signal by the coupling of biochemical and electrochemical interactions. Biosensors are used for the measurement of a small amount of different bodily fluids such as interstitial fluid, saliva, tears, sweat, etc. Electrodes have been integrated or incorporated with traditional-based, polymer-based, or glass-based microfluidics devices. Microfluidics devices provide a high time resolution of low-volume flow rates that require very small internal volume. Generally, PDMS or PDMS-glass hybrid materials are used for the fabrication of microfluidics devices. PDMS-based soft lithography is the most commonly used in microfluidics because of its ability to build very small channels, their low cost, and ease of manufacture. Accordingly, PDMS-based microfluidics devices were widely used in proof-of-concept studies for a wide variety of applications. PDMS-based microfluidics devices have some drawbacks that can cause irregular flow, leakage, or the trapping of air bubbles and bending under pressure. Due to these disadvantages of PDMS-based microfluidic devices, it is difficult to build a safe and precise connection between the other components, so they are unable to be used in biomedical applications (biosensors). Nowadays microfluidics devices are fabricated by the 3D-printing (additive manufacture) method as it has many advantages over

PDMS-based microfluidics devices (Sanati et al. 2019). Various 3D-printing techniques are used to design the microfluidics devices for the fabrication of biosensors. In this chapter, we focused on and discussed the various 3D-printed microfluidic-based miniaturized devices and their implication in biosensors for biomedical application. Figure 6.2(A) depicts the evolution of 3D-printed microfluidics systems for a variety of biomedical applications from 2013 to date. Figure 6.2(B) demonstrates how the total number of research papers in the field of 3D-printed microfluidics has gradually risen over time.

On this journey, Samper et al. fabricated a 3D-printed microfluidics system that can easily be combined with an electrochemical biosensor for the detection of biomarkers like glucose, lactate, and glutamate (Samper et al. 2019). The time resolution of devices is characterized by recording short lactate concentration pulses. The system is used to monitor simultaneous changes in glutamate, glucose, and lactate concentrations in the cerebrospinal fluid dialysate, simulating the physiological response to spreading depolarization events. The device is also used in the ICU to remotely track a patient's brain injury, demonstrating its clinical monitoring ability.

Gowers et al. first created a 3D-printed wearable microfluidics system with FDA-approved clinical microdialysis probes and needle-type biosensors for continuous monitoring of human tissue metabolite levels (Gowers et al. 2015) (Figure 6.3). The authors demonstrated the effectiveness of this 3D-printed microfluidics system as a wearable device for monitoring tissue glucose and lactate levels in cyclists during exercise. The clear changes in glucose and lactate levels suggest that this device has enormous potential for real-time monitoring and assessing athlete training effectiveness.

Erkal et al. used Object Connex 350 multi-material printers to create two 3D-printed devices for electrochemical detection of nitric oxide (NO) and dopamine using acrylate-based polymer material (Erkal et al. 2014). The electrode is housed in commercially available polymer-based fittings in both printed devices so that the different electrode materials (platinum, platinum black, carbon, gold, and silver) can be conveniently applied to the device's threaded receiving port. This provides a module-like style to the experimental design, where the electrodes can be removed, repolished, and reused after biological sample treatment. The first printed device represents a microfluidics platform with a 500 × 500 μm channel and a threaded receiving port to allow integration of either polyetheretherketone (PEEK) nut-encased glassy carbon or platinum black (Pt-black) electrodes for dopamine and NO detection, respectively. For NO gas, the embedded Pt/Pt-black electrode was stated to have a limit of detection of 1 M and a wide linearity range of 7.6–190 μM, whereas, for dopamine, the system showed a detection range of 25–500 μM and a LOD of 500 nM. Kadimisetty et al. fabricated compact, inexpensive, highly sensitive, rapid detection devices and worked with small volume 3D-printed electrochemiluminescent (ECL) immunoassays incorporated with pyrolytic graphite sheet microwell chip nanostructures configured for detection of two protein prostate-specific membrane antigen (PSMA) and prostate-specific antigen (PSA) simultaneously (Kadimisetty et al. 2018). The microfluidic array was fabricated by a stereolithographic 3D-printing technique. A programmable syringe pump is used with this setup to precisely control the reagent flow

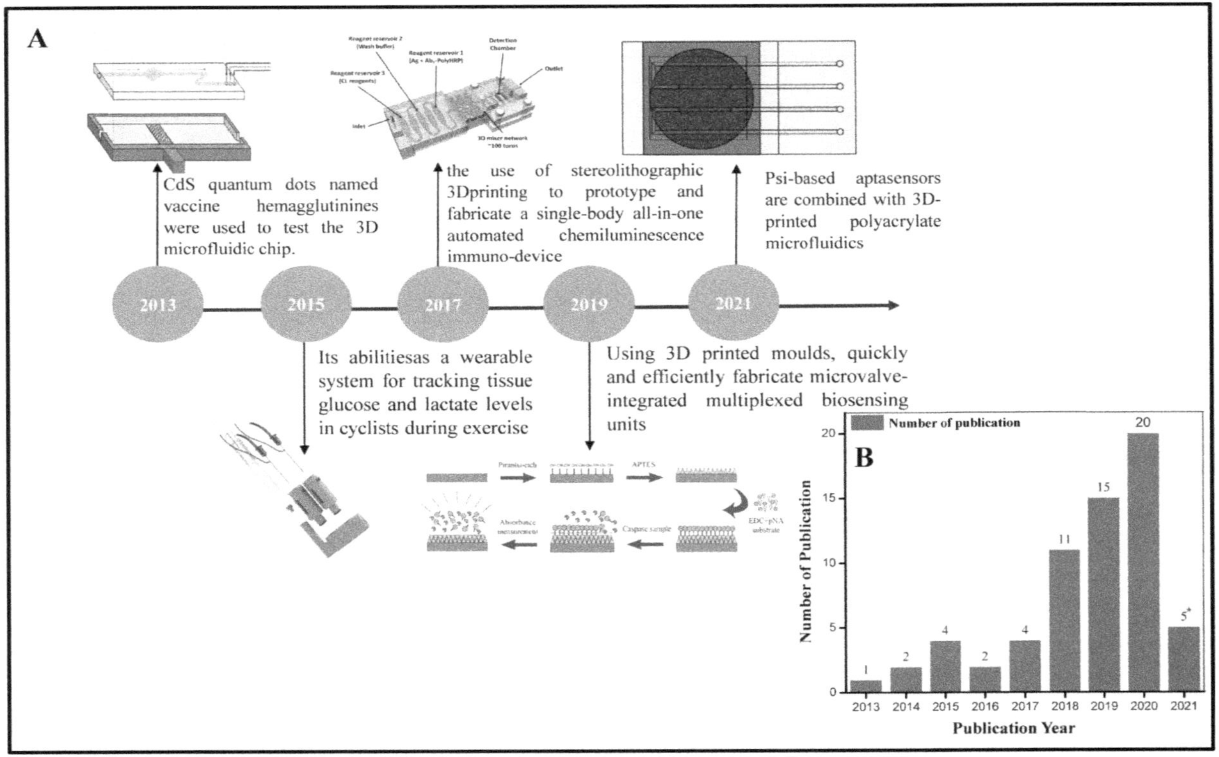

FIGURE 6.2 (A) Timeline for producing 3D-printing microfluidics-based biosensor in 2013 to 2021. Here, we emphasize some examples of 3D-printing microfluidics devices for biomedical applications. (B) The number of articles published on microfluidics and 3D printing annually. Data were obtained from "PubMed" with "3D Printed Microfluidic Biosensors" entered as "subject" in the search box (date: March 16, 2021) (the figure is reproduced with permission of Krejcova et al. 2014, Gowers et al. 2015, Tang et al. 2017, Dang et al. 2019, Arshavsky-Graham et al. 2021).

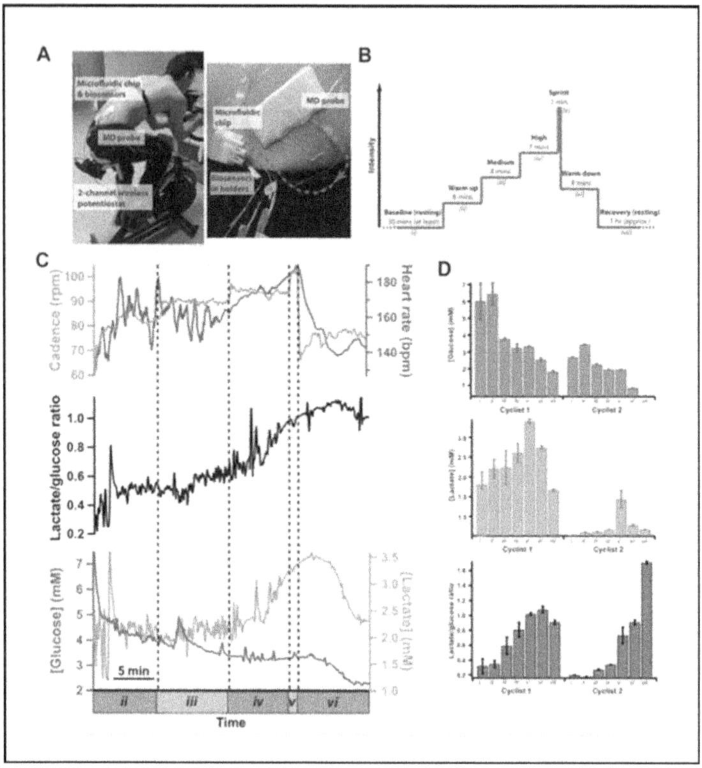

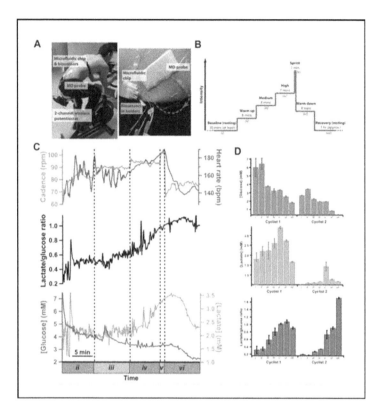

FIGURE 6.3 (A) Picture of a 3D-printed microfluidic system to measure tissue and lactate levels in dialysate during the cycling protocol. (B) Experimental setup. (C) Dialysate glucose and lactate levels were measured during the cycling protocol system's exercise phase. (D) For two separate cyclists, histograms depicting mean dialysate amounts (reproduced with permission from Gowers et al. 2015).

to facilitate incubation and washing without users. Nanostructured surfaces feature antibody-decorated single-wall carbon nanotube forests on pyrolytic graphite sheet (PGS) chip microwells, and sensitivity is increased by massively labeled RuBPY-silica nanoparticles for detection. The assay was cost efficient and fast, with low detection limits and dynamic ranges that could be customized for clinical research. Krejcova et al. reported a new 3D bead-based microfluidics chip developed for rapid, sensitive, and specific detection of hemagglutinin influenza (Krejcova et al. 2014). The polylactide material is used to fabricate a microfluidics chip by the extrusion process of a continuous molten filament at a temperature of 210°C It has been shown that the electrochemical analysis of isolated hemagglutinin using CdS quantum dots is very effective. Due to this fact, the Cd signal has great potential to become an alternative option as a rapid, sensitive, and specific detector of the influenza hemagglutinin or influenza virus, and this 3D microfluidic system may be considered as a promising and powerful platform for the rapid diagnosis of influenza antigens and may be extended for diagnosis of other pathogens. Kadimisetty et al. reported a low-cost, responsive, super-capacitor-powered electrochemical (ECL) protein immunoarray developed by the FDM technique (Kadimisetty et al. 2016). Within 35 min the immunosensor is used to identify three biomarker proteins for cancer in the serum. PLA is used for fabricating the microfluidic array and electrode of a sensor fabricated by hand screen printing carbon graphite ink by using an adhesive-backed vinyl mask template. PSA, PF-4, prostate cancer biomarker proteins, and PSMA in serum were captured on the antibody-coated carbon sensors followed by delivery of detection-antibody-coated Ru(bpy)32+ (RuBPY)-doped silica nanoparticles in a sandwich immunoassay. Supercapacitors are used to power the ECL arrays, when 1.5 V was applied the ECL was generated by electrochemical oxidation of both tri-poylamine (TPrA) and RuBPy on the sensor. By using a charged-coupled device (CCD) camera, the generated ECL was captured from the sensor array. By using an inexpensive solar cell, the supercapacitor was easily photo-recharged between assays. This technology will provide responsive on-site cancer diagnostic tests in resource-limited environments, requiring only moderate-level training.

Bishop et al. successfully manufactured a microfluidics device using 3D-printed SLA techniques and integrated it with a biosensor electrode (Bishop et al. 2015). Acrylate-based resin is used for preparing the 3D-printed microfluidics device and a pencil graphite rod is used. By threaded ports, the electrode is integrated into the printed channels. The transmittance of the cured acrylate resin was sufficient to observe the ECL form of tris (2,2′-bipyridyl) dichlororuthenium (II) hexahydrate ($[Ru(bpy)3]^{2+}$) with tri-*n*-propylamine (TPA) co-reactant. Experiments with multilayer polydiallyldimethylammonium (PDDA)/DNA film-coated electrodes and $[Ru(bpy)3]^{2+}$ in solution suggest the utility of this strategy to ascertain DNA damage or oxidation of guanine bases. The combination of clear 3D-printed flow-cells with reusable, interchangeable electrodes opens the door to new sensing and diagnostic applications at a very low cost. For the "downward" sensing of metabolites or excreted biologically active molecules Ragone et al. fabricated a portable and disposable electrochemical sensor. The rapid detection capability of biomarker alkaline phosphatase (ALP) excreted from colon cancer cell lines has been demonstrated (Ragones et al. 2015). This microfluidics chip consists of a biocompatible substrate composed of an electrochemical cell with two gold electrodes as working and counter electrodes and an Ag/AgCl electrode as quasi reference electrode and it was fabricated using the SLA technique. The electroactivity of working electrodes was verified by cyclic voltammetry of a ferrocyanide/ferricyanide redox reaction. Amperometric in vitro detection of the biomarker alkaline phosphate was successfully demonstrated directly in a cell culture plate from three separate colon cancer lines while preserving their biological environment. Cevenini et al. developed a compact standalone toxicity sensor which is integrated with bioluminescent cells into a smartphone-based system (Cevenini et al. 2016). They fabricated 3D-printed cartridges to incorporate a variety of bioluminescent cells into ready-to-use cartridges and demonstrated the feasibility of accurate detection and quantification of BL signals (Figure 6.4). Human embryonic kidney cells (Hek293T) have been constitutively used to express green-emitting luciferase as sentinel cells, and an Android

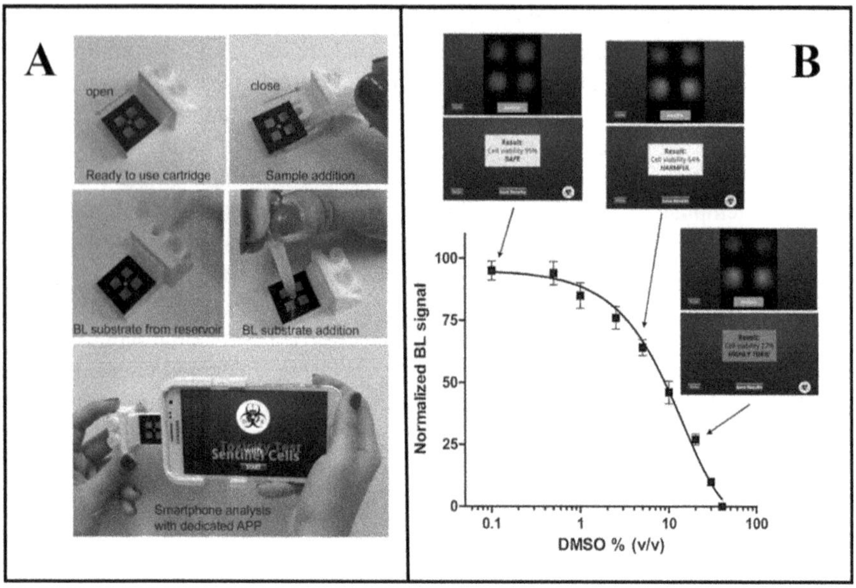

FIGURE 6.4 (A) Picture of 3D-printed smartphone adaptor designed for Samsung Galaxy Note II and easy steps to perform a toxicity test. (B) DMSO toxicity curve obtained with the smartphone and BL image screenshot and the relative result of selected DMSO concentrations corresponding to the three Tox-App warning levels set as "safe" (100–80%), "harmful (79–30%), and "highly toxic" (<30%) (reproduced with permission from Cevenini et al. 2016).

app has been developed to provide a user-friendly environment. The smartphone adapter and mini cartridges were manufactured using 3D-printing technology. Fraser et al. fabricated a microfluidics device with the help of SLA for the diagnosis of malaria (Fraser et al. 2018). They developed a new concept for malaria biosensors in which magnetic beads are coated with aptamers for magnet-guided capture, wash, and detection of the biomarker. The three separate microfluidics chambers are incorporated in the biosensor it was designed to enable such magnet-guided equipment-free colorimetric detection of PfLDH.

Chen et al. proposed a novel 3D-printed microfluidic device functionalized with anti-epithelial cell adhesion molecule (EpCAM) antibodies to separate circulating tumor cells from human blood samples (Chen et al. 2019). These devices were demonstrated to test three kinds of EpCAM-positive cancer cell lines (PC3 prostate cancer, MCF-7 breast cancer, and SW_480 colon cancer) and one type of EpCAM-negative cancer cell line (293T kidney cancer).

Tang et al. used the SLA 3D-printed technique to fabricate a microfluidics device for the diagnosis of cancer biomarker proteins (Tang et al. 2017). The unibody contains (i) an effective 3D network for passive mixing, (ii) three reagent reservoirs, and (iii) an optically transparent detection chamber store in the glass slides and decorated capture antibodies array for observing chemiluminescence output with a CCD camera (Figures 6.5 (A, B)). This low-cost automated 3D-printed microfluidics

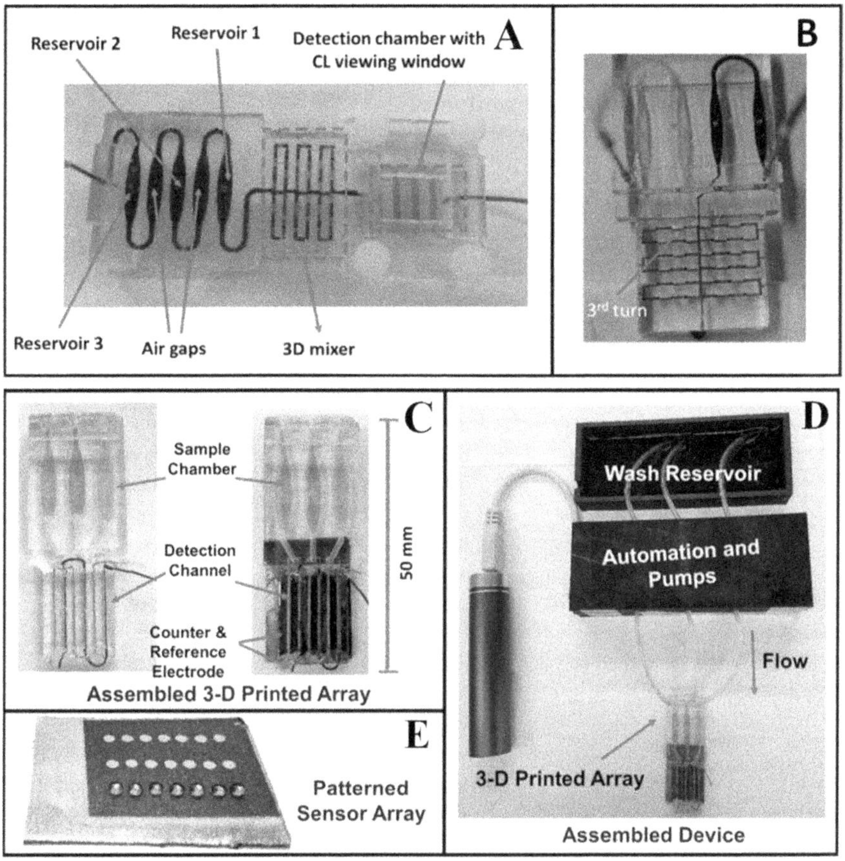

FIGURE 6.5 (A) 3D-printed immunoarray device for unibody CL. (B) Modified 3D-printed system to demonstrate the passive 3D mixing network's capacity. (C) The sample chamber dye solution is seen in a 3D-printed automated genotoxicity screening system without or with a microwell chip and counter electrode wires inserted. (D) The first row of the microwell-patterned pyrolytic graphite detection array contains 1 μL water droplets that are held by the hydrophobic microwell boundaries. A separate sample line feeds each row. Each microwell in the working array includes films of DNA, metabolic enzymes, and RuPVP. (E) Assembled array device with a wash reservoir containing pH 7.4 buffer and electronic microprocessors and micropumps driven by a rechargeable battery and linked to the 3-D printed array below. (From Tang et al. 2107 and Kadimisetty et al. 2017 reproduced with permission.)

device was used for the detection of the cancer biomarker proteins including platelet factor 4(PF-4) and PSA. The sensor device's detection limit is nearly 0.5 pg mL^{-1}, and it has good accuracy for analyzing human serum samples.

Kadimisetty et al. fabricated a 3D-printed microfluidic array developed using the SLA technique for the detection of the genotoxic potential of cigarettes, e-cigarettes, and environmental samples (Kadimisetty et al. 2018) (Figure 6.5(C–E)). Electronic

TABLE 6.1
Summary of 3D-Printing Techniques: Principle, Materials, and 3D-Printed Sensor in the Biomedical Application

Technique	Materials	Application of 3D-Printer Sensor in Biomedical Application	References
FDM	Polylactide and ABS	Biosensor influenza virus, detection of glucose and lactate, smartphone-based toxicity sensor, detection of biomarker cancer	Krejcova et al. (2014), Su et al. (2016), Cevenini et al. (2016), Kadimisetty et al. (2016)
SLA	Pyrolytic graphite sheet, clear resin, acrylate-based resin, clear methacrylate-based resin, acrylate, and epoxy-based mixture and PDMA	Detection of cancer biomarker proteins, detection of genotoxic potential, detection of prostate-specific antigen, detection of biomarker alkaline phosphatase	Tang et al. (2017), Kadimisetty et al. (2017), Bishop et al. (2015), Kadimisetty et al. (2018), Fraser et al. (2018), Ragones et al. (2015)
Inkjet 3D Printing	Casting wax material	Cancer detection	Chen et al. (2019)
PolyJet Process	Acrylate-based polymer	Neurotransmitter detection and measuring oxygen tension in the stream of red blood cells	Erkal et al. (2014)
Ultra 3SP Technique	ABS white	Detection of the level of glucose and lactate, glutamate, glucose, and lactate	Gowers et al. (2015), Samper et al. (2019).

cigarettes are battery-powered devices that vaporize nicotine and were designed as an alternative to tobacco cigarettes, which are extremely harmful to humans.

Su et al. fabricated flow bioreactors using an FDM 3D printer technique and developed a simple method for the functionalization of printed ABS reactors to facilitate the determination of lactate and glucose in a biological sample (Su et al. 2016). Observing the concentration of rat brain extracellular lactate and glucose, this system involved microdialysis (MD) sampling and fluorescence determination in conjunction with a new sample derivatization scheme in which glucose and lactate oxidase was immobilized in ABS flow bioreactors. Table 6.1 summarizes information about the various 3D techniques used in the biosensor for biomedical application.

6.7 CONCLUSION

Printing microfluidics-based biosensors in 3D offers many unique advantages as an evolving material medium for biosensor applications, but there are still problems to be tackled in the future. In this chapter, we discuss the advantages of 3D-printing microfluidic devices for biosensor and biomedical applications. 3D printing is a quick and easy way to create a load-bearing microfluidics system. 3D printing microfluidics enables the development of one-of-a-kind and complex products that are useful to the consumer. All of the features of 3D printing microfluidics can be used to build any shape. This technology has the potential to change not only how researchers collaborate, but also how we interpret experimental design limitations, particularly in 3D-printed microfluidics-based biosensors with spatial control of sensing fluids and sensing properties. Based on the reviewed and collected literature, we are optimistic that the convergence of 3D-printing technology with other advanced techniques, such as microfluidics, will result in a wide variety of benefits and possibilities in medical areas.

REFERENCES

Amin, R., Knowlton, S., Hart, A., Yenilmez, B., Ghaderinezhad, F., Katebifar, S., and Tasoglu, S. 2016. 3D-printed microfluidic devices. *Biofabrication*, 8(2), 022001. doi:10.1088/1758-5090/8/2/022001

Arshavsky-Graham S, Enders A, Ackerman S, Bahnemann J, and Segal E. 2021. 3D-printed microfluidics integrated with optical nanostructured porous aptasensors for protein detection. *Microchim. Acta*, 188(3), 67. doi: 10.1007/s00604-021-04725-0.

Awad, R. H., Habash, S. A., and Hansen, C. J. 2018. 3D printing methods. In *3D Printing Applications in Cardiovascular Medicine*, 11–32. doi:10.1016/b978-0-12-803917-5.00002-x.

Bakhtiar, S. M., Butt, H. A., Zeb, S., Quddusi, D. M., Gul, S., and Dilshad, E. 2018. 3D printing technologies and their applications in biomedical science. In *Omics Technologies and Bio-Engineering*, 167–189. Academic Press. doi:10.1016/b978-0-12-804659-3.00010-5.

Bhalla, N., Jolly, P., Formisano, N., and Estrela, P. 2016. Introduction to biosensors. *Essays Biochem.*, 60(1), 1–8. doi:10.1042/ebc20150001

Bishop, G. W., Satterwhite-Warden, J. E., Bist, I., Chen, E., and Rusling, J. F. 2015. Electrochemiluminescence at bare and DNA-coated graphite electrodes in 3D-printed fluidic devices. *ACS Sens.*, 1(2), 197–202. doi:10.1021/acssensors.5b00156

Bragheri, F., Martinez Vazquez, R., and Osellame, R. 2016. Three-dimensional microfabrication using two-photon polymerization. *Microfluidics*, 310–334. doi:10.1016/b978-0-323-35321-2.00016-9

Castillo-León, J., and Svendsen, W. E. 2015. *Lab-on-a-Chip Devices and Micro-Total Analysis Systems*. Springer. doi:10.1007/978-3-319-08687-3.

Cevenini, L., Calabretta, M. M., Tarantino, G., Michelini, E., and Roda, A. 2016. Smartphone-interfaced 3D printed toxicity biosensor integrating bioluminescent "sentinel cells". *Sens. Actuators B: Chem.*, 225, 249–257. doi:10.1016/j.snb.2015.11.017

Chen, C., Mehl, B. T., Munshi, A. S., Townsend, A. D., Spence, D. M., and Martin, R. S. 2016. 3D-printed microfluidic devices: fabrication, advantages and limitations: a mini review. *Anal. Methods*, 8(31), 6005–6012. doi:10.1039/c6ay01671e

Chen, C.-S., Breslauer, D. N., Luna, J. I., Grimes, A., Chin, W., Lee, L. P., and Khine, M. 2008. Shrinky-Dink microfluidics: 3D polystyrene chips. *Lab Chip*, 8(4), 622. doi:10.1039/b719029h

Chen, Cheng Peng, Benjamin T. Mehl, Akash S. Munshi, Alexandra D. Townsend, Dana M. Spenceb, and R. Scott Martin. 2016. 3D-printed microfluidic devices: fabrication, advantages and limitations: a mini review. *The Royal Society of Chemistry*. 8, 6005–6012. doi: 10.1039/c6ay01671e

Chen, J., Liu, C. Y., Wang, X., Sweet, E., Liu, N., Gong, X., et al. 2019. 3D printed microfluidic devices for circulating tumor cells (CTCs) isolation. *Biosens. Bioelectron.*, 150, 111900. doi:10.1016/j.bios.2019.111900

Convery, Neil, and Nikolaj Gadegaard. 2016. 30 years of microfluidics. *Micro Nano Eng.*, 2(2019), 76–91.

Dang, B. V., Hassanzadeh-Barforoushi, A., Syed, M., Yang, D., Kim, S.-J., Taylor, R. A.… Barber, T. 2019. Microfluidic actuation via 3D-printed molds towards multiplex biosensing of cell apoptosis. *ACS Sens.*, 4, 2181–2189. doi:10.1021/acssensors.9b01057

Erkal, J. L., Selimovic, A., Gross, B. C., Lockwood, S. Y., Walton, E. L., McNamara, S., and Spence, D. M. 2014. 3D printed microfluidic devices with integrated versatile and reusable electrodes. *Lab Chip*, 14(12), 2023–2032. doi:10.1039/c4lc00171k

Fraser, L. A., Kinghorn, A. B., Dirkzwager, R. M., Liang, S., Cheung, Y. W., Lim, B., and Tanner, J. A. (2018). A portable microfluidic aptamer-tethered enzyme capture (APTEC) biosensor for malaria diagnosis. *Biosens. Bioelectron.*, 100, 591–596. doi: 10.1016/j.bios.2017.10.001

Gokhare, Vinod G., Dr. D. N. Raut, and Dr. D. K. Shinde. 2017. A review paper on 3D-printing aspects and various processes used in the 3D-printing. *International Journal of Engineering Research & Technology*, 6, 952–958. doi: 10.17577/IJERTV6IS060409

Gowers, S. A. N., Curto, V. F., Seneci, C. A., Wang, C., Anastasova, S., Vadgama, P.… Boutelle, M. G. 2015. 3D printed microfluidic device with integrated biosensors for online analysis of subcutaneous human microdialysate. *Anal. Chem.*, 87(15), 7763–7770. doi:10.1021/acs.analchem.5b01353

Hagedorn, Y. 2017. Laser additive manufacturing of ceramic components. *Laser Additive Manufacturing*, 163–180. doi:10.1016/b978-0-08-100433-3.00006-3

Hanada, Y., Sugioka, K., Shihira-Ishikawa, I., Kawano, H., Miyawaki, A., and Midorikawa, K. 2011. 3D microfluidic chips with integrated functional microelements fabricated by a femtosecond laser for studying the gliding mechanism of cyanobacteria. *Lab Chip*, 11(12), 2109. doi:10.1039/c1lc20101h

He, Yong, Yan Wu, Jian-zhong Fu, Qing Gao, and Jing-jiang Qiu. 2016. Developments of 3D printing microfluidics and applications in chemistry and biology: A review. *Electroanalysis*, 28, 1658–1678. doi: 10.1002/elan.201600043

http://3dprintingfromscratch.com/common/types-of-3d-printers-or-3d-printing-technologies-overview/

https://all3dp.com/2/what-is-a-dlp-3d-printer-3d-printing-simply-explain

https://www.3dhubs.com/guides/3d-printing/

https://www.3dhubs.com/knowledge-base/introduction-sls-3d-printing/

https://www.autodesk.in/solutions/3d-printing

https://www.sculpteo.com/en/3d-printing/3d-printing-technologies/

https://www.techpats.com/3d-printing-technologies-overview/

Hwang, H. H., Zhu, W., Victorine, G., Lawrence, N., and Chen, S. (2017). 3DPrinting of functional biomedical microdevices via light- and extrusion-based approaches. *Small Methods*, 2(2), 1700277. doi:10.1002/smtd.201700277

Kadimisetty, K., I. M. Mosa, S. Malla, J. E. SatterwhiteWardena, T. Kuhns, R. C. Faria, N. H. Leed, and J. F. Rusling. 2016. 3D-printed supercapacitor-powered electrochemiluminescent protein immunoarray. *Biosens Bioelectron.*, 15(77), 188–193. doi: 10.1016/j.bios.2015.09.017.

Kadimisetty, K., Malla, S., and Rusling, J. F. 2017. Automated 3-D printed arrays to evaluate genotoxic chemistry: E-cigarettes and water samples. *ACS Sens.* 2(5), 670–678. doi:10.1021/acssensors.7b00118

Kadimisetty, K., Spak, A. P., Bhalerao, K. S., Sharafeldin, M., Mosa, I. M., Lee, N. H., and Rusling, J. F. 2018. Automated 4-sample protein immunoassays using 3D-printed microfluidics. *Anal. Methods*, 10(32), 4000–4006. doi:10.1039/c8ay01271g

Klank, H., Kutter, J. P., and Geschke, O. 2002. CO2-laser micromachining and back-end processing for rapid production of PMMA-based microfluidic systems. *Lab Chip*, 2(4), 242. doi:10.1039/b206409j.

Krejcova, L., Nejdl, L., Rodrigo, M. A. M., Zurek, M., Matousek, M., Hynek, D....Kizek, R. 2014. 3D printed chip for electrochemical detection of influenza virus labeled with CdS quantum dots. *Biosens. Bioelectron.*, 54, 421–427. doi:10.1016/j.bios.2013.10.031

Lei, Kin Fong. 2018. *Microfluidics: Fundamentals, Devices, and Applications.* Wiley. ISBN: 978-3-527-34106-1 544.

Liu, Z., Zhang, M., Bhandari, B., and Wang, Y. 2017. 3D printing: Printing precision and application in food sector. *Trends Food Sci. Technol.*, 69, 83–94. doi:10.1016/j.tifs.2017.08.018

Matias, E., and Rao, B. 2015. 3D printing: On its historical evolution and the implications for business. In Portland International Conference on Management of Engineering and Technology (PICMET), Portland, OR, USA. doi:10.1109/picmet.2015.7273052

Matter-Parrat, V., and Liverneaux, P. 2019. Impression 3D en chirurgie de la main. *Hand Surg. Rehabilitat.*, 38, 338–347. doi:10.1016/j.hansur.2019.09.006

Mehrotra, P. 2016. Biosensors and their applications: A review. *J. Oral Biol. Craniofacial Res.*, 6(2), 153–159. doi:10.1016/j.jobcr.2015.12.002

Melchels, F. P. W., Feijen, J., and Grijpma, D. W. (2010). A review on stereolithography and its applications in biomedical engineering. *Biomaterials*, 31(24), 6121–6130. doi: 10.1016/j.biomaterials.2010.04.050

Mpofu, Thabiso Peter, Cephas Mawere, and Macdonald Mukosera. 2014. The impact and application of 3D printing technology. *Int. J. Sci. Res.*, 3, 2148–2152.

Pîrjan, Alexandru, and Dana-Mihaela Petroşanu. n.d. The impact of 3d printing technology on the society and economy. https://ideas.repec.org/a/rau/journl/v7y2013i2p360-370.html

Prabhakar, P., Sen, R. K., Dwivedi, N., Khan, R., Solanki, P. R., Srivastava, A. K., and Dhand, C. 2021. 3D-printed microfluidics and potential biomedical applications. *Front. Nanotechnol.*, 3, 609355. doi: 10.3389/fnano.2021.609355

Ragones, H., Schreiber, D., Inberg, A., Berkh, O., Kosa, G., Freeman, A., et al. 2015. Disposable electrochemical sensor prepared using 3D printing for cell and tissue diagnostics. *Sens. Actuators B: Chem.*, 216, 434–442. doi:10.1016/j. snb.2105.04.065

Samper, I. C., Gowers, S. A. N., Rogers, M. L., Murray, D.-S. R. K., Jewell, S. L., Pahl, C....Boutelle, M. G. 2019. 3D printed microfluidic device for online detection of neurochemical changes with high temporal resolution in human brain microdialysate. *Lab Chip*, 19, 2038–2048. doi:10.1039/c9lc00044e

Sanati, Alireza, Mahsa Jalali, Keyvan Raeissi, Fathallah Karimzadeh, Mahshid Kharaziha, Sahar Sadat Mahshid, and Sara Mahshid. 2019. A review on recent advancements in electrochemical biosensing using carbonaceous nanomaterials. *Microchim. Acta*, 186, 773. doi: 10.1007/s00604-019-3854-2

Savini, A., and Savini, G. G. 2015. A short history of 3D printing, a technological revolution just started. In ICOHTEC/IEEE International History of High-Technologies and Their Socio-Cultural Contexts Conference (HISTELCON), Tel-Aviv, Israel. doi:10.1109/histelcon.2015.7307314

Slapar, Vesna, and dr. Igor Poberaj. 2008. *Microfluidics*. University of Ljubljana.

Su, C. K., Yen, S. C., Li, T. W., and Sun, Y. C. 2016. Enzyme-immobilized 3Dprinted reactors for online monitoring of rat brain extracellular glucose and lactate. *Anal. Chem.*, 88(12), 6265–6273. doi:10.1021/acs.analchem.6b00272

Tang, C. K., Vaze, A., and Rusling, J. F. (2017). Automated 3D-printed unibody immunoarray for chemiluminescence detection of cancer biomarker proteins. *Lab Chip*, 17(3), 484–489. doi:10.1039/c6lc01238h

Tappa, K., and Jammalamadaka, U. 2018. Novel biomaterials used in medical 3d printing techniques. *J. Funct. Biomater.*, 9(1), 17. doi:10.3390/jfb9010017

Tarn, M. D., and Pamme, N. 2014. Microfluidics. *Reference Module in Chemistry, Molecular Sciences and Chemical Engineering*. Elsevier Inc. doi:10.1016/b978-0-12-409547-2.05351-8

Wabuyele, M. B., Ford, S. M., Stryjewski, W., Barrow, J., and Soper, S. A. 2001. Single molecule detection of double-stranded DNA in poly(methylmethacrylate) and polycarbonate microfluidic devices. *Electrophoresis*, 22(18), 3939–3948. doi:10.1002/1522-2683(200110)22:18<3939::aid-elps3939>3.0.co;2-9

Wang, Y.-C., Chen, T., and Yeh, Y.-L. 2018. Advanced 3D printing technologies for the aircraft industry: a fuzzy systematic approach for assessing the critical factors. *Int. J. Adv. Manuf. Technol.*, 105, 4059–4069. doi:10.1007/s00170-018-1927-8

Weisgrab, G., Ovsianikov, A., and Costa, P. F. 2019. Functional 3D Printing for Microfluidic Chips. *Adv. Mater. Technol.*, 1900275. doi:10.1002/admt.201900275

Whitesides, G. M. 2006. The origins and the future of microfluidics. *Nature*, 442(7101), 368–373. doi:10.1038/nature05058

Wu, H., Odom, T. W., Chiu, D. T., and Whitesides, G. M. 2003. Fabrication of complex three-dimensional microchannel systems in PDMS. *J. Am. Chem. Soc.*, 125(2), 554–559. doi:10.1021/ja021045y.

Xia, Y., and Whitesides, G. M. 1998. Soft lithography. *Annu. Rev. Mater. Sci.*, 28, 153–184.

Zhang, Y. 2019. 3D-Printing for microfluidics or the other way around? *Int. J. Bioprint*, 5(2), 192. doi: 10.18063/ijb.v5i2.192.

7 Integrated Biosensors for Rapid and Point-of-Care Biomedical Diagnosis

Sunil Kumar and Rashmi Madhuri

CONTENTS

7.1 INTRODUCTION

Biosensors are integrated devices in which the recognition elements act as biomolecules to bind the analyte and are then followed by a transduction mechanism to read a measurable signal. These are applied in different fields, such as biomedicine, food safety, environmental sensing, and biological warfare, etc. [1]. There are three basic components required for preparation of a biosensor, such as a transducer, a

DOI: 10.1201/9781003033479-7

bioreceptor, and a signal processing device. This device has many essential properties over current analytical devices as it has higher selectivity, a low detection limit, good reversibility, a long lifespan, good response time, and high biocompatibility (Figure 7.1) [2,3]. A bioreceptor is a molecular compound which binds with target compounds through a biochemical reaction. Some examples of bioreceptors are antibodies, nucleic acid, enzymes, phages, and whole cells [4]. The sensitivity of biosensors depends on the type of transducer and the immobilization interaction while specificity depends on the sensitivity of materials [5].

The preparation of integrated biosensors depends on the two types, one is the biological recognition elements and immobilization technique in which the selectivity is based on the target interest, i.e., aptamers and antibodies used for the detection

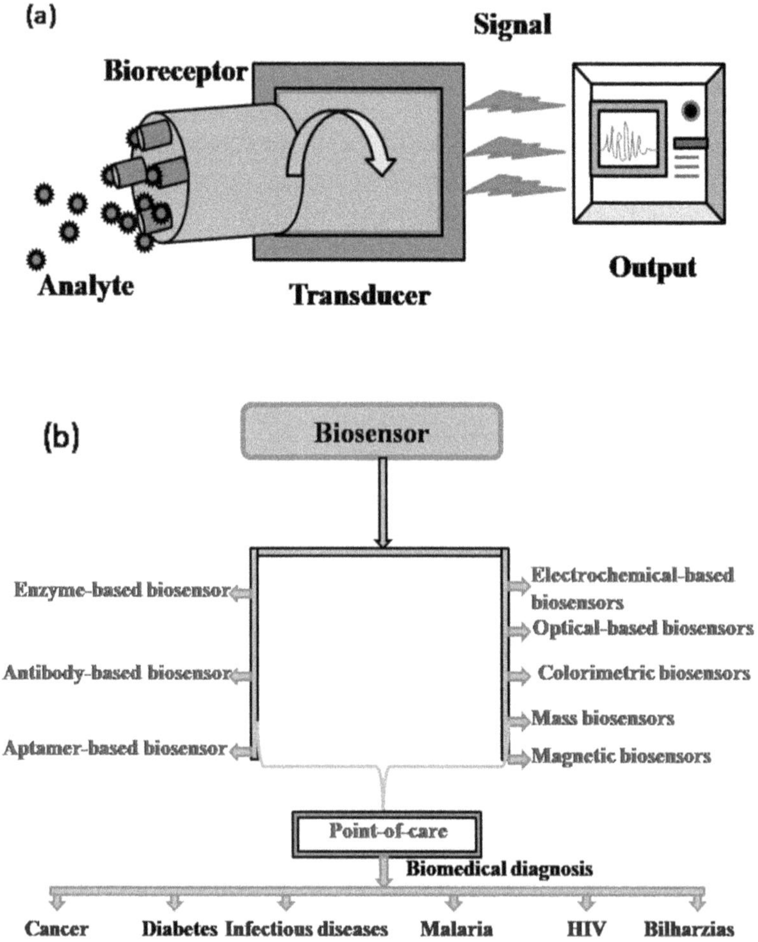

FIGURE 7.1 (a) Schematic representation of biosensor. (b) Schematic illustration of biosensors and their types.

of bacteria and pathogens, but enzymes for catalytic reactions. They are categorized into three types – enzyme-based biosensors, aptamer-based biosensors, and antibody-based biosensors. Second are the types of transducers used. The selection of a transducer is based upon the sensing layer on which the reaction is completed. There are five types – electrochemical-based biosensors, optical-based biosensors, colorimetric biosensors, mass biosensors, and magnetic biosensors (Figure 7.2) [6]. The enzyme is used as a biological catalyst for particular reactions which bind themselves to a specific substrate that apply in the biosensor [7]. Antibodies, applied through vaccines, act as protecting agents against many pathogens [8]. In addition, the aptamer has important properties like its high flexibility, reusability, being easy to fix and regenerate, and having no difference between batches, which is widely applied in the preparation of a sensor [9]. In electrochemical biosensors, a signal is produced by the chemical reaction between the transducer and template molecule which generates an electronic charge in the form of the electrical signal [10]. An optical biosensor depends on the absorbance, surface plasmon resonance, and photoluminescence [11]. A colorimetric biosensor is widely applied because of its easy operation, cost effectiveness, simplicity, speed, and easy measurement [12]. Recently, magnetic materials have been promising materials for the preparation of biosensors for point-of-care (POC) diagnosis [13].

Point-of-care testing (POCT) has the ability to detect the target analyte near to the patient because of easy diagnosis, monitoring, and management. Moreover, it provides quick medical results which are useful for a disease diagnosed at a very early stage in the absence of experts and types of equipment which are useful for global health, in comparison to ordinary techniques including polymerase chain

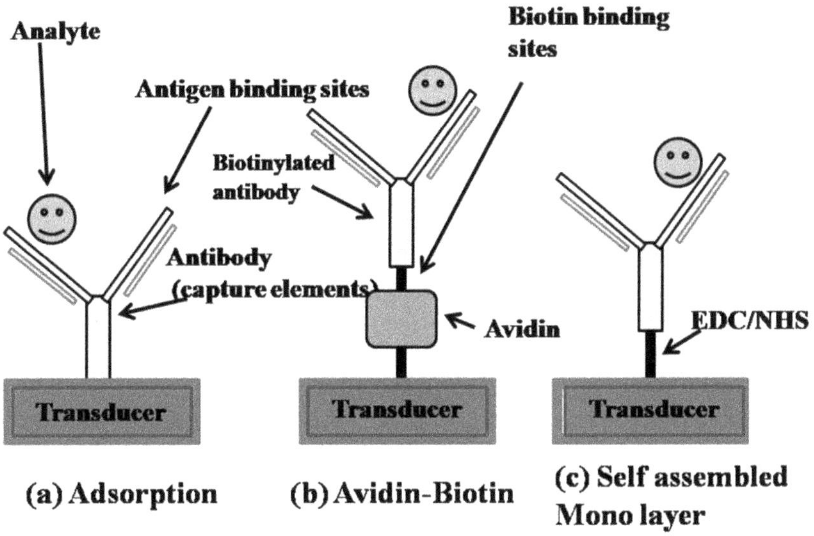

FIGURE 7.2 Schematic illustration of the basic immobilization method, modified from reference [6].

reaction, radioimmunoassay, and enzyme-linked immune-sorbent assay, while some drawbacks such as high cost and complicated instrumentation are recognized [13, 14]. This is to show outstanding advantages for POCT as it is an easily developed, reusable, affordable, sensitive, specific, rapid, user friendly, and effective miniaturized platform. It is categorized into two types –portable hand-held and bench-top systems. A portable hand-held system is prepared by applying a micro assembling process. This prepares simple analysis, assay step, and analysis of a signal automatically. From this process, the estimate of a wide range of analytes is obtained quantitatively or qualitatively. The bench-top system is the essential part of the main lab equipment, which is reduced in both size and rigidity (Table7.1) [15].

The current coronavirus (COVID-19) is a pandemic disease that was declared by the World Health Organization (WHO) on March 11, 2020. The coronavirus or SARS-CoV-19 (severe acute respiratory syndrome coronavirus), originated from Wuhan in China, with around 13 million people infected and approximately 580,000 deaths globally due to this pandemic disease. The main symptoms of infected people with the virus show within 2–14 days, such as fever, shortness of breath, cough, headache, loss of taste and smell, body pain, and sore throat [16, 17]. Recently, Jane Ru Choi has developed a technique for COVID-19 disease. The PoCT was developed by using paper, poly-dimethylsiloxane, and additional flexible materials like carbon nanosheets, film, and textile. From this process, robust PoCT biosensors have been applied for rapid detection of COVID-19 [18]. Additionally, Ruiz-Vega et al. developed a POC nanophotonic biosensor which is used for direct, fast, and specific identification of SARS-COVID-2 virus from affected people in a few minutes. In this technique, two diagnostic assays were involved; one is direct virus detection by antigen recognition and the second is viral genomic identification via hybridization assay. For this reason, this is urgently demanded by clinics and society [19].

7.2 TYPES OF INTEGRATED BIOSENSOR

These are categorized into two types as follows:

7.2.1 Biosensors Categorized Based on the Type of Biological Recognition Element and Immobilization Technique

The immobilization of recognition of the biological elements takes place by using various methods, i.e., adsorption, allurement, covalent interaction, and film imprisonment. Immobilization with covalent bonding is widely applied due to the biosensor's stability and irreversible nature which help to prevent dropping of the bioelements from the surface. These types of biosensors are antibody-modified biosensors, enzyme-modified biosensors, and aptamer-modified biosensors [6].

7.2.1.1 Enzyme-Modified Biosensor

An enzyme-modified biosensor (EMB) is an analytical tool in which enzymes act as recognition elements which bind with substrate acting as a target analyte. A development of EMBs depends on the modification of an enzyme on the surface of the transducer; the resulting enzymes maintain their activity after the modification process. The modification process and suitable biological compounds directly affect the function of an enzyme and the performance of a biosensor, i.e., repeatability, selectivity, sensitivity, stability, and excellent performance [20, 21]. Nanomaterials (NMs) such as metallic oxide nanoparticles (NPs) [22], carbon nanotubes (CNTs) [23], graphene [24], and composite NMs like polymer or NPs [25], carbon nanotube/enzyme biofuel cells [26], and chitosan-modified silver-gold/reduced graphene oxide nanohybrids [27] which are applied as a biological membrane to modified enzymes [20]. Mohammad et al. developed an enzyme in metal-organic-framework (MOF) complex materials which is useful for increasing enzyme stability without disturbing their activities, resulting in this device being applied as a biosensor. Complex compounds like the ZIF-8-GOx and horseradish peroxide (HRP) materials based on polydopamine (PDA) or polyethyleneimine (PEI) contain excellent stability in terms of thermal and acidity which is compared in the absence of ZIF-8. ZIF-8-GOx and HRP in situ exhibits were highly selective for glucose. The proposed materials show sensitivity and limit of detection (LOD) for glucose concentration is 0.00303 Abs/μM and 8 μM [28]. Also, Muguruma et al. prepared an amperometric biosensor which is associated with different compounds like an enzyme, CNTs, and plasma membrane in which CNTs are modified with a layer of cobalt/titanium/chromium materials via microwave plasma-enhanced chemical vapor deposition methods, and after that modified CNTs followed by nitrogen plasma due to this surface having been changed from hydrophobic to hydrophilic, which is why an interaction between CNTs and enzymes takes place. The proposed biosensor shows high sensitivity (38 μA/mM cm^2) along with good linear range as 0.25 to 19 mM. The lower LOD is found to be 34 μM (S/N = 3) [29]. The vertically aligned CNTs were developed, which are applied for the analysis of glucose. The proposed biosensor based on POCT was successfully employed for the analysis of glucose in human blood plasma. Sensitivity is found to be 620 $\mu AmM^{-1}cm^{-2}$ with good linear range (2–426 μM) and detection limit (1.1 μM) [30].

Magar et al. synthesized a biosensor by using choline oxidase coated on the surface of a glassy carbon electrode which was further coated with multiwalled carbon nanotubes (MWCNT) via glutaraldehyde to form a choline oxidase-MWCNT-glassy carbon electrode biosensor under normal conditions. The biosensor exhibited a detection limit of 0.04 nM with a good linear range from 0.1 to 1.0 nM for determination of lead (II) ions[23]. Kou et al. studied the facile, stable, ecofriendly MOFs system which is called smartphone-added biomimetic MOF nanoreactor colorimetric paper (SBMCP).By using this, POCT detected the interior of biomaterials. Therefore, a SBMCP biosensor has the ability to perform colorimetric analysis of glucose and uric acid in diabetes[31]. In this work, we prepared a POC biosensor modified with enzyme tracers and a simple strip reader to form a cost-effective, simple, sensitive biosensor for detection of protein [32].

7.2.1.2 Antibody-Modified Biosensor

Antibodies (Abs) have excellent specificity towards antigens which are generated against any compound. These are widely used for targets in analyzing solution. It is applied successfully to sensitive and rapid detection of different pathogens and toxins. Plant pathogen-specific Ab produced by injecting entire pathogens, particles of the soluble surface component which has been prepared an animal host for polyclonal antibody (pAb)/monoclonal antibody (mAb). pAb and mAb preparation is not an easy process in terms of its quality assured mode which is essential for obtaining an enduring analytical technique. Bacterial-based antibody particles and Fabs, antigen binding fragments of antibodies, are valued due to their specificity and low cost in comparison to mAbs which provide bacterial clones compared to hybridomas [33, 34]. Moreover, a flexible capacitive sensor is developed by using a laser-etching method. The analysis of protein on the surface of flexible compounds was developed for POC detection in which gold is etched on the substrate of polyethylene terephthalate and is important for reducing the damage. They have to judge, based on the amount of the C-reactive protein available in the sample, which can be useful for assessing the risk of cardiac disease. The proposed sensor has the ability to measure protein related to cardiovascular diseases [35]. Faja et al. discovered a novel POC device which is useful for the assessment of specific sepsis biomarkers. The sensor depends on the prepared metal NPs and allows biofunctionalization which also immobilizes with a different type of receptor, for example, antibodies or oligonucleotides. Here, they described two types of biomarker, with a different structure like proteins such as mi-ribonucleic acids, interleukin 6, and C-reactive protein. The limit of detection is 18 μg/mL, 88 μg/mL, and 1 μM (6 μg/mL) for C-reactive protein, interleukin 6, and mi-ribonucleic acids respectively [36]. The MOF is used for preservation of functionality of antibodies conjugated to nanotransducers. A MOF (ZIF-8)-modified biosensor used for preservation of various kinds of biodiagnostic reagents and the proposed biosensor for POC diagnosis in resource-limited settings with drastic environmental circumstances [37].

7.2.1.3 Aptamer-Modified Biosensor

Aptasensors or biosensors, for example ribonucleic acid (RNA) or deoxyribonucleic acid (DNA) aptamers, are used as their recognition elements. The term aptamer is obtained from the Latin word, aptus, meaning "to fit" and the Greek word, meros, meaning "the part." Firstly, the aptamer technique was developed in 1990 by three research group scientists. The selection of RNA with raised enzymatic reactivity which is useful for breakdown of DNA, was analyzed by Robertson and Joyce. Tuerk and Gold developed the method for the selection of a DNA molecule as a selector for T4 RNA polymerase. Ellington and Szostak discovered the technique for a choice of RNA which is suitable for easy binding to organic dyes. Peptide or single-stranded oligonucleotide consisted of DNA and RNA aptamers which formed secondary and tertiary structures after folding under certain conditions such as ionic strength, pH, and temperature, then ultimately formed low or macromolecular compounds, for example cells, cell surface proteins, bacteria, and viruses [38,39]. Dirkzwager et al. developed POC device prototypes for malaria by applying an aptamer-modified colorimetric assay for Pf LDH analysis. In this case, a paper-modified syringe test, as well as a magnetic bead-modified well test, was formed. The syringe test shows a result with higher sensitivity, but it required an extra synthesis process while the well test needed fewer tests, so it is more important for future clinical testing. Therefore, this technique applied for rapid diagnostic tests for malaria [40]. Also, Khan et al. developed a cost-effective inkjet-printed aptamer-based electrochemical (AMB) for the analysis of lysozyme. In this work, inkjet printing plays an important role as disposable sensors and flexible electronics, as well as in wearable sensors, due to their multifunctional character. The strong bonding between CNTs and single-stranded DNA allows it to modify the aptamer on the surface of the working electrode. Therefore, the proposed sensor exhibits a selective limit of detection of 90 ng/mL^{-1} for lysozyme [41]. In this work, electrochemical AMB for the analysis of cardiac biomarkers for brain natriuretic peptide (BNP-32), a good concentration range is from 1 pg/mL to 1 μg/mL in serum. The linear range obtained for cardiac troponin I (cTnI) is 1 pg/mL to 10 ng/mL which is essential for early-stage diagnosis of heart failure[42]. The electrochemical, RNA AMB was used for the analysis of aminoglycoside antibiotics in the blood serum, discovered by Rowe et al. [43].

7.2.2 Different Biosensors Based on the Type of Transducer

The transducer plays a main role in converting this energy into a valuable analytical signal which is used in electronic form [44]. These are described as follows:

7.2.2.1 Electrochemical-Modified Biosensor

Electrochemistry is a powerful device that has the ability to obtain information from enzymatic compounds, redox mechanisms of enzymatic interactions, and metabolic mechanisms. It is applied in different fields including medical, environmental, and pharmaceutical, etc. because of its cost-effective, miniaturized, and simple analysis systems, and it is easy to fabricate. It is divided into three electrochemical techniques such as amperometry, potentiometry, and impedance spectroscopy [45]. The potentiometric sensor is simple and fabricated easily but exhibits limited sensitivity in comparison to the amperometric sensor. It consists of a two or three-electrode system. In two-electrode systems, the working and the reference (counter) electrodes are used but, in three-electrode systems, the working, reference and counter electrodes are used [46]. A biosensor-based electrochemical sensor is comprised of two parts, one is a transducer, and another is recognition elements for example enzymes or enzyme-labeled antibodies [47]. Shabaninejad et al. developed electrochemical-modified biosensors for microRNA detection. The microRNA is called single-stranded RNA molecules, containing 22 nucleotides in length which regulate the biological function by cellular proliferation, and cause death to cancer development and progression. Moreover, the diagnostic value of miRNAs in different diseases has been analyzed. Therefore, the prepared biosensor can be used for the analysis of nucleic acid in the future [48]. Kilic et al. discovered an electrochemical-modified biosensor for the analysis of microRNA, miR21, in breast cancer cells. The prepared biosensor has several important advantages as being accurate, reproducible, robust, and the short time taken. They are also applied to the detection of total RNA, and microRNAs from various cancer tissues or cell lines[49].

7.2.2.2 Optical-Modified Biosensors

A biosensor exhibits two essential properties, like sensitivity and specificity which consist of two parts; one is the biological recognition elements and other is the sensing part. Sensitivity is generated by various kinds of receptors like antibodies, nucleic acid (e.g., aptamer), enzymes, and microbes, but specificity is affected by the interaction between biological recognition elements and sensing parts. The optical properties are produced in biosensors by using different elements such as gold, silver NPs, CNTs, quantum dots, and graphene [50]. There are two detection properties for optical modified biosensors such as label-based and label-free detection. From this, the analyte is detected near to their original shape. They exhibit essential properties like being facile and cost effective also are used for measuring molecular interactions and target quantification [51]. In addition, a mesoporous porous silicon (PSi) is applied for the preparation of the label-free optical biosensor which is used for detection of broad molecule heat shock protein 70 (HSP70). The electrochemical etching method is applied for the preparation of a PSi single-layer platform which is maintained by thermal oxidation. Afterwards, it is modified with anti-HSP70 antibodies via 3-aminopropyltriethoxysilane or glutaraldehyde linking interaction. As a result, this compound exhibits a concentration range of 3000–500,000 ng/mL with a detection limit of 1290 ± 160 ng/mL for detection of HSP70 [52]. Eissa et al. discovered graphene-based label-free voltametric immunosensor for the analysis of the egg white allergen, ovalbumin. The concentration range for the immunosensor is found of 1 pg/mL^{-1} to 0.5 μg/mL^{-1} with a limit of detection of 0.83 pg/mL^{-1}in the presence of a PBS buffer [53]. Hypoxia is a pathological process due to limited supply of oxygen, tissue metabolism affects the structure, and physiological activity results; the hypoxia activates the cancer-producing genes or cancer disease. Nitroreductase (NTR) acts as a biomarker for cellular hypoxia. To overcome this problem, Zhang et al. developed fluorescent biosensor-modified conjugated polymers which are applied for the analysis of NTR and hypoxia diagnosis in tumor cells. The LOD is calculated as 19.7 ng/m^{-1} [54].

7.2.2.3 Colorimetric Biosensors

Colorimetric detection is widely used for a biomolecule due to it having important properties including rapid reading, detectable radiation, and simple processing. The main problem arises in the colorimetric technique; how to convert the invisible signals into a color signal? To overcome this problem, gold NPs are used in surface plasmon resonance (SPR)-based colorimetric biosensors (CB), resulting in the development of color change properties [55]. For multiplex detection, Erickson's group applied two NPs as gold and silver in one solution. Problem arising during the multiplex detection can be resolved by adding multiple signal indicators with colors of different wavelengths. Hao et al. observed pH-resolved CB for multiple targets [56]. Breast cancer in women is a worldwide health problem according to the International Agency for Research on Cancer which reported the annual death among women at around 15% due to breast cancer. Bai et al. developed the sandwich-type Bi_2Se_3-AuNPs biosensor which exhibits good selectivity and sensitivity along with the limit of detection and concentration range of 1018 M, 10–12–10–18 M for detection of breast cancer-associated BRCA1 mutation [57]. In this work, a method is used for the synthesis of "naked" gold NPs modified with Bi_2Se_3 nanosheets (Au@Bi_2Se_3) via simple sonication. The prepared Au@Bi_2Se_3 nanosheets are used as a colorimetric sensor for analysis of cancer biomarkers, this depends on catalytic behavior on the reduction of 4-nitrophenol (4-NP) by sodium borohydride itself and implies a visible color change. Therefore, the colorimetric sensor reported high selectivity and sensitivity for the cancer biomarker like α-fetoprotein (AFP) and prostate-specific antigen (PSA) [55]. Glucose is an essential compound for living organisms due to its good energy source and metabolic intermediation in the preparation of complicated biological compounds. An irregular concentration of glucose in human blood indicates many diseases, like diabetes. Feng has discovered the gold-platinum bimetallic nanocluster-modified biosensor, applied for detection of the glucose level present in serum. The LOD is 2.4 μM with a concentration range of 5–55 μM [58].

7.2.2.4 Mass Biosensors

The mass-sensitive biosensor or surface acoustic wave (SAW) sensor is a powerful device which is applied for detection of different biological molecules such as protein–protein, protein–nucleic acid, or cell–virus interaction. An acoustic wave mass-sensitive detector depends on the quartz crystal microbalance. They are sensitive towards mass and viscoelastic changes at the interface of solid solutions which can indicate mass change at the interface with a resolution of 1 ng/cm^{-2}. When the quartz crystal microbalance is allowed, any unwanted materials on its surface. A stable oscillation is only obtained in the presence of the natural resonance frequency. The natural resonance frequency corresponds to the mass placed on the surface of the QCM electrode which is calculated by using the Sauerbrey equation [59, 60]. Galvan developed a surface plasmon resonance biosensor based on the dielectrophoretic enhanced mass transport for detection of *E. coli* [61].

7.2.2.5 Magnetic Biosensors

The magnetic biosensor provides an actual result which has the ability to detect single magnetic labels. They have many essential properties in comparison to ordinary biosensors including simple handling, rapid analysis, cost effectiveness, are facile, and have a simple fabrication process [62]. The enhancement of biosensor capacity in terms of potential, the magnetic particles are modified with multifunctional materials, resulting in facile separation via magnets and the wide active surface area leads to modification and purification of biological compounds with a magnetic field and therefore decreases the matrix effect. Therefore, the use of magnetic particles in the biosensor increases the signal transduction as well as target recognition, which is why it increases the sensitivity [63]. Xu et al. studied electrochemical biosensors based on magnetic micro/nanoparticles in which the magnetic particles could be modified with transducers via physical adsorption and chemical covalent bonding. This resulted in an increase in the electrode surface area which is a help for the immobilization process [64]. Diagnostics play a critical role in identifying the existence and cause of disease in a human population at an early stage. POCT has several properties – it is easy, simple, fast, and has accurate methods; due to this, selective detection of pathogens can occur. Cortina et al. developed a biosensor for POC sero-diagnosis of infectious diseases [65].

7.3 VARIOUS INTEGRATED BIOSENSORS FOR POC BIOMEDICAL DIAGNOSIS

7.3.1 Biosensors for POC Diagnosis of Cancer

Cancer is one of the most widespread diseases. There are different types of cancer (~200) which affect more than 60 human organs. Almost 90% of cancer-associated deaths occur due to metastasis of the main cancer tumor. Cancer is a genetic disease generated by the disturbance of genes that control different aspects of cellular functions. Two major fields like molecular biology and biochemistry explain about the genes and their products which are involved in the development and progression of cancer cells [66, 67]. Pancreatic cancer causes death, it is in fourth position in the United States and globally. In 2019, about 56,770 patients affected with pancreatic cancers were diagnosed, but 45,750 would die from the disease. As a result, the five-year relative survival rate for all stages is found to be ~8% [68]. There are different types of cancer, occurring in the breast, ovary, lung, colon, bladder, neuron, bone, etc. They caused about 9.6 million deaths globally in 2018 [69]. There are several methods for diagnostic of cancer – ELISAs for detecting CUB domain-containing protein 1 [70], liquid chromatography–tandem mass spectrometry [71] and, immunohistochemistry [72]. These methods are usually time consuming, costly, and labor intensive. Due to this drawback, the POC technique provides the opportunity for complicated examination by nonprofessionals. These technologies are used as a personalized medicine which diagnoses the disease as well as tailoring therapeutics to the individual [73]. Chen et al. developed an electrochemical biosensor for POC use in which the semiconductor manufacturing technology production settings were reformed as a bioreceptor and elevated by using a special interaction. The proposed sensor shows high stability, reproducibility, and impressive accuracy. [74]. Moreover, a premature cancer analysis is essential for the safeguard of metastasis. The nanoplasmonic aptamer sensor was applied to a single step detection of vascular endothelial growth factor-165. The proposed sensor has utility for POC cancer prognostics which have several properties like being simple, cost effective for the test, and needing a small sample volume (Figure 7.3) [75]. Gai et al. developed a redox-free, one-compartment enzyme biofuel cell modified with a cytosensing device for detection, including acute leukemia cells and circulating tumor cells. The proposed biosensor is a POC device for analysis of cancer [76]. Yoo et al. discovered a molybdenum sulfide (MoS_2) biosensor for POC diagnostics for detection of epidermal skin. The great performance of MoS_2 biosensors can detect prostate cancer antigens with a concentration of 1 pg/mL in the absence of a particular surface cure as an anti-PSA modification on the surface of MoS_2. These proposed materials are characterized by applying Kelvin force microscopy and atomic force microscopy, resulting in early detection based on the POC diagnostics for prostate cancer [77].

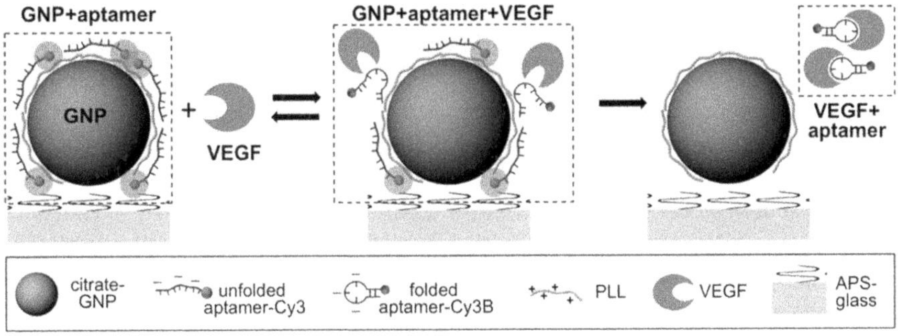

FIGURE 7.3 Schematic representations of detection mechanism of the aptasensor for VEGF165. (a) In the absence of target molecules, VEGF165, unfolded VEGF165 aptamer is electrostatically bound to a positively charged PLL-coated gold nanoparticle (GNP) surface and surface-enhanced fluorescence (SEF) of Cy3B conjugated with the VEGF165 aptamer created by both a metal interaction increasing the radiative fluorescent decay rate of Cy3B and the local surface plasmon resonance (LSPR) enhancing the intensity of an incident light. (b) The interaction of the VEGF165 aptamer to its target induces the reversible conformation change of the aptamer and, consequently, the decreased electrostatic binding force. (c) As a result, the target-binding interaction of the aptamer causes the irreversible detachment of the aptamer from the GNP surface and avoids the SEF effect of Cy_3B.Copy with permission taken from [72].

7.3.2 Biosensors for POC Diagnosis of Diabetes

Diabetes mellitus is a metabolic disorder by chronic hyperglycemia along with the obstruction of fat, carbohydrate, and protein metabolism, leading to defects in insulin production. A diabetes- affected person faces many problems like the failure of different organs, dysfunction, and damage to the organs. Diabetes is categorized as two types, one is Type-1 (T_1) and other is Type-2 (T_2). T_1 is insulin-based diabetes mellitus which reported ~5% of all diagnosed cases of diabetes. T_2 diabetes is reported around 95% of all diagnosed diabetic patients. T_2 diabetes starts with insulin obstruction that means no proper use of insulin by the muscles, liver, and fat tissue [78, 79]. Diabetes mellitus is maintained by regular screening and early detection, and is essential for successful treatment [80]. Zhai et al. illustrated an electro-chemiluminescent biosensor for analysis of I27L genes based on gold NPs functionalized ITO electrode. In this mechanism, this technique is applied to regulate the association of DNA strands by changes in their intensity. The prepared biosensor shows good stability, outstanding selectivity, and excellent repeatability for diabetes with good linear range and detection limits of 1.0×10^{-1} to 1.0×10^{-7} M and 8.1×10^{-12} M [81].Diabetic retinopathy (DR) is a disease of the eye in which infected cases experience blurred vision. This occurs due to abnormal growth in the retinal blood vessels. The amount of β-2-microglobulin present in human tears helps to cure diabetic retinopathy, generated from diabetes mellitus. Therefore, a biosensor has been used for the POC detection of β-2-microglobulin which is important for early diagnosis of diabetic retinopathy [82]. In this method, the quantification of tear fluid may be a precious symbol for systemic glucose observation. A glucose sensor placed under the eyelid provides information about the concentration of basal tear fluid which is related to blood glucose values. This applies to both animals and humans. As a result, a noninvasive biosensor was used for successful detection of tear glucose which helps to cure diabetes (Figure 7.4) [83]. In addition, an electrochemical analyzer or a smartphone-powered medical system is used for the analysis of blood ketone present in blood at the POC after the addition of enzymatic β-hydroxybutyrate which has the ability to diagnose diabetic ketoacidosis and diabetic ketosis acid. The final observed blood ketone obtained for the medical device is good in comparison to the bulky biochemical analyzer. Therefore, the proposed medical smartphone-powered device based on the POC was applied for the diagnosis of blood ketone and was a good solution for mobile medical management [84]. In this case, the POC device is necessary for the detection of many target molecules along with simplicity and good accuracy. A reconfigurable smartphone-interfaced electrochemical lab-on-a-chip was prepared with the help of two working electrodes for the detection of two analytes. The selectivity of a biomarker depends on the coated alginate hydro gels. Alginate film, having either glucose oxidase or lactate oxidase, was coated on each working electrode for at least 4 min. The observation of the sample was completed in less than 1 min. The linear range and sensitivity for glucose and lactate was found to be 0–12 mM, 0–5 mM, and 0.24–0.54 μA/cm^2 mM respectively. This process permitted the detection of T_1 diabetes by using a facile dual analysis device. POC use of this device leads to the adjustability and flexibility of the present system [85].

7.3.3 Biosensors for POC Diagnosis of Infectious Diseases

Infectious diseases (IDs) are a worldwide problem due to a lack of proper health education, treatment regimens, and diagnostic devices. In 2016, 8.2 million

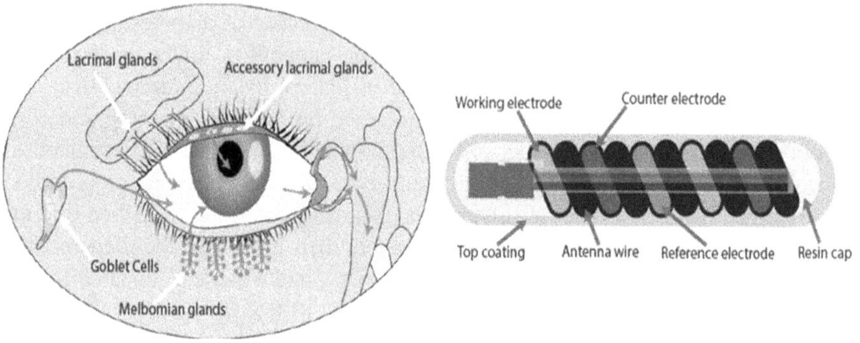

FIGURE 7.4 Schematic representation and properties of NovioSense minimally invasive tear glucose sensor. (a) Illustration showing tear fluid production. (b) Direction of the tear flow and anatomical position of the NovioSense device. (c) Design and structure of the final product showing electronic components. (d) Mechanism of glucose detection where glucose oxidase is used as the enzymatic sensing element. Copy with permission taken from [81].

deaths of people worldwide occurred due to infectious diseases [86]. Various IDs were reported worldwide, including, avian influenza, SARS, Hendra, Nipah, etc. Biosensors are an attractive device to apply for the detection of virus and diseases. In this work, a portable, simple, accurate method for rapid PoC detection of IDs was developed. An electrochemical magnetic micro beads-based biosensor for POC serodiagnosis begins with various microorganisms, including bacteria, viruses, and parasitic protozoa. An electrochemical magnetic microbeads-based biosensor device is used to differentiate infected patients from noninfected patients. As a result, the proposed biosensor was applied for POC serodiagnosis of IDs [65]. Pelaez et al. developed a portable, ecofriendly, and cost-effective biosensor of POC tuberculosis diagnosis. This biosensor is used for early diagnosis of tuberculosis with a good optimum range of 116–175 ng/mL and LOD, and limit of quantification of 0.63 ng/mL and 2.12 ng/mL [87]. Viruses are generated by various IDs in humans and animals, so sensitive and rapid POC tools for diagnosis of viruses are very urgently required. A smartphone-based POC device was used for detection of the avian influenza virus, modified with nanomaterial-enabled colorimetric analysis. Virus capture by chip Au NPs-based colorimetric reaction which allowed detection of viruses with the naked eye with a good LOD of 2.7×10^4 EID50 mL^{-1} but LOD of 8×10^3 EID50 mL^{-1} obtained by a smartphone imaging system with data processing capability. The entire virus capture process was completed in 1.5 h. For an early analysis of the virus, an integrated system is predicted with a POC system, which is shown in Figure 7.5 [88]. In addition, Choi et al. studied that COVID-19 is a global pandemic. A traditional technique like quantitative real-time polymerase chain reaction has been widely applied for detection of COVID-19 but it is labor intensive, time consuming and has an absence of remote settings. To overcome this drawback, POC biosensors, i.e., paper-based and chip-based biosensors are cost effective and easy; due to this, they are used for early diagnosis of COVID-19 [18].

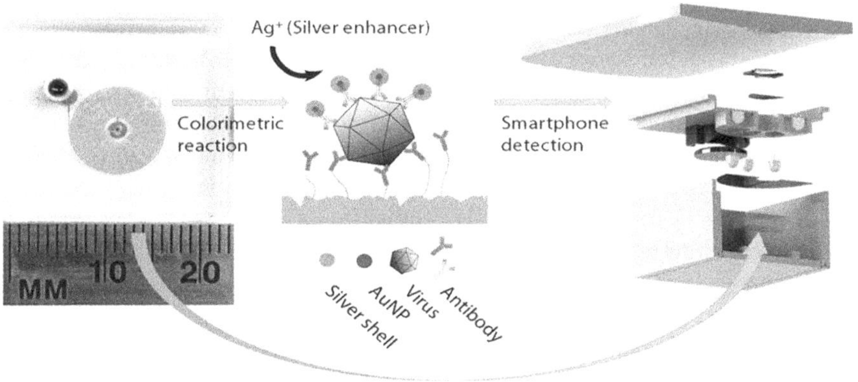

FIGURE 7.5 Synthesis of a nanostructure micro device and the portable smartphone-enabled virus detection system. Copy with permission taken from [86].

7.3.4 Biosensors for PoC Diagnosis of Malaria

Malaria is a mosquito-borne disease which is a serious threat to human health worldwide. This is an infectious disease transmitted from female mosquitoes (genus *Anopheles*) to humans in tropical countries. Single-celled plasmodium parasites in infected mosquitoes are transferred into the human bloodstream through saliva during the mosquito bite. There are many symptoms of malaria, which are recorded as muscle pain, headaches, fever, chills, and fatigue. A total of 429,000 deaths were recorded in the world in 2015, out of which 70% were children. Accordingly, the WHO set a goal for the elimination of malaria from temperate regions by 2030. To achieve eradication of malaria, government, organizations, and health agencies are actively participating in making strategy, teaching health workers, raising awareness, and fundraising. To cure malaria, a vaccine called Mosquirix was developed by the European Union in 2015 which is a significant step toward the eradication of malaria infection. Malaria diagnosis at an early stage is useful for prevention of the disease's spread and provides a cure at the appropriate time to manage the disease as well as the mortality rate. Many methods are involved in malaria diagnosis, including immune-based rapid diagnostic test kits, microscopic analysis, and the nucleic acid-based analytical method. Despite many techniques available, malaria has remained in 91 countries, and is a basic health problem worldwide [89–91]. In this method, a biosensor is developed for the analysis of malaria. DNA aptamer was modified with the *Plasmodium falciparum* lactate dehydrogenase enzyme to improve specificity and sensitivity. A malaria-based biosensor in which the aptamer is modified with magnetic micro beads for magnet-guided encapsulation, cleaning, and analysis of the biomarker. The device is combined with three individual microfluidics chambers which have the ability for magnet-guided equipment-free colorimetric detection of *Plasmodium falciparum* lactate dehydrogenase. For early-stage detection of malaria, a biosensor based on a POC device was developed [92]. Ruiz-Vega et al. synthesized an electrochemical POC device which is used for the rapid, easy, and measurable analysis of *Plasmodium falciparum* lactate dehydrogenase in all blood samples. The magneto-immunoassay is when sample filtration, magnetic-bead cleaning, and electrochemical detection are carried out on a disposable paper electrode microfluidics device. The proposed sensor allowed the *Plasmodium falciparum* lactate dehydrogenase quantization down to 2.47 ng/mL in spiked samples and 0.006–1.5%parasitemias in *Plasmodium*-infected cultured red blood cells and differentiate between malaria patients and a healthy person current parasitemias more than 0.3% [93]. Dirkzwager et al. discovered the aptamer-based POC diagnostic devices for detection of malaria in the presence of 3D printing rapid prototyping. The analytical consequence is detecting the malaria biomarker *Plasmodium falciparum* lactate dehydrogenase from samples through the inherent enzymatic functionality, which provides a blue color in response to *Plasmodium*-positive samples. The paper-based syringe test and magnetic bead-based well test were synthesized by applying 3D printing rapid prototyping. These are applied for successful analysis of recombinant Pf LDH at ng/mL value in the presence of low-volume samples as 20 μL. The syringe test was more sensitive because it involves extra formation key steps, while a good test needs fewer steps, as a result it becomes suitable for future clinical testing. This novel technology based on POC was applied for rapid diagnostic tests for malaria, which is shown in Figure 7.6 [40]. In this method, a MWCNTs@ZnO nanofiber-based chemiresistive biosensor was synthesized which was used for the analysis of a malaria biomarker, histidine-rich protein II (HRP-2). The MWCNTs@ZnO nanofibers were prepared by applying the electrospinning technique and calcinations process. The functional groups stay on the nanofiber surface and are applied to coat histamine-rich protein II antibodies in the absence of surface modification. The sensitivity obtained from this device is 8.29 kΩ/g mL along with the large optimum range of 10 fg/mL–10ng/mL. For rapid detection of malaria, a MWCNTs@ZnO nanofiber-based flexible chemiresistive biosensor based on POC may apply [94]. Geldert et al. also developed a paper-based MoS_2 nanosheet-mediated fluorescence resonance energy transfer (FRET) aptasensor for rapid malaria diagnosis in which the aptamer immobilized to a malarial biomarker and *Plasmodium falciparum* lactate dehydrogenase. A resource area is required after using the POC device, such as those where malaria is frequent [95].

7.3.5 Biosensors for PoC Diagnosis of Human Immunodeficiency Virus (HIV)

Firstly, acquired immunodeficiency syndrome (AIDS), which is spread due to human immunodeficiency virus (HIV), was recorded by the US Centers for Disease Control in 1981. Recently, AIDS acquired a great challenge for a cure because of rapidly rising infection rates. According to the WHO, more than 35 million HIV-infected people were reported in 2013 and around 940,000 people died from HIV-related causes worldwide in 2017. The two types of the virus, HIV and human T-cell lymph tropic virus, act like retroviruses which cause a high mortality potential. HIV viruses are categorized into two types as HIV-1 and HIV-2, classified and based on their primary stage. The HIV-1 virus was generated from simian immunodeficiency virus (SIV) and is found in chimpanzees which live in Central or West Central Africa while the HIV-2 virus was generated from monkey species which are generally found in the coastal range in West Africa. HIV infection in humans is spread via inter-species contamination by the slaughter and consumption of an HIV-virus carrier [96–98]. Two conventional methods were applied for diagnosis of blood-borne pathogens (HIV) like the ELISA and nucleic acid amplification by PCR, but they needed specific reagents, i.e., particular buffers and enzymes, and the basic problems are costly, wide, and complex. To overcome this drawback, mass detection devices modified with piezoelectric materials have the ability to generate surface acoustic waves which are used for the diagnosis of HIV [99]. For early detection, the rapid development of POC diagnostic technologies spread worldwide. This is based on any diagnostic test governed by an external main laboratory at or near the patient [100]. Kaushik et al. developed an electrochemical sensor based on POC for detection of plasma cortisol in HIV. The complete integrated and optimized electrochemical device shows a large cortisol-optimum range as 10 pg/mL^{-1}–500 ng/mL^{-1}, along with LOD, and sensitivity is 10 pg/mL^{-1} and 5.8 μA/pg/mL. The cortisol-selective sensing device was

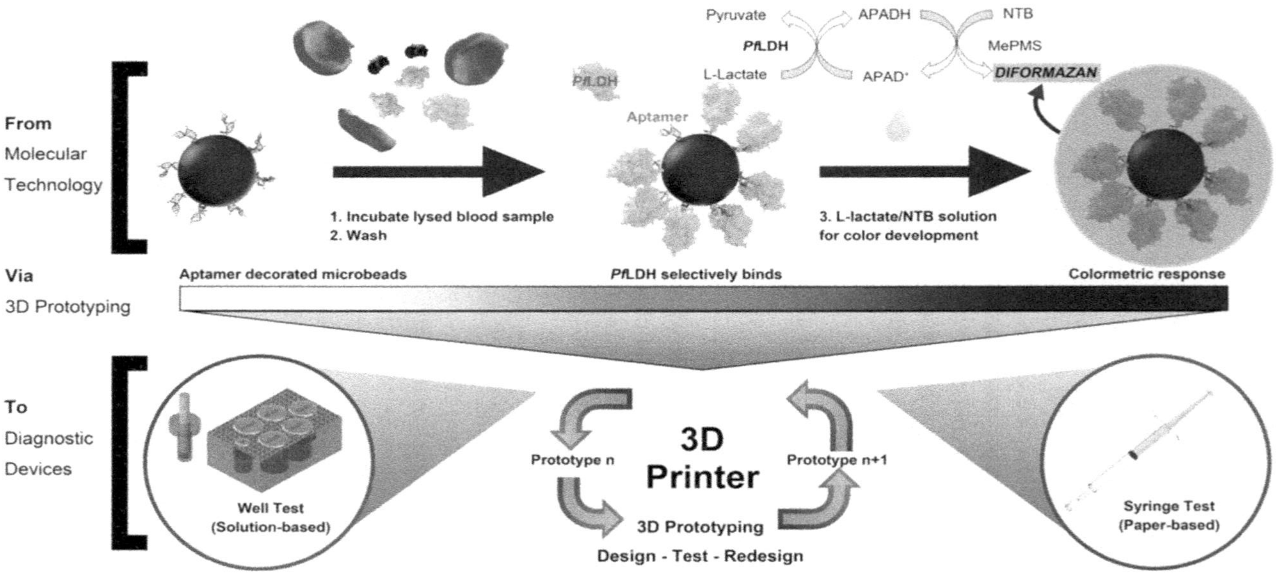

FIGURE 7.6 A 96-well plate-based assay for malaria diagnosis (APTEC) was adapted into a microbead-based assay for enhanced functionality. The assay functions by using aptamer-functionalized micro beads to capture Pf LDH from patient blood samples. After washing, a development reagent is added which, using the enzymatic activity of Pf LDH, produces a blue response for malaria-positive samples. Using 3D printing to aid rapid prototyping and redesign, this microbead-based assay was incorporated into two new prototypes for POC testing: A well test and a syringe test. Copy with permission taken from [91].

applied to determine plasma cortisol in ten samples from HIV patients [101]. Also, an electrochemical DNA biosensor was used for sensitive and selective analysis of HIV-related gene diagnosis. The DNA biosensor depended on the combination of two components as hybrids – double-stranded DNA and protein. The target or probe hybridization comes across the NF-kB protein in the presence of target DNA. They synthesized a sandwich-like DNA biosensor with the addition of horseradish peroxidase (HRP). They formed a sandwich-like ds/NF-B/HRP mixed compound which catalyzed the H_2O_2-mediated oxidation of 3,3,5,5-tetramethylbenzidin (TMB) with changing the color of solution which is why it enhanced the electrochemical current signal. However, no duplex DNA could bind NF-kB protein or HRP to prepare the DNA biosensor in the absence of target DNA. The DNA biosensor was used to detect target DNA as low as 7.05 pM. The proposed biosensor based on POC was used for early diagnosis of HIV, which is explained in Figure 7.7 [102]. In this case, a bioactivated PDMS micro-scale sensor was used for the analysis of human CD^{4+} cells; this acts as a POC device which is used for HIV detection. The surface of the biosensor is immobilized with heparin-based material which increases the water-loving and thrombo-resistance to the PDMS material. In this work, without specific adsorption of CD^{4-} leucocytes it is found to be insignificant, but detection is completed by using $HOECHST^{+}$ cells [103]. Surface acoustic wave biosensors were prepared, leading to very low-cost components obtained in smartphones which were used to diagnose HIV in 133 patient samples. Completion of this process required small, portable, dual-channel biochips and a laboratory precursor, along with reference control coating and minimum complication needed 6 μL plasma. Biochips were approved with 31 plasma samples, taken from an HIV-infected patient and 102 noninfected people. The sensitivity found for individual biomarkers is 100%, i.e., anti-gp41 and 64.5%, i.e., anti-p24 and 100% specificity, obtained in 5 min. For the next generation of this biosensor, it will be connected with POC HIV tests [104].

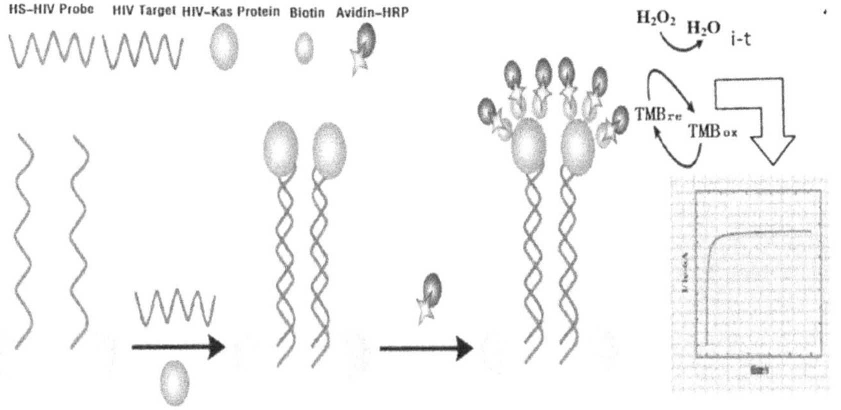

FIGURE 7.7 Schematic illustration of the proposed biosensor. Copy with permission taken from [99].

7.3.6 Biosensors for POC Diagnosis of Bilharzia

Schistosomiasis or bilharzia has infected ~200 million people in the world, of which ~25 million have lost their life. It is caused by five blood fluke species of the genus including *Schistosoma*: *S. japonicum*, *S. intercalatum*, *S. haematobium*, *S. mansoni*, and *S. mekongi*. Schistosomiasis was eliminated in two countries such as Japan the coastal plain of the People's Republic of China in the last two decades via comprehensive multidisciplinary campaigns. The Philippines has more people affected by this endemic disease while Indonesia recorded a lower number of affected people [105]. Accordingly, the WHO formed schistosomiasis control programs in the affected regions in nations in the last decade [106]. There is different immune reactivity of synthetic peptides like P_1, P_2, P_3, P_4, P_5, P_6, and P_7 which is generated by a sequence of amino acids in proteins from different antigenic formations. This process is useful for the diagnosis of *Schistosoma mansoni* [107]. Odundo et al. developed a nanobiosensor which interacts with a nanostrip with immobilized gold NPs, after that they are conjugated with antibodies (bilharzia) which have the potential for the diagnosis of antigen (bilharzia). The proposed biosensor shows a good linear range of 1.13 × 101 ng/mL to 2.3 × 103 ng/mL along with the limit of detection of 8.3887×10^{-2} ng/mL for bilharzia [108].

7.4 CONCLUSION

The main aim of this work is to provide information about the types of integrated biosensor and their essential properties. The preparation of biosensors depends on two basic units, one is recognition elements and second is immobilization techniques. Based on recognition elements, they are categorized into three integrated biosensors such as enzyme-modified biosensors, aptamer-modified biosensors, and antibody-modified biosensors, while the second is based on the types of transducers used. They are categorized into five types as electrochemical-modified biosensors, optical-modified biosensors, colorimetric biosensors, mass biosensors, and magnetic biosensors. The integrated biosensors have some limitations, such as huge cost, bulky size, and requirement for professional operation. To resolve this problem, it has been linked with POC diagnosis to form an integrated biosensor based on POC medical diagnosis which has provided rapid and accurate results. They are useful for diseases (cancer, diabetes, infectious diseases, malaria, human immunodeficiency virus (HIV), and schistosomiasis or bilharzia) diagnosed at a very early stage in the absence of professional experts and laboratory equipment ultimately useful for global health. The integrated biosensor based on POC diagnosis has various important properties like being easily prepared, recyclable, low cost, sensitive, specific, rapid, user friendly, and an effective miniaturized platform. Therefore, we can use these devices to obtaina wide range of analytes quantitatively or qualitatively. It is clear from the data discussed, that the future of integrated biosensors based on POC biomedical diagnosis will be used as a result of the success of emerging technologies in different areas such as biotechnology, biochemistry, biology, materials science, medicine, and physics.

REFERENCES

1. Griffin, G. D., D. N. Stratis-Cullum, and T. E. McKnight. 2019. Biosensors. In *Encyclopedia of Microbiology*, ed. T. M. Schmidt, 542–574. Elsevier.
2. Dang, Y. T., S. Gangadoo, P. Rajapaksha, V. K. Truong, D. Cozzolino, and J. Chapman. 2021. *Biosensors in Food Traceability and Quality. Comprehensive Foodomics*, ed. A. Cifuentes, 308–321. Elsevier.
3. Chakraborty, M., and M. S. J. Hashmi. 2017. *An Overview of Biosensors and Devices. Reference Module in Material Science and Material Engineering*, 1–24. Elsevier.
4. Santoro, K., and C. Ricciardi. 2016. Biosensors. In *Encyclopedia of Food and Health, Food Science*, 430–436. Elsevier.
5. Ye, Y., J. Ji, Z. Sun, P. Shen and X. Sun. 2019. Recent advances in electrochemical biosensors for antioxidant analysis in food stuff. *Trends Anal. Chem.* 122:115718–115759.
6. Luka, G., A. Ahmadi, H. Najjaran, E. Alocilja, M. D. Rosa, K. Wolthers, A. Malki, H. Aziz, A. Althani, and M. Hoorfar. 2015. Microfluidics integrated biosensors: A leading technology towards lab-on-a-chip and sensing applications. *Sensors* 15, no. 12:30011–30031.
7. Pérez, J. A. C., J. E. Sosa-Hernández, S. M. Hussain, M. Bilal, R. Parra-Saldivar, and H. M. N. Iqbal. 2019. Bioinspired biomaterials and enzyme-based biosensors for point-of-care applications with reference to cancer and bio-imaging. *Biocatal. Agric. Biotechnol.* 17:168–176.
8. Guo, Z., J. R. Wilson, I. A. York, and J. Stevens. 2018. Biosensor-based epitope mapping of antibodies targeting the hemagglutinin and neuraminidase of influenza A virus. *J. Immunol. Methods* 461:23–29.
9. Xiang, W. Q. Lv, H. Shi, B. Xie, and L. Gao. 2020. Aptamer-based biosensor for detecting carcinoembryonic antigen. *Talanta* 214:120716–120733.
10. Kumar, H., N. Kumari, and R. Sharma. 2020. Nanocomposites (conducting polymer and nanoparticles) based electrochemical biosensor for the detection of environment pollutant: Its issues and challenges. *Environ. Impact Assess. Rev.* 85: 106438–106452.
11. Tereshchenko, A., M. Bechelany, R. Viter, V. Khranovskyy, V. Smyntyna, N. Starodub, and R. Yakimova. 2016. Optical biosensors based on ZnO nanostructures: Advantages and perspectives. A review. *Sens. Actuators B* 229:664–677.
12. Kal-Koshvandi A. T., 2020. Recent advances in optical biosensors for the detection of cancer biomarker α-fetoprotein (AFP). *Trends Anal. Chem.* 128:115920–115967.
13. Xianyu, Y., Q. Wang, and Y. Chen, 2018. Magnetic particles-enabled biosensors for point-of-care testing. *Trends Anal. Chem.* 106:213–224.
14. Vashist, S. K. 2017. Point-of-care diagnostics: Recent advances and trends. *Biosensor* 74, no. 4:1–4.
15. Syedmoradi, L., M. Daneshpour, M. Alvandipour, F. A. Gomez, H. Hajghassem, and K. Omidfar. 2017. Point of care testing: The impact of nanotechnology. *Biosens. Bioelectron.* 87:373–387.
16. Parihar, A., P. Ranjan, S. K. Sanghi, A. K. Srivastava, and R. Khan. 2020. Point-of-care biosensor-based diagnosis of COVID-19 holds promise to combat current and future pandemics. *ACS Appl. Bio Mater.* 3:7326–7343.
17. Suleman S., S. K. Shukla, N. Malhotra, S. D. Bukkitgar, N. P. Shetti, R. Pilloton, J. Narang, Y. N. Tan, T. M. Aminabhavi. 2021. Point of care detection of COVID-19: Advancement in biosensing and diagnostic methods. *Chem. Eng. J.* 414:128759–128773.
18. Choi, J. R. 2020. Development of point-of-care biosensors for COVID-19. *Front Chem.* 8:517.
19. Ruiz-Vega, G., M. Soler and L. M Lechuga. 2020. Nanophotonic biosensors for point-of-care COVID-19 diagnostics and coronavirus surveillance. *J. Phys. Photon.* 3:011002–011010.

20. Rahimi, P. and Y. Joseph. 2019. Enzyme-based biosensors for choline analysis: A review. *Trends Anal. Chem.* 110:367–374.
21. Songa, E. A. and J. O. Okonkwo. 2016. Recent approaches to improving selectivity and sensitivity of enzyme-based biosensors for organophosphorus pesticides: A review. *Talanta* 155:289–304.
22. Bai, Y-H. Y. Du, J-J. Xu, and H-Y. Chen. 2007. Choline biosensors based on a bi-electrocatalytic property of MnO_2 nanoparticles modified electrodes to H_2O_2. *Electrochem. Commun.* 9, no. 10:2611–2616.
23. Magar, H. S., M. E. Ghica, M. N. Abbas, and C. M. Brett., 2017. Highly sensitive choline oxidase enzyme inhibition biosensor for lead ions based on multiwalled carbon nanotube modified glassy carbon electrodes. *Electroanalysis* 29, no. 7:1741–1748.
24. Peña-Bahamonde, J., H. N. Nguyen, S. K. Fanourakis, and D. F. Rodrigues. 2018. Recent advances in graphene-based biosensor technology with applications in life sciences. J. *Nanobiotechnology* 16, no. 1:75.
25. Zhang, Z., J. Wang, X. Wang, Y. Wang, and X. Yang. 2010. A sensitive choline biosensor with supramolecular architecture. *Talanta* 82, no. 2:483–487.
26. Holzinger, M., A. L. Goff, and S. Cosnier, 2012. Carbon nanotube/enzyme biofuel cells. *Electrochimica Acta* 82:179–190.
27. Nabih, S., and S. S. Hassn. 2020. Chitosan-capped Ag-Au/rGO nanohybrids as promising enzymatic amperometric glucose biosensor. *J. Mater. Sci.-Mater. Electron.* 31:3352–13361
28. Mohammad, M., A. Razmjou, K. Liang, M. Asadnia, and V. Chen. 2018. Metal-organic-framework-based enzymatic microfluidic biosensor via surface patterning and biomineralization. *ACS Appl. Mater. Interfaces* 11, no. 2:1807–1820.
29. Muguruma, H., T. Hoshino, and Y. Matsui., 2011. Enzyme biosensor based on plasma-polymerized film-covered carbon nanotube layer grown directly on a flat substrate. *ACS Appl. Mater. Interfaces* 3, no. 7:2445–2450.
30. Azimi, S., A. Farahani, and H. Sereshti. 2020. Plasma-functionalized highly aligned CNT-based biosensor for point of care determination of glucose in human blood plasma. *Electroanalysis* 32, no. 2:394–403.
31. Kou, X., L. Tong, Y. Shen, W. Zhu, L. Yin, S. Huang, F. Zhu, G. Chen, and G. Ouyang. 2020. Smartphone-assisted robust enzymes@ MOFs-based paper biosensor for point-of-care detection. *Biosens. Bioelectron.* 156:112095–112104.
32. Kawde, A-N., X. Mao, H. Xu, Q. Zeng, Y. He, and G. Liu. 2010. Moving enzyme-linked immunosorbent assay to the point-of-care dry-reagent strip biosensors. *Am. J. Biomed. Sci.* 2, no.1:23–32.
33. Ueda, H., and J. Dong. 2014. from fluorescence polarization to Quenchbody: Recent progress in fluorescent reagentless biosensors based on antibody and other binding proteins. *Biochim. Biophys. Acta* 1844, no. 11: 1951–1959.
34. Conroy, P. J., S. Hearty, P. Leonard, and R. J. O'Kennedy. 2009. Antibody production, design and use for biosensor-based applications. *Semin. Cell Dev. Biol.*, 20, no. 1:10–26.
35. Sivashankar, S., C. Sapsanis, U. Buttner, and K. N. Salama. 2015. Flexible low-cost cardiovascular risk marker biosensor for point-of-care applications. *Electron. Lett.*, 51, no. 22:1746–1748.
36. Fabri-Faja, N., O. Calvo-Lozano, P. Dey, R. A. Terborg, M-C. Estevez, A. Belushkin, and F. Yesilköy. 2019. Early sepsis diagnosis via protein and miRNA biomarkers using a novel point-of-care photonic biosensor. *Anal. Chim. Acta* 1077:232–242.
37. Wang, C., L. Wang, S. Tadepalli, J. J. Morrissey, E. D. Kharasch, R. R. Naik, and S. Singamaneni. 2018. Ultrarobust biochips with metal–organic framework coating for point-of-care diagnosis. *ACS Sens.* 3, no. 2:342–351.

38. Hianik, T. 2018. Aptamer based biosensors. Surface science and electrochemistry. In *Encyclopedia of Interfacial Chemistry*, 11–19.
39. Razmi, N., B. Baradaran, M. Hejazi, M. Hasanzadeh, J. Mosafer, A. Mokhtarzadeh, and M. Guardia. 2018. Recent advances on aptamer-based biosensors to detection of platelet-derived growth factor. *Biosens. Bioelectron.* 113:58–71.
40. Dirkzwager, R. M., S. Liang, and J. A. Tanner. 2016. Development of aptamer-based point-of-care diagnostic devices for malaria using three-dimensional printing rapid prototyping. *ACS Sens.* 1, no. 4:420–426.
41. Khan, N. I., A. G. Maddaus, and E. Song. 2018. A low-cost inkjet-printed aptamer-based electrochemical biosensor for the selective detection of lysozyme. *Biosensors* 8, no. 1:1–18.
42. Grabowska, I., N. Sharma, A. Vasilescu, M. Iancu, G. Badea, R. Boukherroub, S. Ogale, and S. Szunerits. 2018. Electrochemical aptamer-based biosensors for the detection of cardiac biomarkers. *ACS Omega* 3, no. 9:12010–12018.
43. Rowe, A. A., E. A. Miller, and K. W. Plaxco. 2010. Reagent less measurement of amino glycoside antibiotics in blood serum via an electrochemical, ribonucleic acid aptamer-based biosensor. *Anal. Chem.* 82, no. 17:7090–7095.
44. Ali, J., J. Najeeb, M. A. Ali, M. F. Aslam, and A. J. J. B. B. Raza. 2017. Biosensors: Their fundamentals, designs, types and most recent impactful applications: A review. *J. Biosens. Bioelectron.* 8, no. 1:1–9.
45. A-Angeles, G., G. A. Á-Romero, and A. Merkoçi. 2018. *Electrochemical Biosensors: Enzyme Kinetics and Role of Nanomaterials. Surf. Sci. Electrochem.* 140–155.
46. Liu, C. C. 2012. Electrochemical based biosensors. *Encyclopedia of Interfacial Chemistry: Surface Science and Electrochemistry* 2, no. 3:269–272.
47. Mongra, A. C., A. Kaur, and R. K. Bansal. 2012. Review study on electrochemical-based biosensors. *Measurements* 2, no. 2:743–749.
48. Shabaninejad, Z., F. Yousefi, A. Movahedpour, Y. Ghasemi, S. Dokanehiifard, S. Rezaei, R. Aryan, A. Savardashtaki, and H. Mirzaei. 2019. Electrochemical-based biosensors for microRNA detection: Nanotechnology comes into view. *Anal. Biochem.* 581:13349–113361.
49. Kilic, T., S. N. Topkaya, D. O. Ariksoysal, M. Ozsoz, P. Ballar, Y. Erac, and O. Gozen. 2012. Electrochemical based detection of microRNA, mir21 in breast cancer cells. *Biosens. Bioelectron.* 38, no. 1:195–201.
50. Shao, B., and Z. Xiao. 2020. Recent achievements in exosomal biomarkers detection by nanomaterials-based optical biosensors: A review. *Anal. Chim. Acta.* 1114:74–84.
51. Chen, Y., J. Liu, Z. Yang, J. S. Wilkinson, and X. Zhou. 2019. Optical biosensors based on refractometric sensing schemes: A review. *Biosens. Bioelectron.* 144: 111693–117047.
52. Maniya, N. H., and D. N. Srivastava. 2020. Fabrication of porous silicon based label-free optical biosensor for heat shock protein 70 detection. *Mater. Sci. Semicond. Process.* 115:105126–105135.
53. Eissa, S., L. L'Hocine, M. Siaj, and M. Zourob. 2013. A graphene-based label-free voltammetric immunosensor for sensitive detection of the egg allergen ovalbumin. *Analyst* 138, no. 15:4378–4384.
54. Zhang, Z., Q. Feng, M. Yang, and Y. Tang. 2020. A ratiometric fluorescent biosensor based on conjugated polymers for sensitive detection of nitroreductase and hypoxia diagnosis in tumor cells. *Sens. Actuators, B* 318:128257–128265.
55. Xiao, L., A. Zhu, Q. Xu, Y. Chen, J. Xu, and J. Weng. 2017. Colorimetric biosensor for detection of cancer biomarker by au nanoparticle-decorated Bi_2Se_3 nanosheets. *ACS Appl. Mater. Interfaces* 9, no. 8:6931–6940.

56. Hao, N., J. Lu, Z. Zhou, R. Hua, and K. Wang. 2018. A pH-resolved colorimetric biosensor for simultaneous multiple target detection. *ACS Sensors* 3, no. 10:2159–2165.
57. Bai, Y., H. Li, J. Xu, Y. Huang, X. Zhang, J. Weng, Z. Li, and L. Sun. 2020. Ultrasensitivity colorimetric biosensor for BRCA1 mutation based on multiple signal amplification strategy. *Biosens. Bioelectron.* 166:112424–112442.
58. Feng, J., P. Huang, and F-Y. Wu. 2017. Gold–platinum bimetallic nanoclusters with enhanced peroxidase-like activity and their integrated agarose hydrogel-based sensing platform for the colorimetric analysis of glucose levels in serum. *Analyst* 142, no. 21: 4106–4115.
59. Hoß, S. G., and G. Bendas. 2017. Mass-sensitive biosensor systems to determine the membrane interaction of analytes. In *Antibiotics*, 145–157. Humana Press.
60. Kabay, G., G. K. Can, and M. Mutlu. 2017. Amyloid-like protein nanofibrous membranes as a sensing layer infrastructure for the design of mass-sensitive biosensors. *Biosens. Bioelectron.* 97:285–291.
61. Galvan, D. D., V. Parekh, E. Liu, E-L.Liu, and Q. Yu. 2018. Sensitive bacterial detection via dielectrophoretic-enhanced mass transport using surface-plasmon-resonance biosensors. *Anal. Chem.* 90, no. 24:14635–14642.
62. Will, I., A. Ding, and Y. Xu. 2015. Proximity effect of magnetic permalloy nanoelements used to induce AMR changes in magnetic biosensor nanowires at specific receptor sites. *J. Magn. Magn. Mater.* 388, 5–9.
63. X., Yunlei, Q. Wang, and Y. Chen. 2018. Magnetic particles-enabled biosensors for point-of-care testing. *Trends Anal. Chem.* 106:213–224.
64. Xu, Y., and E. Wang. 2012. Electrochemical biosensors based on magnetic micro/nano particles. *Electrochim. Acta* 84:62–73.
65. Cortina, M. E., L. J. Melli, M. Roberti, M. Mass, G. Longinotti, S. Tropea, P. Lloret. 2016. Electrochemical magnetic microbeads-based biosensor for point-of-care serodiagnosis of infectious diseases. *Biosens. Bioelectron.* 80:24–33.
66. Khanmohammadi, A., A. Aghaie, E. Vahedi, A. Qazvini, M. Ghanei, A. Afkhami, A. Hajian, and H. Bagheri. 2020. Electrochemical biosensors for the detection of lung cancer biomarkers: A review. *Talanta* 206:120251–120265.
67. Aoki, K., N. Komatsu, E. Hirata, Y. Kamioka, and M. Matsuda. 2012. Stable expression of FRET biosensors: A new light in cancer research. *Cancer Sci.* 103, no. 4:614–619.
68. Qian, L., Q. Li, K. Baryeh, W. Qiu, K. Li, J. Zhang, and Q. Yu. 2019. Biosensors for early diagnosis of pancreatic cancer: A review. *Transl. Res.* 213:67–89.
69. Sadighbayan, D., K. Sadighbayan, M. R. T-Kia, A. Y. Khosroushahi, and M. Hasanzadeh. 2019. Development of electrochemical biosensors for tumor marker determination towards cancer diagnosis: Recent progress. *Trends Anal. Chem.* 118:73–88.
70. Chen, Y., B. S. Harrington, K. C. Lau, L. J. Burke, Y. He, M. Iconomou, J. S. Palmer, B. Meade, J. W. Lumley, and J. D. Hooper. 2017. Development of an enzyme-linked immunosorbent assay for detection of CDCP1 shed from the cell surface and present in colorectal cancer serum specimens. *J. Pharm. Biomed. Anal.* 139:65–72.
71. Qian, X., Q. Zhan, L. Lv, H. Zhang, Z. Hong, Y. Li, H. Xu, Y. Chai, L. Zhao, and G. Zhang. 2016. Steroid hormone profiles plus α-fetoprotein for diagnosing primary liver cancer by liquid chromatography tandem mass spectrometry. *Clin. Chim. Acta* 457:92–98.
72. Kim, M. S., T. Kim, S-Y. Kong, S. Kwon, C. Y. Bae, J. Choi, C. H. Kim, E. S. Lee, and J-K. Park. 2010. Breast cancer diagnosis using a micro fluidic multiplexed immunohisto chemistry platform. *PloS One* 5, no. 5:10441–10453.
73. Soper, S. A., K. Brown, A. Ellington, B. Frazier, G. G-Manero, V. Gau, S. I. Gutman. 2006. Point-of-care biosensor systems for cancer diagnostics/prognostics. *Biosens. Bioelectron.* 21, no. 10:1932–1942.

74. Chen, L-C., E. Wang, C.-S. Tai, Y-C. Chiu, C.-W. Li, Y-R. Lin, T-H. Lee, C-W. Huang, J-C. Chen, and W. L. Chen. 2020. Improving the reproducibility, accuracy, and stability of an electrochemical biosensor platform for point-of-care use. *Biosens. Bioelectron.* 155:112111–112119.
75. Cho, H., E-C. Yeh, R. Sinha, T. A. Laurence, J. P. Bearinger, and L. P. Lee. 2012. Single-step nanoplasmonic VEGF165 aptasensor for early cancer diagnosis. *ACS Nano* 6, no. 9:7607–7614.
76. Gai, P., R. Song, C. Zhu, Y. Ji, W. Wang, J-R. Zhang, and J-J Zhu. 2015. Ultrasensitive self-powered cytosensors based on exogenous redox-free enzyme biofuel cells as point-of-care tools for early cancer diagnosis. *Chem. Commun.* 51, no. 94:16763–16766.
77. Yoo, G., H. Park, M. Kim, W. G. Song, S. Jeong, M. H. Kim, H. Lee. 2017. Real-time electrical detection of epidermal skin MoS_2 biosensor for point-of-care diagnostics. *Nano Res.* 10, no. 3:767–775.
78. Kanchi, S., D. Sharma, K. Bisetty, and V. N. Nuthalapati. 2015. Diabetes and its effects: Statistics and biosensors. *J. Env. Anal Chem.* 2:111–116.
79. Sabu, C., T. K. Henna, V. R. Raphey, K. P. Nivitha, and K. Pramod. 2019. Advanced biosensors for glucose and insulin. *Biosens. Bioelectron.* 141:111201–111222.
80. Usman, F., J. O. Dennis, A. Y. Ahmed, F. Meriaudeau, O. B. Ayodele, and A. A. Rabih. 2018. A review of biosensors for non-invasive diabetes monitoring and screening in human exhaled breath. *IEEE Access* 7:5963–5974.
81. Zhai, S., C. Fang, J. Yan, Q. Zhao, and Y. Tu. 2017. A label-free genetic biosensor for diabetes based on AuNPs decorated ITO with electro-chemiluminescent signaling. *Anal. Chim. Acta* 982:62–71.
82. Maity, S., S. Ghosh, T. Bhuyan, D. Das, and D. Bandyopadhyay. 2020. Microfluidic immunosensor for point-of-care-testing of beta-2-microglobulin in tear. *ACS Sustain. Chem. Eng.* 8, no. 25:9268–9276.
83. Kownacka, A. E., D. Vegelyte, M. Joosse, N. Anton, B. J. Toebes, J. Lauko, I. Buzzacchera. 2018. Clinical evidence for use of a noninvasive biosensor for tear glucose as an alternative to painful finger-prick for diabetes management utilizing a biopolymer coating. *Biomacromolecules* 19, no. 11 (2018): 4504–4511.
84. Guo, J., 2017. Smartphone-powered electrochemical dongle for point-of-care monitoring of blood β-ketone. *Anal. Chem.* 89, no. 17:8609–8613.
85. Márquez, A., J. Aymerich, M. Dei, R. R-Rodríguez, M. V-Carrera, J. P-Delgado, P. G-Gómez. 2019. Reconfigurable multiplexed point of Care System for monitoring type 1 diabetes patients. *Biosens. Bioelectron.* 136:38–46.
86. Li, Z., L. Leustean, F. Inci, M. Zheng, U. Demirci, and S. Wang. 2019. Plasmonic-based platforms for diagnosis of infectious diseases at the point-of-care. *Biotechnol. Adv.* 37, no. 8:107440–107463.
87. Peláez, E. C., M. C. Estevez, A. Mongui, M-C. Menéndez, C. Toro, O. L. H-Sandoval, J. Robledo, M. J. García, P. D. Portillo, and L. M. Lechuga. 2020. Detection and quantification of HspX antigen in sputum samples using plasmonic biosensing: Toward a real point-of-care (POC) for tuberculosis diagnosis. *ACS Infect. Dis.* 6, no. 5:1110–1120.
88. Xia, Y., Y. Chen, Y. Tang, G. Cheng, X. Yu, H. He, G. Cao, H. Lu, Z. Liu, and S-Y. Zheng. 2019. Smartphone-based point-of-care microfluidic platform fabricated with a ZnO nanorod template for colorimetric virus detection. *ACS Sensors* 4, no. 12:3298–3307.
89. Ragavan, K. V., S. Kumar, S. Swaraj, and S. Neethirajan. 2018. Advances in biosensors and optical assays for diagnosis and detection of malaria. *Biosens. Bioelectron.* 105:188–210.
90. Yan, S. L. R., F. Wakasuqui, and C. Wrenger. 2020. Point-of-care tests for malaria: Speeding up the diagnostics at the bedside and challenges in malaria cases detection. *Diagn Microbiol. Infect. Dis.* 98:115122–115154.

91. Krampa, F. D., Y. Aniweh, P. Kanyong, and G. A. Awandare. 2020. Recent advances in the development of biosensors for malaria diagnosis. *Sensors* 20, no. 3:799–820.
92. Fraser, L. A., A. B. Kinghorn, R. M. Dirkzwager, S. Liang, Y-W. Cheung, B. Lim, S. C-C. Shiu. 2018. A portable microfluidic aptamer-tethered enzyme capture (APTEC) biosensor for malaria diagnosis. *Biosens. Bioelectron.* 100:591–596.
93. Ruiz-Vega, G, Arias-Alpízar, K, de la Serna, E, Borgheti-Cardoso, LN, Sulleiro, E, Molina, I, Fernàndez-Busquets, X, Sánchez-Montalvá, A, Del Campo, FJ, Baldrich, E. 2020. Electrochemical POC device for fast malaria quantitative diagnosis in whole blood by using magnetic beads, Poly-HRP and microfluidic paper electrodes. *Biosens. Bioelectron.* 150:111925–111949.
94. Panigrahi, A. K., V. Singh, and S. G. Singh. 2017. A multi-walled carbon nanotube-zinc oxide nanofiber based flexible chemiresistive biosensor for malaria biomarker detection. *Analyst* 142, no. 12:2128–2135.
95. Geldert, A., and C. T. Lim. 2017. Paper based MoS_2 nanosheet-mediated FRET aptasensor for rapid malaria diagnosis. *Sci. Rep.* 7, no. 1:1–8.
96. Fatin, M. F., A. R. Ruslinda, M. M. Arshad, K. K. Tee, R. M. Ayub, U. Hashim, A. Kamarulzaman, and S. C. Gopinath. 2016. HIV-1 Tat biosensor: Current development and trends for early detection strategies. *Biosens. Bioelectron.* 78:358–366.
97. Mozhgani, S-Hamidreza, H. A. Kermani, M. Norouzi, M. Arabi, and S. Soltani. 2020. Nanotechnology based strategies for HIV-1 and HTLV-1 retroviruses gene detection. *Heliyon* 6, no. 5:4048–4064.
98. Farzin, L., M. Shamsipur, L. Samandari, and S. Sheibani. 2020. HIV biosensors for early diagnosis of infection: The intertwine of nanotechnology with sensing strategies. *Talanta* 206:120201–120215.
99. Bisoffi, M., V. Severns, D. W. Branch, T. L. Edwards, and R. S. Larson. 2013. Rapid detection of human immunodeficiency virus types 1 and 2 by use of an improved piezoelectric biosensor. *J. Clin. Microbiol.* 51, no. 6:1685–1691.
100. Zhang, S., Z. Li, and Q. Wei. 2020. Smartphone-based cytometric biosensors for point-of-care cellular diagnostics. *Nanotechnology Precis. Eng.* 3, no. 1: 32–42.
101. Kaushik, A., A. Yndart, R. D. Jayant, V. Sagar, V. Atluri, S. Bhansali, and M. Nair. 2015. Electrochemical sensing method for point-of-care cortisol detection in human immunodeficiency virus-infected patients. *Int. J. Nanomed.* 10:677–686.
102. Guo, Y., J. H. Chen, and G. Chen. 2013. A label-free electrochemical biosensor for detection of HIV related gene based on interaction between DNA and protein. *Sens. Actuators, B* 184:113–117.
103. Thorslund, S., R. Larsson, F. Nikolajeff, J. Bergquist, and J. Sanchez. 2007. Bioactivated PDMS microchannel evaluated as sensor for human CD^{4+} cells-The concept of a point-of-care method for HIV monitoring. *Sens. Actuators, B* 123, no. 2:847–855.
104. Gray, E. R., V. Turbé, V. E. Lawson, R. H. Page, Z. C. Cook, R. B. Ferns, E. Nastouli. 2018. Ultra-rapid, sensitive and specific digital diagnosis of HIV with a dual-channel SAW biosensor in a pilot clinical study. *NPJ Digit. Med.* 1, no. 1:1–8.
105. Olveda, D. U., Y. Li, R. M. Olveda, A. K. Lam, D. P. McManus, T. N. Chau, D. A. Harn, G. M. Williams, D. J. Gray, and A. G. Ross. 2014. Bilharzia in the Philippines: Past, present, and future. *Int. J. Infect. Dis.* 18:52–56.
106. Barsoum, R. S., G. Esmat, and T. El-Baz. 2013. Human schistosomiasis: Clinical perspective. *J. Adv. Res.* 4, no. 5:433–444.
107. de Oliveira, E. J., H. Y. Kanamura, K. Takei, R. D. C. Hirata, L. C. P. Valli, N. Y. Nguyen, I. C. Rodrigues, A. R. Jesus, and M. H. Hirata. 2008. Synthetic peptides as an antigenic base in an ELISA for laboratory diagnosis of schistosomiasis mansoni. *Trans. R. Soc. Trop. Med. Hyg.* 102, no. 4:360–366.
108. Odundo, J., N. Noah, D. Andala, J. Kiragu, and E. Masika. 2018. Development of an electrochemical nano-biosensor for rapid and sensitive diagnosis of bilharzia in Kenya. *S. Afr. J. Chem.* 71, no. 1:127–134.

8 Paper-Based Microfluidics Devices with Integrated Nanostructured Materials for Glucose Detection

Abhinav Sharma, Wejdan S. AlGhamdi, Hendrik Faber, and Thomas D. Anthopoulos

CONTENTS

DOI: 10.1201/9781003033479-8

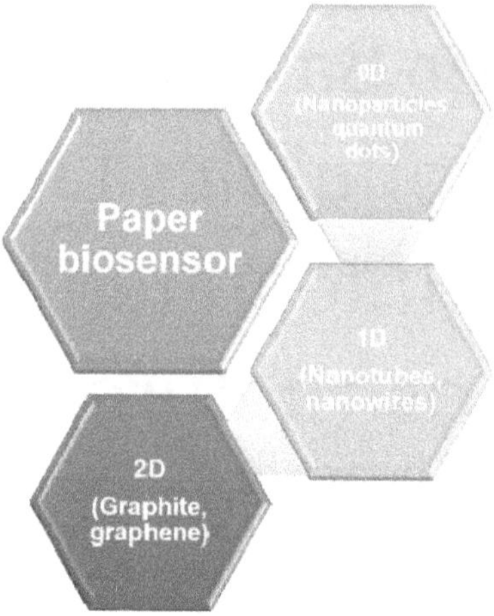

SCHEME 8.1 Schematic representation of nanostructured electrode materials integrated paper-based biosensors for glucose detection.

8.1 INTRODUCTION

8.1.1 Microfluidics Paper-Based Analytical Devices (μPADs)

The paper-based device is a natural platform to develop simple, cost-effective, disposable, and flexible microfluidics point-of-care (POC) diagnostics. They can be classified into the dipstick test, lateral flow assays (LFAs), and microfluidics paper-based analytical devices (μPADs). The commercial dipstick test was easy and simple to measure the glucose, protein, and other biomarkers in blood and urine samples; however, inaccuracy and long-time analysis are the main disadvantages. Afterward, LFAs have been widely studied over the last few decades for the rapid diagnosis of biomarkers and pathogens. In the LFAs, the sample flows horizontally towards a detection zone, while in the vertical flow immunoassays (VFAs), sample flow is directed vertically to a detection line; vertical flow serves more rapid detection using a low sample volume. In both assays, the labeled antibodies are used to capture and detect a biomolecule through a colorimetric or fluorescent reading. LFAs became a standard assay for clinical diagnostics such as pregnancy tests and blood coagulation screening (Kuswandi and Ensafi 2020). Despite the high potential of LFAs, there are several drawbacks regarding analytical performances, such as their accuracy, low sensitivity, and simultaneous multiple biomarker detection.

Since the Whitesides group first proposed μPADs in 2007 they have attracted considerable attention to develop and improve the performances of paper devices. These are ideal candidates for POC diagnostics, where centralized medical facilities are not available, and the cost of diagnosis is a major concern, especially in developing countries. These devices are also well suited to the food industry, biomedical, and environmental monitoring applications (Martinez et al. 2007), due to their several advantages such as low-cost fabrication, low sample volume, and they are portable, disposable, and biodegradable (Xie et al. 2019). The μPADs eliminate the problems of LFAs via creating microfluidics channel networks by patterning paper, which supports the direct flow to different regions of the device without the need for external pumps used in most traditional microfluidics devices and their ability to multiplex detection. Many different techniques, including wax printing (Lu et al. 2010), screen printing (Dungchai et al. 2011), laser-cutting (Chitnis et al. 2011), photolithography (Martinez et al. 2008), plasma treatment (Li et al. 2008), and chemical vapor-phase deposit (Lam et al. 2017) have been reported to fabricate hydrophilic/hydrophobic microstructures on paper substrates to construct μPADs. Despite the quick manufacture and high resolution of hydrophobic patterns and microfluidic channels on paper, some of these methods show limitations, such as the need for expensive instruments, multistep processing, and trained personnel. To avoid these limitations, some groups have developed simple origami techniques to construct μPADs. While μPADs may lack the high mechanical robustness compared to other materials (i.e., glass, silicon, polymers), they have other benefits over a paper substrate (Nge et al. 2013). The hydrophilic and porous nature of μPADs makes it possible to efficiently modify them with electrode material and easily diffuse a target analyte via capillary flow action without the need of external pumps or power supplies; they have the ability to store analytes within the fiber network and a high surface-area-to-volume ratio that can accelerate electron transfer, resulting in improved detection limits in the colorimetric method (Yang et al. 2017). Moreover, these devices can be quickly printed and modified to make a complex design and geometry using the origami technique (Liu and Crooks 2011). As a result of these benefits, paper devices were used in applications ranging from spot tests for metals (West 1945), paper chromatography (Fisher et al. 1948, Han et al. 2018), lateral flow immunoassays (Anfossi et al. 2018), and vertical flow immunoassay (Chen et al. 2019).

Moreover, μPADs can be implemented with other techniques such as electrochemistry (Fava et al. 2019), chemiluminescence (Yu et al. 2011), and fluorescence (Durán et al. 2016). Among them, the integration of μPADs with electrochemical detection approaches has shown to provide very useful, portable, accurate, and robust detection systems for glucose. Despite this substantial growth, much innovative work is needed to be a practical implementation of μPADs in future developments of POC diagnostics in clinical settings. This chapter explores the significant importance of the different nanostructured electrode materials used in the design and modifies the μPADs for glucose detection. In recent years, μPAD technology has developed significantly, especially in biomedical applications. Nanomaterial-integrated paper devices have evolved as an ideal platform for the desired analytical and clinical bio-sensing applications. Accordingly, the nanostructured materials integrated for μPADs have several advantages; they support the paper substrate and act as a working electrode that can accelerate electron transfer, they improve the electroanalytical performances, ease, and fast immobilization of the bioreceptor and analyte. Moreover, it can be utilized to enhance the conduction of charges (electronic and ionic) within the 3D ordered structure of cellulose paper (Ge et al. 2014).

8.1.2 Glucose Detection Techniques

Glucose monitoring from blood is of great importance in the diagnosis and treatment of diabetes. It is a metabolic disease that causes an abnormal blood sugar level, which consequently activates several metabolic complications. This disease has no cure so far. Hence, patients diagnosed with diabetes consistently need to monitor their blood glucose levels to avoid complications. In clinical medicine, only a glucometer is used to monitor the glucose levels from the patients' blood (Tonyushkina and Nichols 2009). In recent years, various types of paper-based glucose sensors that include electrochemical and optical methods have been reported. The conventional electrochemical biosensor consists of three separate electrodes – working electrode (WE), counter electrode (CE), and reference electrode (RE). The WE and CE material must be a chemically stable conductive material, such as carbon, gold, platinum, etc. The CE provides electron flow between the WE and CE, and closes the current circuit in the cell. The CE surface area must be much larger than WE, to avoid the kinetic limit of the process. The conductive Ag/AgCl ink is made of standard RE, which produces constant potential in the whole cell and balances the WE reaction. The electrochemical biosensor can be divided into amperometric, potentiometric, conductometric, and coulometric transducers, which transform the reaction of bioreceptor molecules into the measurable signal (Grieshaber et al. 2008).

To fabricate electrochemical microfluidics paper-based analytical devices (EμPADs), all three electrodes (WE, CE, RE) are patterned on paper using a printing technique. Electrochemical detection integrated with a μPAD plays an important role in glucose detection due to the advantage of low cost, high sensitivity and selectivity, minimal sample preparation, and short response time. In the enzymatic glucose sensor, the WE modified with glucose oxidase (GOx) is the most commonly used enzyme, to improve the selectivity of the sensor, and redox mediator [potassium ferricyanide/ferrocyanide ($K_3/K_4[Fe(CN)_6]$), ferrocene carboxylic acid (FCA), and Prussian blue (PB), etc.], which catalyzes glucose oxidation to produce hydrogen peroxide (H_2O_2) and gluconic acid simultaneously. It exhibits excellent performance in terms of sensitivity and selectivity. Colorimetric detection is the widely used method for μPADs due to the advantages of visual readout, speed, straightforward operation, and superior stability (Tian et al. 2018). GOx and horseradish peroxidase (HRP) are the commonly used bienzyme system to catalyze the reaction between glucose and the color indicator (chromic agent) in μPADs. The catalytic reaction of glucose in the presence of the enzyme results in H_2O_2 and gluconic acid. The HRP then catalyzes the reaction of H_2O_2 with a color indicator and generates a visual color change. Identifying an appropriate color indicator is one of the crucial steps in the advancement of μPADs for glucose concentration determination. Potassium iodide (i.e., KI) was one of the commonly used color indicators for the detection of glucose concentrations in colorimetric sensors. Other different organic dyes such as tetramethylbenzydine (TMB), o-phenylenediamine (OPD), 2,2′-azino-bis(3-ethylbenzothioazol ine-6-sulfonic acid) (ABTS), and nanoparticles were used as color indicators for a colorimetric signal of glucose.

8.2 NANOSTRUCTURED ELECTRODE-INTEGRATED µPADS FOR GLUCOSE DETECTION

Nanomaterials (NMs) emerge as promising candidates in electronic devices due to their remarkable electrical, optical, mechanical, and thermal properties (Ge et al. 2014, Pirzada and Altintas 2019, Su et al. 2017). They were developed in the early 1900s with the evolution of quantum mechanics that predicts the unique behavior of materials, as they are sized below 100 nm. Based on the nanoscale dimensions, NMs are classified into 0D (i.e., nanoparticles, quantum dots), 1D (i.e., nanowires, nanorods, nanotubes, nanobelts, and nanofibers), and 2D (i.e., graphene, boron nitride, metal dichalcogenides) (Pokropivny and Skorokhod 2008). NMs are engineered in laboratories using different strategies (i.e., top-down and bottom-up methods) and integrated into electronic devices for biosensing and bioimaging applications (Batool et al. 2019, Pirzada and Altintas 2019, Rani et al. 2019). Recently, biosensor researchers have tended to design low-cost and simple devices that are suitable for POC diagnosis. In this regard, µPADs have received great attention in the last decade and have shown a good capability to monitor different biomolecules (Boobphahom et al. 2020).

The major challenges for the development of high-performance paper-based biosensors are lack of robustness due to low mechanical strength, low electroanalytical performances, mainly in terms of sensitivity and stability; most paper-based sensors tend to rapidly lose their performance after storage for a few days and thus limit the lifetime of the sensor. The biosensor selectivity of the target analyte is mainly determined by the biorecognition element, while the transducer greatly influences the sensitivity of the biosensor. To overcome these limitations, modifying the paper surfaces with the conductive electrode materials in a way that improves the µPAD's performance and stability is still a difficult task. Many research groups have integrated various metals (Núnez-Bajo et al. 2017, Parrilla et al. 2017), metal oxide nanowires (Li et al. 2015) and nanoparticles (Cinti et al. 2018, Deng et al. 2014, Evans et al. 2014, Núnez-Bajo et al. 2018), carbon nanomaterials (Amor-Gutiérrez et al. 2019, Jia et al. 2018, González-Guerrero et al. 2017), and quantum dots (QDs) (Durán et al. 2016, Yuan et al. 2012) to develop paper-based glucose biosensors. In biosensing applications, NMs were used to modify the working electrode, which helps to accelerate the electron transfer to enhance sensitivity and stability. A combination of NMs with a paper-based device shows great potential for monitoring biomolecules. This chapter discusses the most common electrode materials used in the development of µPADs for glucose detection, including carbon-based materials (carbon nanotubes, carbon ink, graphene ink, and graphite), nanowires (ZnONWs), nanoparticles (AuNPs, CeO_2NPs, Fe_3O_4NPs, PBNPs, SiO_2NPs, SnO_2NPs, etc.) and quantum dots (CdSe/ZnS, CdTe).

As an overview of relevant data, Table 8.1 collects values from selected literature reports, and Figure 8.1 provides a graphical representation of the linear detection ranges that were established, broken down by electrode category and test medium. It can be observed that the majority of studies were targeting carbon-based materials as electrodes in paper-based µPADs and made use of various buffer solutions as

TABLE 8.1
Comparative Study of Microfluidics Paper-Based Devices (µPADs) using Different Nanostructured Electrode Materials for Glucose Detection

Electrode Materials	Biofunctionalization	Sensing Type	Linear Range	LOD	References
Carbon ink	GOx	CA	5–20 mM (urine)	0.35 mM	(Zhao et al. 2013)
Carbon ink	GOx	CA	1–20 mM (serum)	1 mM	(Li et al. 2017)
Carbon ink	GOx/HRP	CA	0.3–15 mM (buffer)	0.12 mM	(Amor-Gutiérrez et al. 2017)
Carbon ink	GOx/HRP	CA	0.5–15 mM (buffer)	0.4 mM	(Amor-Gutiérrez et al. 2019)
Carbon ink	GOx	CA	0.2–22.2 mM (urine)	0.22 mM	(Nie et al. 2010)
Carbon ink	GOx	EC	2–20 mM (urine)	2 mM	(Li et al. 2015)
Carbon ink	CTS/CBNPs/GOx	CA	0.1–40 mM (buffer) 2.5–10 mM (urine)	0.03 mM (buffer) 2.5 mM (urine)	(Fava et al. 2019)
Carbon ink	GOx	CA	5–17.5 mM (buffer)	5 mM	(Punjiya et al. 2018)
Carbon ink	GOx	CA	0.5–10 mM (buffer), 0.25–0.75 mM (urine)	0.35 mM (buffer), 0.25 mM (urine)	(Yao and Zhang 2016)

(Continued)

TABLE 8.1 (CONTINUED)
Comparative Study of Microfluidics Paper-Based Devices (μPADs) using Different Nanostructured Electrode Materials for Glucose Detection

Electrode Materials	Biofunctionalization	Sensing Type	Linear Range	LOD	References
Carbon ink	GOx	CA, CM	CA; 1–40 mM CM; 0–10 mM (buffer)	CA; 0.32 mM CM; 1.3 mM	(Wu et al. 2019)
Carbon ink/PB	GOx	CA	2–100 mM (buffer)	0.21 mM	(Dungchai et al. 2009)
MWCNTs	GDH/Os(dmobpy)PVI (anode), BOx/Os(bpy)PVI (cathode)	CA	2.5–30 mM (buffer)	2.5 mM	(González-Guerrero et al. 2017)
CNT/PB	GOx	Amperometric	0.02–1.4 mM (CSF)	0.39 mM	(Qin et al. 2018)
GO	TMB/GOx/HRP	CM	0.05–1 mM (buffer), 0.1–0.75 mM (saliva)	0.02 mM (buffer) 0.1 mM (saliva)	(Jia et al. 2018)
Graphite ink	GDH	Coulometric	0.5–50 Mm (buffer)	0.33 mM	(Lamas-Ardisana et al. 2018)
PB/graphite ink	Nafion/GOx/ CTS/Nafion	Amperometric	0–1.9 mM (buffer), 0.4–1.5 mM (sweat)	5 μM (buffer) 0.4 mM (sweat)	(Cao et al. 2019)
Graphite pencil	GOx/FCA	CA	1–12 mM (buffer)	0.05 mM	(Li et al. 2016)
PEDOT:POS/ graphite ink	GOx	Optical	0.5–4.5 mM (urine)	0.5 mM	(Mohammadifar et al. 2019)

(Continued)

TABLE 8.1 (CONTINUED)
Comparative Study of Microfluidics Paper-Based Devices (μPADs) using Different Nanostructured Electrode Materials for Glucose Detection

Electrode Materials	Biofunctionalization	Sensing Type	Linear Range	LOD	References
Graphene/PFLO	GOx/AuNPs	Amperometric	0.1–1.5 mM (buffer)	0.081 mM	(Gokoglan et al. 2017)
PEGME/graphite	ABTS/GOx/HRP	CM	0.25–1.25 mM (buffer)	0.25 mM	(Mitchell et al. 2015)
Au	GOx/HRP	CA	0.1–15 mM (buffer)	0.11 mM	(Núnez-Bajo et al. 2017)
Pt	CTS/GOx	CA	2–10 mM (buffer)	2 mM	(Witkowska Nery et al. 2016)
Pt	Nafion/GOx /Nafion	Potentiometric	0.1–3 mM serum)	0.03 mM	(Parrilla et al. 2017)
Pt	Nafion/GOx /Nafion	Potentiometric	0.3–3 mM (buffer, serum)	0.2 mM (buffer), 0.1 mM (serum)	(Cánovas et al. 2017)
ZnONWs/Carbon ink	GOx/Nafion	CA	0-15 mM (buffer), 5.1–15.1 mM (serum)	94.7 μM (buffer), 115.2 μM (serum)	(Li et al. 2015)
AuNPs/PEDOT	GOx	CA	0–20 mM (buffer)	0.1 mM	(Määttänen et al. 2013)
AuNPs/Au	GOx	LSV	0.01–5 mM (buffer)	6 μM	(Núnez-Bajo et al. 2018)
PBNPs	GOx	CA	0–25 mM (buffer)	0.17 mM	(Cinti et al. 2018)
SnO_2	GOx	Conductometric	0.5–12 mM (buffer)	0.5 mM	(Mahadeva and Kim 2011)

(Continued)

TABLE 8.1 (CONTINUED)
Comparative Study of Microfluidics Paper-Based Devices (µPADs) using Different Nanostructured Electrode Materials for Glucose Detection

Electrode Materials	Biofunctionalization	Sensing Type	Linear Range	LOD	References
Fe_3O_4, MWCNT, GO	TMB/GOx/HRP	CM	0.05–1 mM (Fe_3O_4 and MWCNT), 0–1 mM (GO) (PBS)	43 µM (Fe_3O_4), 62 µM (MWCNT), and 18 µM (GO)	(Figueredo et al. 2016)
SiO_2 NPs	KI/GOx/HRP	CM	0.5–10 mM (PBS)	0.5 mM	(Evans et al. 2014)
GO/SiO_2/CeO_2	OPD/GOx	CM	0.5–30 mM (PBS)	0.2 mM (PBS)	(Deng et al. 2014)
CeO_2 NPs	APTS/CTS/GOx	CM	0.5–100 mM (PBS) 1–35 mM (serum)	0.5 mM (PBS) 1 mM (serum)	(Ornatska et al. 2011)
CdSe/ZnS QDs	GOx	Fluorescence	0.27–11 mM (buffer)	0.027 µM	(Durán et al. 2016)

ABTS: 2,2′-azino-bis(3-ethylbenzothioazoline-6-sulfonic acid), APTS: aminopropyltrimethoxysilane, AuNPs: gold nanoparticles, BOx: bilirubin oxidase, CA: chrono-amperometry, CB NPs: carbon black nanoparticles, CeO_2 NPs: cerium oxide nanoparticles, CM: colorimetric, CNT: carbon nanotube, CSF: cerebrospinal fluid, CTS: chitosan, CV: cyclic voltammetry, EC: electrochemical, EDC: *N*-(3-dimethylaminopropyl)-*N*′-ethylcarbodiimide hydrochloride, FCA: ferrocenecarboxylic acid, Fe_3O_4 NPs: iron oxide nanoparticles, GDH: glucose dehydrogenase, GO: graphene oxide, GOx: glucose oxidase, HRP: horseradish peroxidase, KI: potassium iodide, LSV: linear sweep voltammetry, Na_3PO_4 buffer: sodium phosphate buffer, MWCNT: multiple-walled carbon nanotube, NHS: *N*-hydroxysuccinimide, OD: O-dianisidine dihydrochloride, Os(dmobpy)PVI: $[Os(4,4'\text{-dimethoxy-}2,2'\text{-bipyridine})_2(\text{poly-vinylimidazole})_{10}Cl]^+$, Os(bpy)PVI: $[Os(2,2'\text{-bipyridine})_2(\text{poly-vinylimidazole})_{10}Cl]^+$, OPD: o-phenylenediamine, PB: prussian blue, PBNPs: Prussian blue nanoparticles, PEDOT:PSS: poly(3,4-ethylenedioxythiophene) polystyrene sulfonate, PEGME: poly(ethylene glycol) methyl ether, PFLO: poly(9,9-di-(2-ethylhexyl)-fluorenyl-2,7-diyl)-end capped with 2,5-diphenyl-1,2,4-oxadiazole, Pt: platinum, SnO_2 NPs: tin oxide nanoparticles, SPEs: screen-printed electrodes, TMB: tetramethylbenzydine, ZnO NWs: zinc oxide nanowires, QDs: quantum dots

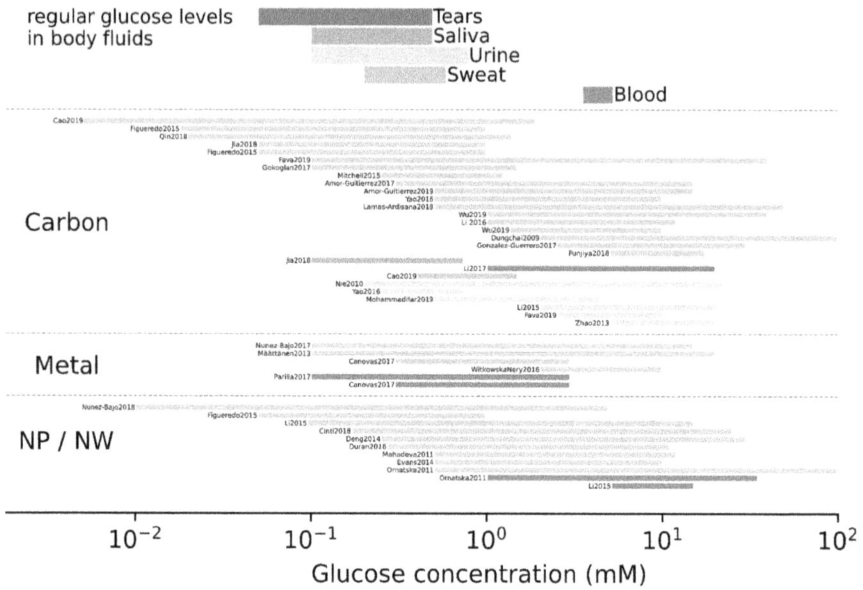

FIGURE 8.1 Overview of the glucose detection ranges of paper-based glucose sensors based on the values listed in Table 8.1. As a comparison, the top section shows the range of common glucose concentrations found in the various body fluids in healthy humans. Below, each entry from the table is drawn as a horizontal bar, indicating the linear range of detection that was determined in the study. The data in the plot is grouped by electrode categories (carbon materials, metals, and NPs/NWs). Additionally, the color of each bar represents the test medium, i.e., light blue for buffer solutions, green for saliva, yellow for urine, and orange for sweat. Note that blood and serum as the test medium are both combined into one color (red). Also, in studies where the lower limit of the linear range was stated as zero, the limit of detection (LOD) was used in this graph instead.

the test medium. The sensing of glucose in actual bodily fluids was carried out in considerably fewer cases, and no study used tears as the test medium. In the majority of cases, the linear detection range of the reported μPADs was large enough to be relevant for the glucose concentrations within the human body.

8.2.1 Carbon Nanomaterials

Carbon nanomaterials offer attractive opportunities to improve biosensor performance based on their ability to form different structures in the nanoscale. These nanostructures exhibit excellent properties such as high surface area, electrical conductivity, mechanical flexibility, and biocompatibility (Kour et al. 2020, Wang and Dai 2015). The incorporation of carbon nanomaterials into paper-based biosensors has been rapidly growing (Ge et al. 2014, Koo et al. 2016) with a wide range of nanostructures including single-walled carbon nanotubes (SWCNTs) (Shen et al. 2019), multiple-walled carbon nanotubes (MWCNTs) (Wang et al. 2018), graphene oxide (GO) (Jia et al. 2018), reduced graphene oxide (rGO) (Fan et al. 2019), and carbon quantum-dots (QD) (Tian et al. 2017).

8.2.1.1 Carbon Nanotubes

Carbon nanotubes (CNTs) are carbon allotropes with a cylindrical and hollow nanostructure that consists of sheets of graphene rolled into a cylindrical shape with a length of micrometers and a diameter of hundreds of nanometers (Tasis et al. 2006). Moreover, it is also considered to be a derivative of both carbon fibers and fullerenes (C_{60}) with 60 carbon atoms arranged in a cage-like fused-ring (Simon et al. 2019). Iijima first discovered CNTs in 1991 using the arc discharge method that applies a direct current arc voltage between two carbon electrodes under an inert atmosphere (He or Ar gas). The high temperature (3000–4000°C) between the electrodes causes sublimation of carbon electrodes as the growth of a large quantity of CNTs, which is successfully observed by transmission electron microscopy (Iijima 1991). Later on, several methods were proposed to synthesize CNTs, such as the laser ablation method (Chrzanowska et al. 2015) that uses a laser pulse to evaporate the graphite target in an electrical furnace at 1200°C, and the chemical vapor deposition (CVD) method (Öncel and Yürüm 2006) provide a high quality of CNTs via pyrolysis of hydrocarbon gases such as methane, ethylene at high temperature. To date, CVD is the most efficient way to obtain high purity CNTs with unique electronic properties (De Volder et al. 2013), good tensile strength (up to 63 GPa) (Yu et al. 2000), high chemical and thermal stability, which have the great potential used in various applications such as drug delivery, medicine, biosensors, industrial, and biomedical applications (He et al. 2013, Simon et al. 2019, Sireesha et al. 2018).

CNTs are categorized, based on the number of carbon layers, into (i) SWCNTs, typically occurring as hexagonal bundles, made of a single graphene layer with diameter ranges from 0.4–2 nm, (ii) MWCNTs, in which many SWCNTs are nested within each other. The number of nanotubes in a MWCNT varies from three to more than 20, while the diameter of both the inner nanotube and the outermost nanotube differs, from 2 nm to over 100 nm. MWCNTs have a hollow helical tubular structure in sp^2 hybridization that exhibits a large surface area and higher catalytic activity. The outer walls of MWCNTs can be modified with functional groups such as hydroxides (O-H), carboxylic acids (-COOH), or amides ($-CONH_2$) to improve their solubility and ability to attach the bioreceptors. However, the larger the number of walls, the higher the risk of defects, which leads to poor efficiency compared to SWCNTs. The yield, growth dimensions, and other unique properties of CNTs could be effectively tailored by the catalyst, gas precursor, flow rate, and temperature (Eatemadi et al. 2014). The existence of a catalytic agent is not mandatory for MWCNTs, whereas during the preparation of individual SWCNTs, catalytic agents (such as cobalt, yttrium, nickel, iron, etc.) are required.

For μPADs, MWCNTs can modify the porous structure of the paper (WE) to enhance its electronic conductivity and electrocatalytic activity (Wang et al. 2012). The porous nature of the paper allows efficient immobilization of a sensing material and easy diffusion of a target analyte. González-Guerrero et al. (2017) developed a flexible, portable, compact paper-based enzymatic microfluidic glucose/O_2 fuel cell that operates with a small sample volume (~35 μL), which is easily obtained by pricking the finger. The energy generated from the paper fuel cells was evaluated with different glucose concentrations from the whole blood sample. (Figure 8.2(a)) shows the carbon paper electrodes (20 mm^2) from Fuel Cell Earth (type TG-H-060) are modified by drop casting using a mixture of enzymatic ink, [which consists of the corresponding enzymes (glucose dehydrogenase (GDH) or bilirubin oxidase (BOx), redox mediator [Os(4,4′-dimethoxy-2,2′-bipyridine)$_2$(poly-vinylimidazole)$_{10}$Cl]$^+$ (Os(dmobpy) PVI at anode and [Os(2,2′-bipyridine)$_2$(poly-vinylimidazole)$_{10}$Cl]$^+$ Os(bpy)PVI) at cathode, crosslinker poly(ethylene glycol) diglycidyl ether (PEGDGE) and MWCNT]. The selection of enzymes for glucose oxidation at the anode, the GDH enzyme is preferably used due to its stability in the presence of oxygen in the solution (Tsujimura et al. 2006), bilirubin oxidase (BOx) enzyme is used for the electrical reduction of oxygen to water at the cathode (Ivanov et al. 2010). To construct the paper-based fuel cell, carbon paper electrodes (anode and cathode) are combined with a paper sample absorption substrate and enclosed within a plastic microfluidic casing. The selection of redox mediator is based on matching the appropriate redox potential that can enhance electron transfer between the enzyme and electrode. In comparison with ferrocene (Fc) mediators and its derivatives (Cass et al. 1984), osmium-based redox polymers have also been widely used in enzymatic reactions (Mano and Heller 2003). The main advantage of using osmium mediators is the flexibility of their redox potential, which can be tailored conveniently by changing the ligands attached to its central metal atom (Gallaway and Barton 2008), allowing a broad voltage range for both anode and cathode in glucose/O_2 enzymatic fuel cells (Osadebe et al. 2015). This combination of enzymes and mediators allowed

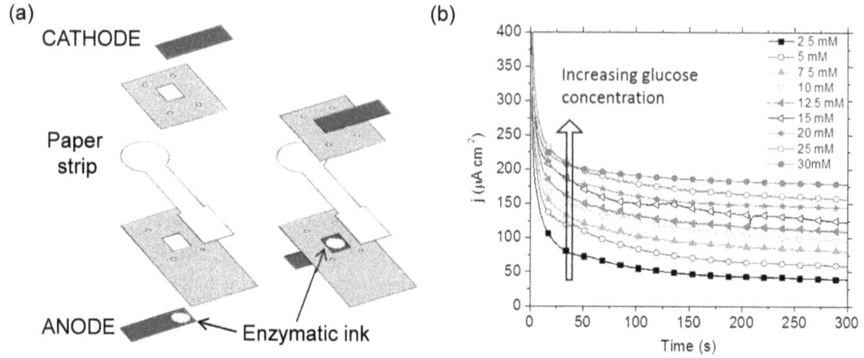

FIGURE 8.2 (a) The components of the device consist of paper strip (L-35 mm, W-5 mm) have a circular reception zone (D-10 mm), and the absorption region (10 x 5 mm); and the two carbon paper electrodes (5 x 15 mm^2). A pair of transparent pressure-sensitive adhesive (PSA) layers (thickness-75 μm, dimensions-30 × 15 mm^2 (bottom) and 17 × 15 mm^2 (top)) were added to the assembly to maintain the elements together. The PSA layers also define the active electrode area (4 × 5 mm^2) exposed to the paper substrate. (b) Chronoamperograms of enzymatic paper-based fuel cell. (Reprinted from González-Guerrero et al. 2017. Biosensors and Bioelectronics 90: 475–480, with permission from Elsevier.)

the fuel cell to be operated within one buffer solution at neutral pH. The paper fuel cells were performed at a fixed voltage (0.45 V) with different glucose concentrations (2.5–15 mM), showed a linear power output response with power densities ranging (from 20 to 90 μW cm^{-2}) (Figure 8.2(b)). The devices showed high variability, which is normal for a manual fabrication; however, they are still promising for a self-powered glucose sensor. Qin et al. (2018) proposed a simple, cost-effective carbon nanotube modified paper as a working electrode (CNTPE) for the detection of glucose in brain microdialysate. The CNTPE combined the advantages of both the electrochemical property of CNTs and the diversity function of the paper. CNTPE was directly fabricated through drop cast of CNT dispersion onto the filter paper and dried at 60°C, and CNT-coated paper was cut into 3-mm-wide strips. Scanning electron microscope (SEM) images of the CNTPE and Prussian blue (PB) were uniformly electrodeposited at the CNTPE, and enzyme (GOx) immobilized on PB/CNTPE, dried at room temperature. The GOx/PB/CNTPE exhibited good amperometric response towards the glucose concentrations (20–1.4 mM) in artificial cerebrospinal fluid (aCSF), showed the linear response, and detected the glucose concentration (0.39 ± 0.02 mM) in microdialysate. The low potential of PB towards the reduction of H_2O_2-enabled GOx/PB/CNTPE to interfere free from other electrochemical species in the brain. Therefore, CNTPE has remarkable specificity against the other electroactive species. Hence, this approach has great potential applications in the detection of neurochemicals for understanding the function of the brain.

8.2.1.2 Carbon Ink

Carbon ink is commonly made from lampblack or soot and a binding agent, which keeps holding carbon particles in suspension and sticking strongly to the paper. It can be printed using screen printing, dipping, and drawing with a ballpoint pen to make the working electrode onto the paper. Among the printing techniques, screen printing is considered the most versatile and widely used to fabricate μPADs. Carbon ink provides a high surface-area-to-volume ratio, good electrical conductivity, and excellent thermal and chemical stability. Moreover, unlike conventional conductive ink (Ag), carbon inks are cost effective, very resistant to abrasion, scratching, and flexing, and make them suitable candidates for the fabrication of printed-paper-based biosensors.

Carbon ink was first used for the fabrication of EμPADs through screen printing (Dungchai et al. 2009). Afterwards, it has recently been of interest for electrode fabrication on paper (Cate et al. 2015). However, other researchers employed it using different strategies like a pencil (Li et al. 2016a), pen on paper (Russo et al. 2011), stencil-printing process (Taleat et al. 2014). The printing method requires a stencil or a mask for patterning the electrochemical cell in order to separate the three electrodes (WE, CE, RE), whereas drawing directly with pencil/pen or following a pattern previously designed by software can replace using the masks, but the geometry has to be carefully optimized to obtain adequate analytical features.

Dungchai et al. (2009) were the first to report EμPADs for glucose detection. Photolithography was used to define the hydrophilic patterns on filter paper using a photoresist (SU-8 3025), and electrodes (WE and CE) are screen printed using carbon ink containing PB (C2070424D2) and Ag/AgCl ink as the RE and make contact pads to connect the potentiostat. The enzyme (GOx) modified WE successfully detected the different glucose concentrations in buffer solution and serum samples as well. This approach is an easy-to-use, inexpensive, and portable alternative for POC monitoring. Amor-Gutiérrez et al. (2017) developed a paper-based maskless platform for an amperometric enzymatic biosensor. It consists of a three-electrode configuration as carbon ink is used for the WE, and gold-plated pins for RE and CE. To prepare the WE, hydrophobic wax-defined paper area (d-4 mm) was coated with diluted carbon ink and dried overnight at room temperature. The carbon-coated electrodes are connected with a gold-plated connector header in a specific way (Figure 8.3(a)); the one in the center assists as a connection for the WE, and is placed below the paper, and the other two are located at the upper side and connected to the RE and CE to avoid short circuits in between. Using a gold-plated connector as a RE and CE forms a low-cost standard electronic connection to a potentiostat and works as a clip in order to support the paper between electrodes. The mixture of bienzyme (GOx/HRP) and potassium ferrocyanide ($K_4[Fe(CN)_6]$) as a redox mediator (Biscay et al. 2011) was deposited onto the paper working electrode (on the opposite side of the carbon ink). To optimize different concentrations of enzymes, the modified WE was combined with the screen-printed carbon electrode (SPCE), which integrates a silver pseudo-reference electrode and a carbon auxiliary electrode. In comparison, one integrated SPCE can be used for multiple (~10) measurements, while the paper WE is disposable and different for each measurement, but its connector head can operate multiple times without affecting the signal. The chronoamperometric measurements of paper-based biosensors were recorded by adding a small volume of different glucose concentrations (0.3–15 mM). This paper-based biosensor achieved good linearity with LOD of 0.12 mM, showed good reproducibility, stability, and specificity by being able to quantify glucose in real food samples. To scale up the production of this method, Amor-Gutierrez et al. developed an advanced multiplex (eight-channel platform) paper-based electrochemical device. They constructed eight electrochemical cells due to the commercial availability of eight-channel micropipettes as well as multichannel connectors and multipotentiostats (Figure 8.3(b)). For the easy sampling step of the microfluidics approach, a glass-fiber pad (sampler) is immersed in a container with the glucose solution, which flows by capillarity until it reaches the working electrode. This device shows a linear range (0.5–15 mM) of glucose concentrations with good precision (Amor-Gutiérrez et al. 2019). Li et al. (2017) developed a pen-on-paper (POP)-based microfluidics electrochemical platform for glucose detection in human serum. The POP approach is based on two custom-designed pens, which are made of a ballpoint pen filled with wax to write microfluidics channels and carbon ink to write electrodes on paper, which are integrated with the heating unit. In this process, the empty ballpoint pen was filled with liquid melted wax through pipetting prior to being cleaned properly by water and dried at room temperature, then the ball was pen inserted into a customized conducting tube surrounded by a heating coil wire, which were connected to a power supply and a thermostatic controller to transfer the heat uniformly to the wax inside the pen

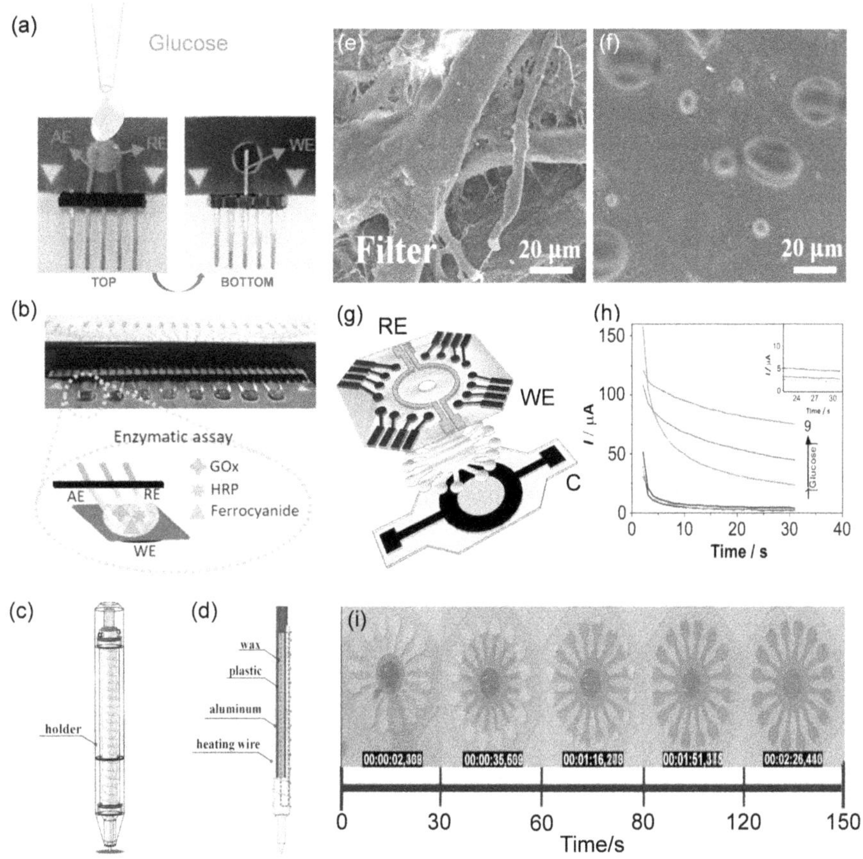

FIGURE 8.3 (a) A carbon-coated paper electrode (top and bottom) clipped with three gold-plated pins (reprinted from Amor-Gutiérrez et al. 2017. Biosensors and Bioelectronics 93: 40–45. with permission from Elsevier). (b) Schematic of an array of electrodes connected to the multipotentiostat; paper electrodes modified with enzymes, connected with gold-plated pins used as WE, CE, RE (reprinted from Amor-Gutiérrez et al. 2019. Biosensors and Bioelectronics 135: 64–70, with permission from Elsevier). (c) The inner structure of the wax pen and (d) custom-designed holder encompassed the conducting tube. SEM images of the wax patterns on a filter paper, (e) blank paper, and (f) wax-coated paper (reprinted from Li et al. 2017. Biosensors and Bioelectronics 98: 478–485, with permission from Elsevier). (g) The device was assembled by overlapping of each component's layer (WE, RE layer, micro-fluidic channel, CE layer), and the double-sided adhesive used to seal the multi-µEPaD. (h) Chronoamperograms recorded toward different glucose concentrations (0.1 to 40 mM) in 0.1 mol L^{-1} NaCl. (i) shows the microfluidic pattern (hexadecagon), and Allura red dye solution (1.0×10^{-3} mol L^{-1}) added in the sample spot and homogeneous elution of in the microfluidic pathway (reprinted from Fava et al. 2019. Biosensors and Bioelectronics 203: 280–286, with permission from Elsevier).

(Figure 8.3(c)). A custom-designed holder encompassed the conducting tube, as shown in (Figure 8.3(d)), which both are made of aluminum to make it handy for the users because it is a good thermal conductor and is lightweight. The fabrication of the conductive-ink pen was similar to that of the wax pen using a diluted carbon ink, which is the mixture of the carbon ink (Electrodage 423ss) and the diluent (SBJ-5) with the mass ratio of 1:0.6. This simple ink pen can write electrodes on paper without any extra pressure-assisted devices. Particularly, the wax pen is capable of realizing a one-step fabrication of wax channels on paper, as the melted wax penetrates into paper during the writing process without any post treatment. To verify wax coating on paper, the SEM image shows the melted wax filled in the pores of filter paper, result in the uniform coated on filter paper (Figure 8.3(e, f)). A fully written POP approach was successfully developed for the paper-based microfluidics biosensor for glucose detection in serum samples. It provides a promising perspective for POC applications; however, this approach mainly relies on handwriting, which makes it difficult to realize mass production. Li et al. (2015) demonstrate a simple approach to draw the electrodes on paper using a pressure-assisted ballpoint pen. It is composed of a ballpoint pen that was connected to the tip of the pressure-assisted device. The ballpoint pen is then filled with carbon (Electrodage 423ss) and Ag ink to write the WE, CE, and RE, respectively, using the PMMA mask. The proposed EμPAD was successfully employed to quantitatively analyze glucose in artificial urine. Punjiya et al. (2018) proposed a 3D-PAD integrated with a custom-designed potentiostat for glucose sensing. The device is fabricated using wax-printing, simple folding, and laser cutting via a cleanroom-free process without the need for expensive equipment. The WE and CE are screen printed with a carbon ink (applied ink solutions, C-200) and RE with Ag/AgCl ink. The device contains a hollow 3D analyte reservoir structure constructed from paper folding and double-sided adhesive tape to confine the analyte on the top of the SPEs, which allows a more uniform SPE top surface as the electroactive area without the need for external analyte containment. The WEs are functionalized with an enzyme (GOx), and CA measurement was carried out in a redox mediator with adding the glucose concentrations (5–17.5 mM) in PBS. The device shows the linear response with a sensitivity (0.34 μA mM^{-1}). The 3D-ePAD demonstrates a low-cost, versatile, self-contained system suitable for POC diagnostic devices. Fava et al. (2019) developed a multiplex microfluidics electrochemical paper-based device (multi-EμPaD) using all three screen-printed carbon electrodes for glucose detection in urine. First, microfluidics channels were produced using a simple craft cutter printer to cut the filter paper in the desired patterns. It consists of 16 replicate channels circularly distributed around to the sample spot, which provides a radial elution, with a homogeneous distribution of the solute into detection spots where the sample is analyzed. Afterward, the screen-printed carbon electrodes (WE, CE, RE) are deposited using carbon ink (Gwent Group, UK) via a vinyl adhesive mask attached to the polyester substrate. In the case of RE, Ag/AgCl ink was applied to the carbon electrodes to make an electrical contact. However, the counter electrode has different geometry and design and assembled with the sheet containing electrodes (WE, RE) and microfluidic channels in a way to keep in contact with all 16 individual microfluidics channels and electrodes (Figure 8.3(g)).

The WEs were previously modified with the mixture of redox mediator (ferrocene-carboxylic acid; FCA), chitosan (CTS), carbon black nanoparticles (CBNPs, type VXC-72R), and enzyme (GOx). The CTS was used to make stable aqueous dispersions of CBNPs, that produced the homogeneous and stable films on the electrode surface, and CBNPs are added to the CTS film in order to ensure the electrical contact, which enhanced a fast electron transfer rate and GOx enzyme and electrodes. This analytical device has a linear response with a glucose concentration range from 0.1 to 40 mM, with LOD of 0.03 mM (Figure 8.3(h)). To demonstrate the homogeneous flow behavior from the sample injection point to sensing points, Allura red dye solution was injected, and the complete microfluidics device filling time was less than 2.5 min (Figure 8.3(i)). This linear range of the multi-EμPaD covers the range of glucose concentration in human urine. Nie et al. (2010, Zhao et al. (2013), and Yao and Zhang (2016) deposited all three carbon electrodes (WE, CE, RE) using screen-printed carbon ink (E3456) on wax-patterned paper, which is able to detect the glucose concentration in artificial urine. These approaches have the potential to detect the glucose level directly from urine samples of diabetic patients. Wu et al. (2019) proposed 3D-μPADs containing three working detection zones for multiple colorimetric and electrochemical detections in parallel. The 3D-μPAD is prepared through a combination of thin adhesive films and paper folding, which avoids the use of cellulose powders, and the complex folding sequence simultaneously permits assays in several layers. For the electrochemical method, the limit of detection (LOD) is found at 0.32 mM, which is lower than the colorimetric method (1.35 mM). The proposed devices are cost effective, rapid, and suitable for multiplex assays.

8.2.1.3 Graphene

Graphene is a one-atom-thick 2D sheet with a close-packed honeycomb lattice of sp^2 hybridization. Since Geim and Novoselov discovered graphene in 2004, it has been widely studied in biosensing applications due to excellent electrical conductivity, high carrier mobility, low charge-transfer resistance, and wide electrochemical potential (Geim and Novoselov 2007, Ren et al. 2018). Graphene also exhibits good mechanical flexibility, which makes it an ideal replacement for indium tin oxide (ITO) or other rigid electrodes in futuristic flexible electronics. The biocompatibility and high surface area help to increase the surface loading of the target enzyme and bioreceptors. These excellent graphene properties make it a suitable material for multifunctional biosensors to detect various biomolecules, including DNA, protein, bacteria, and glucose (Coroş et al. 2020, Suvarnaphaet and Pechprasarn 2017). In addition, graphene-based electrodes exhibit superior performance compared to other carbon-based electrodes in terms of their electro-catalytic activity. The electrodes act as signal amplifiers that improve the performance of the enzymatic biosensors by allowing a direct charge transfer between enzyme and graphene electrode (Filip et al. 2015, Shao et al. 2010).

To date, several approaches have been studied to construct novel biosensors using a single or multilayers of graphene such as mechanical exfoliation (Novoselov et al. 2004), chemical vapor deposition (CVD) (Li et al. 2009), thermal annealing of SiC (Berger et al. 2006), and the arc discharge method (Subrahmanyam et al. 2009).

Among these methods, CVD is one of the most widely used to produce large-area graphene sheets; however, it is a high-temperature process, limits the choices of substrates, and makes it costly to scale up the production. Many research groups have modified graphene to improve the electrical and optical properties for biosensing by combining it with various types of polymers such as polyaniline (Feng et al. 2011), polypyrrole (Alwarappan et al. 2010), polyallylamine (Zhang et al. 2014), poly(vinyl alcohol) (Su et al. 2013) and poly(diallyl dimethylammonium chloride) (Jia et al. 2013). Moreover, rGO functionalized with hydrophilic polymers can prevent aggregation in aqueous media, and some polymers (i.e., poly(diallyldimethylammonium chloride) (PDDA), polyethylenimine (PEI)) act as reducing agents to reduce GO to rGO simultaneously. Other forms of graphene-like reduced graphene oxide and graphene-based ink are of particular interest due to their advantages of being processed in solution, which facilitates a cost-effective biosensor fabrication (Labroo and Cui 2014, Krishnan et al. 2019). In comparison, solution processing of graphene oxide (GO) is an efficient approach to produce large quantities and facilitate the preparation of graphene-based thin films, which are compatible with various substrates such as paper and plastic.

Jia et al. (2018) reported a simple and efficient approach to modified GO with μPADs, coupled with smartphone-based colorimetric detection to quantify glucose from artificial saliva. For μPADs fabrication, a photolithography technique was used instead of wax printing to make a more precise pattern of various μPADs sizes. Aqueous GO solution (5 $mg.mL^{-1}$) was pipetted on the detection zones of μPADs and dried at room temperature; SEM images show the GO warped on filter paper. The chromogenic agent (3,3′,5,5′-tetramethylbenzidine; TMB) and the mixture of enzymatic solution (GOx/HRP) were pipetted to the detection zones and dried at room temperature. A constant glucose concentration (0.5 mM) was tested with and without coated GO paper electrodes, the assay with the GO coating showed a much better color uniformity within the detection zone. The μPADs were detected glucose concentrations (0.1–0.75 mM) in artificial saliva with LOD of 0.1 mM, which are appropriate for analyzing glucose concentrations in oral fluid.

8.2.1.4 Graphite Ink

Graphite ink, a dispersion of graphene flakes of a variety of thicknesses in solvents, can be easily patterned on different substrates via spin coating (Becerril et al. 2008), dip coating (Wang et al. 2008), spraying/stencil (Huang et al. 2016), screen printing (Huang et al. 2015), and inkjet printing (Torrisi et al. 2012) techniques. Graphene ink has many advantages over other metal-based conductive inks such as silver and copper. In comparison, graphene ink has multifunctional properties, is transparent, and cost effective. Moreover, it is difficult to recycle silver, which means that every single-use electronic product must be disposed of, but graphene ink is recycled and could be used to produce more ecofriendly printed electronics.

Lamas-Ardisana et al. (2018) reported a bulk production of EμPADs for glucose detection. First, hydrophobic barriers are patterned onto the paper using diluted UV screen-printing ink (Ultraswitch UVSW/UVV6), immediately cured the ink into the UV oven. This UV ink mixture minimizes the clogging of the ink and makes the passage of the ink smoother through the paper surface than other solid wax and polymer solutions. Graphite ink (WE, CE) and Ag/AgCl ink (RE) was screen printed on UV-patterned paper. Figures 8.4(a, b) show the SEM image of UV ink and graphite ink on wax-printed paper. The microfluidic channels of the device were modified with the solutions redox mediator ($K_3[Fe(CN)_6]$) and glucose dehydrogenase (GDH) enzyme. The sensor showed the wide linear range (0.5–50 mM) with LOD (0.33 mM) and sensitivity (0.426 C/M cm^2) and analyzed the biosensor in real samples. Cao et al. (2019) reported a three-dimensional paper-based microfluidic electrochemical integrated device (3D-PMED) for real-time monitoring of glucose from sweat. The 3D-PMED was fabricated by wax screen printing to make patterns on cellulose paper; the pre-patterned paper was then folded four times to form five stacked layers: The sweat collector, the vertical and transverse channel, the electrode, and the sweat evaporators shown in Figure 8.4(c). The WE and CE were screen printed using Prussian blue/graphite ink (C2070424P2) and the RE (Ag/AgCl) (Figure 8.4(d)). Prussian blue is an efficient redox mediator for selective detection of hydrogen peroxide (H_2O_2) in the presence of oxygen and other interference species at very low potential. Moreover, PB is the most effective catalyst for electroreduction of H_2O_2. Therefore, the printed biosensor was modified with PB, resulting in a decrease of the applied potential to avoid the penetration of electrochemical interferences. As the low level of glucose in sweat (0.25–1.5 mM), this glucose sensor showed good linearity for glucose concentrations (0–1.9 mM) with a low detection limit (5 μM) (Figure 8.4(e)). Moreover, the 3D-PMED was integrated on human skin to evaluate the glucose from sweat during exercise. Amperometric current decreased gradually with glucose in sweat ranging from 1.5 to 0.4 mM until the end of cycling. That also indicated that the 3D-PMED exhibited a good performance and could be readily used for on-body measurement. Mohammadifar et al. coated the PEDOT:PSS/graphite ink to develop a robust paper-based glucose sensor that can directly transform the electrochemical enhanced signal to distinct visual quantification utilizing light-emitting diodes (Mohammadifar et al. 2019).

8.2.1.5 Graphite Pencil

A graphite pencil is an everyday tool, essentially a nanocomposite of fine graphite powders with intercalated clay particles that can easily make traces on a paper by drawing using a gentle force. While drawing, friction between the pencil and the paper deposit the graphite particles on the paper. Compared to other commonly used methods to fabricate paper-based devices, using a pencil drawing is the simplest approach to integrate electrodes on paper with relatively good conductivity; there is no critical requirement to prepare the ink and rapid fabrication without using sophisticated equipment, which has the great convenience for practical use in resource-limited areas.

Pencil drawing on paper allows an easy, green, and fast method to fabricate electrode prototypes and easily change the designs and geometries. Li et al. reported fully drawn POP to detect glucose (Li et al. 2016b) with three electrodes designed directly on paper by drawing with an ordinary 6B pencil. Figure 8.4(f) shows that the SEM image of the bare paper contains a large amount of cellulose fiber, which

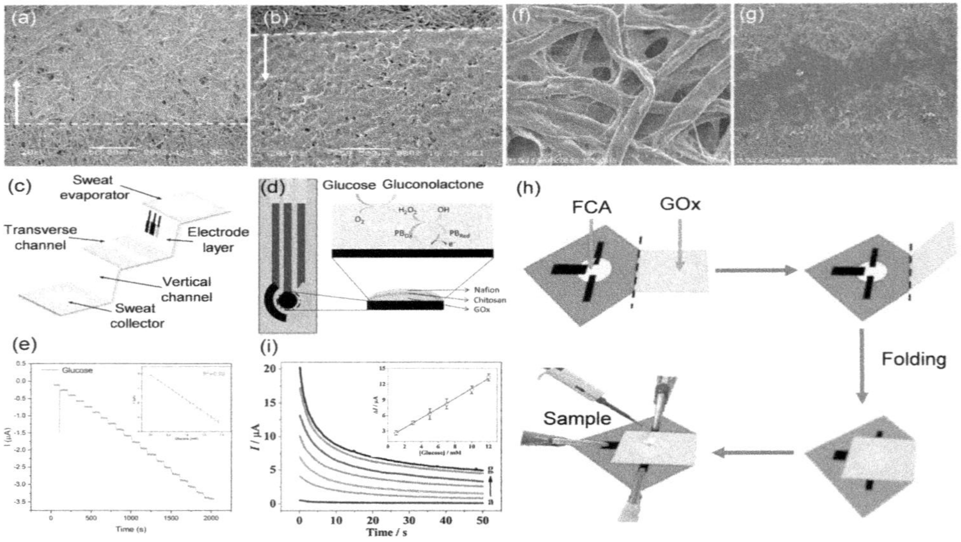

FIGURE 8.4 (a) SEM images with UV ink (25% diluted), (b) and graphite ink coated onto the chromatographic paper, discontinuous lines, and arrows indicate the limits and surfaces of the printed areas. (Reprinted from Lamas-Ardisana et al. 2018. Biosensors and Bioelectronics 109: 8–12, with permission from Elsevier.) (c) Schematic illustration of the 3D-PMED has five layers. A three-electrode electrochemical sensor was attached to the electrode layer by double-side adhesive tape. (d) Schematic diagram of a glucose sensor, working electrodes modified with the enzyme (GOx), which catalyzes the oxidation of glucose to gluconolactone and generates H_2O_2 simultaneously, while H_2O_2 was electrochemically reduced directly on the surface of the PB modified carbon electrode. The chitosan and nafion layer is coated to remove interferents present in sweat. (e) Amperometric current response with addition of 0.1 mM glucose concentration (0 to 1.9 mM) in 0.01 M PBS, inset figure shows the calibration curve (from Cao, et al. 2019. RSC Advances 9: 5674–5681, with permission from Elsevier). (f) SEM images of the paper substrate, (g) graphite deposition on paper, (h) schematic illustration of functionalization steps of the paper device with the mediator and enzyme and representation of combining both zones together (detection and enzyme). (i) Chronoamperometric curves response with glucose concentrations (1–12 mM), inset figure shows the calibration plot (reprinted from Li, et al. 2016. Sensors and Actuators, B: Chemical 231: 230–238, with permission from Elsevier).

results in a porous and rough surface. While drawing, this paper surface is favorable for the adhesion of the graphitic particles, allows them to form a continuous carbon particle path (Figure 8.4(g)). The device comprises two zones separated by a central crease: The first zone is for detection where electrodes are fully drawn with a pencil and ferrocenecarboxylic acid (FCA) introduced as the electron transfer mediator for the catalytic oxidation of glucose, and the second zone for the enzyme (GOx) immobilization, which is a circle of hydrophobic barrier on the paper. The final form of combining the two zones is by folding the device on its central crease (Figure 8.4(h)). An amperometric signal was recorded by adding a glucose sample at a certain concentration on the detection zone, achieving a linear range of 1–12 mM, and LOD of 0.05 mM with good reliability and reproducibility (Figure 8.4(i)). Moreover, a μPAD demonstrates its analytical performance in human blood samples. Along with fresh paper devices that can also be fabricated without any equipment, the fully drawn origami devices offer a cheap, flexible, portable, disposable, and practical tool in resource-limited areas. Gokoglan et al. (2017) reported a disposable paper-based electrochemical biosensor using graphene modified with a conducting polymer for glucose detection. First, graphene flakes produced using the solution process of graphite power, dispersed in dimethylformamide (DMF), and brush-painted onto the paper substrates (10 × 10 cm). A conducting polymer, poly(9,9-di-(2-ethylhexyl)-fluorenyl-2,7-diyl)-end capped with 2,5-diphenyl-1,2,4-oxadiazole (PFLO) over the graphene-coated paper, and then immobilized the enzyme (GOx) with gold nanoparticles (AuNPs) via physical adsorption. AuNPs are preferable to use in a wide range of biosensor applications as it is a good stable conductor that provides a direct charge transfer between the active site of the enzyme and electrode. The graphene/PFLO/AuNPs/GOx modified paper electrode has an amperometric response that is highly sensitive for glucose and successfully determines the glucose concentration in commercial beverages. This disposable biosensor showed a linear response for glucose concentrations (0.1 to 1.5 mM) with LOD of 0.081 mM. Mitchell et al. (2015) demonstrated custom-made reagents (75% PEGME and 25% graphite) pencils for rapid and solvent-free deposition of reagents onto μPADs. It has a sample zone, a reagent zone, a test zone, and a waste zone, all connected by a single channel. The 2,2′-azino-bis (3-ethylbenzothioazoline-6-sulfonic acid) (ABTS) as the electron donor dye deposited in the sample zone, and enzymes (GOx/HRP) were deposited in the reagent zone of the devices. The quantitative colorimetric assay developed a blue-green color when glucose concentrations (0.25 to 1.25 mM) were present in the sample. The devices show a longer shelf life for sensitive reagents compared to solution-based deposition and do not impact the accuracy or precision results of enzymatic colorimetric assays. The reagent pencils provide a modern alternative to configure diagnostic POC tests without advanced equipment being required.

8.2.2 Metal Electrodes (Au, Pt)

For bioelectronics, gold (Au) is the most suitable choice for the electrodes because it is highly conductive, highly stable when used in aqueous or biological conditions, and easy to functionalize using thiol groups. Au electrodes can be deposited on the paper substrates using several techniques such as physical vapor deposition (PVD), electron beam (e-beam) evaporation, sputtering, and spray deposition. Ultrathin gold films (less than 10 nm) were fabricated on a latex-coated paper substrate by PVD, which provides mechanical stability and good adhesion without adding other adhesive layers. Robust, flexible, highly conductive (near the conductivity of bulk gold) paper electrodes can be achieved with a low material consumption (Ihalainen et al. 2015). The electrochemical biosensor was a three-gold electrode system (thickness; ~200 nm) deposited through a mask using sputtering. Au, as a reference electrode, shows a cathodic shift of the potential compared to the saturated calomel electrode (SCE) and Ag/AgCl electrode. The WE of the thin film gold has an electrochemical behavior similar to the conventional gold electrode that forms a gold oxide on the electrode surface (Shiroma et al. 2012). Núnez-Bajo et al. (2017) developed a robust platform using gold-sputtered paper (WE) integrated with reusable metal wires (as a RE and CE). The thickness of the Au layer was controlled by the time (240 s) and the intensity of the discharge (20 mA). The bienzymes (GOx/HRP) were immobilized by adsorption on the Au surface, and ferrocyanide was added as an electron-transfer mediator, which electrochemically reduced to produce a current within 30 s. A chronoamperometric signal recorded the current changes induced by the enzymatic glucose reaction. Moreover, glucose was analyzed in real food samples with good accuracy, which proves the feasibility of the developed system being built for different food industrial applications.

A catalytic platinum (Pt) surface was the first material used for glucose sensing by Leland Clark that started the era of biosensing technology in 1962 (Clark and Lyons 1962). Pt electrodes have been used based on monitoring the change in the redox potential that produces a redox-active substance, H_2O_2, during the enzymatic reaction of the glucose. Canovas et al. developed a fully integrated, compact, portable, and wireless paper-based potentiometric system for the detection of glucose in biological fluids (i.e., serum and whole blood) (Cánovas et al. 2017). The WE (platinum-coated paper) and RE (Ag/AgCl) are fitted with the sample module that consists of a hydrophilic plastic mask placed over the electrodes in order to create a channel with a smaller measuring volume of 25 μL. The sputtered platinum-coated paper acts as a redox-sensitive surface to oxidize the glucose with the same strategy that Clark used. A nafion layer was coated on WE to efficiently entrap the GOx enzyme that enhances the stability and sensitivity of the glucose sensor. Moreover, it increases the selectivity by restricting the effect of redox-active anions due to the negatively charged nature of Nafion. This property acts as a permselective barrier against the interference of anions such as ascorbate and urate, which are typical redox-active species found in biological media. It has been shown that the sensitivity improvement is linked to the Donnan potential, which is generated because of the ion-exchange capacity of Nafion. The volume of nafion influences the reproducibility and sensitivity of the sensors. The potentiometric cell is connected to a portable wireless data transmission device. The whole sensor system allows in situ measurements with a procedure of less than 2 min, including a two-point calibration, washing, and sample measurement. The data was recorded wirelessly by Bluetooth using a cellphone application and which created a real-time plot on the screen. The analytical device exhibited a good sensitivity (−107.1 ± 7.2 mV), (−95.9 ± 4.8 mV) per decade in PBS and artificial serum, respectively, with a linear range of 0.3–3 mM. The wireless system was also validated with whole blood and serum samples from diabetic patients, and the test results were compared with the conventional colorimetric method and commercial glucometer, which opened up new opportunities in telemedicine and POC testing. Parrilla et al. (2017) also used the same cell configuration for a sensitive potentiometric-based enzymatic biosensor, which is able to detect glucose in serum. WitkowskaNery et al. (2016) proposed a paper-based electrochemical device composed of a bioactive channel, Pt working electrode, and pencil-drawn pseudo-reference electrode. Different immobilization methods and architectures were analyzed to obtain the maximum performance of a paper-based assay. They created a paper-based device with wax-printed borders to make a hydrophobic barrier and comprise a buffer-filled sponge providing the wicking solution. A channel was immobilized with an enzyme (GOx), and a wicking pad guaranteed constant movement of the fluid. A pencil-drawn pseudo reference electrode was placed between the device and the wicking pads, and a Pt working electrode (100 mm diameter) was positioned at the end of the channel. In order to perform the measurement, glucose concentrations were dissolved in water and spotted in various distances (every 5 mm) along the channel, and the time to obtain the maximal response of the hydrogen peroxide on the electrode was measured. The study included computer modeling as well as electrochemical experiments and led to reduced time of analysis (from ~30 to 15 min) without loss of sensitivity. The sensor could quantify glucose (2.0–10.0 mM) in the clinical range with good linearity and repeatability and showed the limit of detection was 2.0 mM.

8.2.3 Nanowires (ZnO)

Among the one-dimensional nanostructures, zinc oxide nanowires (ZnO NWs) are widely studied in various biosensing applications because of their high surface-to-volume ratio with a fast electron transfer rate, chemical stability, and biocompatibility (Napi et al. 2019, Izyumskaya et al. 2017). ZnO is a semiconductive material with a direct wide bandgap (3.37 eV), high binding exciton energy (60 meV) at room temperature (Lu et al. 2008), high optical gain, and unique piezoelectric properties. The crystal structures of ZnO are cubic zinc blende, rock salt, and hexagonal wurtzite, which is the most common and stable one at ambient conditions. Owing to its excellent properties, ZnO NWs are an ideal candidate for other applications such as gas sensors (Tiwale 2015), solar cells (Consonni et al. 2019), and light-emitting diodes (Rahman 2019).

Li et al. (2015) reported the first EμPAD that could detect the glucose from human serum using ZnO NWs. A biosensor designed via direct growth of ZnO NWs on a carbon ink-coated WE using the hydrothermal process. ZnO NWs show a high enzyme-capturing efficiency that leads to high sensitivity and a low limit of detection. For enzymatic glucose sensing, an electron mediator ($K_3/K_4[Fe(CN)_6]$) is usually needed to reduce the working voltage and improve the current response. However, due to the high conductivity of ZnO NWs, electron mediators are no longer required, which is an advantage also because electron mediators are typically light-sensitive materials, which can cause a decline in biosensor stability. The fabrication of a ZnO-NW EμPAD, is where the WE (bottom layer) formed on a paper substrate using wax and stencil printing of carbon-ink, and the top layer consists of a hydrophilic reaction zone, which has patterned CE (carbon ink), and a RE (Ag/AgCl) on top. In this paper, the WE is designed separately from the rest of the device, which provides two advantages: First, it helps to grow ZnO NWs easily and selectively on the carbon working electrode. Second, it allows them to achieve a large-sized WE without being limited to the sizes of the remaining two electrodes and the area of the reaction zone. This strategy established good contact between ZnO NWs on the WE and the reaction zone that leads to enhance the sensitivity of the EμPAD. Figure 8.5(a, b) shows SEM images of carbon ink-coated WE surface and high density of ZnO NWs grown on carbon ink-coated WE. An enzyme (GOx) was then immobilized on ZnO NWs/WE and measured the cyclic voltammetry (CV) response. As a control experiment, they prepared the same set of samples without the ZnO NWs layer on WE. The results show an increase in the CV current by increasing the glucose concentrations, which is expected due to the enzyme-catalyzed oxidation of the glucose. The response with the presence of ZnO NWs was much higher than the non-ZnO NWs devices, which

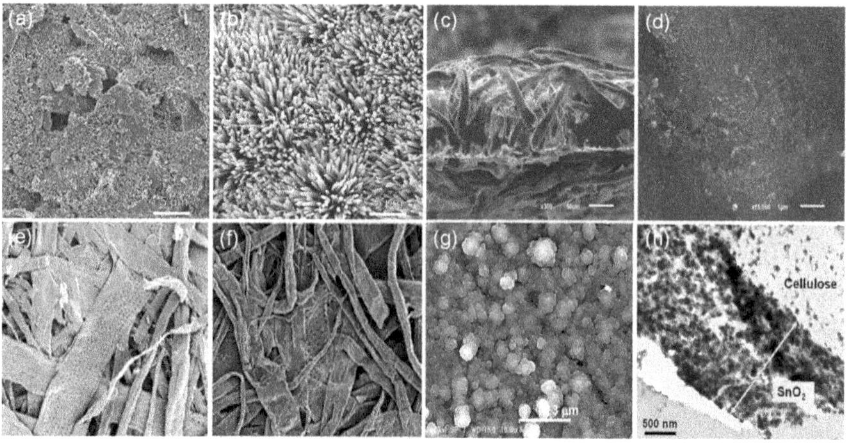

FIGURE 8.5 (a) SEM image of carbon ink-coated paper surface before the growth of ZnO NWs. (b) High density of ZnO NWs grown on carbon ink-coated WE (reprinted from Li et al. 2015. Microsystems and Nanoengineering 1: 15014). (c) Cross-section the unmodified Au-sputtered paper. (d) AuNPs modified with Au-sputtered paper obtained electrodepositing gold for 6000 s. The average diameter (56 ± 16 nm) of AuNPs and clusters (170 ± 44 nm) are also found in some regions (reprinted from Núnez-Bajo et al. 2018. Talanta 178: 160–165, with permission from Elsevier). (e) Bare filter paper, (f) and PBNP-containing filter paper (reprinted from Cinti et al. 2018. Talanta 187: 59–64, with permission from Elsevier). (g) Cellulose–SnO_2 hybrid nanocomposite (24 h deposition time), (h) TEM image of a uniform and homogeneous deposition of SnO_2 coating on the cellulose surface (reprinted from Mahadeva et al. 2011. Sensors and Actuators, B: Chemical 157: 177–182, with permission from Elsevier).

confirms the lower electrical potential that is required for the device to generate an electrochemical response. This ability comes from the low reduction potential of the ZnO NWs that is close to the oxidation potential of the glucose. To apply a whole blood sample, a filter paper was integrated on top of the reaction zone to filter the blood cells. EμPADs detect the glucose concentrations range (0 to15 mM) and achieved the highest sensitivity (2.88 $\mu A{\cdot}mM^{-1}{\cdot}cm^{-2}$), with a LOD of 94.7 μM, which covers the clinical range of glucose concentrations. This device also tested glucose in human serum samples, achieved a linear range (5.1–15.1 mM) and sensitivity (2.36 $\mu A{\cdot}mM^{-1}{\cdot}cm^{-2}$). EμPADs integrated with ZnO NWs provide a facile and inexpensive platform to enhance the sensitivity of the biosensor.

8.2.4 Nanoparticles (NPs)

Nanoparticles (NPs) are solid colloidal particles that range between 10 and 1000 nm in size. They can exhibit significantly different chemical and physical properties to their larger material counterparts. They can be divided into different groups based on their properties, shapes, or sizes, e.g., fullerenes, metal NPs, ceramic NPs, and polymeric NPs. Among them, metal NPs, such as Ag, Au, Pt, and Pd NPs have attracted much interest in developing biosensors because of their unique properties such as large surface-to-volume ratio, high surface reaction activity, and strong adsorption ability to immobilize the desired biomolecules. AuNPs, in particular, were widely used to construct biosensors because of their excellent optical and electronic properties, signal transduction, and biocompatibility. Different kinds of NPs were incorporated into paper microfluidics biosensors, which greatly improve their sensitivity and also provide novel detection methods. Therefore, NP-based paper microfluidics biosensors set a new trend in developing portable, sensitive biosensing devices and show great promise in many research areas, especially for the diagnosis of the most common diseases and environmental applications.

Määttänen et al. (2013) developed fully inkjet-printed gold nanoparticles (AuNPs) as WE and CE, and silver nanoparticles (AgNPs) as a RE in which an Ag/AgCl layer was deposited electrochemically on a coated paper substrate. The resistivity of the printed Au electrodes (1.6×10^{-7} Ωm) is higher than the one of bulk gold (2.2×10^{-8} Ωm), due to the possible contaminants from the ambient atmosphere, porosity, and organic residues remain during IR-sintering; both affect the electron transfer kinetics. PDMS was used to create reaction wells that confine the electrolyte solution over the electrodes. A SAM layer was formed on the WE using 1-octadecanethiol (ODT) solution. Although the thiolation time for printed gold electrodes was significantly shorter than required for forming a fully oriented SAM on gold electrodes (Love et al. 2005). The poly(3,4- ethylenedioxythiophene) (PEDOT) has been used as entrapment support for enzymes to maintain its electroactivity and stability during sensing. The PEDOT-GOx films were electropolymerized on the WE surface using the galvanostatic and cyclic potential sweep method. The paper/Au/PEDOT-GOx chips showed a distinct current response by changing the glucose concentration that was also selective since no increase in current was recorded for a high concentration of mannose. Núnez-Bajo et al. (2018) integrated AuNPs on

Au-sputtered paper electrodes for non-enzymatic glucose detection. The porous gold structure deposited using sputtering on paper. The AuNPs are synthesized on Au-sputtered paper used as a WE by electrochemical reduction of $AuCl_4$ without any reducing agents. Using potentiostatic/galvanostatic method, a constant current (−100 μA) was applied on the electrode surfaces to produces the reduction of Au(III) to Au(0) for different times (1500–9000 s) in $AuCl_4$ solution (10.15 mM in HCl). An optimized time (6000 s) was appropriate to generate uniform gold nanostructures on paper electrodes with high and precise signals. Figure 8.5(c, d) shows the Au-sputtered paper and AuNPs connected to the Au surface on the cellulosic fibers. The electrochemical behavior of AuNPs integrated paper electrodes was tested by CV using a redox mediator [[$K_4Fe(CN)_6$] in PBS]. The current intensity of AuNP-nanostructured electrodes improved ten-fold compared to unmodified Au-sputtered paper electrodes. The linear sweep voltammogram (LSV) was recorded with glucose solutions (0.01 to 5 mM) in 0.1 M NaOH, and glucose oxidation occurs at the lowest potential (+0.080 V) on the Au electrode. A linear response was obtained for glucose concentrations, with a LOD of 6 μM. In addition, the sensor was used to precisely analyze glucose levels in real food samples. Cinti et al. (2018) proposed a facile and simple approach to synthesized Prussian blue nanoparticles (PBNPs) on paper and developed disposable diagnostic strips for glucose detection. To synthesize PBNPs, the wax-patterned filter paper is impregnated with the mixture (0.05 M $K_3Fe(CN)_6$/ $FeCl_3$) in water for 1.5 h. The PBNPs are formed when the reaction is carried out within the structure of the filter paper, with no specific stimuli (e.g., pH, voltage, reducing agent, etc.). This is possible because of the presence of reducing agents in mineral fillers allows for an auto-reduction of the precursor. Figure 8.5(e, f) shows the homogeneous dispersion of PBNPs onto the filter paper. Usually, PB is formed by the electrochemical reduction of a solution that contains both ferric species or hexac yanoferrate(II)/hexacyanoferrate(III) ions. The screen-printed carbon and Ag/AgCl inks are used to deposit the WE, CE, and RE, respectively. A PB-based paper sensor is immobilized with an enzyme (GOx) solution, shows a linear response (0 to 25 mM) upon adding the glucose concentration leads to an increase of the cathodic current due to the electrocatalytic reduction of the enzymatically produced H_2O_2. In addition, the selectivity of the PB-based paper biosensor has been evaluated in the presence of common potential blood interferents (i.e., ascorbic acid, uric acid, and acetaminophen). This proposed work has the potential to detect other biological analytes and heavy metals. Mahadeva and Kim (2011) reported cellulose integrated SnO_2 particles used for a glucose biosensor. SnO_2 is a known electrical semiconductor that is optically transparent in the visible spectrum with a wide band gap of 3.6 eV at room temperature. A porous SnO_2 layer was grown on regenerated cellulose films via liquid phase deposition technique with varying deposition time. As the deposition time increased the SnO_2 particles grew uniformly all over the surface, and particles were connected each other to form a layer on the cellulose film as shown in Figure 8.5(g, h). The enzyme (GO_x) was immobilized into a cellulose–SnO_2 hybrid nanocomposite via covalent bonding to detect the glucose. This nanocomposite can be an inexpensive, flexible, and disposable glucose biosensor. Figueredo et al. (2016) modified μPADs using three different nanomaterials such as Fe_3O_4 nanoparticles

(MNPs), MWCNT, and GO to enhance the color signal quality (Figure 8.6(a)). These nanomaterials differ from each other regarding their shape and surface-to-volume ratio. They can act as supports for enzyme immobilization to enhance the analytical signal for glucose. The mixture solution of (TMB/GOx/HRP) modified μPADs enhanced the color signal for detection of glucose at low concentrations (Figure 8.6(b)). The limit of detection values achieved for glucose with μPADs integrated with MNPs, MWCNT, and GO were 43, 62, and 18 μM, respectively. The analytical performance of the nanomaterial-modified μPADs are also tested in urine samples, GO-μPAD shows the best results with recovery values (83 to 109%) obtained in a considerable glucose concentration range (0.3–0.8 mM), in comparison to MWCNT-μPADs (76 to 96%), and μPADs-MNPs (40 to 71%). The integration of μPADs with carbon-based nanomaterials and magnetic NPs has great potential to be explored in bioanalytical applications. Evans et al. (2014) integrated μPADs with silica nanoparticles (SiO_2 NPs) to improve the color intensity and uniformity in colorimetric assay (Figure 8.6(c–e)). The devices are fabricated using a CO_2 laser engraver and then immersed in a suspension containing the APTES-modified SiO_2 NPs, which increases the adsorption capacity of selected enzymes and prevents the washing away effect that creates color gradients in the colorimetric measurements. The SiO_2 NPs trapped within the structure of the cellulose served as solid support to immobilize the enzyme solution (KI/GOx/HRP) responsible for the colorimetric reaction. This platform was successfully applied to the qualitative analysis of glucose (7.33 mM) in urine samples. Deng et al. (2014) synthesized hybrid GSC ($GO/SiO_2/CeO_2$) nanosheets used as an alternative chromogenic probe which obtained high intrinsic peroxidase activity due to the combination of GO and CeO_2. GSC nanosheets are able to catalyze the oxidation reaction of different peroxidase substrates such as TMB, ABTS, and o-phenylenediamine (OPD) in the presence of H_2O_2. GSCs have a superior catalytic activity toward the H_2O_2-mediated oxidation of OPD. This assay exhibited excellent analytical performance for glucose detection in human serum and urine (Figure 8.6(f)). Ornatska et al. (2011) used redox-active ceria nanoparticles (CeO_2 NPs) as a chromogenic indicator for colorimetric assay, without using either the organic dye or the peroxidase enzyme. To construct the sensor, CeO_2 NPs and GOx were immobilized onto the silane-modified paper using a silanization procedure, which stabilized the CeO_2 NPs onto the filter paper (Figure 8.6(g)). The visible color change from white-yellowish to dark orange in the presence of glucose is due to the change of the oxidation state (Ce^{3+}/Ce^{4+}) and formation of surface complexes onto the NP surface, induced by the enzymatically produced H_2O_2. The assay shows sensitivity for glucose detection, and it is robust, inexpensive, and performs successfully in human serum samples (Figure 8.6(h)). The results show that these NPs can be used in the production of reliable colorimetric bioassays.

8.2.5 Quantum Dots

Quantum dots (QDs) are semiconductor nanocrystals with atoms from groups II–VI (e.g., CdTe, CdSe, CdS, ZnS, ZnSe, or ZnTe), III–V (GaN, GaP, GaAs, InP, InN), IV–VI (PbSe, PbS or PbTe) I-III-VI_2 (e.g., $CuInS_2$ or $AgInS_2$) or IV elements

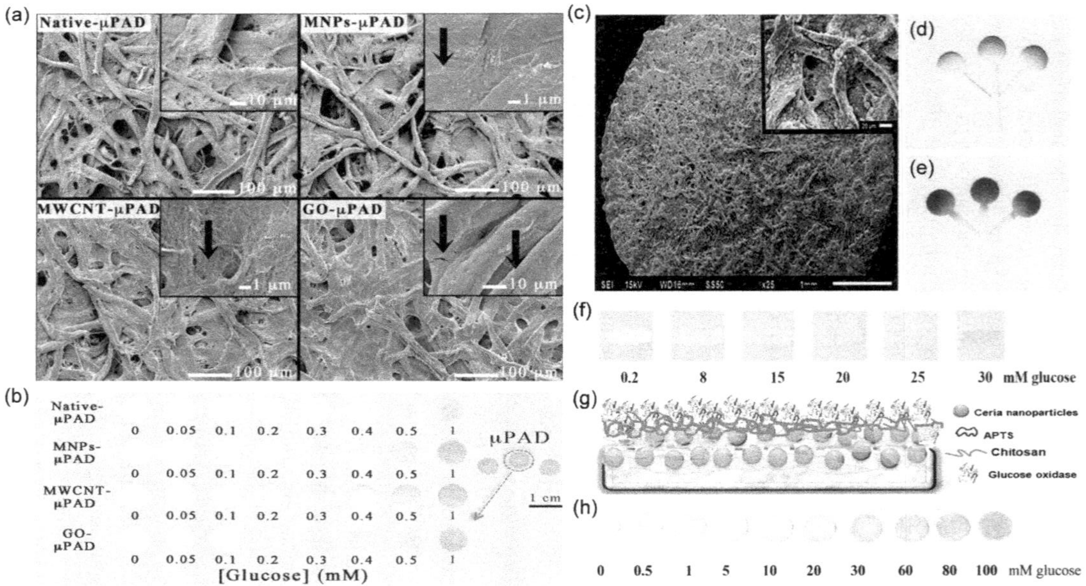

FIGURE 8.6 (a) FESEM images of bare and treated devices with Fe_3O_4 (MNPs), MWCNT, and GO; arrows indicate the presence of nanomaterials over the paper fibers (inset figure). (b) Colorimetric assay for bare μPAD, MNP-μPAD, MWCNT-μPAD, and GO-μPAD with glucose concentration (0–1 mM) (reprinted from Figueredo et al. 2016. ACS Applied Materials and Interfaces 8: 11–15, with permission from the American Chemical Society). (c) SEM of the detection zone of a μPAD after the deposition of SiO_2 nanoparticles, optical images of colorimetric assay of (d) bare (without SiO_2), and (e) modified (with SiO_2) papers (reprinted from Evans et al. 2014. Analyst 139: 5560–5567, with permission from the Royal Society of Chemistry). (f) The colorimetric responses of GSCs bioactive paper to glucose concentrations are 0.2 mM, 8 mM, 15 mM, 20 mM, 25 mM, 30 mM, and 50 mM (from left to right) (reprinted from Deng et al. 2014. Biosensors and Bioelectronics 52: 324–329, with permission from Elsevier). (g) Ceria-based paper bioassay with multilayer sequence of the immobilization of enzyme onto silanazed paper. (h) Colorimetric response of the paper sensors (reprinted from Ornatska et al. 2011. Analytical Chemistry 83: 4273–4280, with permission from the American Chemical Society).

(e.g., Si, C, or Ge). The metal chalcogenide (i.e., MX M = Cd, Zn and X = S, Se, Te) QDs are unique nanostructures that show quantum confinement and size-dependent photoemission characteristics, including absorbance and photoluminescence (PL) (Hildebrandt et al. 2017, Wegner and Hildebrandt 2015, Juan Zhou et al. 2015). The most common synthesized QDs contain highly toxic elements such as cadmium, lead, and mercury, but more effort to design less toxic or non-toxic QDs has been increased recently by using elements such as Ag, Zn, Mn, and Cu (Xu et al. 2016). Since the size of QDs is 1–10 nm in diameter with 100 to 1000 atoms, the surface-to-volume ratio of a QD is relatively large, which allows efficient functionalization of biomolecules (Murphy 2002). Moreover, QDs show several significant advantages over most of the organic fluorophore dyes and proteins, such as high quantum yield, narrow emission bands with broad excitation spectra, and high brightness (large molar extinction coefficients) (Bruchez et al. 1998, Chan and Nie 1998). QDs of different sizes or compositions can be excited by a single light source to produce emission colors with a minimal spectral overlap over a wide spectral range, making them particularly attractive for multiplex imaging (Liu et al. 2010). In contrast to organic dyes, most QDs are highly resistant to photo-bleaching, which allows them to be used for long-term in vitro and in vivo imaging (Montalti et al. 2015, Wegner and Hildebrandt 2015). More specifically, these nanocrystals can be engineered to emit at wavelengths between 450 and 1500 nm by changing their size, shape, and composition (Vanmaekelbergh et al. 2015). These unique optical and fluoresce properties of QDs were first discovered early in 1980, but it was not until 1998 when Bruchez et al. (1998) and Chan and Nie (1998) groups used QDs for the first time as new fluorescent probes for biological detection and imaging. Since then, major improvement has been made in the usage of QDs as luminescent labels for bioimaging, biophotonic, and medical applications (Chinnathambi and Shirahata 2019). A major advantage of QDs for biosensing is their size tunability that directly affects their optical properties, which greatly facilitates the detection of different targets by simply using QDs with different sizes. There are many examples that show sensitive detection of various biomolecules such as DNA (Wang et al. 2018), miRNA (Hu et al. 2018), viruses (Nguyen et al. 2020, Yang et al. 2020), proteins (Qiu et al. 2017), cells (Tsuboi and Jin 2017), enzymes (Petryayeva and Algar 2013).

QDs were initially synthesized as colloidal dispersions in nonpolar organic media; surface modification is critical to improving their bioconjugation, biocompatibility, dispersibility in water and biological media, which can be further used in biosensing applications. Different modification methods were used to achieve these properties, such as surface cap exchange, encapsulation in silica shells or micelles, and using hydrophilic/amphiphilic molecules. To develop QD-labeled probes, it can be functionalized with a wide variety of bio-receptors (e.g., antibodies, DNA, proteins, peptides, and vitamins) through established conjugation techniques (Biju 2014). Therefore, QDs are commonly used as fluorescent labeling and chemiluminescent emitters for the development of a variety of fluorescence, bioluminescence, chemiluminescence, electrochemiluminescence, and electrochemical biosensors (Yao et al. 2017). Additional to these optical properties, the QDs have unique photoelectrochemical activity (Yue et al. 2013). The excitation of QDs by light, when immobilized on a conductive electrode, will cause an electron transfer reaction between the QDs and the electrode to produce a photocurrent signal. The QD-based photoelectrochemical biosensors evolved later than the optical biosensors, but they were rapidly developed due to their high sensitivity without intervention of costly instruments (Freeman et al. 2013, Shu and Tang 2017).

Only a few reports are published about integrating QD in paper devices. Yuan et al. (2012) reported a facile method for the generation of polymer QD-enzyme hybrid film for the development of glucose biosensors. The glutathione-capped CdTe QDs and enzyme (GOx) were used to spontaneously encapsulate poly-(diallyldimethylammonium chloride) (PDDA) via electrostatic interaction to form PDDA-enzyme-QDs hybrid films on paper substrates. The hybrid mixture was dropped onto chromatography paper dried at room temperature. The PDDA QD hybrid material is considered as a good candidate for test paper preparations as the polymer offers the immobilization of both QDs and enzymes in the paper. The obtained PDDA QD enzyme hybrids feature both high fluorescence and biorecognition. The paper sensor detected the glucose concentration range (0 to 10 mM). A prepared paper-based glucose sensor showed relatively high stability, as both the fluorescence of the QDs and the activity of the enzyme encapsulated in the hybrids were maintained well for at least four weeks. This hybrid material possesses great potential in the development of portable sensing devices. Durán et al. (2016) fabricated a QD-modified paper-based assay for glucose detection. First, the CdSe/ZnS core/shell QD was synthesized using a chemical route. Circular paper sheets were uniformly loaded with CdSe/ZnS QDs and enzyme (GOx), which displayed strong fluorescence under a UV lamp (365 nm excitation). After 20 min of exposure, glucose oxidized by GOx and produced H_2O_2, which etched the surface of the QDs and changed their optical properties, resulting in a decrease in their fluorescence intensity. The changes in QD fluorescence can be easily seen by either the naked eye with a portable smartphone camera or a fluorospectrometer. In this work, paper sheets were photographed using a digital camera at an excitation of 365 nm. Different parameters were designed to improve the glucose reaction, including sample pH, the number of QDs, and enzymes (GOx). The paper-based assay showed a sigmoidal response for glucose concentration (5–200 $mg.dL^{-1}$) and (LOD–5 $\mu g{\cdot}dL^{-1}$), demonstrating their potential use for biomedical applications.

8.3 CONCLUSION AND FUTURE ASPECTS

Nanomaterials have become essential components of bioanalytical devices, as they significantly boost overall efficiency in terms of sensitivity, lower the detection limit, and enable the detection of a variety of biomolecules. The specific properties of nanostructures also provide alternatives to conventional transduction methods. Additionally, combining novel nanomaterials with μPADs may open the window for the development of paper-based electronic devices for biosensing applications. The recent development in nanomaterials integrated into μPADs was demonstrated for various biosensing applications. These μPADs are gaining more popularity than the conventional substrates in terms of their cost, flexibility, availability, ease of modification, disposability, and ease of use. In addition, porous structure and high absorbance of the paper are useful features for strong adhesion of nanomaterials and bioreceptor immobilization. Despite all of these benefits, the commercialization of paper-based electronic devices faces a number of fundamental and realistic problems. For instance, nanomaterials-embedded μPADs currently have less stability and longevity than those of conventional devices. Besides this, the inherent flexibility of paper may interfere with the electrical conductivity and non-specific adsorption of other analytes, which may generate false signals during measurement. The number of different nanomaterials with their own unique properties to integrate with μPADs still leaves lots of room to explore. Thus, the development of robust, sensitive, and accurate μPADs would be very helpful to commercialize POCT. Novel nanomaterials, the advancements in fabrication techniques, and appropriate structural designs may result in improved properties, accuracy, and sensitivity of the μPADs.

REFERENCES

Alwarappan, S., C. Liu, A. Kumar, and C. Z. Li. 2010. Enzyme-doped graphene nanosheets for enhanced glucose biosensing. *Journal of Physical Chemistry C* 114: 12920–12924.

Amor-Gutiérrez, O., E. Costa Rama, A. Costa-García, and M. T. Fernández-Abedul. 2017. Paper-based maskless enzymatic sensor for glucose determination combining ink and wire electrodes. *Biosensors and Bioelectronics* 93: 40–45.

Amor-Gutiérrez, O., E. Costa-Rama, and M. T. Fernández-Abedul. 2019. Sampling and multiplexing in lab-on-paper bioelectroanalytical devices for glucose determination. *Biosensors and Bioelectronics* 135: 64–70.

Anfossi, L., F. Di Nardo, S. Cavalera, C. Giovannoli, and C. Baggiani. 2018. Multiplex lateral flow immunoassay: An overview of strategies towards high-throughput point-of-need testing. *Biosensors* 9: 1–14.

Batool, R., A. Rhouati, M. H. Nawaz, A. Hayat, and J. L. Marty. 2019. A review of the construction of nano-hybrids for electrochemical biosensing of glucose. *Biosensors* 9: 1–19.

Becerril, H. A., J. Mao, Z. Liu, R. M. Stoltenberg, Z. Bao, and Y. Chen. 2008. Evaluation of solution-processed reduced graphene oxide films as transparent conductors. *ACS Nano* 2: 463–470.

Berger, C., X. Wu, N. Brown, C. Naud, X. Li, Z. Song, D. Mayou, et al. 2006. Electronic confinement and coherence in patterned epitaxial graphene. *Science* 312: 1191–1196.

Biju, V. 2014. Chemical modifications and bioconjugate reactions of nanomaterials for sensing, imaging, drug delivery and therapy. *Chemical Society Reviews* 43: 744–764.

Biscay, J., E. C. Rama, M. B. G. García, J. M. P. Carrazón, and A. C. García. 2011. Enzymatic sensor using mediator-screen-printed carbon electrodes. *Electroanalysis* 23: 209–214.

Boobphahom, S., M. N. Ly, V. Soum, N. Pyun, O. S. Kwon, N. Rodthongkum, and K. Shin. 2020. Recent advances in microfluidic paper-based analytical devices toward high-throughput screening. *Molecules* 25: 2970.

Bruchez, M., M. Moronne, P. Gin, S. Weiss, and A. P. Alivisatos. 1998. Semiconductor nanocrystals as fluorescent biological labels. *Science* 281: 2013–2016.

Cánovas, R., M. Parrilla, P. Blondeau, and F. J. Andrade. 2017. A novel wireless paper-based potentiometric platform for monitoring glucose in blood. *Lab on a Chip* 17: 2500–2507.

Cao, Q., B. Liang, T. Tu, J. Wei, L. Fang, and X. Ye. 2019. Three-dimensional paper-based microfluidic electrochemical integrated devices (3D-PMED) for wearable electrochemical glucose detection. *RSC Advances* 9: 5674–5681.

Cass, A. E. G., G. Davis, G. D. Francis, H. Allen, O. Hill, W. J. Aston, I. J. Higgins, E. V. Plotkin, L. D. L. Scott, and A. P. F. Turner. 1984. Ferrocene-mediated enzyme electrode for amperometric determination of glucose. *Analytical Chemistry* 56: 667–671.

Cate, D. M., J. A. Adkins, J. Mettakoonpitak, and C. S. Henry. 2015. Recent developments in paper-based microfluidic devices. *Analytical Chemistry* 87: 19–41.

Chan, W. C. W., and S. Nie. 1998. Quantum dot bioconjugates for ultrasensitive nonisotopic detection. *Science* 281: 2016–2018.

Chen, P., M. Gates-Hollingsworth, S. Pandit, A. Park, D. Montgomery, D. AuCoin, J. Gu, and F. Zenhausern. 2019. Paper-based vertical flow immunoassay (VFI) for detection of bio-threat pathogens. *Talanta* 191: 81–88.

Chinnathambi, S., and N. Shirahata. 2019. Recent advances on fluorescent biomarkers of near-infrared quantum dots for in vitro and in vivo imaging. *Science and Technology of Advanced Materials* 20: 337–355.

Chitnis, G., Z. Ding, C. L. Chang, C. A. Savran, and B. Ziaie. 2011. Laser-treated hydrophobic paper: An inexpensive microfluidic platform. *Lab on a Chip*, 11:1161–1165.

Chrzanowska, J., J. Hoffman, A. Małolepszy, M. Mazurkiewicz, T. A. Kowalewski, Z. Szymanski, and L. Stobinski. 2015. Synthesis of carbon nanotubes by the laser ablation method: Effect of laser wavelength. *Physica Status Solidi (B)* 252: 1860–1867.

Cinti, S., R. Cusenza, D. Moscone, and F. Arduini. 2018. Paper-based synthesis of prussian blue nanoparticles for the development of whole blood glucose electrochemical biosensor. *Talanta* 187: 59–64.

Clark, L. C., and C. Lyons. 1962. Electrode systems for continuous monitoring in cardiovascular surgery. *Annals of the New York Academy of Sciences* 102: 29–45.

Consonni, V., J. Briscoe, E. Kärber, X. Li, and T. Cossuet. 2019. ZnO nanowires for solar cells: A comprehensive review. *Nanotechnology* 30: 362001–362041.

Coroş, M., S. Pruneanu, and R. I. Stefan-van Staden. 2020. Review: Recent progress in the graphene-based electrochemical sensors and biosensors. *Journal of The Electrochemical Society* 167: 037528.

De Volder, M. F. L., S. H. Tawfick, R. H. Baughman, and A. J. Hart. 2013. Carbon nanotubes: Present and future commercial applications. *Science* 339: 535–539.

Deng, L., C. Chen, C. Zhu, S. Dong, and H. Lu. 2014. Multiplexed bioactive paper based on GO@SiO_2@CeO_2 nanosheets for a low-cost diagnostics platform. *Biosensors and Bioelectronics* 52: 324–329.

Dungchai, W., O. Chailapakul, and C. S. Henry. 2009. Electrochemical detection for paper-based microfluidics. *Analytical Chemistry* 81: 5821–5826.

Dungchai, W., O. Chailapakul, and C. S. Henry. 2011. A low-cost, simple, and rapid fabrication method for paper-based microfluidics using wax screen-printing. *Analyst* 136: 77–82.

Durán, G. M., T. E. Benavidez, Á. Ríos, and C. D. García. 2016. Quantum dot-modified paper-based assay for glucose screening. *Microchimica Acta* 183: 611–616.

Eatemadi, A., H. Daraee, H. Karimkhanloo, M. Kouhi, N. Zarghami, A. Akbarzadeh, M. Abasi, Y. Hanifehpour, and S. W. Joo. 2014. Carbon nanotubes: Properties, synthesis, purification, and medical applications. *Nanoscale Research Letters* 9: 1–13.

Evans, E., E. F. Moreira Gabriel, T. E. Benavidez, W. K. Tomazelli Coltro, and C. D. Garcia. 2014. Modification of microfluidic paper-based devices with silica nanoparticles. *Analyst* 139: 5560–5567.

Fan, Y., S. Shi, J. Ma, and Y. Guo. 2019. A paper-based electrochemical immunosensor with reduced graphene oxide/thionine/gold nanoparticles nanocomposites modification for the detection of cancer antigen 125. *Biosensors and Bioelectronics* 135: 1–7.

Fava, E. L., T. A. Silva, T. M. do Prado, F. C. de Moraes, R. C. Faria, and O. Fatibello-Filho. 2019. Electrochemical paper-based microfluidic device for high throughput multiplexed analysis. *Talanta* 203: 280–286.

Feng, X. M., R. M. Li, Y. W. Ma, R. F. Chen, N. E. Shi, Q. L. Fan, and W. Huang. 2011. One-step electrochemical synthesis of graphene/polyaniline composite film and its applications. *Advanced Functional Materials* 21: 2989–2996.

Figueredo, F., P. T. Garcia, E. Cortón, and W. K. T. Coltro. 2016. Enhanced analytical performance of paper microfluidic devices by using Fe3O4 nanoparticles, MWCNT, and graphene oxide. *ACS Applied Materials and Interfaces* 8: 11–15.

Filip, J., P. Kasák, and J. Tkac. 2015. Graphene as signal amplifier for preparation of ultrasensitive electrochemical biosensors. *Chemical Papers* 69: 112–133.

Fisher, R. B., D. S. Parsons, and G. A. Morrison. 1948. Quantitative paper chromatography. *Nature* 161: 764–765.

Freeman, R., J. Girsh, and I. Willner. 2013. Nucleic acid/quantum dots (QDs) hybrid systems for optical and photoelectrochemical sensing. *ACS Applied Materials and Interfaces* 5: 2815–2834.

Gallaway, J. W., and S. A. C. Barton. 2008. Kinetics of redox polymer-mediated enzyme electrodes. *Journal of the American Chemical Society* 130: 8527–8536.

Ge, X., A. M. Asiri, D. Du, W. Wen, S. Wang, and Y. Lin. 2014. Nanomaterial-enhanced paper-based biosensors. *Trends in Analytical Chemistry* 58: 31–39.

Geim, A. K., and K. S. Novoselov. 2007. The rise of graphene. *Nature Materials* 6: 183–191.

Gokoglan, T. C., M. Kesik, S. Soylemez, R. Yuksel, H. E. Unalan, and L. Toppare. 2017. Paper based glucose biosensor using graphene modified with a conducting polymer and gold nanoparticles. *Journal of the Electrochemical Society* 164: G59–G64.

González-Guerrero, M. J., F. J. del Campo, J. P. Esquivel, D. Leech, and N. Sabaté. 2017. Paper-based microfluidic biofuel cell operating under glucose concentrations within physiological range. *Biosensors and Bioelectronics* 90: 475–480.

Grieshaber, D., R. MacKenzie, J. Vörös, and E. Reimhult. 2008. Electrochemical biosensors-sensor principles and architectures. *Sensors* 8: 1400–1458.

Han, S. M., Y. W. Kim, Y. K. Kim, J. H. Chun, H. B. Oh, and S. H. Paek. 2018. Performance characterization of two-dimensional paper chromatography-based biosensors for biodefense, exemplified by detection of bacillus anthracis spores. *Biochip Journal* 12: 59–68.

He, H., L. A. Pham-Huy, P. Dramou, D. Xiao, P. Zuo, and C. Pham-Huy. 2013. Carbon nanotubes: Applications in pharmacy and medicine. *BioMed Research International* 2013: 1–12.

Hildebrandt, N., C. M. Spillmann, W. Russ Algar, T. Pons, M. H. Stewart, E. Oh, K. Susumu, S. A. Díaz, J. B. Delehanty, and I. L. Medintz. 2017. Energy transfer with semiconductor quantum dot bioconjugates: A versatile platform for biosensing, energy harvesting, and other developing applications. *Chemical Reviews* 117: 536–711.

Hu, J., M. H. Liu, and C. Y. Zhang. 2018. Integration of isothermal amplification with quantum dot-based fluorescence resonance energy transfer for simultaneous detection of multiple microRNAs. *Chemical Science* 9: 4258–4267.

Huang, X., T. Leng, X. Zhang, J. C. Chen, K. H. Chang, A. K. Geim, K. S. Novoselov, and Z. Hu. 2015. Binder-free highly conductive graphene laminate for low cost printed radio frequency applications. *Applied Physics Letters* 106: 203105.

Huang, X., K. Pan, and Z. Hu. 2016. Experimental demonstration of printed graphene nanoflakes enabled flexible and conformable wideband radar absorbers. *Scientific Reports* 6: 4–11.

Ihalainen, P., A. Määttänen, M. Pesonen, P. Sjöberg, J. Sarfraz, R. Österbacka, and J. Peltonen. 2015. Paper-supported nanostructured ultrathin gold film electrodes: Characterization and functionalization. *Applied Surface Science* 329: 321–329.

Iijima, S. 1991. Helical microtubules of graphitic carbon. *Nature* 354: 56–58.

Ivanov, I., T. Vidaković-Koch, and K. Sundmacher. 2010. Recent advances in enzymatic fuel cells: Experiments and modeling. *Energies* 3: 803–846.

Izyumskaya, N., A. Tahira, Z. H. Ibupoto, N. Lewinski, V. Avrutin, Ü. Özgür, E. Topsakal, M. Willander, and H. Morkoç. 2017. Review: Electrochemical biosensors based on ZnO nanostructures. *ECS Journal of Solid State Science and Technology* 6: Q84–Q100.

Jia, L., J. Liu, and H. Wang. 2013. Preparation of poly(diallyldimethylammonium chloride)-functionalized graphene and its applications for H2O2and glucose sensing. *Electrochimica Acta* 111: 411–418.

Jia, Y., H. Sun, X. Li, D. Sun, T. Hu, N. Xiang, and Z. Ni. 2018. Paper-based graphene oxide biosensor coupled with smartphone for the quantification of glucose in oral fluid. *Biomedical Microdevices* 20: 1–9.

Koo, Y., V. N. Shanov, and Y. Yun. 2016. Carbon nanotube paper-based electroanalytical devices. *Micromachines* 7: 1–9.

Kour, R., S. Arya, S.-J. Young, V. Gupta, P. Bandhoria, and A. Khosla. 2020. Review: Recent advances in carbon nanomaterials as electrochemical biosensors. *Journal of the Electrochemical Society* 167: 037555.

Krishnan, S. K., E. Singh, P. Singh, M. Meyyappan, and H. S. Nalwa. 2019. A review on graphene-based nanocomposites for electrochemical and fluorescent biosensors. *RSC Advances* 9: 8778–8781.

Kuswandi, B., and A. A. Ensafi. 2020. Perspective: Paper-based biosensors: Trending topic in clinical diagnostics developments and commercialization. *Journal of the Electrochemical Society* 167: 037509.

Labroo, P., and Y. Cui. 2014. Graphene nano-ink biosensor arrays on a microfluidic paper for multiplexed detection of metabolites. *Analytica Chimica Acta* 813: 90–96.

Lam, T., J. P. Devadhasan, R. Howse, and J. Kim. 2017. A chemically patterned microfluidic paper-based analytical cevice (C-MPAD) for point-of-care diagnostics. *Scientific Reports* 7: 1188.

Lamas-Ardisana, P. J., G. Martínez-Paredes, L. Añorga, and H. J. Grande. 2018. Glucose biosensor based on disposable electrochemical paper-based transducers fully fabricated by screen-printing. *Biosensors and Bioelectronics* 109: 8–12.

Li, W., D. Qian, Y. Li, N. Bao, H. Gu, and C Yu. 2016a. Fully-drawn pencil-on-paper sensors for electroanalysis of dopamine. *Journal of Electroanalytical Chemistry* 769: 72–79.

Li, W., D. Qian, Q. Wang, Y. Li, N. Bao, H. Gu, and C. Yu. 2016b. Fully-drawn origami paper analytical device for electrochemical detection of glucose. *Sensors and Actuators, B: Chemical* 231: 230–238.

Li, X., J. Tian, T. Nguyen, and W. Shen. 2008. Paper-based microfluidic devices by plasma treatment. *Analytical Chemistry* 80: 9131–9134.

Li, X., W. Cai, J. An, S. Kim, J. Nah, D. Yang, R. Piner, et al. 2009. Large-area synthesis of high-quality and uniform graphene films on copper foils. *Science* 324: 1312–1314.

Li, X., C. Zhao, and X. Liu. 2015. A paper-based microfluidic biosensor integrating zinc oxide nanowires for electrochemical glucose detection. *Microsystems and Nanoengineering* 1: 15014.

Li, Z., F. Li, J. Hu, W. H. Wee, Y. L. Han, B. Pingguan-Murphy, T. J. Lu, and F. Xu. 2015. Direct writing electrodes using a ball pen for paper-based point-of-care testing. *Analyst* 140: 5526–5535.

Li, Z., F. Li, Y. Xing, Z. Liu, M. You, Y. Li, T. Wen, Z. Qu, X. Ling Li, and F. Xu. 2017. Pen-on-paper strategy for point-of-care testing: Rapid prototyping of fully written microfluidic biosensor. *Biosensors and Bioelectronics* 98: 478–485.

Liu, H., and R. M. Crooks. 2011. Three-dimensional paper microfluidic devices assembled using the principles of origami. *Journal of the American Chemical Society* 133: 17564–17566.

Liu, J., S. K. Lau, V. A. Varma, R. A. Moffitt, M. Caldwell, T. Liu, A. N. Young, et al. 2010. Molecular mapping of tumor heterogeneity on clinical tissue specimens with multiplexed quantum dots. *ACS Nano* 4: 2755–2765.

Love, J. C., L. A. Estroff, J. K. Kriebel, R. G. Nuzzo, and G. M. Whitesides. 2005. Self-assembled monolayers of thiolates on metals as a form of nanotechnology. *Chemical Reviews* 103: 1103–1169.

Lu, F., W. Cai, and Y. Zhang. 2008. ZnO hierarchical micro/nanoarchitectures: Solvothermal synthesis and structurally enhanced photocatalytic performance. *Advanced Functional Materials* 18: 1047–1056.

Lu, Y., W. Shi, J. Qin, and B. Lin. 2010. Fabrication and characterization of paper-based microfluidics prepared in nitrocellulose membrane by Wax printing. *Analytical Chemistry* 82: 329–335.

Määttänen, A., U. Vanamo, P. Ihalainen, P. Pulkkinen, H. Tenhu, J. Bobacka, and J. Peltonen. 2013. A low-cost paper-based inkjet-printed platform for electrochemical analyses. *Sensors and Actuators B: Chemical* 177: 153–162.

Mahadeva, S. K., and J. Kim. 2011. Conductometric glucose biosensor made with cellulose and tin oxide hybrid nanocomposite. *Sensors and Actuators B: Chemical* 157: 177–182.

Mano, N., and A. Heller. 2003. A miniature membraneless biofuel cell operating at 0.36 V under physiological conditions. *Journal of the Electrochemical Society* 105: A1136–A1138.

Martinez, A. W., S. T. Phillips, M. J. Butte, and G. M. Whitesides. 2007. Patterned paper as a platform for inexpensive, low-volume, portable bioassays. *Angewandte Chemie: International Edition* 46: 1318–1320.

Martinez, A. W., S. T. Phillips, and G. M. Whitesides. 2008. Three-dimensional microfluidic devices fabricated in layered paper and tape. *Proceedings of the National Academy of Sciences of the United States of America* 105: 19606–19611.

Mitchell, H. T., I. C. Noxon, C. A. Chaplan, S. J. Carlton, C. H. Liu, K. A. Ganaja, N. W. Martinez, C. E. Immoos, P. J. Costanzo, and A. W. Martinez. 2015. Reagent pencils: A new technique for solvent-free deposition of reagents onto paper-based microfluidic devices. *Lab on a Chip* 15: 2213–2220.

Mohammadifar, M., M. Tahernia, and S. Choi. 2019. An equipment-free, paper-based electrochemical sensor for visual monitoring of glucose levels in urine. *SLAS Technology* 24: 499–505.

Montalti, M., A. Cantelli, and G. Battistelli. 2015. Nanodiamonds and silicon quantum dots: Ultrastable and biocompatible luminescent nanoprobes for long-term bioimaging. *Chemical Society Reviews* 44: 4853–4921.

Murphy, C. J. 2002. Peer reviewed: Optical sensing with quantum dots. *Analytical Chemistry* 74: 520 A–526 A.

Napi, M. L. M., S. M. Sultan, R. Ismail, K. W. How, and M. K. Ahmad. 2019. Electrochemical-based biosensors on different zinc oxide nanostructures: A review. *Materials* 12: 1–34.

Nge, P. N., C .I. Rogers, and A. T. Woolley. 2013. Advances in microfluidic materials, functions, integration, and applications. *Chemical Reviews* 113: 2550–2583.

Nguyen, A. V. T., T. D. Dao, T. T. T. Trinh, D. Y. Choi, S. T. Yu, H. Park, and S. J. Yeo. 2020. Sensitive detection of influenza a virus based on a CdSe/CdS/ZnS quantum dot-linked rapid fluorescent immunochromatographic test. *Biosensors and Bioelectronics* 155: 112090.

Nie, Z., C. A. Nijhuis, J. Gong, X. Chen, A. Kumachev, A. W. Martinez, M. Narovlyansky, and G. M. Whitesides. 2010. Electrochemical sensing in paper-based microfluidic devices. *Lab on a Chip* 1: 477–483.

Novoselov, K. S., A. K. Geim, S. V. Morozov, D. Jiang, Y. Zhang, S. V. Dubonos, I. V. Grigorieva, and A. A. Firsov. 2004. Electric field in atomically thin carbon films. *Science* 306: 666–669.

Núnez-Bajo, E., M. C. Blanco-López, A. Costa-García, and M. Teresa Fernández-Abedul. 2017. Integration of gold-sputtered electrofluidic paper on wire-included analytical platforms for glucose biosensing. *Biosensors and Bioelectronics* 91: 824–832.

Núnez-Bajo, E., M. C. Blanco-López, A. Costa-García, and M. T. Fernández-Abedul. 2018. In situ gold-nanoparticle electrogeneration on gold films deposited on paper for non-enzymatic electrochemical determination of glucose. *Talanta* 178: 160–165.

Öncel, Ç., and Y. Yürüm. 2006. Carbon nanotube synthesis via the catalytic CVD method: A review on the effect of reaction parameters. *Fullerenes, Nanotubes and Carbon Nanostructures* 14: 17–37.

Ornatska, M., E. Sharpe, D. Andreescu, and S. Andreescu. 2011. Paper bioassay based on ceria nanoparticles as colorimetric probes. *Analytical Chemistry* 83: 4273–4280.

Osadebe, I., P. Conghaile, P. Kavanagh, and D. Leech. 2015. Glucose oxidation by osmium redox polymer mediated enzyme electrodes operating at low potential and in oxygen, for application to enzymatic fuel cells. *Electrochimica Acta* 182: 320–326.

Parrilla, M., R. Cánovas, and F. J. Andrade. 2017. Paper-based enzymatic electrode with enhanced potentiometric response for monitoring glucose in biological fluids. *Biosensors and Bioelectronics* 90: 110–116.

Petryayeva, E., and W. R. Algar. 2013. Proteolytic assays on quantum-dot-modified paper substrates using simple optical readout platforms. *Analytical Chemistry* 85: 8817–8825.

Pirzada, M., and Z. Altintas. 2019. Nanomaterials for healthcare biosensing applications. *Sensors* 19: 5311.

Pokropivny, V. V., and V. V. Skorokhod. 2008. New dimensionality classifications of nanostructures. *Physica E: Low-Dimensional Systems and Nanostructures* 40: 2521–2525.

Punjiya, M., C. H. Moon, Z. Matharu, H. Rezaei Nejad, and S. Sonkusale. 2018. A three-dimensional electrochemical paper-based analytical device for low-cost diagnostics. *Analyst* 143: 1059–1064.

Qin, H., Z. Zhu, W. Ji, and M. Zhang. 2018. Carbon nanotube paper-based electrode for electrochemical detection of chemicals in rat microdialysate. *Electroanalysis* 30: 1022–1027.

Qiu, Z., J. Shu, and D. Tang. 2017. Bioresponsive release system for visual fluorescence detection of carcinoembryonic antigen from mesoporous silica nanocontainers mediated optical color on quantum dot-enzyme-impregnated paper. *Analytical Chemistry* 89: 5152–5160.

Rahman, F. 2019. Zinc oxide light-emitting diodes: A Review. *Optical Engineering* 58: 010901.

Rani, R., K. Sethi, and G. Singh. 2019. Nanomaterials and their applications in bioimaging. In *Nanotechnology in the Life Sciences*, Plant Nanobionics. Springer, Cham, 429–450.

Ren, S., P. Rong, and Q. Yu. 2018. Preparations, properties and applications of graphene in functional devices: A concise review. *Ceramics International* 44: 11940–11955.

Russo, A., B. Y. Ahn, J. J. Adams, E. B. Duoss, J. T. Bernhard, and J. A. Lewis. 2011. Pen-on-paper flexible electronics. *Advanced Materials* 23: 3426–3430.

Shao, Y., J. Wang, H. Wu, J. Liu, I. A. Aksay, and Y. Lin. 2010. Graphene based electrochemical sensors and biosensors: A review. *Electroanalysis* 22: 1027–1036.

Shen, Y., T. T. Tran, S. Modha, H. Tsutsui, and A. Mulchandani. 2019. A paper-based chemiresistive biosensor employing single-walled carbon nanotubes for low-cost, point-of-care detection. *Biosensors and Bioelectronics* 130: 367–373.

Shiroma, L. Y., M. Santhiago, A. L. Gobbi, and L. T. Kubota. 2012. Separation and electrochemical detection of paracetamol and 4-aminophenol in a paper-based microfluidic device. *Analytica Chimica Acta* 725: 44–50.

Shu, J., and D. Tang. 2017. Current advances in quantum-dots-based photoelectrochemical immunoassays. *Chemistry: An Asian Journal* 12: 2780–2789.

Simon, J., E. Flahaut, and M. Golzio. 2019. Overview of carbon nanotubes for biomedical applications. *Materials* 12: 624.

Sireesha, M., V. Jagadeesh Babu, A. S. Kranthi Kiran, and S. Ramakrishna. 2018. A review on carbon nanotubes in biosensor devices and their applications in medicine. *Nanocomposites* 4: 36–57.

Su, H., S. Li, Y. Jin, Z. Xian, D. Yang, W. Zhou, F. Mangaran, F. Leung, G. Sithamparanathan, and K. Kerman. 2017. Nanomaterial-based biosensors for biological detections. *Advanced Health Care Technologies* 3: 19–29.

Su, X., J. Ren, X. Meng, X. Ren, and F. Tang. 2013. A novel platform for enhanced biosensing based on the synergy effects of electrospun polymer nanofibers and graphene oxides. *Analyst* 138: 1459–1466.

Subrahmanyam, K. S., L. S. Panchakarla, A. Govindaraj, and C. N. R. Rao. 2009. Simple method of preparing graphene flakes by an arc-discharge method. *Journal of Physical Chemistry C* 113: 4257–4259.

Suvarnaphaet, P., and S. Pechprasarn. 2017. Graphene-based materials for biosensors: A review. *Sensors* 17: 2161.

Taleat, Z., A. Khoshroo, and M. Mazloum-Ardakani. 2014. Screen-printed electrodes for biosensing: A review (2008–2013). *Microchimica Acta* 181: 865–891.

Tasis, D., N. Tagmatarchis, A. Bianco, and M. Prato. 2006. Chemistry of carbon nanotubes. *Chemical Reviews* 106,: 1105–1136.

Tian, T., Y. Bi, X. Xu, Z. Zhu, and C. Yang. 2018. Integrated paper-based microfluidic devices for point-of-care testing. *Analytical Methods* 10: 3567–3581.

Tian, X., H. Peng, Y. Li, C. Yang, Z. Zhou, and Y. Wang. 2017. Highly sensitive and selective paper sensor based on carbon quantum dots for visual detection of TNT residues in groundwater. *Sensors and Actuators, B: Chemical* 243: 1002–1009.

Tiwale, N. 2015. Zinc oxide nanowire gas sensors: Fabrication, functionalisation and devices. *Materials Science and Technology* 31: 1681–1697.

Tonyushkina, K., and J. H. Nichols. 2009. Glucose meters: A review of technical challenges to obtaining accurate results. *Journal of Diabetes Science and Technology* 3: 971–980.

Torrisi, F., T. Hasan, W. Wu, Z. Sun, A. Lombardo, T. S. Kulmala, G. W. Hsieh, et al. 2012. Inkjet-printed graphene electronics. *ACS Nano* 6: 2992–3006.

Tsuboi, S., and T. Jin. 2017. Bioluminescence resonance energy transfer (BRET)-coupled annexin V-Functionalized quantum dots for near-infrared optical detection of apoptotic cells. *ChemBioChem* 18: 2231–2235.

Tsujimura, S., S. Kojima, K. Kano, T. Ikeda, M. Sato, H. Sanada, and H. Omura. 2006. Novel FAD-dependent glucose dehydrogenase for a dioxygen-insensitive glucose biosensor. *Bioscience, Biotechnology and Biochemistry* 70: 654–659.

Vanmaekelbergh, D., L. K. Van Vugt, H. E. Bakker, F. T. Rabouw, B. De Nijs, R. J. A. Van Dijk-Moes, M. A. Van Huis, P. J. Baesjou, and A. Van Blaaderen. 2015. Shape-dependent multiexciton emission and whispering gallery modes in supraparticles of CdSe/multishell quantum dots. *ACS Nano* 9: 3942–3950.

Wang, P., L. Ge, M. Yan, X. Song, S. Ge, and J. Yu 2012. Paper-based three-dimensional electrochemical immunodevice based on multi-walled carbon nanotubes functionalized paper for sensitive point-of-care testing. *Biosensors and Bioelectronics* 32: 238–243.

Wang, X., L. Zhi, and K. Müllen. 2008. Transparent, conductive graphene electrodes for dye-sensitized solar cells. *Nano Letters* 8: 323–327.

Wang, Y., J. Luo, J. Liu, X. Li, Z. Kong, H. Jin, and X. Cai. 2018. Electrochemical integrated paper-based immunosensor modified with multi-walled carbon nanotubes nanocomposites for point-of-care testing of 17β-estradiol. *Biosensors and Bioelectronics* 107: 47–53.

Wang, Z., and Z. Dai. 2015. Carbon nanomaterial-based electrochemical biosensors: An overview. *Nanoscale* 7: 6420–6431.

Wang, Z. Y., L. J. Wang, Q. Zhang, B. Tang, and C. Y. Zhang. 2018. Single quantum dot-based nanosensor for sensitive detection of 5-methylcytosine at both CpG and non-CpG sites. *Chemical Science* 9: 1330–1338.

Wegner, K. D., and N. Hildebrandt. 2015. Quantum dots: Bright and versatile in vitro and in vivo fluorescence imaging biosensors. *Chemical Society Reviews* 44: 4792–4834.

West, P. W. 1945. A selective spot test for copper. *Industrial and Engineering Chemistry Analytical Edition* 17: 740–741.

WitkowskaNery, E., M. Santhiago, and L. T. Kubota. 2016. Flow in a paper-based bioactive channel – study on electrochemical detection of glucose and uric acid. *Electroanalysis* 28: 2245–2252.

Wu, Y., Y. Ren, L. Han, Y. Yan, and H. Jiang. 2019. Three-dimensional paper based platform for automatically running multiple assays in a single step. *Talanta* 200: 177–185.

Xie, L., X. Zi, H. Zeng, J. Sun, L. Xu, and S. Chen. 2019. Low-cost fabrication of a paper-based microfluidic using a folded pattern paper. *Analytica Chimica Acta* 1053: 131–138.

Xu, G., S. Zeng, B. Zhang, M. T. Swihart, K. T. Yong, and P. N. Prasad. 2016. New generation cadmium-free quantum dots for biophotonics and nanomedicine. *Chemical Reviews* 116: 12234–12327.

Yang, Y., E. Noviana, M. P. Nguyen, B. J. Geiss, D. S. Dandy, and C. S. Henry. 2017. Paper-based microfluidic devices: Emerging themes and applications. *Analytical Chemistry* 89: 71–91.

Yang, Y. B., Y. D. Tang, Y. Hu, F. Yu, J. Y. Xiong, M. X. Sun, C. Lyu, et al. 2020. Single virus tracking with quantum dots packaged into enveloped viruses using CRISPR. *Nano Letters* 20: 1417–1427.

Yao, J., L. Li, P. Li, and M. Yang. 2017. Quantum dots: From fluorescence to chemiluminescence, bioluminescence, electrochemiluminescence, and electrochemistry. *Nanoscale* 9: 13364–13383.

Yao, Y., and C. Zhang. 2016. A novel screen-printed microfluidic paper-based electrochemical device for detection of glucose and uric acid in urine. *Biomedical Microdevices* 18: 1–9.

Yu, J., L. Ge, J. Huang, S. Wang, and S. Ge. 2011. Microfluidic paper-based chemiluminescence biosensor for simultaneous determination of glucose and uric acid. *Lab on a Chip* 11: 1286–1291.

Yu, M. F., O. Lourie, M. J. Dyer, K. Moloni, T. F. Kelly, and R. S. Ruoff. 2000. Strength and breaking mechanism of multiwalled carbon nanotubes under tensile load. *Science* 287: 637–640.

Yuan, J., N. Gaponik, and A. Eychmü. 2012. Application of polymer quantum dot-enzyme hybrids in the biosensor development and test paper fabrication. *Analytical Chemistry* 84: 50047–50052.

Yue, Z., F. Lisdat, W. J. Parak, S. G. Hickey, L. Tu, N. Sabir, D. Dorfs, and N. C. Bigall. 2013. Quantum-dot-based photoelectrochemical sensors for chemical and biological detection. *ACS Applied Materials and Interfaces* 5: 2800–2814.

Zhang, Z., S. Liu, Y. Shi, Y. Zhang, D. Peacock, F. Yan, P. Wang, L. He, X. Feng, and S. Fang. 2014. Label-free aptamer biosensor for thrombin detection on a nanocomposite of graphene and plasma polymerized allylamine. *Journal of Materials Chemistry B* 2: 1530–1538.

Zhao, C., M. M. Thuo, and X. Liu. 2013. A microfluidic paper-based electrochemical biosensor array for multiplexed detection of metabolic biomarkers. *Science and Technology of Advanced Materials* 14: 054402.

Zhou, Juan, Y. Yang, and C. Y. Zhang. 2015. Toward biocompatible semiconductor quantum dots: from biosynthesis and bioconjugation to biomedical application. *Chemical Reviews* 115: 11669–11717.

9 Microfluidics Devices as Miniaturized Analytical Modules for Cancer Diagnosis

Niraj K. Vishwakarma, Parul Chaurasia, Pranjal Chandra, and Sanjeev Kumar Mahto

CONTENTS

9.1 INTRODUCTION

Cancer is a chronic disease that can develop in almost any organ or tissue of the human body when abnormal cells grow uncontrollably.[1] Breast, colorectal, lung, liver, cervical, prostate, and thyroid cancers in humans are the most common types of cancer. Cancer is the second leading cause of death globally after accidental death; recently it was estimated that in 2018, 9.6 million people died globally from various types of cancer.[2] The mortality rate in cancer patients decreases dramatically if cancers can be detected at an early stage.[3] In contrast, due to the localized and smaller size of the tumor at a primary site, early-stage cancer detection is quite crucial for improved diagnosis.[4–7] Various technologies have been developed for pre-screening, prognostic, diagnostic, therapy assessment, and monitoring of cancer.[8–10] In addition, several types of direct-tumor-biopsy are also known for diagnosis of cancer in humans. Moreover, in the case of solid tumors, cancer tissue is extracted using an image-assisted core needle biopsy followed by tissue interrogation by histopathologists.[11] On the another hand, direct-tumor-biopsy provokes the risk of bleeding and inflammation, therefore, it can be an alternative route for the cancer to spread.

DOI: 10.1201/9781003033479-9

Furthermore, tissue biopsy cannot be performed for many cancers such as lung and brain cancer, as tumor heterogeneity is an additional challenge.[1, 12–15] The time to process results in tumor biopsy takes several days and sometimes weeks.[16] Overall, traditional tumor biopsy can only be used for static tumors at a specific time while it does not reveal dynamic changes that occur during cancer treatment.

Cancer cells usually break off from the primary tumor at the initial stages of malignant progression, circulate in the blood or lymph vessels and invade through the interstitial extracellular matrix to a distant organ, thereby establishing a secondary tumor.[17–20] The process is known as cancer metastases.[21, 22] Over 90% of mortality in cancer patients occurs due to metastases. Therefore, most of the diagnostic techniques have been designed to reproduce and study the complexity of metastases.

The concentration of break-off circulating tumor cells (CTCs) in the blood of cancer patients is often found in only up to 10 100 cells mL^{-1} while red blood cells (erythrocyte) and white blood cells (leukocyte) exist in 10^9 and 10^6 cell mL^{-1}, respectively.[21, 23] CTCs reveal a wealth of information about the primary tumor and mutations occurring in tumor tissue, allowing for targeted therapeutic approaches.[24] In addition, the analysis of CTCs can also express the mechanism of treatment resistance.[25, 26] Therefore, CTC analyses are considered as a real-time "liquid biopsy" for patients with cancer.[27] On the other hand, the isolation and analysis of CTCs require exceptionally sensitive and specific analytical instrumentation.[28] Several approaches have been developed in the past decade for CTC isolation and analysis that take advantage of their physical (density, size, and electric charges)[8, 29, 30] and biological properties[27] (cell viability and cell surface protein expression). Although CTC isolation can be performed by flow cytometry, while it requires a large amount of sample and has a too-low throughput which limits their feasibility. Whereas the enzyme-linked immunosorbent assay (ELISA) method requires a high sample volume, long analysis time, and is unable to measure multiple proteins.[31] Liquid chromatography-mass spectrometry (LC-MS)-based proteomics is a realistic tool for biomarker discovery, however it is technically complex for clinical diagnostics and too expensive.[3, 24] On another hand, fluorescence, electrochemiluminescence (ECL), and surface plasmon resonance (SPR) measurement technologies are available which have detection limits of 1–100 pg mL^{-1} and is capable of measuring up to ten proteins per sample.[28, 32] Nevertheless, instrument cost and sampling cost per single-use limit their applications and require trained professionals for operation. CellSearchsystem (Menarini Silicon Biosystems, Italy) is renowned as an FDA-approved commercial product for CTC analysis, however, its high cost, manual process, and false-positive/false-negative limit its applications.[29, 32]

Microfluidics-based CTCs isolation technologies have come to the fore in the last two decades, where the convenience of such technologies is prevailing over conventional batch techniques.[22, 33–40] Microfluidics provides miniaturized devices in a very small area and is capable of fast CTC isolation with high efficiencies as well as automation. It can be integrated easily into multiple workflows. The microfluidics technologies have many advantages over conventional methods such as reducing the size of equipment, allowing for parallelization, and eliminating complex protocols, which can enable complete lab-on-a-chip devices for CTCs isoalation.[20, 41–44]

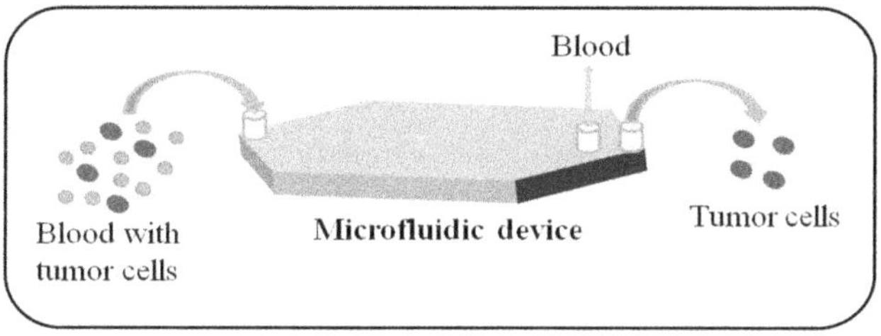

FIGURE 9.1 Schematic illustration of microfluidics approach for isolation of tumor cells.

A general schematic diagram of microfluidics-based CTCs isolation is shown in Figure 9.1.

This chapter focuses on the microfluidics technologies used for the isolation and analysis of CTCs. The current technologies that are outlined and discussed are based on exploiting the physical and biological properties of CTCs. The main technologies for CTC detection that are discussed are cell-affinity chromatography,[45–55] immunomagnetic-based cell sorting,[56–61] and methods that differentiate based on cellular biophysics such as cell size,[29,62–65] and dielectrophoresis (DEP).[66–70] We discuss the performance and capability of these systems in terms of throughput, purity, yield, cell viability, and the capability for on-chip post-processing after cancer cell capture.

9.2 MICROFLUIDICS APPROACHES FOR CANCER DETECTION

9.2.1 Cell-Affinity MicroChromatography (CAMC)

Cell-affinity micro-chromatography (CAMC), developed from affinity chromatography, is a technique based on antigen-antibody interactions.[1, 7] It is a method in which cancer cells from a heterogeneous cell population can be captured selectively. A high-affinity ligand (e.g., antibodies), which can bind selectively to a certain cell type (e.g., antigen), is immobilized on a solid support.[71–74] The method has recently been implemented in microfluidics devices for capturing different types of cancer cells. However, the method requires highly selective and high-affinity antibodies for each type of cancer cell. The selection of ligands is critical for cell-affinity separations which comprise proteins,[74] antibodies,[75] and high-affinity aptamers.[76] The high-affinity DNA-aptamers can also be used as high-affinity ligands in a microfluidic channel. A simple schematic diagram of cell-affinity micro-chromatography is shown in Figure 9.2(A).

An anti-epithelial cell adhesion molecule (anti-EpCAM) is known as the most commonly employed antibody for CTCs isolation via antibody-based cell capture.[77–79] In such a system, the isolated CTCs are stained with fluorescent markers for enumeration and counting. However, the method is restricted to limited capture capacity because of the small surface area for ligating antibodies into the microfluidic channel. Nevertheless, various types of microfluidics channels have been developed for better ligation. The first antibody-based CAMC system was reported by Du et al.[80] Exploiting this system, cervical cancer cells were captured using α6-integrin

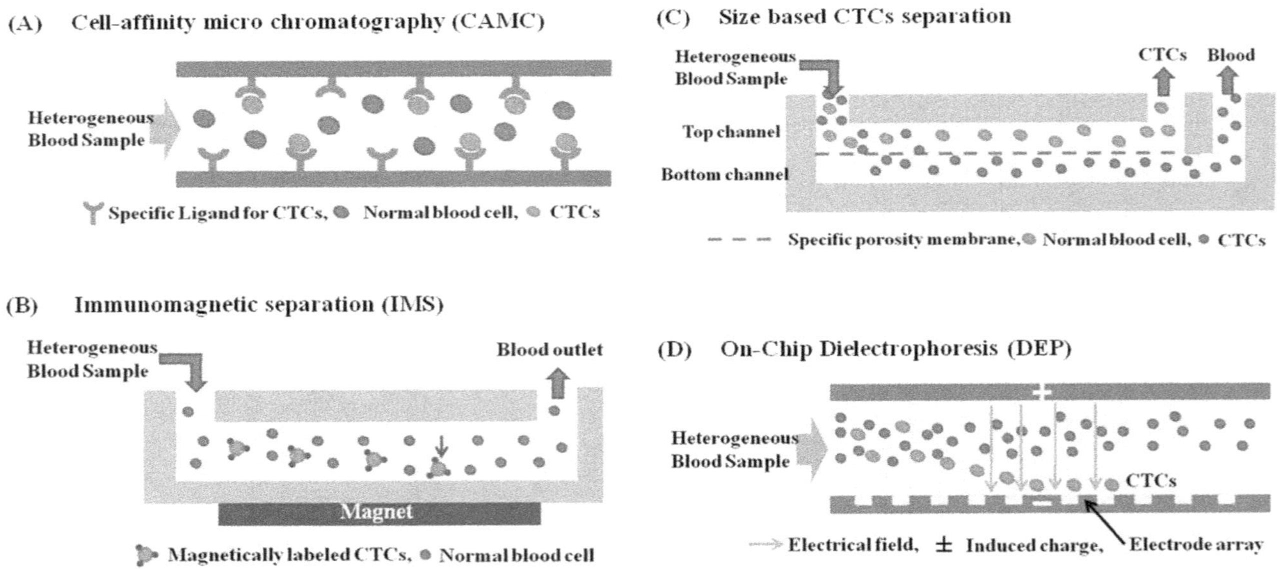

FIGURE 9.2 Schematic illustration of different types of microfluidics technologies used for the isolation and analysis of CTCs. (A) Cell-affinity micro chromatography (CAMC). (B) Immunomagnetic separation (IMS). (C) Size-based CTCs separation. (D) On-chip dielectrophoresis (DEP).

as a capture antibody bound to the channel surface. The capture rate of tumor cells was controlled by the flow rate and concentration of antibodies used to the microfluidic channel surface. The system demonstrated a cancer cell recovery rate >30%, including 5% normal cells, with cell line mixtures of human cervical stroma, normal human glandular epithelial, and cervical cancer cells with up-regulated α6-integrin cell surface receptors.

CTC capture has been demonstrated by utilizing a packed-bed system with a high surface area in a microfluidic channel containing a weir.[81] For efficient capture of CTCs, avidin-functionalized beads were encumbered into the microfluidics device which generated a high surface-to-volume ratio. The isolation efficacy of the system was performed by infusing breast cancer cells present in whole blood pretreated with a biotinylated anti-EpCAM antibody. While flowing, the biotinylated cells come into contact with avidin-functionalized packed bed and bound, yielded captured cancer cells, which were then enumerated by washing. Overall capture efficiency of 40% was determined for multiple blood donors (ranging from 35% up to 70%).

Immunoaffinity-based selective CTCs capture of two phenotypically similar cells was performed by patterning regions of alternating-affinity ligands.[82] In this strategy, cells with strong affinity can be separated into one affinity-patterned region and cells with lower affinity levels can be captured in a second affinity region of the microchip. Adopting a similar strategy, Launiere et al. reported a multi-protein patterned microchip for improving CTC capture by patterning regions of alternating-adhesive proteins.[83] Two proteins, anti-EpCAM and E-selectin, have been used among which anti-EpCAM provides the specificity for CTCs capture and E-selectin bind with unwanted impurities in the flow stream, thereby increases CTC capture efficiency. As a result, this method reduces the capture of normal blood cells up to 82% with a high capture efficiency of tumor cells up to 1.9 times higher than surface coated with anti-EpCAM alone.

A three-dimensional microfluidics chip was developed by Li et al. for high-purity cell separations utilizing a negative selection of background cells.[84] The device was efficient in enriching target cells without labeling or the need to elute captured cells. Therefore, this approach is suitable for applications where subsequent analyses of target cells are required, such as genetic analysis and subculture. Using this system, the separation purities were achieved up to 92–96%, which is comparatively high with straight-channel systems (77% purity). The flexible nature of the system allows parallelization for massive analysis.

9.2.2 Immunomagnetic Separation (IMS)

Magnetic nanoparticles (MNPs) get close to a biological entity of interest because of their comparable dimensions with biological particles.[59, 85] Magnetic bead-based cell separation techniques can overcome the limitations of CAMC in which recovery of captured target cells is difficult. CAMC usually requires acidic solution to recover captured cells which can denatur the cells. Immunomagnetic separation (IMS) is a technique that is used for separation and enrichment of specific cells/CTCs from a heterogeneous suspension. It involves the coupling of biological macromolecules such as CTCs with supermagnetic iron oxide (Fe_3O_4) nanoparticles embedded as core in a polymer shell. The Fe_3O_4 beads can be controlled and stabilized under the influence of the external magnetic field, While the magnetic memory will be lost when the external magnetic field is detached, resulting in a re-dispersion of magnetic beads. When Fe_3O_4 beads are grafted with biological molecules, e.g., antibodies, it would be capable of identifying and capturing target cells such as CTCs. In contrast to classical IMS, wherein cell capture and separation followed by washing is required, microfluidics-based IMS overcomes these limitations. A schematic diagram of immunomagnetic separation (IMS) is shown in Figure 9.2(B).

Utilizing microfluidics-based IMS, Liu et al. fabricated a hexagonal array of nickel micropillars for isolation of CTCs from a population of normal blood cells.[56] The micropillars were designed in such a way that they can trap superparamagnetic beads and be fabricated onto the bottom of the inner channel. It also avoids the agglomeration of magnetic beads inside the channel. A549 cancer cells spiked in populated normal blood cells were infused in the device and captured over pre-trapped magnetic beads.

Further, Sun et al. proposed a strategy comprised of a combination of magnetic beads and nonmagnetic beads of IMS working as opposite selection criteria. The device was characterized by the isolation of tumor-initiating cells from breast cancer cells (SUM149).[57] $CD44^+/CD24^-$cells were isolated in a microfluidic channel by favorably mixing with nonmagnetic beads coated with anti-CD24 antibody followed by mixing with magnetic beads coated with anti-CD44 antibody under the influence of a magnetic field. Consequently, two IMS beads produced a higher tumor sphere formation of 1.62% compared with 1.16% of one IMS bead using only $CD44^+$ criterion with the same initial cell number.

Low abundance, small size, and heterogeneity of CTCs signify major engineering challenges for their capturing. Shields IV et al. advanced a microfluidics device to enumerate cancer cells from normal blood followed by staining and analysis.[61] The microfluidics device is comprised of three modules, viz., acoustic standing wave, magnetic field gradient, and an array of micromagnets which results in the alignment of cells, deflection of magnetically labeled cells from unlabeled cells, followed by the capturing of labeled cells (cancer cells), respectively. The first two modules separated magnetically labeled cells with purity of >85%, whereas the third module was capable of capturing with accuracies of >80%. Therefore, the device is efficient for the enumeration of CTCs which can be extended for application of single-cell biology and immunology.

A microfluidics-based IMS chip was presented for CTCs detection with a very fast flow rate of 10 mL h^{-1}.[86] The microchip was fabricated by attaching a microchannel-patterned polydimethylsiloxane (PDMS) slab with a glass coverslip. The MNP-labeled cancer cells containing blood sample flowed closely above the magnetic array which become deposited at the bottom wall of glass coverslip of the device. Therefore, the deposited cells could be analyzed directly through a fluorescence microscope. In the proposed microchip, the number of nanoparticles used was 25% fewer than the amount used in the commercial CellSearch system. To label cancer cells, anti-EpCAM antibodies, labeled Fe_3O_4 MNPs, were introduced into the blood samples. Thereafter, the blood sample was flowed through the microchip to capture labeled cells. Further, cells were stained with DAPI, anti-CD45, and fluorescent-labeled anti-cytokeratin. The device can detect cancer cells from a very low number of tumor cells to a blood cell ratio (about 1:10^9), while the device was capable of capturing tumor cells with capture rates of ~90% and ~86% for COLO205 and SKBR3 cells, respectively.

There have been reports published concerning a cell-sorting method based on cell size differences under the influence of biocompatible ferrofluids. Zhao et al. demonstrated ferrohydrodynamic cell separation of low-concentration CTCs spiked with RBC-lysed blood.[60] The method is label free and utilizes magnetic buoyance force which exerted differently on cells depending on their size. The method implies a laminar flow that was capable of enriching CTCs from patients' blood with a high throughput of ~6 mL h^{-1} and a high rate of recovery of ~92.9%. To process a significant amount of blood, systematic optimization of ferrofluid properties and optimal magnetic field and its gradient was determined through a validated analytical model. As a result, cancer cells including H1299 lung cancer and MCF-7 breast cancer with recovery rates of 92.3 ± 3.6% and 94.7 ± 4.0%, respectively, have recovered with ~100 cancer cells per mL spike ratio.

9.2.3 Size-Based Cancer Cell Detection and Separation

The CTCs size (17–52 μm) are considered to be larger than the human blood components, i.e., white blood cells (7–15 μm) and red blood cells (6–8 μm).[1, 62] Therefore, size-based CTC isolation can be performed as a consequence of morphological differences between CTCs and normal blood cells.[1, 7] Size-based isolation of CTCs through microfluidics techniques is one of the challenging and interesting fields. In conventional microfiltration, a common difficulty such as clogging and adhesion of the blood sample is very general. Furthermore, the gathering of cells on the microfilter increases fluid driving pressure that can denature captured cells. On other hand, size-based separation does not require specific biomarkers such as labeling of cells and ligands, in addition, it requires limited components for microdevice fabrication. Moreover, size-based CTCs sorting via microchips enables single-cell analysis or cell culture inside the chip. The key point of size-based separation is a high flow rate which is directly proportional to the high-throughput capturing of CTCs. Finally, the cell viability or gene expression profile of isolated CTCs does not reduce during size-based isolation, consequently, it enables off-chip cellular and molecular characterizations. A microfluidic diagram of size-based cancer cell separation is shown in Figure 9.2(C).

Mohamed et al. developed an on-chip microfabricated sieving microfilters device for size-based separation of CTCs.[87] The device comprises four successively narrower regions of widths 20 μm, 15 μm, 10 μm, and 5 μm with constant channel depth (20 μm) and ~1800 channels per region. The narrower region performs as a barricade for larger cells while allowing smaller cells to continue passing. As cells flow across the device, they combated each region and bigger cells stopped at a specific gap width that prohibited passage. For experimental purposes, a neuroblastoma cell mixed with whole blood was preferred. When mixed cells infused into the device, all other cells except neuroblastoma migrated to the output while neuroblastoma cells were preserved at a 10 μm-wide region barrier.

Using cell size as a biomarker, further, Hur et al. presented a methodology that allows label-free and high-throughput CTCs isolation from a heterogeneous blood sample.[88] The approach employs selective isolation of larger cancer cells in parallel expansion-contraction trapping reservoirs utilizing the irretrievable migration of particles into microvortices developed in reservoirs. Larger CTCs spiked in normal blood were successfully separated in trapping reservoirs with processing rates 7.5×10^6 cells s^{-1}. The processing parameters, operational flow rate, and cell diameter for the trapping of CTCs in microvortices was also determined experimentally. The observed cell recovery rates were ~23% and ~10% for MCF-7 and HeLa cells, respectively. The proposed method would be useful for research and clinical applications wherein in vitro culture is required.

Furthermore, Chung et al. reported a microfluidic cell sorter (μFCS) for the isolation and comprehensive analysis of CTCs in whole blood, processing with large volumes of samples at high flow rates without clogging or pressure buildup.[65] The μFCS has a physical barrier of modified weir-style which capture size-based target cells and successively allows in situ tumor identification. Owing to a modified weir, the system was able to achieve a $>10^4$ enrichment ratio at a flow rate of 20mL h^{-1} and allow facile retrieval of the captured cells. The methodology can be further extended to enrich CTCs of a broader size range. For the passage of human blood cells through the weir, the optimal height gap was observed ~10 μm under which >99% blood cells passed. The captured cells can be examined in situ for comprehensive and multifaceted evaluation.

A crossflow filtration technique was developed by Li et al. using a high porosity PDMS microfiltration membrane.[64] This chip is a kind of membrane-assisted dual-channel[89] microfilter which allows penetration of particles of a certain size (depending on the pore size of PDMS membrane) under flow conditions while carrying away bigger particles. For favorable filtration of target cells without clogging the PDMS membrane, the porosity of the PDMS membrane can be maintained as high as 30% with a surface area of 3 × 3 cm. The crossflow filtration allows the passing of undesired RBCs across the membrane while the continuous main flow carries away CTCs with WBCs. The crossflow filtration chip attained a sample throughput of ~1 mL h^{-1} for the processing of undiluted blood samples while maintaining high cell purity (93.5 ± 0.5%) and a high recovery rate (27.4 ± 4.9%) for WBCs.

The working principle of Dean flow fractionation based on particle inertia, is another separation technique based on cell size for CTC isolation from the blood.[90] In this technique, a curvilinear channel is used to accelerate particles in the radial direction and curvature adopted by cells depending on their size. The influence of centrifugal acceleration in the radial direction results in the formation of two symmetrical counter-rotating vortices (top and bottom) across the channel cross-section which are known as Dean vortices.[91] Using this technique, initially, a mixture of 7.32 μm and 1.9 μm polystyrene particles have been separated successfully.[92]

Applying Dean drag and inertial focusing forces together upon cells of different sizes, Lim et al. separated CTCs from normal blood.[93] Channel dimensions were designed in such a way that only larger cells, CTCs, experience inertial focusing, while smaller cells (RBCs and leukocytes) can be affected by the Dean drag. The microchannel consists of a two-inlet and two-outlet circular microchannel (width = 500 μm and height = 160 μm) with a total length of 10 cm. A sheath flow (phosphate buffer solution) was infused with a high flow rate through the inner inlet together with a diluted blood sample from the outer inlet. All the cells (including CTCs) initiate migration along with the Dean vortex and move towards the inner channel. As a result, CTCs (experience strong inertial lift forces) confined to the inner wall whereas blood cells (influence by Dean drag) continue flowing along with the Dean vortex and confine towards the outer wall. The performance of the system was authenticated using CTCs spiked into normal blood after that preliminary clinical test with healthy and lung cancer patients' blood samples.

Furthermore, Warkiani et al. presented a circular microfluidics technique with a trapezoidal cross-section outlet for CTC enrichment from clinically significant blood in a label-free and ultrafast manner.[94] Similar to the above discussion, smaller particles (hematologic components) influenced by inertial force and rotational flow and come closer to outer wall whereas larger particles (CTCs) experienced the Dean vortex preferably and approach toward the inner wall. More than 80% of cancer cells have been successfully isolated and recovered using a single spiral microchannel with one inlet and two outlets.

Advancing microfluidics devices based on Dean drag and inertial focusing Lim et al. further presented an improved and high-throughput multiplexed version of the circular microchannel (three circular microchannels stacked together). The device validated high sensitivity by the consistent detection of CTCs (breast cancer samples: 12–1275 CTCs mL^{-1}, lung cancer samples: 10–1535 CTCs mL^{-1}) from 100% clinically significant blood of patients (n = 56).[95] The multiplexed device was fabricated by stacking three individual devices together through manual alignment and oxygen plasma bonding. Under the influence of inertial lift force CTCs approaches close the inner wall of the microchannel, whereas WBCs with platelets experience Dean drag forces and migrate near the outer wall.

For high purity separation outcome of CTCs, Zhou et al. demonstrated a novel multi-flow microfluidics device based on size-dependent inertial migration of cells.[96] Rotational-induced force, a kind of inertial force, applied on the cells is the leading force for size-based isolation of CTCs in the multi-flow microfluidics device. The device was proficient in providing >87% purity of separation outcome of unlabeled cancer cells with a recovery rate of >93% at clinically appropriate concentrations of cells (2 cells/mL and above). Likewise, the device detected six CTCs out of eight non-small cells of lung-cancer patients, while none for five healthy control subjects.

9.2.4 On-Chip Dielectrophoresis (DEP)

Dielectrophoresis (DEP) is an electrokinetic method that uses the polarization of dielectric particles such as cells under nonuniform electrical field force.[67, 97] The technique is commonly utilized in microfluidics for separation of cells and biomolecular diagnosis.

Recently, DEP is being used for the isolation of different types of CTCs from blood, and therefore, anticipated as a molecular biomarker.[7, 97, 98] There have been several reports that deal with CTCs separation from normal blood using microfluidics DEP devices.[1, 7, 99, 100] DEP's working mechanism mainly depends on the dielectric properties of cells (surface charge and surface area) and works under an applied electrical field gradient. Therefore, surface charge differences between CTCs and normal blood cells (neutral) are mainly responsible for CTCs separation through DEP. When the cells are placed in the nonuniform electric field they start moving in a certain direction depending on the magnitude of electric field and charge developed on the cells.[97] It can also be applied for cell manipulation and characterization of cell dielectric properties. A general schematic diagram of on-chip DEP is shown in Figure 9.2(D).

As a result of asymmetry in surface charge, surface area, and the size of different cells, they experience dissimilar DEP responses. Therefore, cells can be separated by choosing a certain electric field frequency which lies in between the crossover frequencies of different cells.[101–104] Cells with lower crossover frequency will be attracted to high electrical magnitude and vice versa. This difference discriminates against different types of cells. DEP is known as the most promising separation technique due to its high efficiency, high throughput, high selectivity, and low sampling cost.[97, 99]

Firstly, Becker et al. were able to sort human leukemia cells at a rate of about 10^3 cells s^{-1} using DEP based dielectric affinity column. The proposed method was able to manage a sorting rate than a conventional fluorescent activated cell sorter. The leukemia cells containing normal blood cells were eluted through a DEP column, resulting in leukemia cell retention by microelectrode arrays while normal blood cells freely passed across the column.[104] Subsequently, removal of the DEP field yielded the release and assortment of cancer cells from the microelectrode. Using a similar dielectric column with distinction DEP forces Becker et al. further demonstrated separation of several different types of cancer cell from human blood.[105] Separation of epithelial cancer cells (MDA-231) was performed with the device. The recovery rate was ~95%.[103]

An insulator-based DEP (iDEP) microfluidic chip made of PDMS was developed by the researchers. The DEP microfluidic chip, associated with hydrodynamic techniques, was utilized for plasma separation from normal human blood.[106] The channel of the chip was composed of a set of dead-end branches at each side of the main channel and fitted with two Pt electrodes at the inlet and outlet points. The device has an optical real-time monitoring system that monitors pre- or post-processing components of plasma. After filling microchannel with a 2 μL droplet of fresh blood via capillary force, an electric field was generated into the device using Pt electrodes. DEP trapping of RBCs started through the cross-junctions formed at the channel which prevented more RBCs entering into the channel and thus working like a sieve. Importantly, an iDEP device can be used with blood samples prior to dilution and offer formation of plasma with a purity of 99%.

A technique known as dielectrophoretic field flow fractionation (DEP-FFF) has been established for cell separation efficiently via electrically controlled discrimination. The technique separates cells based on their density and surface morphology.[107–109] Gascoyne et al. implemented this technique for continuous flow separation of CTCs such as breast cancer cells from normal T-lymphocytes as well as from $CD34^+$ hematopoietic stem cells.[69, 110–112] The microelectrodes (50 μm width) of device was fabricated using standard photolithography on glass substrates (50 × 50 mm) and a Teflon spacer was introduced between the electrode plates and glass plate. The separation channel dimensions were height 0.42 × width 25 × length 388 mm. A thin PDMS chamber was energized with alternating current signals to separate the cells. Cells were levitated against DEP and sedimentation forces to different equilibrium heights depending on cell density and surface charge. Thereby, cells were transported at differing velocities through the microchannel and eventually got separated. Cells were counted by an in-line flow cytometer, which exited from the bottom outlet port of the chamber.

A constant electric field can be generated by maintaining identical widths and gaps between electrodes, therefore, possessing high separation efficiency. A DEP microfluidics device was developed for continuous separation of a heterogeneous mixture of blood cells of MDA-MB-231-labeled protein and cancer cells.[113] A planar inter-digitated transducer electrode was located on the bottom of the microchannel that slightly protruded into the microchannel from one of the sidewalls. The device allowed separation along the lateral direction (independent of effect of gravity), which resulted in faster separation than in DEP-FFF devices. The device does not require pre-focusing of cells as well as minimizing the adverse effect of the electric field on cells. The microdevice is comprised of two parts: Glass substrate and PDMS. A glass substrate was used to make an electrode, a sandwich layer of Cr/Au was coated on it. There were 20 pairs of electrodes with ≈39 μm widths and ≈36 μm of gaps, protruding dimension ≈2 μm to ≈3 μm. In the second part, the microfluidics structure was formed using PDMS. The channel width was ≈50 μm and it was enlarged to ≈250 μm. Cells entering the microdevice were deflected by the DEP force which was generated by the electrodes. Dissimilar cells were deflected to different lateral locations inside the microchannel by DEP force.

Fluid inertia with finite inertial force can be an effective approach for manipulation and separation of fluid contents.[114] In conventional microfluidics technologies, inertial effects are almost negligible and fluid speed is very low. However, fluid inertia cannot be neglected at a much higher fluid speed. Inertial microfluidics offers excellent advantages such as precise manipulation, high throughput, robustness, and simplicity.[114–116] A variety of inertial microfluidics devices has been developed for various applications including isolation of CTCs.[32, 93, 94] Many of them lack the tunability to adapt to different particle samples and flexibility due to the fixed microchannel structure and dimensions.

A hybrid technique combining DEP and the inertial system was developed by Li et al.[117] The system combined the benefits of both methods while overcoming their limitations. A serpentine PDMS channel (width 200 μm × depth 40 μm × length 700 μm) was introduced above the pre-patterned Ti (50 nm)/Pt (150 nm) zigzag electrode on a glass slide which generated DEP force within the bottom region of the microchannel. The gap and width of the electrodes were both 20 µm. The particles moving close to top region of microchannel cannot be adjusted instantaneously, which hampers particle separation by DEP modification. To push particles to get close to the electrode and remove the particle hindrance, a top sheath downward flow at the inlet of the samples flow was introduced that pushed all the particles towards the bottom region of the microchannel. The main advantage of the technique is that the voltage adjustment can be done in order to separate different size particles without a redesign of the layout of the chip. Therefore, the separation efficiency was improved by the implementation of the top sheath flow.

Acoustophoresis is a non-contacting and label-free technology for cell manipulation with improved bioanalytical and clinical applications. The integration of DEP with acoustophoresis remains a promising tool for automated cell separation systems where the washing of cells is essential. Cetin et al. developed a platform for particle washing followed by separation.[68] A low-conductivity buffer solution was used to wash particles which prevents the adverse effects of Joule heating developed via DEP. To create acoustic waves for acoustophoresis, a piezoelectric material was utilized which developed standing waves along the channel. The alignment of piezoelectric materials was achieved with a unique mold. The experiments are performed using polystyrene particles.

A bipolar electrode (BPE), a high-throughput and scalable wireless array, was developed by Anand et al. and subsequently tested on MDA-MB-231 cells.[99] The device helped in capturing single cells and their selective separation. The BPE device was able to remove ohmic contact developed in individual electrode wires. The CTCs were captured across the parallel microchannels consisting of micro-pockets with dimensions closely matching the dimension of targeted cells, which allowed easy incorporation of cells. Further, Henslee et al. used cDEP to separate the same cells from a heterogeneous mixture.[118] The heterogeneous mixture was comprised of early, intermediate, and late-stage breast cancer cells MCF-10A, MDA-MB-231, and MCF-7, respectively. Furthermore, the device was improved to be utilized with clinically relevant samples.[119]

A liquid electrode DEP device was developed for continuous separation of particles of different conductivity. The device is advantageous in terms of its low cost because there is no use of expensive metal electrodes; additionally, it does not require a high voltage field for operation.[120] To complete the circuit, an ionic liquid was used instead of metal electrodes in the PDMS device. The device performance was measured by continuous separation of PC-3 human prostate cancer cells. DEP force separates the particles in the electrode region due to the fact that different particles have different electrical properties and sizes. Ionic liquid and cells do not contact each other directly, and cells are collected in the culture medium. Moreover, the device is also capable of separating human breast cancer cells (MDA-MB-231) with high purity from stem cells (ADSCs). There are no metal electrodes, hence the fabrication is simpler and inexpensive.

Furthermore, an optically induced DEP (ODEP) was developed to separate PC-3 human prostate cancer cells from leucocytes. The ODEP device was integrated with in-line fluorescent microscopic imaging.[121] The device is flexible and user friendly and capable of modifying electrode layout. The electrode layout is designed with a computer interface and anticipated onto the microchannel through a digital projector. The device was capable of isolating CTCs with purity of 100%, which is not achievable with other DEP techniques. The device performs on the basis of light configurations for the efficient manipulation of leukocytes and PC-3 cells. The recovery rate achieved was ~41.5%.

To increase throughput and recovery rates, Lee et al. proposed a platform that combined negative DEP (n-DEP) force, gravitational forces and drag forces to sort K562 cells.[122] In addition, fabricating the components was minimized to fabricate the device with a simplified structure which minimizes the chance of leaking during experiments. To obtain greater proficiency and high throughput, iterated electrode pairs and a meso-sized channel was introduced into the device. Two electrode arrays were used to increase cells sorting efficacy. The arrays of electrodes were further modified by depositing five sub-electrodes which generated five repeated n-DEP force barriers. The device succeeded in achieving a throughput of 17,000 cells min^{-1}, separation efficiency of 94.74 ± 0.77%, with a 49.42% recovery rate.

Ovarian and cervical cancers are common tumors of female reproductive organs.[123] Cervical cancer develops by persistent growth of abnormal tissue in the cervix of the uterus. The common cell line used for cervical cancer cell is HeLa cells.[124] It is generally derived from an aggressive cervical adenocarcinoma. Mouse ovarian surface epithelial (MOSE) cells are commonly used for ovarian cancer research. MOSE can be originated from mouse ovarian cancer. A contactless DEP was developed for off-chip assortment and analysis.[68] The device is comprised with thin PDMS membrane which separates suspended cells from electrodes and provides improved cell viability. Further, researchers compared a contactless DEP device with a device comprised with pillars of cells (diameter 20 μm) in terms of cell viability and efficacy of same cell line. The trapping area on each pillar was maintained in such a way that only one or two cells could be trapped. The device was capable of sorting 10^6 cells h^{-1} and achieved optimum cell viabilities of 71% in untrapped and 81% in trapped MOSE cells.

Cheng et al. presented 3D lateral DEP (LDEP) microfluidics chip for high-throughput isolation of AS2-GFP, lung cancer cells, from diluted blood.[125] The microfluidics chip is allowed to focus the suspended samples to the channel side-walls with the help of hydrodynamic sheath flow. It allowed the particles to be in a queue through DEP-induced lateral displacements. Under a determined electrical field, larger cells (cancer cells) experienced higher LDEP force which pushed them to flow in the middle of the microchannel. On the other hand, smaller cells (blood cells) experienced lower LDEP force, resulting in a flow of cells to a distance of 200 μm away from the sidewall. With a 6 cm-long LDEP channel and 20 μL min^{-1} flow rate, the device was capable of achieving a recovery rate of 85% from a sample containing 0.001% cancer cells. The device throughput was amplified by increasing the channel length, resulting in a higher particle residence time in the induced region. The presented design allows the continuous fractionation of particles on the basis of differences in their critical negative DEP strengths.

Different cell lines can also be characterized based on their intrinsic electrical properties. Agah et al. presented a technique which was able to discriminate two closely related breast cancer cell lines (LCC1 and LCC9), known as off-chip passivated-electrode insulator-based DEP (OπDEP).[126] The presented technique also assessed the sensitivity of the cell line through an anti-cancer agent "Obatoclax." LCC1 and LCC9 showed different DEP responses under identical experimental conditions. Applying the photolithography technique, electrodes were patterned on a 500 μm-thick glass substrate. The electrode surface was then modified with chrome/gold by physical vapor deposition. The dimensions of electrodes were 1000 μm wide and had 600 μm horizontal spacing. The patterned electrodes were attached with a PDMS mold with an array of insulator pillars inside the channel. After the experiment on the mentioned cell lines, it was concluded that LCC1 showed higher crossover frequency (700 kHz) than LCC9 (100 kHz) with trapping efficiency 30–40% and 40–60%, respectively. When exposed to the anti-cancer agent Obatoclax, both cell lines exhibited dose-dependent shifts in DEP crossover frequency and trapping efficiency.

9.3 OUTLOOK FOR MICROFLUIDICS APPROACHES FOR CANCER DETECTION

In contrast to conventional processes wherein a large number of samples is required for analysis, the microfluidics technique needs a very small amount of materials owing to its specific features, such as miniaturization and automation. Due to confined space and high surface area, the analysis rate is quite a lot higher and can be integrated with other analytical techniques. There have been several techniques reported for the isolation and analysis of CTCs, which are otherwise not possible through the conventional technique reported so far. Microfluidics is the only technique which makes "liquid biopsy" possible. This chapter summarizes microfluidics technologies and their working principles and experimental results for CTC separation and analysis. Detection and isolation of CTCs using microfluidics is a noninvasive technique that can be repeated nearly limitlessly, facilitating the real-time monitoring of tumor progression. CTC analysis provides a wealth of information including early cancer detection, development, progression, and treatment as well as cancer subtypes and gene mutations. Various microfluidics technologies are described, so far, for the separation and analysis of CTCs from whole blood. Moreover, in contrast to the conventional method, these strategies have a great advantage in terms of noninvasive diagnosis by liquid biopsy and improved comfort of patients by avoiding unnecessary invasive sampling of a tumor. Microfluidics systems continue to serve as a platform for many cell-separation approaches and often drive new innovations due to the advantages of integration, ease of fabrication, and flow control. In particular, the microfluidics platform that allows high CTCs recovery rate and high sample throughput in short experimental time can be considered as ideal. Despite the recent microfluidics technologies available, the development of a single device capable of simultaneously achieving high throughput in a short time, high target cancer cell recovery, and high purity still remains challenging. Low-cost and POC diagnostic tools and technologies remain a necessity for cancer management and reduction in mortality rates.

ACKNOWLEDGMENTS

We gratefully acknowledge the financial support from Scheme for Transformational and Advanced Research in Sciences (STARS), Ministry of Human Resource Development, Government of India, through a research grant (STARS/APR2019/BS/583/FS). NKV acknowledges Council of Scientific and Industrial Research (CSIR) (SRA Pool No. 13(9077-A)/2019), Government of India, for the Research Fellowships.

REFERENCES

1. Kulasinghe, A. 2018. The use of microfluidic technology for cancer applications and liquid biopsy. *Micromachines* 9:397.
2. American Cancer Society. 2018. *Annual Report.*
3. Etzioni, R. 2003. The case for early detection. *Nat. Rev. Cancer* 3:243–252.
4. Bannasch, P. 1992. *Cancer Diagnosis: Early Detection.* New York: Springer.
5. Elwood, J. M. and S. B. Sutcliffe. 2010. *Cancer Control.* Oxford: Oxford University Press.
6. Verma, M. 2003. *New York Academy of Sciences and National Cancer Institute (U.S.), Division of Cancer Prevention, Epigenetics in Cancer Prevention: Early Detection and Risk Assessment.* New York: New York Academy of Sciences.
7. Chen, J. 2012. Microfluidic approaches for cancer cell detection, characterization, and separation. *Lab Chip* 12:1753–1767.
8. Zhang, J. Z. 2013. Microfluidics and cancer: Are we there yet?. *Biomed. Microdevices* 15:595–609.
9. Chandra, P. 2017. *Next Generation Point-of-care Biomedical Sensors Technologies for Cancer Diagnosis.* ed. Y. N. Tan, and S. P. Singh, New York: Springer.
10. Purohit, B. 2019. Cancer cytosensing approaches in miniaturized settings based on advanced nanomaterials and biosensors. *Nanotechnol. Modern Animal Biotechnol. Concepts Appl.* 133–147.
11. Radhakrishna, S. 2013. Needle core biopsy for breast lesions: An audit of 467 needle core biopsies. *Indian J. Med. Paediatr. Oncol.* 34:252–256.
12. Ilie, M. 2016. Pros: Can tissue biopsy be replaced by liquid biopsy? *Transl. Lung Cancer Res.* 5:420–423.
13. Welch, H. G. 2010. Overdiagnosis in cancer. *J. Natl. Cancer Inst.* 102:605–613.
14. Loughran, C. F. 2011. Seeding of tumour cells following breast biopsy: A literature review. *Br. J. Radiol.* 84:869–874.
15. Yates, L. R. 2012. Evolution of the cancer genome. *Nat. Rev. Genet.* 13:795–806.
16. Sacher, A. G. 2016. Prospective validation of rapid plasma genotyping for the detection of EGFR and KRAS mutations in advanced lung cancer. *JAMA Oncol.* 2:1014–1022.
17. Chambers, A. F. 2002. Dissemination and growth of cancer cells in metastatic sites. *Nat. Rev. Cancer* 2:563–572.
18. Steeg, P. S. 2006. Tumor metastasis: Mechanistic insights and clinical challenges. *Nat. Med.* 12:895–904.
19. Wirtz, D. 2011. The physics of cancer: The role of physical interactions and mechanical forces in metastasis. *Nat. Rev. Cancer* 11:512–522.

20. Huang, Y. L. 2017. Microfluidic modeling of the biophysical microenvironment in tumor cell invasion. *Lab Chip* 17:3221–3233.
21. Ma, Y.-H. V. 2018. A review of microfluidic approaches for investigating cancer extravasation during metastasis. *Microsys. & Nanoeng.* 4:17104.
22. Portillo-Lara, R. 2016. Microengineered cancer-on-a-chip platforms to study the metastatic microenvironment. *Lab Chip* 16:4063–4081.
23. Wang, S. 2011. Highly efficient capture of circulating tumor cells by using nanostructured silicon substrates with integrated chaotic micromixers. *Angew. Chem. Int. Ed.* 50:3084–3088.
24. Dagogo-Jack, I. 2018. Tumour heterogeneity and resistance to cancer therapies. *Nat. Rev. Clin. Oncol.* 15:81–94.
25. Alix-Panabières, C. 2014. Challenges in circulating tumour cell research. *Nat. Rev. Cancer* 14:623–631.
26. den Toonder, J. 2011. Circulating tumor cells: The grand challenge. *Lab Chip* 11:375–377.
27. Alix-Panabières, C. 2013. Circulating tumor cells: Liquid biopsy of cancer. *Clin. Chem.* 59:110–118.
28. Bardelli, A. 2017. Liquid biopsies, what we do not know (yet). *Cancer Cell* 31:172–179.
29. Poudineh, M. 2018. Profiling circulating tumour cells and other biomarkers of invasive cancers. *Nat. Biomed. Eng.* 2:72–84.
30. Alix-Panabières, C. 2014. Technologies for detection of circulating tumor cells: Facts and vision. *Lab Chip* 14:57–62.
31. Espinosa, O. A. 2018. Accuracy of Enzyme-Linked Immunosorbent Assays (ELISAs) in Detecting Antibodies against Mycobacterium leprae in Leprosy Patients: A Systematic Review and Meta-Analysis. *Can. J. Infect. Dis. Med. Microbiol.* 2018: 9828023.
32. Ozkumur, E. 2013. Inertial focusing for tumor antigen–dependent and –independent sorting of rare circulating tumor cells. *Sci. Transl. Med.* 5:179ra47.
33. Heng, Y. 2018. A review of microfluidic approaches for investigating cancer extravasation during metastasis. *Microsys. & Nanoeng.* 4:17104.
34. Chandra, P. 2011. Separation and simultaneous detection of anticancer drugs in a microfluidic device with an amperometric biosensor. *Biosens. Bioelectron.* 28:326–332.
35. Mahato, K. 2018. Shifting paradigm of cancer diagnoses in clinically relevant samples based on miniaturized electrochemical nanobiosensors and microfluidic devices. *Biosens. Bioelectron.* 100:411–428.
36. Samatov, T. R. 2015. Modelling the metastatic cascade by in vitro microfluidic platforms. *Prog. Histochem. Cytochem.* 49:21–29.
37. Guo, Q.-r. 2021. Multifunctional microfluidic chip for cancer diagnosis and treatment. *Nanotheranostics* 5:73–89.
38. Su, W. 2019. Integrated microfluidic device for enrichment and identification of circulating tumor cells from the blood of patients with colorectal cancer. *Dis. Markers* 2019:8945974
39. Ruzycka, M. 2019. Microfluidics for studying metastatic patterns of lung cancer. *J. Nanobiotechnol.* 17:71.
40. Skardal, A. 2016. A reductionist metastasis-on-a-chip platform for in vitro tumor progression modeling and drug screening. *Biotechnol. Bioeng.* 113:2020–2032.
41. Reátegui, E. 2015. Tunable nanostructured coating for the capture and selective release of viable circulating tumor cells. *Adv. Mater.* 27:1593–1599.
42. Jiang, X. 2017. Microfluidic isolation of platelet-covered circulating tumor cells. *Lab Chip* 17:3498–3503.
43. Poddar, S. 2019. Low density culture of mammalian primary neurons in compartmentalized microfluidic devices. *Biomed. Microdevices* 21:67.

44. Mahto, S. K. 2009. Multicompartmented microfluidic device for characterization of dose-dependent cadmium cytotoxicity in BALB/3T3 fibroblast cells. *Biomed. Microdevices* 11:401–411.
45. Nagrath, S. 2007. Isolation of rare circulating tumour cells in cancer patients by microchip technology. *Nature* 450:1235–1239.
46. Adams, A. A. 2008. Highly efficient circulating tumor cell isolation from whole blood and label-free enumeration using polymer-based microfluidics with an integrated conductivity sensor. *J. Am. Chem. Soc.* 130:8633–8641.
47. Maheswaran, S. 2008. Detection of mutations in EGFR in circulating lung-cancer cells. *N. Engl. J. Med.* 359:366–377.
48. Cheung, L. S. L. 2009. Detachment of captured cancer cells under flow acceleration in a bio-functionalized microchannel. *Lab Chip* 9:1721–1731.
49. Gleghorn, J. P. 2010. Capture of circulating tumor cells from whole blood of prostate cancer patients using geometrically enhanced differential immunocapture (GEDI) and a prostate-specific antibody. *Lab Chip* 10:27–29.
50. Wang, S. T. 2009. Three-dimensional nanostructured substrates toward efficient capture of circulating tumor cells. *Angew. Chem. Int. Ed.* 48:8970–8973.
51. Thierry, B. 2010. Herceptin functionalized microfluidic polydimethylsiloxane devices for the capture of human epidermal growth factor receptor 2 positive circulating breast cancer cells. *Biomicrofluidics* 4:32205.
52. Stott, S. L. 2010. Isolation of circulating tumor cells using a microvortex-generating herringbone-chip. *Proc. Natl. Acad. Sci. U. S. A.* 107:18392–18397.
53. Kurkuri, M. D. 2011. Plasma functionalized PDMS microfluidic chips: Towards point-of-care capture of circulating tumor cells. *J. Mater. Chem.* 21:8841–8848.
54. Dharmasiri, U. 2011. High-throughput selection, enumeration, electrokinetic manipulation, and molecular profiling of low-abundance circulating tumor cells using a microfluidic system. *Anal. Chem.* 83:2301–2309.
55. Zheng, X. 2011. A high-performance microsystem for isolating circulating tumor cells. *Lab Chip* 11:3269–3276.
56. Liu, Y.-J. 2007. A micropillar-integrated smart microfluidic device for specific capture and sorting of cells. *Electrophoresis* 28:4713–4722.
57. Sun, C. 2017. Immunomagnetic separation of tumor initiating cells by screening two surface markers. *Sci. Rep.* 7:40632.
58. Zborowski, M. 2011. Rare cell separation and analysis by magnetic sorting. *Anal. Chem.* 83:8050–8056.
59. Bankó, P. 2019. Technologies for circulating tumor cell separation from whole blood. *J. Hematol. Oncol.* 12:48.
60. Zhao, W. 2017. Label-free ferrohydrodynamic cell separation of circulating tumor cells. *Lab Chip* 17:3097–3111.
61. Shields IV, C. W. 2016. Magnetic separation of acoustically focused cancer cells from blood for magnetographic templating and analysis. *Lab Chip* 16:3833–3844.
62. Hao, S.-J. 2018. Size-based separation methods of circulating tumor cells. *Adv. Drug Delivery Rev.* 125:3–20.
63. Geislinger, T. M. 2015. Hydrodynamic and label-free sorting of circulating tumor cells from whole blood. *Appl. Phys. Lett.* 107:203702.
64. Li, X. 2014. Continuous-flow microfluidic blood cell sorting for unprocessed whole blood using surface-micromachined microfiltration membranes. *Lab Chip* 14:2565–2575.
65. Chung, J. 2012. Cell sorting: Microfluidic cell sorter (μFCS) for on-chip capture and analysis of single cells. *Adv. Healthcare Mater.* 4:432–436.
66. Alazzam, A. 2011. Interdigitated comb-like electrodes for continuous separation of malignant cells from blood using dielectrophoresis. *Electrophoresis* 32:1327–1336.

67. Chan, J. Y. 2018. Dielectrophoresis-based microfluidic platforms for cancer diagnostics. *Biomicrofluidics* 12:011503.
68. Cetin, B. 2016. An integrated acoustic and dielectrophoretic particle manipulation in a microfluidic device for particle wash and separation fabricated by mechanical machining. *Biomicrofluidics* 10:014112.
69. Wang, X.-B. 2000. Cell separation by dielectrophoretic field-flow-fractionation. *Anal Chem.* 72:832–839.
70. Yang, F. 2010. Dielectrophoretic separation of colorectal cancer cells. *Biomicrofluidics* 4:013204.
71. Gao, Y. 2013. Recent advances in microfluidic cell separations. *Analyst* 138:4714–4721.
72. Pallela, R. 2016. An amperometric nanobiosensor using a biocompatible conjugate for early detection of metastatic cancer cells in biological fluid. *Biosens. Bioelectron.* 85:883–890.
73. Choudhary, M. 2016. CD59 targeted ultrasensitive electrochemical immunosensor for fast and noninvasive diagnosis of oral cancer. *Electroanalysis* 28:2565–2574.
74. Green, J. V. 2009. Microfluidic enrichment of a target cell type from a heterogenous suspension by adhesion-based negative selection. *Lab Chip* 9:2245–2248.
75. Phillips, J. A. 2009. Enrichment of cancer cells using aptamers immobilized on a microfluidic channel. *Anal. Chem.* 81:1033–1039.
76. Wang, K. 2009. Differential mobility cytometry. *Anal. Chem.* 81:3334–3343.
77. Lacroix, M. 2006. Significance, detection and markers of disseminated breast cancer cells. *Endocr. Relat. Cancer* 13:1033–1067.
78. Pantel, K. 2008. Detection, clinical relevance and specific biological properties of disseminating tumour cells. *Nat. Rev. Cancer* 8:329–340.
79. Sleijfer, S. 2007. Circulating tumour cell detection on its way to routine diagnostic implementation?. *Eur. J. Cancer* 43:2645–2650.
80. Du, Z. 2006. Microfluidic-based diagnostics for cervical cancer cells. *Biosens. Bioelectron.* 21:1991–1995.
81. Kralj, J. G. 2012. A simple packed bed device for antibody labelled rare cell capture from whole blood. *Lab Chip* 12:4972–4975.
82. Vickers, D. A. L. 2012. Separation of two phenotypically similar cell types via a single common marker in microfluidic channels. *Lab Chip* 12:3399–3407.
83. Launiere, C. 2012. Channel surface patterning of alternating biomimetic protein combinations for enhanced microfluidic tumor cell isolation. *Anal. Chem.* 84: 4022–4028.
84. Li, P. 2011. Negative enrichment of target cells by microfluidic affinity chromatography. *Anal. Chem.* 83:7863–7869.
85. Ferreira, M. M. 2016. Circulating tumor cell technologies. *Mol. Oncol.* 10:374–394.
86. Hoshino, K. 2011. Microchip-based immunomagnetic detection of circulating tumor cells. *Lab Chip* 11:3449–3457.
87. Mohamed, H. 2004. Development of a rare cell fractionation device: Application for cancer detection. *IEEE Trans. NanoBiosci.* 3:251–256.
88. Hur, S. C. 2011. High-throughput size-based rare cell enrichment using microscale vortices. *Biomicrofluidics* 5:022206.
89. Ramanjaneyulu, B. T. 2018. Towards versatile continuous-flow chemistry and process technology via new conceptual microreactor systems. *Bull. Korean Chem. Soc.* 39:757–772.
90. Johnston, I. D. 2014. Dean flow focusing and separation of small microspheres within a narrow size range. *Microfluid Nanofluid* 17:509–518.
91. Dean, W. 1928. LXXII. The stream-line motion of fluid in a curved pipe (Second paper). *Philos. Mag. Ser.* 5:673–695.

92. Bhagat, A. A. S. 2008. Continuous particle separation in spiral microchannels using dean flows and differential migration. *Lab Chip* 8:1906–1914.
93. Hou, H. W. 2013. Isolation and retrieval of circulating tumor cells using centrifugal forces. *Sci. Rep.* 3:1259.
94. Warkiani, M. E. 2014. Slanted spiral microfluidics for the ultra-fast, label-free isolation of circulating tumor cells. *Lab Chip* 14:128–137.
95. Khoo, B. L. 2014. Clinical validation of an ultra high-throughput spiral microfluidics for the detection and enrichment of viable circulating tumor cells. *PLoS One* 9:e99409.
96. Zhou, J. 2019. Isolation of circulating tumor cells in non-small-cell-lung-cancer patients using a multi-flow microfluidic channel. *Microsyst. Nanoeng.* 5:8.
97. Khoshmanesh, K. 2011. Dielectrophoretic platforms for bio-microfluidic systems. *Biosens. Bioelectron.* 26:1800–1814.
98. Zhang, H. 2019. DEP-on-a-chip: Dielectrophoresis applied to microfluidic platforms. *Micromachines* 10:423.
99. Li, M. 2017. High-throughput selective capture of single circulating tumor cells by dielectrophoresis at a wireless electrode array. *J. Am. Chem. Soc.* 139:8950–8959.
100. Peter, R. C. 2014. Isolation of circulating tumor cells by dielectrophoresis. *Cancers* 6:545–579.
101. Pethig, R. 1996. Dielectrophoresis: Using inhomogeneous AC electrical fields to separate and manipulate cells. *Crit. Rev. Biotechnol* 16:331–348.
102. Fuhr, G. and G. A. Neil. 1996. *Electromanipulation of Cells*, 259–328. Boca Raton, FL: CRC Press/Taylor & Francis.
103. Becker, F. F. 1995. Separation of human breast cancer cells from blood by differential dielectric affinity. *Proc. Natl. Acad. Sci. U.S.A* 92:860–864.
104. Becker, F. F. 1994. The removal of human leukaemia cells from blood using interdigitated microelectrodes. *J. Phys. D: Appl. Phys.* 27:2659–2662.
105. Gascoyne, P. R. C. 1997. Dielectrophoretic separation of cancer cells from blood. *IEEE Trans. Ind. Gen. Appl* 33:670–678.
106. Mohammadi, M. 2015. Hydrodynamic and direct-current insulator-based dielectrophoresis (H-DC-iDEP) microfluidic blood plasma separation. *Anal. Bioanal. Chem.* 407:4733–4744.
107. Huang, Y. 1997. Introducing dielectrophoresis as a new force field for field-flow fractionation. *Biophys. J.* 73:1118–1129.
108. Wang, X.-B. 1998. Separation of polystyrene beads using dielectrophoretic/gravitational field-flow-fractionation. *Biophys. J.* 74:2689–2701.
109. Yang, J. 1999. Cell separation on microfabricated electrodes using dielectrophoretic/gravitational field-flow fractionation. *Anal. Chem.* 71:911–918.
110. Racilla, E. 1998. Detection and characterization of carcinoma cells in the blood. *Proc. Natl. Acad. Sci. U.S.A.* 95:4589–4594.
111. Terstappen, L. W. M. M. 1998. Flowcytometry: Principles and feasibility in transfusion medicine. Enumeration of epithelial derived tumor cells in peripheral blood. *Vox. Sang* 74 Supplement 2:269–274.
112. Peck, K. 1998. Detection and quantitation of circulating cancer cells in the peripheral blood of lung cancer patients. *Cancer Res.* 58:2761–2765.
113. Alazzam, A. 2017. Novel microfluidic device for the continuous separation of cancer cells using dielectrophoresis. *J. Sep. Sci.* 40:1193–1200.
114. Carlo, D. D. 2009. Inertial microfluidics. *Lab Chip* 9:3038–3046.
115. Zhang, J. 2016. Fundamentals and applications of inertial microfluidics: A review. *Lab Chip* 16;10–34.
116. Martel, J. M. 2014. Inertial focusing in microfluidics. *Annu. Rev. Biomed. Eng.* 16:371–396.

117. Zhang, J. 2018. Tunable particle separation in a hybrid dielectrophoresis (DEP)-inertial microfluidic device. *Sens. Actuators B Chem.* 267:14–25.
118. Henslee, E. A. 2011. Selective concentration of human cancer cells using contactless dielectrophoresis. *Electrophoresis* 32:2523–2529.
119. Zhu, Y. 2013. Ultrasensitive and selective electrochemical diagnosis of breast cancer based on a hydrazine-Au nanoparticle-aptamer bioconjugate. *Anal. Chem.* 85:1058–1064.
120. Sun, M. 2016. Continuous on-chip cell separation based on conductivity-induced dielectrophoresis with 3D self-assembled ionic liquid electrodes. *Anal. Chem.* 88:8264–8271.
121. Chiu, T.-K. 2016. Application of optically-induced-dielectrophoresis in microfluidic system for purification of circulating tumour cells for gene expression analysis- Cancer cell line model. *Sci. Rep.* 6:32851.
122. Lee, D. 2016. Negative dielectrophoretic force based cell sorter with simplified structure for high reliability. *Int. J. Precis. Eng. Manuf.* 17:247–251.
123. Markovic, N. 2008. *What Every Woman Should Know About Cervical Cancer.* Springer Netherlands, Imprint, Dordrecht: Springer.
124. Kniss, D. A. 2014. Discovery of HeLa cell contamination in HES cells: Call for cell line authentication in reproductive biology research. *Reprod. Sci.* 21:1015–1019.
125. Cheng, I.-F. 2015. Antibody-free isolation of rare cancer cells from blood based on 3D lateral dielectrophoresis. *Lab Chip* 15:2950–2959.
126. Soltanian-Zadeh, S. 2017. Breast cancer cell obatoclax response characterization using passivated-electrode insulator-based dielectrophoresis. *Electrophoresis* 38:1988–1995.

10 Analytical Devices with Instrument-Free Detection Based on Paper Microfluidics

Sasikarn Seetasang and Takashi Kaneta

CONTENTS

10.1 INTRODUCTION: BACKGROUND

As a solution to the ever-increasing demand for point-of-need analysis, μPADs promise to satisfy the ASSURED criteria that was suggested by the World Health Organization. An analytical device must be affordable, sensitive, specific, user friendly, rapid and robust, equipment free, and deliverable. Among them, the user-friendly, equipment-free, and deliverable features must be achieved to allow practical analysis outside equipped laboratories. Therefore, several types of μPADs that do not require large instruments have been examined.

The key features of the μPADs, which satisfy the user-friendly, equipment-free, and deliverable criteria, include the detection scheme and the sampling method. In the first report of the μPADs, a colorimetric reaction was employed for semi-quantitative

DOI: 10.1201/9781003033479-10

analysis by judging the color intensity using the naked eye. Subsequently, image processing software was used to measure either the intensity or the hue of the primary colors (red, green, and blue (RGB)) in order to quantify a target analyte according to its customary production of a colored product via several types of chemistry such as complex formation, precipitation formation, redox reactions, and enzymatic reactions. Colorimetry has been the most popular detection scheme in μPADs because of its simplicity, compatibility with small or miniaturized devices, and easy availability. Other methods such as electrochemistry, absorption spectrometry, fluorometry, and chemiluminescence measurement generally require relatively large devices such as a potentiostat, a light source, and a photodetector in addition to a personal computer (PC) that can interface with the devices. Some of these devices can be made transportable by developing a miniaturized system, but extra electric power supplies are needed to operate them.

In this chapter, we focus on the colorimetric detection schemes that are the simplest and most user friendly because this method is based on a color change that is intuitive for users. In early research, scanners and digital cameras served as the primary image acquisition devices to allow easy control of illumination conditions to ensure the capture of reproducible images. In the development of image acquisition and processing, several developments have been reported using smartphones, whereas the naked eye is the primary instrument used to realize quantitative analysis via μPADs, which are considered to be instrument free. Here, we discuss colorimetric methods that are potentially applicable to chemical on-site analyses.

10.2 COLORIMETRIC MEASUREMENT VIA TRANSPORTABLE SMALL DEVICES

Colorimetric detection is achieved by chemical reactions that take place in detection zones that contain reagents pre-deposited on hydrophilic areas of the μPADs. In early studies, μPADs provided only positive/negative answers or semi-quantitative analysis to estimate a rough concentration of the analyte by comparing the intensity of the color of the standard with that of the sample. Due to the hard work and diligence of scientists, several techniques and technologies have been developed and combined with μPADs for capturing and processing images. These methodologies have enhanced accuracy and provided precise quantitative answers. Electronic image capture devices such as scanners, cameras, and smartphones have been employed to record the intensity of reflected light from the detection zones. Quantitative results are obtained by processing captured images using software such as ImageJ or Adobe Photoshop® or via the development of applications for converting image pixels to RGB, hue saturation values (HSV), cyan, magenta, yellow, and black (CMYK), or greyscale values. The extract color values are utilized to construct a calibration curve to determine the relationship between the amount of the analyte and the color intensity.

In the laboratory, scanners are the most reliable device for image acquisition due to a stable light source with no influence from ambient light. Therefore, the colorimetric method requires a scanner and a PC with image processing software for precise and accurate quantification. These devices are transportable for on-site analysis and point-of-care (POC) testing because they need no extra electric power supply. Scanners can be powered via the USB port of a PC that can be charged prior to use outside the laboratory. An example of a scanner and a PC is shown in Figure 10.1(I). Ogawa and Kaneta (2016) used these devices to demonstrate the field analysis of iron ions in hot spring water. There are two problems associated with the use of a scanner. The first involves the need to dry the μPADs completely before image capture since wetted μPADs can influence the reflected light on the scanner surface. Thus, the color intensity of the μPADs must be stable at least during the time needed to dry them and to scan their images. The second involves repeated scanning, as the glass on the scanner can become contaminated by colored products, toxic reagents, and/or bio substrates. Although scanners prevent errors caused by ambient light, these two problems should be carefully avoided.

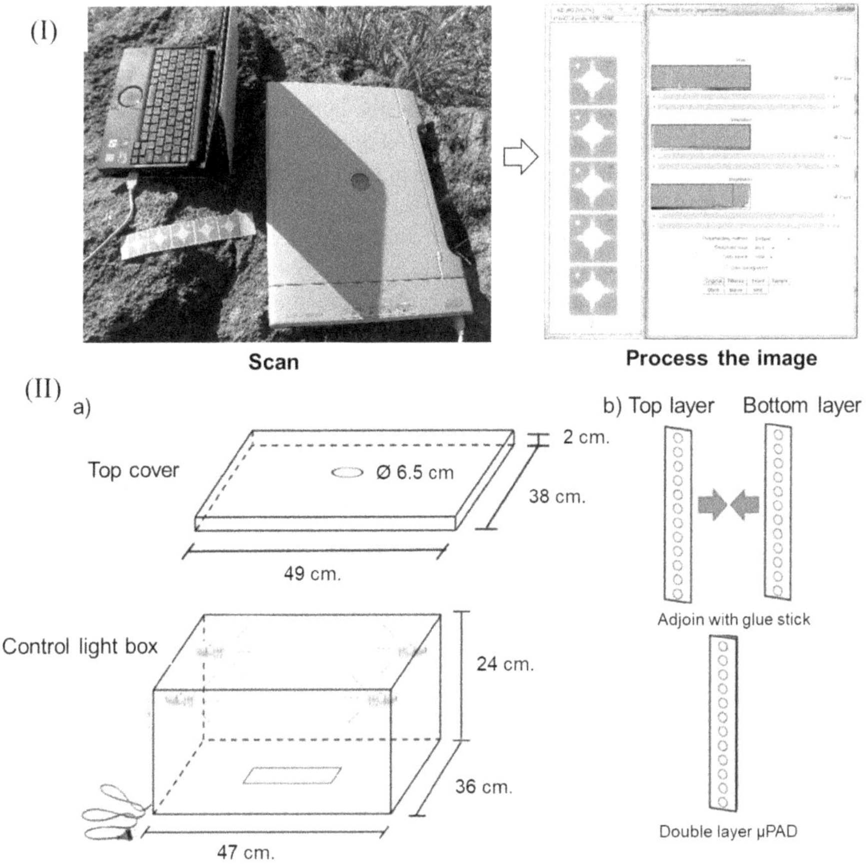

FIGURE 10.1 (I) On-site measurements of iron (III) using a scanner before data analysis by ImageJ software (Ogawa and Kaneta 2016) and (II) a homemade control light box and double layer μPAD fabrication, reprint adapted from Meelapsom et al. (2016).

High-performance digital cameras also can be useful. Several functions such as optical zoom, macro feature, optical or sensor-shift image stabilization, and high-quality filters allow researchers to produce photos with good quality and resolution. In addition, camera sensors allow clear images even under conditions of low light. All of these features are accurately and precisely controllable with manual settings for professional photographers or automatic settings for beginners. Additionally, cameras can capture photos at any time of the reaction without contact with a μPAD, which improves the issue of contamination. However, cameras require precise regulation of illumination to prevent errors caused by disturbance from ambient light, which means an external photo studio or light source could be required to control the consistency and brightness of the environment. A controlled light box can perform the function of a photo studio, as displayed in Figure 10.1(II). The featured homemade light studio is equipped with four light sources that can control light consistency (Meelapsom et al. 2016). A controlled-light box was employed for the determination of mercury (II) on a double-layer paper device while data processing required the installation of ImageJ software on a laptop. As shown in Figure 10.1(II), the controlled-light box was developed for portability and precise image capture of a μPAD, but on-site analysis was difficult due to the large size of the box and the need for a plug-in power supply for the light-emitting diodes (LEDs). Therefore, miniaturization of the photo studio and replacing the plug-in power supply with small batteries is an alternative method for POC analysis.

Even though scanners and cameras are transportable, a PC is still required for capturing and processing images. Therefore, a device that could both capture images and perform data analysis would be preferable for on-site analysis. In addition, size and weight remain important parameters to realize convenience. Therefore, smartphones are alternative devices that can play the dual roles of image capture and processing. Although smartphones are intrinsically communication tools, functions that allow point-of-need analysis include a fundamentally equipped digital camera and software for image processing. Furthermore, high-performance smartphones are small and powered by a built-in rechargeable battery, which equates to portability that is better than a PC with a scanner. However, with the introduction of a phone as an image capture device in 2008 (Martinez et al. 2008), researchers encountered problems in focusing at close distances and lighting conditions, which reduced both the accuracy and precision of the analytical results.

The problems with focus were solved by attaching a macro lens to the camera phones to enhance magnification and reduce the focal distance. The lightning conditions, which also were a major obstacle, were highlighted by Choi et al. (2015). They compared three types of image-capture devices including a scanner, a microscope, and a smartphone for producing images of a developed paper-based 3D microfluidics device. They found that the photos taken by the smartphone under ambient light caused errors, they attributed it to the light intensity and levels of light contrast of the images that were lower than those of either the scanner or the microscope, which resulted in poor sensitivity. To improve sensitivity and precision, cover boxes were added with/without light sources. To solve the problems of focus and illumination, the design for pattern recognition was altered, the colors of paper devices were rescaled, and processing software was downloaded to smartphones.

10.2.1 Combination of Additional Cover Boxes with/without Light Sources

The use of an additional light source increases light intensity and produces brighter photographs with improved contrast. The additional light sources that are normally incorporated with μPADs use LEDs and UV lights such as a mercury lamp. Even though a smartphone has an LED as a standard feature, external light sources need a power supply for on-site analysis. Hence, many portable devices such as a smartphone with a built-in flashlight, the screen light of a smartphone, and LEDs powered by small batteries have been introduced to accomplish on-site analyses. Improving the contrast in photos depends on the position of the light source, as noted by Li et al. (2019). When the μPADs were placed on a flat LED emitting white light, the light penetrated the paper substrate and provided better contrast between the boundary of the white border and the colored detection zone.

An additional device that can be used to control the conditions of illumination is an enclosed box that excludes ambient light. Different designs, such as a cover box and a smartphone case, have been employed for shielding ambient light, controlling light stability, and fixing the focal distance between the μPADs and a smartphone during a photo shoot. The materials used for these boxes include Styrofoam, normal paper, flute board, paper cardboard, and poly(methyl methacrylate). In addition to the appropriate selection for the color of the box, some additional components such as a light diffuser, a prism, and reflection film also improve the conditions of light reflection, saturation, and distribution. A dark room also shields ambient light; however, it is unavailable for on-site analysis. Han et al. (2020) fabricated a smartphone-based reader with simple opto-mechanical parts for accurate quantification. The smartphone-based reader showed the consistency of the images obtained in indoor, outdoor, and darkroom environments.

10.2.2 Design of Paper Devices with Pattern Recognition

Methods for pattern recognition such as the use of barcodes or QR codes on μPADs solves the problem of light dependency during image capture since the result produces either a specific pattern or a perfect code rather than subjective color intensities of the detection zones. The principle is based on the depletion of some portions in the code, which are invalidated before use. The depleted portions contain a pre-deposited colorimetric reagent that appears when prompted by a reaction with an analyte, which then validates the code. Thus, this type of device gives a positive or negative result. After the reaction, a smartphone installed with a barcode reader application recognizes the code and displays the result, as illustrated in Figure 10.2(I) (Guo et al. 2015). On the other hand, the QR code design reports the results in the opposite way, and a colored product disturbs and invalidates the QR code if it is positive (see Figure 10.2(II)) (Russell et al. 2017). Although the code-based μPADs can detect an analyte while avoiding the problem of fluctuating illumination, the results are only qualitative or semi-quantitative. Therefore, code-based μPADs will require further development to achieve a fully quantitative analysis that would be more beneficial for gaining accurate and precise output without the need to control the lighting conditions.

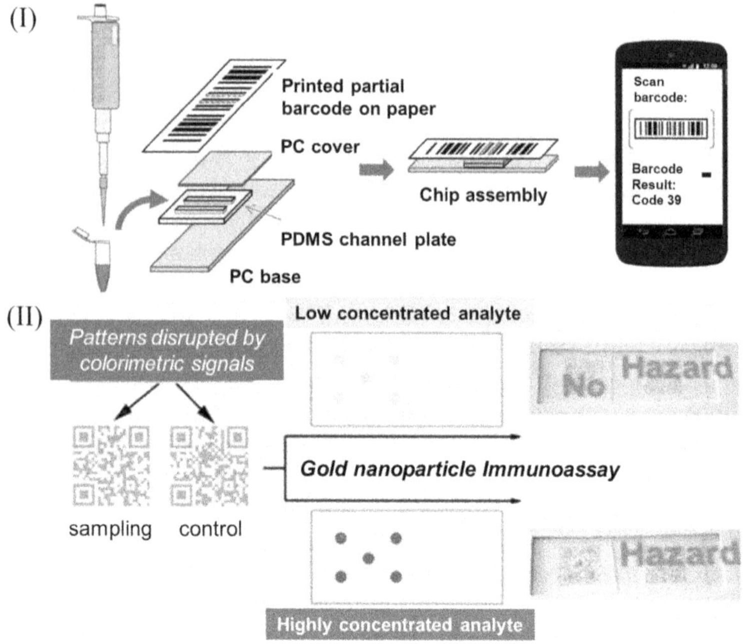

FIGURE 10.2 (I) Design and preparation steps of the colorimetric barcode assay before reading the results via a smartphone, and (II) design of the colorimetric QR code assay consisting of μPADs with five reaction spots covered by transparent film printed with two QR codes. Reprint adapted from Russell et al. (2017) and Guo et al. (2015).

10.2.3 Design of Paper Devices with Color Rescaling

A preprinted color chart with known color intensities is prepared on μPADs or on an additional paper sheet which plays the role of a color reference guide that is used to correct the white balance and the background in a captured image for data processing. Since the color chart and the detection zone adhere to the same light illumination conditions, the color chart corrects the color intensities of the detection zones, i.e., it can normalize the different color intensities caused by various lighting conditions. Some researchers have used black ink printed on a paper substrate and/or the white color of a paper substrate as a reference in order to simplify the printing of an additional color chart (Lopez-Ruiz et al. 2014, Hong and Chang 2014, Kong et al. 2019). Conversely, others have printed more colors such as an RBG chart and/or a multi-color chart that covers the wavelengths of their targeted colors (Shen et al. 2012, Huang et al. 2018, Chen et al. 2019). The color chart is located at the center of the μPADs or near the detection zone because it is preferable to subject the color chart and the detection zones to the same lighting conditions.

10.2.4 Development of Software/Applications

Methodologies or applications using algorithms with the intention of correcting or normalizing color intensity have been developed to improve the accuracy of results from various photos captured under different lighting conditions. The developed methods include data processing with/without reference points. The development of such methods has utilized the normalization of three reference spots and a fast-Fourier transform-based pre-processing scheme such as that reported by McCracken et al. (2016), or as an application reported by Chen et al. (2019) that works in the manner of human eyes using white balance correction algorithms to eliminate bright light. Another example is the application that subtracts the background values from the RGB values of the analyte signal and converts the remaining values to the corresponding concentrations according to a preload polynomial fitting curve constructed as a function of either the hue (H) or saturation (S) values from the standard color chart under different lighting conditions (Hong and Chang 2014). Other color correction algorithms that may be useful for developing smartphone applications were reported by Finlayson et al. (2001). In that paper, ways to estimate an unknown illuminant in a captured scene were proposed using a correlation matrix involving three steps: Characterization of possible illuminants, the building of a probability distribution for each light, and encoding these distributions in the columns of the correlation matrix.

10.3 COLORIMETRIC DETECTION AND QUANTIFICATION VIA AN INSTRUMENT-FREE READOUT

Colorimetric detection has shown great potential for developing an instrument-free approach because of the ability to determine a result using only the naked eye. Conventional applications with paper substrates use test paper strips and immunochromatography (lateral flow immunoassay) that produces qualitative or semi-quantitative results based on the color change in the device. However, some analyses such as pH, glucose, and contaminated metals require more quantitative answers than a simple yes/no. Therefore, a color chart/scale has been widely developed for estimating the concentration by comparing the obtained color on the test strip using a provided color chart as a reference. This type of test strip has been successfully applied to daily life diagnoses such as that seen in ketosis test strips, glucose test strips, and diabetes test strips.

Fortunately, many attempts have challenged the limitations of qualitative and semi-quantitative analyses since the first μPADs were reported in 2007. Many researchers have proposed many strategies to achieve instrument-free quantitative analyses using output signals transduced by colored distance, reaction time, zone counting, and text-based color changes. These μPADs have achieved levels of simple operation and precise quantitative analysis without requiring bulky equipment, which is the ultimate goal for POC analysis that could benefit people who live in remote areas without sophisticated instruments and technicians.

10.3.1 Distance-Based Method

A distance-based approach provides simple signal readouts for quantification that uses only a ruler or a calibration mark. An example of a common distance-based readout is a thermometer, which consists of a circular sample zone connected with a microfluidic channel containing reagents for a colorimetric reaction.

The thermometer-like channel is surrounded by a hydrophobic barrier fabricated by screen printing, wax printing, flexographic printing, inkjet patterning, and stamping. Some devices use tape, folded paper (origami), a paper bridge, a valve system, or a pseudo-hydrophobic barrier for disconnection between the sample zone and the microfluidic channel to control the incubation time of the reaction before measuring the distance. Quantitative results were obtained by measuring the distance of the colored channel, which depends on the amount of the analyte that forms precipitates with the deposited colorimetric reagent. A calibration curve plots the colored distance as a function of the analyte concentration.

In 1986, Vaughan et al. used chromatography paper instead of normal filter paper to report the fundamental concept of the distance-based readout. In 2013, Henry's group introduced the distance-based strategy to μPADs (Cate et al. 2013). The concept of the distance-based method consists of a hydrophobic barrier, a calibration mark printed along a straight channel, and a circular zone for sample introduction, as shown in Figure 10.3(I). The first demonstration of a μPAD with a distance-based

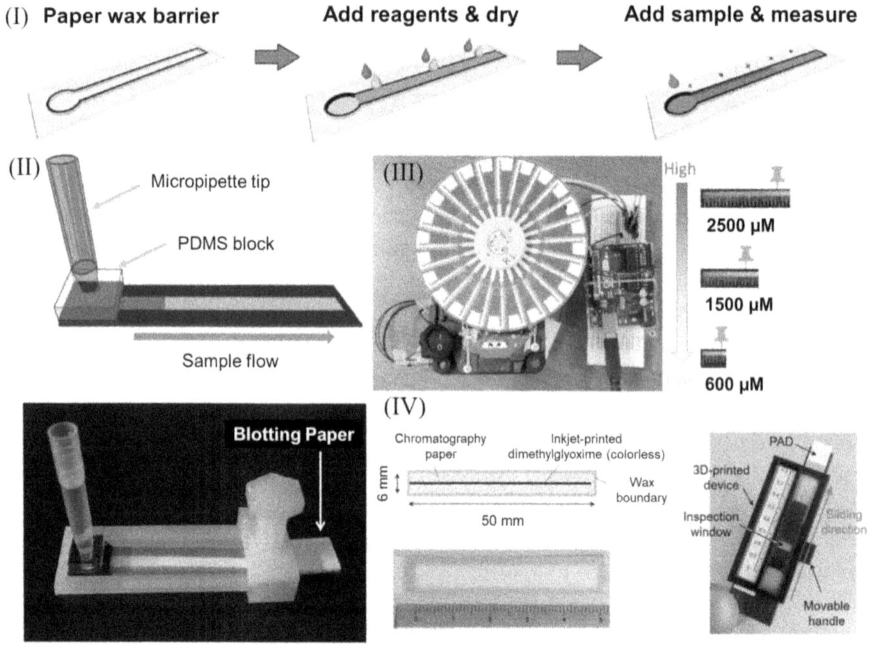

FIGURE 10.3 (I) Operation concept of the first reported distance-based method consisting of printing a hydrophobic barrier on filter paper, patterning reagents, and adding a sample for analysis. (II) Design of μPADs for large-volume introduction with a sample reservoir and a 3D printing holder. (III) CD-PADs with a rotary platform (DVD drive and microcomputer) for an accelerated capillary flow rate and the distance-based results of nickel (II) analysis. (IV) Outline and dimension of single-line μPADs (0.4 mm width, 5 mm length) and a 3D printing holder with calibrated marking for a distance readout. Reprint adapted from Shimada and Kaneta (2018, Maejima et al. (2020), Yamada et al. (2018), and Cate et al., (2013).

readout showed the versatility of the approach by applying three chemistries that included an enzymatic reaction, metal complexation, and nanoparticle aggregation for glucose, nickel, and glutathione analysis, respectively. The results confirmed the simplicity of the concept for quantitative analysis without additional instruments, and the distance-based readout became one of the methods appropriate for ASSURED criteria as well as being applicable to many analytes. Table 10.1 lists the publications that have used the distance-based readout in various applications, which indicates the practical nature of the technique.

Although the distance-based method provides characteristics suitable for POC analysis with accurate quantitative results in a wide range of applications, improvements in sensitivity and swiftness remain challenging. For example, a design for a large volume of sample introduction using spontaneous evaporation and a blotting paper method (Figure 10.1 (I)) significantly reduced the limits of detection for Iron (II) to 20 µg L^{-1}, which approximates that of inductively coupled plasma-optical emission spectrometry (Shimada and Kaneta 2018). However, technicians using this method struggled with the flow of the solution because of difficulties in accelerating the flow rate when introducing a large volume of solution into the channel. An interesting improvement that sped up the solution movement was reported by Citterio's group (Maejima et al. 2020). They employed a CD reader that acted as a centrifuge to accelerate the rate of the capillary flow and successfully reduced the analysis time to 1.5 min per analysis (Figure 10.3(II)). Another idea that shortened the analysis time was a reduction of the channel width for the reaction (Yamada et al. 2018) (Figure 10.3(III)). A channel width of 0.4 mm shortened the analysis time nine-fold from that of a previous report of a channel with the same length (5 cm) but with a width of 2.6 mm (Cate et al. 2015). In the µPADs with the narrow channel, the sample was introduced by dipping it into the sample solution without the need for a micropipette. Such a dip-and-read approach fits well with the ASSURED criteria, because a precise volume is unnecessary, which eliminates the operational skill needed when using a micropipette.

10.3.2 Time-Based Method

A time-based readout is an instrument-free approach that uses time as an indicator for the appearance of a color product or for the duration of the solution movement from the injection zone to the detection zone. The quantitative results are obtained from the relationship between the analyte concentration and the time measurement needed to complete the reaction. The main reaction used for the time-based readout is depolymerization or polymerization to convert hydrophobic properties into hydrophilic versions or vice versa via hydrogen peroxide (H_2O_2). In the first demonstration of a time-based readout, a hydrophobic polymer layer was sandwiched by hydrophilic paper substrates. The hydrophobic polymer layer was depolymerized by H_2O_2, which resulted in a pathway between the hydrophilic paper substrates. The time needed to depolymerize the hydrophobic layer depends on the concentration of H_2O_2, which is correlated with the time needed for the sample solution to flow from the injection zone to the detection zone. When the reaction is combined with

TABLE 10.1
Distance-Based Method Used with Various Applications

Analyte	Sample Matrix	LOD	Linear Range	Ref.
Glucose, Ni^{2+} and glutathione	Glucose: Serum Ni^{2+}: Combustion ash Glutathione: Serum	Glucose: ~20 mg dL^{-1} Ni^{2+}: Not reported Glutathione: Not reported	Glucose: 11–270 mg dL^{-1} Ni^{2+}: 0.792 µg L^{-1} Glutathione: 0.12–2.0 nmol	Cate et al. (2013)
H_2O_2	Hair bleach and Anti-infective solution	Not reported	50–300 mM	Sameenoi et al. (2014)
Ni^{2+}, Cu^{2+}, and Fe^{2+}	Airborne Particulate matter	<u>Single channel</u> Ni^{2+}: 0.1 µg Cu^{2+}: 0.1 µg Fe^{2+}: 0.05 µg <u>Multi-channel</u> Ni^{2+}: 1.0 µg Cu^{2+}: 5.0 µg Fe^{2+}: 1.0 µg	<u>Single channel</u> Ni^{2+}: 0.10–10 µg Cu^{2+}: 0.10–17 µg Fe^{2+}: 0.05–7.0 µg <u>Multi-channel</u> Ni^{2+}: 5.0–55 µg Cu^{2+}: 5.0–65 µg Fe^{2+}: 1.0–65 µg	Cate et al. (2015)
Cocaine	Urine	3.8 µmol L^{-1}	10–400 µmol L^{-1}	Wei et al. (2016)
Cu^{2+}	Drinking water	1 mg L^{-1}	1–6 mg L^{-1}	Pratiwi et al. (2017)
Hg^{2+}	Whitening cream	0.93 µg mL^{-1}	1–30 µg mL^{-1}	Cai et al. (2017)
Carcinoembryonic antigen	Human serum	2 ng mL^{-1}	0–40 ng mL^{-1}	Chen et al. (2018)
Single-stranded DNA oligomer	Not reported	10 nmol L^{-1}	10 nmol L^{-1}–1 µmol L^{-1}	Kalish et al. (2017)
Adenosine	Not reported	20 µmol L^{-1}	0–200 µmol L^{-1}	Tian et al. (2017)
Genomic DNA	Not reported	4.14×10^3 copies µL^{-1}	7.88×10^3–7.88×10^6 copies µL^{-1}	Hongwarittorrn et al. (2017)
Salmonella typhimurium bacteria	Starling bird fecal and whole milk	Pure *S. typhimurium*: 10^2 CFU mL^{-1} Fecal: 10^5 CFU mL^{-1} Milk: 10^3 CFU mL^{-1}	Not reported	Srisa-Art et al. (2018)
Ni^{2+}	Welding fume	Not reported	0.2–1 mmol L^{-1}	Yamada et al. (2018)
Fe^{3+}	Tap water and natural water	20 µg L^{-1}	20–1000 µg L^{-1}	Shimada and Kaneta (2018)
Pb^{2+}	Gunshot residue	Not reported	50–500 mg L^{-1}	Buking et al. (2018)

(Continued)

TABLE 10.1 (CONTINUED)
Distance-Based Method Used with Various Applications

Analyte	Sample Matrix	LOD	Linear Range	Ref.
K^+	Serum	Not reported	0.1–5.0 mmol L^{-1}	Gerold et al. (2018)
EGCG, GA, CA, Q, AA, and VA[1]	Tea	EGCG: 4.0 µmol L^{-1} GA: 5.0 µmol L^{-1} CA: 6.0 µmol L^{-1} Q: 6.0 µmol L^{-1} AA: 8.0 µmol L^{-1} VA: 8.0 µmol L^{-1}	EGCG: 0.02–0.10 µmol L^{-1} GA: 0.08–1.00 µmol L^{-1} CA: 0.04–1.00 µmol L^{-1} Q: 0.40–10.00 µmol L^{-1} AA: 0.10–4.00 µmol L^{-1} VA: 0.01–0.08 µmol L^{-1}	Piyanan et al. (2018)
Cyanide	Seawater, drinking water, tap water and wastewater	10 µg L^{-1}	0.05–1 mg L^{-1}	Khatha et al. (2019)
Ag^+	Water	0.25 µmol L^{-1}	2.5–30 µmol L^{-1}	Fu et al. (2019)
Ca^{2+}	Water	0.05 mmol L^{-1}	0.05–5 mmol L^{-1}	Shibata et al. (2019)
K^+	Serum	Not reported	1–6 mmol L^{-1}	Soda et al. (2019)
Boric acid	Eye drops	0.3 mmol L^{-1}	0.3–3 mmol L^{-1}	Hashimoto and Kaneta (2019)
Single-stranded DNA	Plant leaves of sour orange	Not reported	Not reported	Kalish et al. (2020)
PDA-antiCD81[2]	COLO1, MDA-MB-231, and HuR-KO1 cell media	COLO1: 4.7×10^5 particles mL^{-1} MDA-MB-231: 5.2×10^5 particles mL^{-1} HuR-KO1: 2.4×10^5 particles mL^{-1}	10^6–10^{10} particles mL^{-1}	Chutvirasakul et al. (2020a)
Cu^{2+}	Soil	1 mg L^{-1}	Not reported	Guan and Sun (2020)
Ni^{2+}	Not reported	95.0 µmol L^{-1}	100–5000 µmol L^{-1}	Maejima et al. (2020)
Albumin and creatinine	urine	Not reported	Albumin: 0–1000 mg L^{-1} Creatinine: 0–3000 mg L^{-1}	Hiraoka et al. (2020)
DDAC, BAC and CPC[3]	Fumigation solution	DDAC: 40 µmol L^{-1} BAC: 20 µmol L^{-1} CPC: 80 µmol L^{-1}	DDAC: 0.25–1.2 mmol L^{-1} BAC: 0.25–2.0 mmol L^{-1} CPC: 0.35–2.0 mmol L^{-1}	Chutvirasakul et al. (2020b)

[1] EGCG = Gallate equivalent, GA = Gallic acid, CA = Caffeic acid, Q = Quercetin, AA = Ascorbic acid, and VA = Vanilic acid
[2] PDA-antiCD81 = polydiacetylene conjugated with antiCD81
[3] DDAC = didecyldimethylammonium chloride, BAC = benzyldimethyltetradecyl ammonium chloride, and CPC = cetylpyridinium chloride

another enzymatic reaction that produces H_2O_2, the time-based readout is applicable to several target analytes, which can be demonstrated by the enzymatic reaction of glucose. In that reaction, the time indicating the end of the reaction was visualized by the appearance of the colored dye deposited on a layered paper substrate, which could be judged by the naked eye. Two reagents including 4-(4,4,5,5-tetramethyl-1,3,2-dioxaborolan-2-yl)benzyl(4-nitrophenyl)carbamate (TDBNC) and 3,3,5′,5′-tetra methylbenzidine (TMB) have been employed for this purpose. The mechanisms for converting the hydrophobic properties of TDBNC and TMB to hydrophilic properties are illustrated in Figures 10.4(I) and 10.4(II), respectively.

To improve the analytical performance of the time-based readout Lewis et al. (2013), suggested two interesting strategies: (i) changing the substituent to a methoxy group at the ortho-position (Figure 10.4(I)(1a, 1b) in order to increase the conversion rate from hydrophobic to hydrophilic and (ii) increasing the degree of polymerization to expand the dynamic range (Figure 10.4(I)(1c, 1f). These two strategies could improve limit of detection, sensitivity, and dynamic range.

In the application of the time-based readout, H_2O_2 production is usually the rate-determining step although the wicking rate of the microfluidic constituents in the paper substrate could also affect the time measurement. Several parameters such as paper characteristics (material, pore size, and thickness) and humidity influence the wicking rate. The characteristics of paper can be controlled to obtain precise results

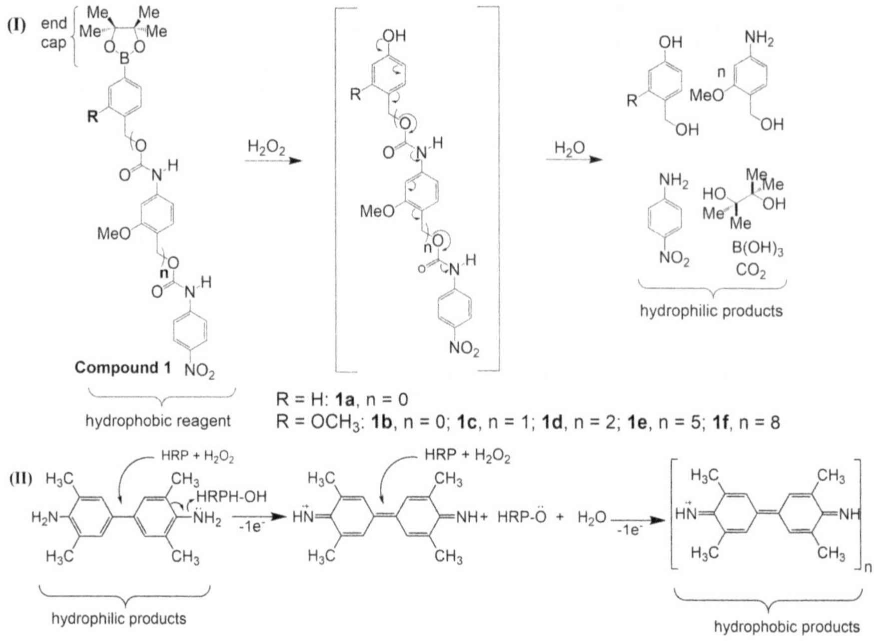

FIGURE 10.4 Mechanisms of (I) depolymerization of TDBNC and (II) polymerization of TMB for changing hydrophobic into hydrophilic products and hydrophilic into hydrophobic products, respectively. Reprint adapted from Lewis et al. (2013) and Nurjayadi et al. (2016).

via selection of the appropriate paper, but humidity is an issue that is difficult to control during experiments. Thus, Noh and Philips (2010) introduced an integrated fluidic timer that enables automatic calibration for accurate time monitoring even under different conditions of humidity. The fluidic timer zone consists of paraffin wax as a timer controller and the use of dye as a signaling material. Although the amount of wax varied the signaling times when reading the results, the integrated fluidic timer on μPADs has the following advantages: (i) self-calibration permits use under conditions of fluctuating humidity, (ii) the measurement automatically starts the reaction without the need for starting, stopping, and resetting operations that require a stopwatch, (iii) simultaneous observations in an external timer and the appearance of color spots are unnecessary, which is more convenient for users, and (iv) neither batteries nor electronic devices are required. However, a problem that remains when using the integrated fluidic timer is that wax thickness must be prepared so as to fit within the required reaction time. In other words, the timer is not applicable to unpredictable reaction times and the wax thickness must be optimized for individual reactions. Moreover, the timer itself plays the role of a stopwatch, which signals the end of the reaction for other μPADs but is not a chemical sensor of the analyte. Thus, an electronic device and software are required to capture the images and to extract the color intensity of the detection zone. Another type of timer μPAD was fabricated by printing an incomplete QR code, in which the missing part is fulfilled via colorimetric reaction. The product of the colorimetric reaction completes a valid QR code after a certain time and suggests the exact timing for photos for every experiment (Wang et al. 2018). These two ideas are interesting for further development to obtain better accuracy and precision in the time-based readout.

In conclusion, the time-based readout provides precise quantitative results under well-controlled humidity conditions. When using a mechanical timer, the operator must simultaneously observe the timer along with the reaction progress. Therefore, if the reaction occurs on the level of seconds or a few minutes, careful attention must be paid to avoid an error in reading the time. Since at least three reagents, H_2O_2, polymer, and dye, are needed for the time-based readout devices, multi-layer devices are necessary, and this leads to a more complicated fabrication and alignment of the μPADs.

10.3.3 Counting-Based Method

The μPADs with counting-based readouts consist of many reaction zones and/or detection zones that contain different amounts of a reagent. The detection zones change color depending on the concentration of the analyte, and the number of the detection zones representing the color change gives the concentration of the analyte. The obtained concentrations are given as a specified range with resolution determined by the concentration increment of the deposited reagent in each detection, or reaction zone. This method is a significant improvement over detection schemes that use only the naked eye for quantitative results without constructing a calibration curve. Many detection zones are needed, however, to accomplish a wide range of measurements with high resolution.

The first introduction of the counting-based method was adapted from the time-based method reported by Lewis et al. (2012). The μPADs were fabricated by stacking eight layers of paper pieces with 16 reaction channels for the determination of H_2O_2. The principle of the detection scheme is based on the transition of a polymer from hydrophilic to hydrophobic properties in the response to H_2O_2 similar to the time-based readouts described in Section 10.3.2. Fabrication of the μPADs requires the stacking of several layers to allow monitoring of the reaction, as displayed in Figure 10.5(I), which could complicate the fabrication process. Therefore, many studies have introduced new designs of μPADs that allow a counting-based readout without the requirement of multiple layers. For example, Zhang et al. (2014) employed the same principle with respect to a transformation of solubility properties, but they achieved the determination of H_2O_2 via two-dimensional μPADs consisting of 24 detection regions, as illustrated in Figure 10.5(II). Karita and Kaneta (2014) also developed two-dimensional μPADs for titrations, as shown in Figure 10.5(III–IV). The μPADs consisted of ten reaction and detection zones containing various amounts of a primary standard substance and a constant amount of an indicator, respectively. For the titration of a base, the μPADs were prepared by pre-depositing different amounts of a primary standard for acidic potassium hydrogen phthalate, which was pre-deposited in different concentrations in the reaction zones while the detection zones contained a constant amount of a pH indicator. Therefore, the detection zones changed the color of the pH indicator when the reaction zones contained amounts of the acid that were less than that of the base in the sample. Namely, when the sample containing the base was added to the μPADs, the endpoint was found at the detection zone neighboring the reaction zone that contained an amount of the acid equivalent to the amount of the base in the sample. For the titration of acids, the acid in the reaction zones was replaced by a primary standard of basic sodium carbonate. This idea was also applied to the chelate titration of magnesium and calcium, as shown in Figure 10.5(IV) (Karita and Kaneta 2016).

10.3.4 Text-Based Method

A text-based readout shows the results directly by displaying the alphabetical designation, the number, or the symbol (+/–) preprinted on the μPAD and represents either the presence of the analyte or its concentration. Hence, personal errors are minimized during data analysis that excludes the construction of a calibration curve. Even though a calibration curve is unnecessary, this advantage sacrifices the attainment of a fully quantitative result, and the obtained results are limited to a semi-quantitative nature. In addition, all expected results must be pre-patterned on the paper device during the fabrication process. This strategy is more pragmatic when the experiment must be performed in a remote area where trained operators and sophisticated instruments are unavailable. Until recently, however, only a few applications have been reported such as for type ABO and rhesus (RhD) identification (Li et al. 2012). In this report, all alphabetic designations and symbols for blood types (A, B, AB, and O) and rhesus (Rh+ and Rh–) were pre-screened by grouping antibodies that were specific to each of the blood types. When a blood sample is

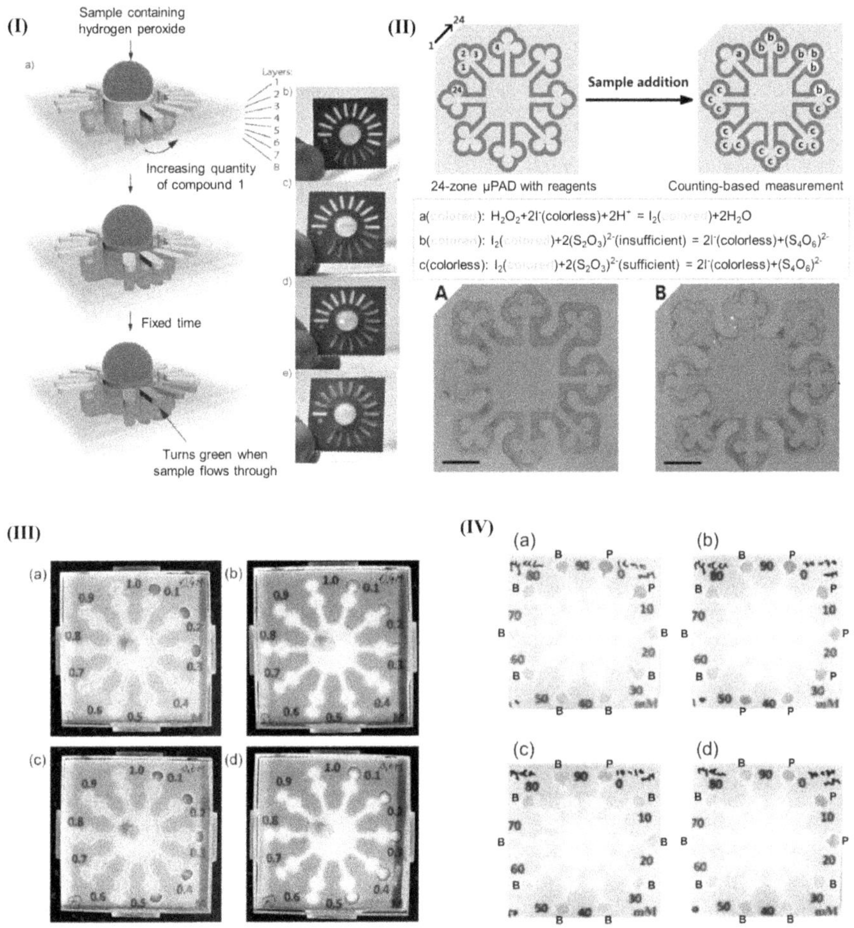

FIGURE 10.5 Design of multi-dimension μPADs for different purposes (I) H_2O_2 determination using the conversion of 4-(4,4,5,5-tetramethyl-1,3,2-dioxaborolan-2-yl)benzyl (4-nitrophenyl)carbamate, (II) H_2O_2 determination using the conversion of 3,3,5′,5′-tetramethylbenzidine, (III) acid-base titration at different concentrations of NaOH, and (IV) chelate titration at different concentrations of calcium. Reprint adapted from Karita and Kaneta (2014, 2016), Lewis et al. (2012), and Zhang et al. (2014).

introduced to the μPADs, the hemagglutination reaction will occur causing agglutination if the antibodies match the antigens in the red blood cells. Therefore, a saline solution cannot wash out the text pattern. Conversely, if the antibodies mismatch there are no agglutination results, and the flow of a saline solution easily removes red blood cells from the μPADs, as shown in Figure 10.6(I).

A semi-quantitative example of the text-based strategy was also demonstrated for human serum albumin analysis where text and numbers were preprinted by a reagent ink of tetrabromophenol blue (TBPB) to indicate the concentration of albumin (Alb)

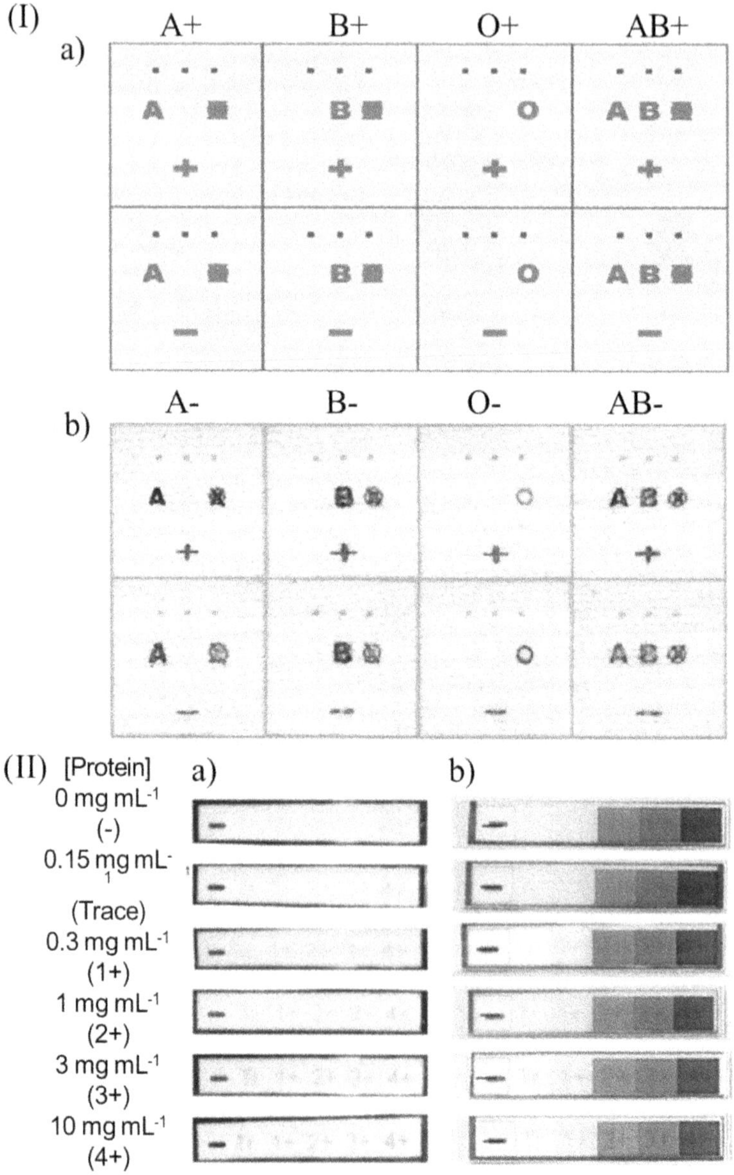

FIGURE 10.6 (I) A schematic of the expected text patterns and the actual test of all eight ABO RhD blood types, and (II) the response text patterns regarding protein concentrations on the original paper device (a) and on the paper device composed of the filter layer with five different colors corresponding to the different color intensities of an TBPB-Alb complex. Reprint adapted from Yamada et al. (2017) and Li et al. (2012).

TABLE 10.2
Advantages and Disadvantages for Colorimetric μPADs

Quantification Approach	Advantage	Disadvantage
1. Transportable small devices		
• Scanner	• Output of quantitative results • Facility in light consistency	• Requirement for trained personnel for data analysis • Requirement for data analysis program • Requirement of drying μPADs before scanning • Possibility of contamination from repeated scanning
• Camera/phone	• Output of quantitative results • Facility in operations • Acquisition of images at any time of reaction • Transportability for on-site analysis	• Requirement forlight consistency • Requirement for adjustment in focal distance • Requirement for trained personnel for data analysis • Requirement for data analysis program or developed application
2. Instrument-free readout		
• Distance-based method	• Easiness of data readout • No requirement of trained personnel for data analysis • Enhancement of sensitivity with introduction of large volume sample	• Long analysis time • Insufficient sensitivity • Requirement of complicated design for incubation reaction
• Time-based method	• No instrument except for timer/watch for data readout • Suitability for the reaction that involves H_2O_2 • Precise quantitative results under well controlled humidity	• Necessity to monitor progress in reaction and time • Long analysis time • Necessity of at least two layers of μPADs • Necessity of humidity control
• Counting-based method	• Ease of data readout • No necessity for calibration curve • No requirement for trained personnel for data analysis	• Necessity for many reaction zones • Low sensitivity and narrow working range
• Text-based method	• Ease of data readout • No necessity for calibration curve • No requirement for trained personnel for data analysis	• Long analysis time • Output of semi-quantitative results • Necessity of pre-patterning for all expected results

(Yamada et al. 2017). The complex formation between Alb and TBPB generated a color change in the numbers from yellow to blue when the μPAD was covered with a filter sheet that displayed five different colors corresponding to the different color intensities of a TBPB-Alb complex. Thus, if the color of the detection zone is stronger than the color of the filter, the number appears to report the concentration of Alb, as illustrated in Figure 10.6(II). Users see the results as the appearance of a written number, so the text-based readout is more useful and practical for minimizing user-dependent errors that take place due to the different perspectives of individual technicians.

10.4 CONCLUSIONS

The μPADs with colorimetric reactions permit quantitative and semi-quantitative analysis using the different readouts that include image capture with small devices and naked-eye detection. The advantages and disadvantages of the small data acquisition devices and the colorimetry-based μPADs with naked-eye detection are summarized in Table 10.2. Many researchers have been challenged to solve the problems, as discussed in this chapter. A suitable readout method should be selected as considered in terms of (i) property of the colorimetric reaction and product, (ii) analysis time, selectivity, and sensitivity, (iii) application field, and (iv) targeted users. For example, naked-eye detection is excellent with less skill in operation and detection, while image processing is more sensitive and precise as expected. Conversely, small image capture devices provide higher sensitive and precise results, although the data processing is skillful. Therefore, the users can choose the detection scheme that is appropriate for their aims.

REFERENCES

Buking, S., P. Saetear, W. Tiyapongpattana, K. Uraisin, P. Wilairat, D. Nacapricha, and N. Ratanawimarnwong. 2018. Microfluidic paper-based analytical device for quantification of lead using reaction band-length for identification of bullet hole and its potential for estimating firing distance. *Anal Sci* 34 (1):83–89.

Cai, L., Y. Fang, Y. Mo, Y. Huang, C.Xu, Z. Zhang, and M. Wang. 2017. Visual quantification of Hg on a microfluidic paper-based analytical device using distance-based detection technique. *AIP Adv* 7 (8):085214.

Cate, D. M., W. Dungchai, J. C. Cunningham, J. Volckens, and C. S. Henry. 2013. Simple, distance-based measurement for paper analytical devices. *Lab Chip* 13 (12): 2397–2404.

Cate, D. M., S. D. Noblitt, J. Volckens, and C. S. Henry. 2015. Multiplexed paper analytical device for quantification of metals using distance-based detection. *Lab Chip* 15 (13):2808–2818.

Chen, C.-A., P.-W. Wang, Y.-C. Yen, H.-L. Lin, Y.-C. Fan, S.-M. Wu, and C.-F. Chen. 2019. Fast analysis of ketamine using a colorimetric immunosorbent assay on a paper-based analytical device. *Sens Actuators B Chem* 282:251–258.

Chen, Y., W. Chu, W. Liu, and X. Guo. 2018. Distance-based carcinoembryonic antigen assay on microfluidic paper immunodevice. *Sens Actuators B Chem* 260:452–459.

Choi, S., S.-K. Kim, G.-J. Lee, and H.-K. Park. 2015. Paper-based 3D microfluidic device for multiple bioassays. *Sens Actuators B Chem* 219:245–250.

Chutvirasakul, B., N. Nuchtavorn, L. Suntornsuk, and Y. Zeng. 2020a. Exosome aggregation mediated stop-flow paper-based portable device for rapid exosome quantification. *Electrophoresis* 41 (5–6):311–318.

Chutvirasakul, B., N. Nuchtavorn, M. Macka, and L. Suntornsuk. 2020b. Distance-based paper device using polydiacetylene liposome as a chromogenic substance for rapid and in-field analysis of quaternary ammonium compounds. *Anal Bioanal Chem* 412 (13):3221–3230.

Finlayson, G. D., S. D. Hordley, and P. M. Hubel. 2001. Color by correlation: a simple, unifying framework for color constancy. *IEEE Trans Pattern Anal Mach Intell* 23 (11):1209–1221.

Fu, G., Y. Zhu, K. Xu, W. Wang, R. Hou, and X. Li. 2019. Photothermal microfluidic sensing platform using near-infrared laser-driven multiplexed dual-mode visual quantitative readout. *Anal Chem* 91 (20):13290–13296.

Gerold, C. T., E. Bakker, and C. S. Henry. 2018. Selective distance-based K(+) quantification on paper-based microfluidics. *Anal Chem* 90 (7):4894–4900.

Guan, Y., and B. Sun. 2020. Detection and extraction of heavy metal ions using paper-based analytical devices fabricated via atom stamp printing. *Microsyst Nanoeng* 6 (1):14.

Guo, J., J. X. H. Wong, C. Cui, X. Li, and H.-Z. Yu. 2015. A smartphone-readable barcode assay for the detection and quantitation of pesticide residues. *Analyst* 140 (16):5518–5525.

Han, G.-R., H. J. Koo, H. Ki, and M.-G. Kim. 2020. Paper/soluble polymer hybrid-based lateral flow biosensing platform for high-performance point-of-care testing. *ACS Appl Mater Interfaces* 2020 12 (31):34564–34575.

Hashimoto, Y., and T. Kaneta. 2019. Chromatographic paper-based analytical devices using an oxidized paper substrate. *Anal Methods* 11 (2):179–184.

Hiraoka, R., K. Kuwahara, Y.-C. Wen, T.-H. Yen, Y. Hiruta, C.-M. Cheng, and D. Citterio. 2020. Paper-based device for naked eye urinary albumin/creatinine ratio evaluation. *ACS Sens* 5 (4):1110–1118.

Hong, J. I., and B. Y. Chang. 2014. Development of the smartphone-based colorimetry for multi-analyte sensing arrays. *Lab Chip* 14 (10):1725–1732.

Hongwarittorrn, I., N. Chaichanawongsaroj, and W. Laiwattanapaisal. 2017. Semi-quantitative visual detection of loop mediated isothermal amplification (LAMP)-generated DNA by distance-based measurement on a paper device. *Talanta* 175:135–142.

Huang, J.-Y., H.-T. Lin, T.-H. Chen, C.-A. Chen, H.-T. Chang, and C.-F. Chen. 2018. Signal amplified gold nanoparticles for cancer diagnosis on paper-based analytical devices. *ACS Sens* 3 (1):174–182.

Kalish, B., J. Luong, J. Roper, C. Beaudette, and H. Tsutsui. 2017. Distance-based quantitative DNA detection in a paper-based microfluidic device. In IEEE 12th International Conference on Nano/Micro Engineered and Molecular Systems (NEMS), Los Angeles, CA, USA, 337–341.

Kalish, B., J. Zhang, H. Edema, J. Luong, J. Roper, C. Beaudette, R. Echodu, and H. Tsutsui. 2020. Distance and microsphere aggregation-based DNA detection in a paper-based microfluidic device. *SLAS Technol* 25 (1):58–66.

Karita, S., and T. Kaneta. 2014. Acid–base titrations using microfluidic paper-based analytical devices. *Anal Chem* 86 (24):12108–12114.

Karita, S., and T. Kaneta. 2016. Chelate titrations of Ca^{2+} and Mg^{2+} using microfluidic paper-based analytical devices. *Anal Chim Acta* 924:60–67.

Khatha, P., T. Phutthaphongloet, P. Timpa, B. Ninwong, K. Income, N. Ratnarathorn, and W. Dungchai. 2019. Distance-based paper device combined with headspace extraction for determination of cyanide. *Sensors* 19 (10):2340.

Kong, T., J. B. You, B. Zhang, B. Nguyen, F. Tarlan, K. Jarvi, and D. Sinton. 2019. Accessory-free quantitative smartphone imaging of colorimetric paper-based assays. *Lab Chip* 19 (11):1991–1999.

Lewis, G. G., M. J. DiTucci, and S. T. Phillips. 2012. Quantifying analytes in paper-based microfluidic devices without using external electronic readers. *Angew Chem Int Ed* 51 (51):12707–12710.

Lewis, G. G., J. S. Robbins, and S. T. Phillips. 2013. Point-of-care assay platform for quantifying active enzymes to femtomolar levels using measurements of time as the readout. *Anal Chem* 85 (21):10432–10439.

Li, F., Y. Hu, Z. Li, J. Liu, L. Guo, and J. He. 2019. Three-dimensional microfluidic paper-based device for multiplexed colorimetric detection of six metal ions combined with use of a smartphone. *Anal Bioanal Chem* 411 (24):6497–6508.

Li, M., J.Tian, M. Al-Tamimi, and W. Shen. 2012. Paper-based blood typing device that reports patient's blood type "in writing". *Angew Chem Int Ed* 51 (22):5497–5501.

Lopez-Ruiz, N., V. F. Curto, M. M. Erenas, F. Benito-Lopez, D. Diamond, A. J. Palma, and L. F. Capitan-Vallvey. 2014. Smartphone-based simultaneous pH and nitrite colorimetric determination for paper microfluidic devices. *Anal Chem* 86 (19):9554–9562.

Maejima, K., Y. Hiruta, and D. Citterio. 2020. Centrifugal paperfluidic platform for accelerated distance-based colorimetric signal readout. *Anal Chem* 92 (7):4749–4754.

Martinez, A. W., S. T. Phillips, E. Carrilho, S. W. Thomas, H. Sindi, and G. M. Whitesides. 2008. Simple telemedicine for developing regions: camera phones and paper-based microfluidic devices for real-time, off-site diagnosis. *Anal Chem* 80 (10):3699–3707.

McCracken, K. E., S. V. Angus, K. A. Reynolds, and J.-Y. Yoon. 2016. Multimodal imaging and lighting bias correction for improved μPAD-based water quality monitoring via smartphones. *Sci Rep* 6 (1):27529.

Meelapsom, R., P. Jarujamrus, M. Amatatongchai, S. Chairam, C. Kulsing, and W. Shen. 2016. Chromatic analysis by monitoring unmodified silver nanoparticles reduction on double layer microfluidic paper-based analytical devices for selective and sensitive determination of mercury(II). *Talanta* 155:193–201.

Noh, H., and S. T. Phillips. 2010. Fluidic timers for time-dependent, point-of-care assays on paper. *Anal Chem* 82 (19):8071–8078.

Nurjayadi, M., D. Apriyani, U. Hasan, I. Santoso, F. Kurniadewi, I. R. Kartika, K. Agustini, F. Puspasari, D. Natalia, and W. Mangunwardoyo. 2016. Immunogenicity and specificity of anti recombinant protein Fim-C-*Salmonella typhimurium* antibody as a model to develop typhoid vaccine. *Procedia Chem* 18:237–245.

Ogawa, K., and T. Kaneta. 2016. Determination of iron ion in the water of a natural hot spring using microfluidic paper-based analytical devices. *Anal Sci* 32 (1):31–4.

Piyanan, T., A. Athipornchai, C. S. Henry, and Y. Sameenoi. 2018. An instrument-free detection of antioxidant activity using paper-based analytical devices coated with nanoceria. *Anal Sci* 34 (1):97–102.

Pratiwi, R., M. P. Nguyen, S. Ibrahim, N. Yoshioka, C. S. Henry, and D. H. Tjahjono. 2017. A selective distance-based paper analytical device for copper(II) determination using a porphyrin derivative. *Talanta* 174:493–499.

Russell, S. M., A. Doménech-Sánchez, and R. de la Rica. 2017. Augmented reality for real-time detection and interpretation of colorimetric signals generated by paper-based biosensors. *ACS Sens* 2 (6):848–853.

Sameenoi, Y., P. N. Nongkai, S. Nouanthavong, C. S. Henry, and D. Nacapricha. 2014. One-step polymer screen-printing for microfluidic paper-based analytical device (μPAD) fabrication. *Analyst* 139 (24):6580–6588.

Shen, L., J. A. Hagen, and I. Papautsky. 2012. Point-of-care colorimetric detection with a smartphone. *Lab Chip* 12 (21):4240–4243.

Shibata, H., Y. Hiruta, and D. Citterio. 2019. Fully inkjet-printed distance-based paper microfluidic devices for colorimetric calcium determination using ion-selective optodes. *Analyst* 144 (4):1178–1186.

Shimada, Y., and T. Kaneta. 2018. Highly sensitive paper-based analytical devices with the introduction of a large-volume sample via continuous flow. *Anal Sci* 34 (1):65–70.

Soda, Y., D. Citterio, and E. Bakker. 2019. Equipment-free detection of K^+ on microfluidic paper-based analytical devices based on exhaustive replacement with ionic dye in ion-selective capillary sensors. *ACS Sens* 4 (3):670–677.

Srisa-Art, M., K. E. Boehle, B. J. Geiss, and C. S. Henry. 2018. Highly sensitive detection of salmonella typhimurium using a colorimetric paper-based analytical device coupled with immunomagnetic separation. *Anal Chem* 90 (1):1035–1043.

Tian, T., Y. An, Y. Wu, Y. Song, Z. Zhu, and C. Yang. 2017. Integrated distance-based origami paper analytical device for one-step visualized analysis. *ACS Appl Mater Inter* 9 (36):30480–30487.

Vaughan, L.M, G. Milavetz, E. Ellis, S.J Szefler, K. Conboy, M.M Weinberger, S. Tillson, J. Jenne, M.B Wiener, T. Shaughnessy, and J. Carrico. 1986. Multicentre evaluation of disposable visual measuring device to assay theophylline from capillary blood sample. *Lancet* 327 (8474):184–186.

Wang, T., G. Xu, W. Wu, X. Wang, X. Chen, S. Zhou, and F. You. 2018. A novel combination of quick response code and microfluidic paper-based analytical devices for rapid and quantitative detection. *Biomed Microdevices* 20 (3):79.

Wei, X., T. Tian, S. Jia, Z. Zhu, Y. Ma, J. Sun, Z. Lin, and C. J. Yang. 2016. Microfluidic distance readout sweet hydrogel integrated paper-based analytical device (μDiSH-PAD) for visual quantitative point-of-care testing. *Anal Chem* 88 (4):2345–2352.

Yamada, K., K. Suzuki, and D. Citterio. 2017. Text-displaying colorimetric paper-based analytical device. *ACS Sens* 2 (8):1247–1254.

Yamada, K., D. Citterio, and C. S. Henry. 2018. Dip-and-read" paper-based analytical devices using distance-based detection with color screening. *Lab Chip* 18 (10):1485–1493.

Zhang, Y., C. Zhou, J. Nie, S. Le, Q. Qin, F. Liu, Y. Li, and J. Li. 2014. Equipment-free quantitative measurement for microfluidic paper-based analytical devices fabricated using the principles of movable-type printing. *Anal Chem* 86 (4):2005–12.

11 Micromixers and Microvalves for Point-of-Care Diagnosis and Lab-on-a-Chip Applications

Aarathi Pradeep and T. G. Satheesh Babu

CONTENTS

11.1 MICROMIXERS FOR LAB-ON-A-CHIP APPLICATIONS

Micromixers are miniaturized mixing devices that have at least one of their characteristic dimensions in the micrometer scale, typically capable of handling microliters to milliliters of sample volume. While eddy diffusions and turbulence can achieve macroscale mixing, it is not possible at the microscale due to the large viscous effects owing to the small size of the micromixer. Because of the low Reynolds number associated with microfluidics devices, mixing in the micro-regime is mainly due to molecular diffusions and induced chaotic advection. Molecular diffusion results from the random motion of molecules, mainly due to concentration gradients of chemical potential. Chaotic advection is an induced phenomenon for microfluidic mixing by modifying the microchannel geometry and fluid motion. An efficient

DOI: 10.1201/9781003033479-11

micromixer should have rapid mixing with a smaller mixing path and increased area of contact between the two mixing streams.

11.1.1 Principle of Micromixing

Mixing is the process of homogenizing the concentration of different fluidic streams such that there is a uniform distribution of concentration in the final stream. In a typical microfluidics device, the flow is characterized by a low Reynolds number, often less than 100. It means that the flow is laminar and streamlined, which means that mixing between streams is primarily due to diffusion.

Diffusion is a slow process, and to achieve mixing based only on diffusion requires a long contact time between the diffusing streams. It is the movement of particles from an area of high concentration to a low concentration until fluid homogeneity is achieved. This phenomenon is governed by Fick's law of diffusion, which states that the diffusion flux is proportional to the concentration gradient existing between the streams. Mathematically, for a one-dimensional diffusion, this is given as

$$J = -D\frac{\partial C}{\partial x} \tag{11.1}$$

where, J is the diffusive flux (mol m^{-2} s^{-1}), D is the diffusion coefficient (m^2 s^{-1}), and $\left(\frac{\partial C}{\partial x}\right)$ is the concentration gradient of the diffusing species existing in the x direction.

11.1.2 Mixing Efficiency in Microchannels

One of the most widely used methods for quantifying mixing efficiency is by calculating the mixing index (MI) of the micromixer. It is given as,

$$MI = \left(1 - \sqrt{\frac{1}{N}\sum_{i=1}^{N}\left(\frac{C_i - \bar{C}}{\bar{C}}\right)}\right) \times 100\% \tag{11.2}$$

where N refers to the total number of sample points, C_i refers to the normalized concentration at the inlets and $\bar{C}$ refers to the normalized expected concentration. MI varies from 0 to 100, with 0 representing an unmixed solution and 100 representing a thoroughly mixed homogeneous solution.

11.1.3 Classification of Micromixers

Micromixers are generally classified as active or passive micromixers based on their energy requirement (Figure 11.1). In the case of active micromixers, an external energy source is used to mix the fluidic streams, whereas in passive micromixers, micromixing is achieved by molecular diffusions and chaotic advection. The flow

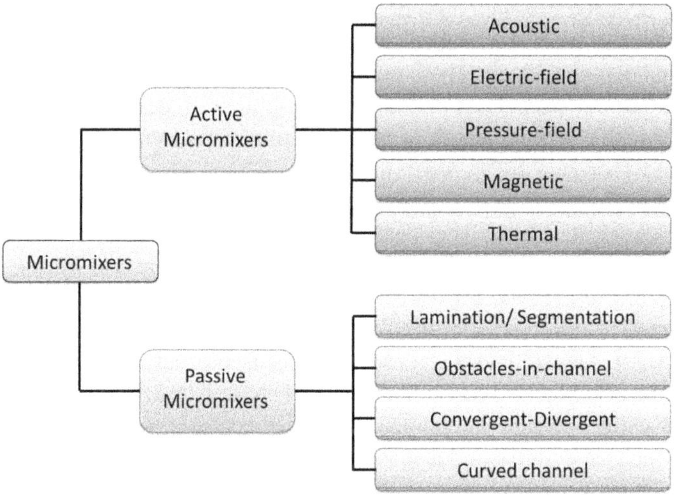

FIGURE 11.1 Schematic representation of the classification of micromixers.

geometry is designed such that laminar perturbations occur as the fluid flows through the channels, leading to enhanced diffusion between the streams. Although active micromixers have higher mixing efficiency, passive micromixers have advantages like ease of fabrication and integration with other microfluidics elements.

11.1.3.1 Active Micromixers

An active micromixer refers to a microfluidics device that employs an external energy source for mixing different fluids. This external energy is used to create disturbances in the fluid or to drive moving parts within the device to promote mixing. The following section briefly explains the different active micromixers based on different energy sources.

11.1.3.1.1 Acoustic Field-Based Micromixers

In acoustic-driven micromixers, the energy from acoustic waves is transferred to the fluid medium to induce mixing. The perturbations from the waves propagate along with the fluid. A commonly employed technique uses acoustically vibrating microstructures with sharp edges that can vibrate in response to fluctuations in pressure in the acoustic waves, thereby inducing convective mixing.

Rife et al. demonstrated acoustic micromixing using an array of piezoelectric actuators (Rife et al. 2000). The actuator array was deposited on a silicon substrate and had a resonant frequency around 50 MHz with an attenuation length of 8.3 mm. This actuator was attached to a polymethyl methacrylate (PMMA) with the microchannels. On actuation, the acoustic energy was transferred to the microchannel enabling the complete mixing of fluids. The efficiency is that mixing depends on the switching and resonant modes of the piezoelectric material (PZT). Jang et al. investigated this. By using a 6 × 6 mm lead-zirconate-titanate PZT layer on a silicon membrane, acoustic mixing could be achieved in less than 30 s (Jang et al. 2007). The piezoelectric layer was placed on top of a circular mixing chamber of 6 mm diameter and depth of 30 μm. These micromixers have been employed for the synthesis of different nanoparticles. For example, Rasouli et al. demonstrated an acoustic micromixer with a mixing length of only 25.4 μm for the synthesis of nanoparticles (Rasouli and Tabrizian 2019). A mixing time of 0.8 ms was achieved for a flow rate of 116 μL min^{-1} with an input voltage of 40 V_{pp}. This micromixer was developed by placing a piezoelectric transducer near a microchannel with sharp edges. The sharp edges and trapped bubbles in the microchannel functioned as a vibrating boundary in response to the acoustic pressure waves generated by the piezoelectric disc. Li et al. reported an ultrafast acoustic micromixer for the synthesis of nanoparticles (Le et al. 2020). An oscillatory electric field was used to actuate a piezoelectric disc coupled to the substrate. Mixing efficiency of 91% could be achieved in 4.1 ms. Other methods of using acoustic waves include surface acoustic waves and bulk acoustic waves in which advection is induced by traveling bulk or surface waves, respectively (Bernassau et al. 2014, Collins et al. 2016).

11.1.3.1.2 Electric Field-Based Micromixers

Micromixers based on electric fields mostly rely on electrohydrodynamic disturbances (EHD) (Figure 11.2). In these micromixers, the motion of electrically charges fluid particles in response to an electric field (AC or DC) used to disturb the laminarity of the flow. The mixing efficiency is dependent on the applied voltage. The fluids are mostly immiscible as at least one of the fluid streams is a dielectric.

These micromixers typically have electrodes whose potential and frequency can be varied to achieve the micromixing. An active micromixer employing EHD was demonstrated by Ould El Moctar et al. (2003). It consisted of titanium electrodes placed perpendicular to the microchannel. Flow disturbances were induced by switching the potential and frequency of the electrodes. The mixing efficiency was

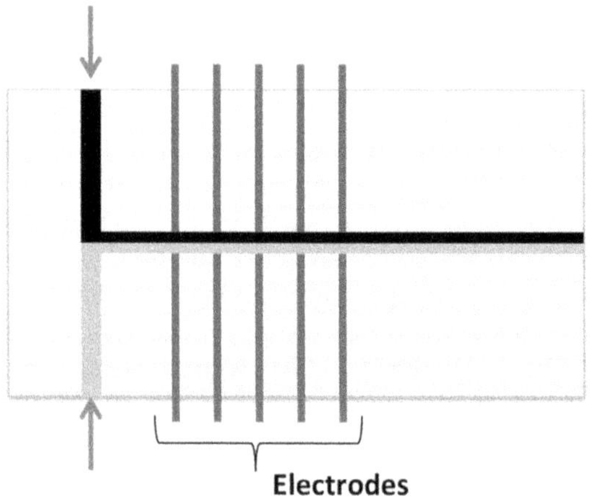

FIGURE 11.2 An electrode configuration for active micromixing by EHD.

proportional to the square of the electric field, implying a direct relation to the EHD force. Such EHD-based active micromixers have been used to synthesize polymer nanoparticles (Lee et al. 2019). They have also been employed for mixing in hydrodynamic flow-focusing devices. Zahn et al. had demonstrated the mixing of water and phenol using this technique (Zahn and Reddy 2006). The microchannel was fabricated in polydimethylsiloxane (PDMS), and the floor of the microchannel was patterned with chromium/gold electrodes. The flow disturbances were generated by varying the frequencies from 250 kHz to 10 MHz, and the root mean square values from 0 to 45 V.

Apart from EHD, electrokinetic instabilities are also used to enhance mixing in electric field-based micromixers. Electrokinetic instabilities are caused by ion transport due to the presence of charge at the solid–liquid interface. This includes mixing based on electroosmosis, electrophoresis, and dielectrophoresis (DEP). Lee et al. demonstrated the development of a DEP-based micromixer in which micromixing was achieved by a folding and stretching mechanism (Lee et al. 2001). Polarizable polystyrene beads were used to disturb the flow in response to a nonuniform AC electric field. A micromixer that employs periodic low-frequency electrokinetic driving force was demonstrated by Fu et al. (2005). Using an electric field with a strength 100 V m^{-1} and a periodic switching frequency of 2 Hz, a mixing efficiency of 95% could be achieved in 1 mm.

11.1.3.1.3 Pressure Field-Based Micromixers

In pressure field-based micromixers, micromixing is achieved by pulsating the flow with a velocity or pressure perturbation mechanism using external actuators (Fu et al. 2013). The design primarily consists of the main channel and one or more side channels. The fluid in the main channel is stirred by pulsing velocity through side channels. This creates flow disturbances within the central channel by stretching and folding fluids. The pressure disturbances are sometimes achieved by employing two micropumps which are operated in an out-of-phase manner.

11.1.3.1.4 Magnetic Field-Based Micromixers

In this class of micromixers, mixing is achieved using magnetic sources like a permanent magnet, electromagnet, and magnetic stirrer with a larger focus on magnetohydrodynamics and magnetic stirring.

- Active micromixers employing permanent magnets usually have a magnetic fluid stream (e.g., ferrofluid) and a nonmagnetic stream. The permanent magnet is placed near the microchannel and provides a nonuniform external force for mixing.
- Electromagnetic micromixers achieve mixing using Lorentz force generated by applying alternating voltages to induce an alternating current. This leads to the formation of multiple micro streams, thereby enhancing mixing efficiency.
- Magnetic stirring micromixers employ small magnetic stirrers or beads to achieve mixing between different fluids.

Veldurthi et al. studied the performance of a micromotor actuated by a magnetic field to load rifampicin drug into TiO_2 nanoparticles with a maximum mixing efficiency of 90% (Veldurthi et al. 2015). The micromotor, with a dimension of 1 × 0.2 × 0.3 mm, was developed by dispersing Fe_2O_3 nanoparticles in PDMS. The microrotor was actuated in a magnetic field resulting in the mixing of fluid streams. Mensing et al. demonstrated the use of a stainless-steel blade for an active magnetic micromixing device (Mensing et al. 2004). The blade was actuated by placing the device on a magnetic stir plate, resulting in the mixing of fluids. Micromixers using embedded micromagnets and macro electromagnets have been used for developing micromixers. Cao et al. had demonstrated ferrofluid-based magnetic micromixer to obtain a high mixing efficiency of 97% in 8 s in a short length of 600 μm (Cao et al. 2015). Here, efficient mixing was attributed to the static gradient magnetic fields and alternating uniform magnetic fields generated by micromagnets and macro electromagnets.

11.1.3.1.5 Thermal Field-Based Micromixers

Thermal field micromixers use thermal energy, thermal bubbles, or electrothermal effects to enhance mixing efficiency by increasing the diffusivity of the fluids. The kinetic energy of the molecules is increased, thereby increasing the molecular diffusion. Tan demonstrated an efficient micromixer that employs the periodic boiling principle of inkjet technology using microheaters (Tan 2019). Meng et al. demonstrated a thermal micromixer using AC electrothermal actuation that can produce a mixing efficiency of 100% using four arc electrodes (Meng et al. 2018).

11.1.3.2 Passive Micromixers

Passive micromixers are micromixers with modified geometry such that as the fluid flows through the microfluidics channels, the laminarity of the flow is disturbed leading to enhanced mass transport. They can be classified as mixing based on molecular diffusions and mixing based on chaotic advection.

- Mixing based on molecular diffusions alone is achieved by increasing the contact area between the fluid streams and by reducing the microchannel thickness (reduced diffusion path). This makes use of the higher concentration gradient existing in microchannels with reduced lateral dimensions. Lamination and segmentation techniques are commonly employed for micromixing based on molecular diffusions.
- Mixing based on chaotic advection focus on introducing laminar perturbations and improving the mass transport in the transversal direction. Modification of the geometry of the channel, like curved channels, channels with sharp turns, obstacles-in-channels, converging-diverging channels, and unsymmetrical channels are commonly used to induce chaotic advection in micromixers.

11.1.3.2.1 Lamination Mixers

The most common type of lamination micromixers includes parallel lamination and serial lamination. In parallel lamination streams, the inlet streams are split into two or more sub-streams (n) and then rejoined in a single channel. As faster diffusion can be achieved in channels with reduced diameter, the mixing efficiency of the micromixer is enhanced by a factor of n^2. Bessoth et al. demonstrated a parallel lamination micromixer with 32 sub-streams that could achieve complete mixing in 15 ms (Bessoth et al. 1999).

A serial lamination micromixer also works by splitting and rejoining the fluid streams. Here, the rejoined streams are split further and rejoined in a serial pattern to enhance faster mixing. With "n" splitting and rejoining steps, this technique leads to 4^{n-1} improvement in micromixing. Rigler et al. demonstrated a wedge-shaped inlet to create four laminated fluid layers that achieve fast mixing based on vertical diffusion. The channels are further modified with Tesla structures, zig-zag channels, or serpentine channels to enhance mixing.

11.1.3.2.2 Obstacles-in-Channel Mixers

This class of micromixers typically employs obstacles in the fluid flow path to generate vortices and chaotic advection that can lead to faster mixing and reduction in mixing length. The obstacles are placed either on the inner walls of the microchannel or in the flow path. Bhagat et al. compared the efficiency of mixing particles by incorporating Tesla structures and obstructions in the channel (Bhagat and Papautsky 2008). It was seen from their studies that an obstruction-based micromixer was efficient in mixing particles. A passive micromixer with baffles in the main channel was developed by Chung et al. (2006). The baffles aided in the formation of recirculation zones within the microchannel, thus enhancing mixing quality.

One of the breakthrough developments in an obstacles-based passive micromixer was the staggered herringbone micromixer developed by Stroock et al. (2002). The design consisted of pairs of slanted ridges, also called the staggered herringbone structure. The micromixer could generate counter-rotating vortices and secondary flows leading to chaotic advection in the microchannel. The interface between the fluidic streams could be significantly increased, leading to faster mixing.

11.1.3.2.3 Convergent–Divergent Microchannels as Micromixers

In convergence–divergence-based micromixers, vortices are generated when abrupt changes occur in the channel geometry leading to sudden expansion or contraction. A passive micromixer with converging–diverging side walls with sinusoidal variations was described by Afzal and Kim (2012). The micromixer was found to be efficient with a Reynolds number (Re) in the range of $10 < Re < 70$. These micromixers usually incorporate obstacles to enhance mixing further. The effect of placing triangular obstacles in convergent–divergent channels was studied by Heshmatnezhad et al., and it was concluded that obstacles in convergent–divergent walls enhance mixing (Heshmatnezhad et al. 2017). This class of micromixers also includes asymmetric microchannels that are designed to enhance micromixing. A passive micromixer with asymmetric rhombic sub-channels was proposed by Hossain and Kim (2014). It was seen from their analysis that the rhombic angles and the width of the sub-channels influenced the mixing performance.

11.1.3.2.4 Curved Channels as Micromixers

Curved-channel micromixers are highly efficient in high Reynolds number flow conditions and are characterized using the Dean number. It was reported in the literature that secondary flow consisting of two and four vortices appear for Dean number <150 and >150, respectively (Cheng et al. 1976). Microchannels with sharp 90° turns are useful in introducing chaotic advection in microchannels. Therefore, increasing the number of turns increases the mixing efficiency. Spiral microchannels with expansion–contraction parts were studied by Mehrdel et al. and it had performed better than spiral microchannels with a uniform cross-section (Mehrdel et al. 2018). Balasubramaniam et al. studied the effect of varying cross-section and hydraulic diameter in spiral microchannels to improve mixing efficiency (Balasubramaniam et al. 2017). It was seen from their analysis that the cross-section of the spiral geometry is an important factor contributing to the strength of Dean vortices which determines the mixing efficiency.

11.1.4 Applications of Micromixers

The applications of micromixers are vast and diverse. They have been widely used for the concentration and detection of samples for biological processes, typically for sample concentration, chemical synthesis, and as chemical reactors.

An integrated programmable microfluidic cell array using micromixers for combinatorial drug screening was developed by Kim et al. (2012). Pradeep et al. developed a passive microfluidic mixing system using meander-shaped micromixers for the analysis of glucose concentration in a sample using nonenzymatic glucose sensors that work in an alkaline medium (Pradeep et al. 2016). An integrated device capable of performing immunoassay for the detection of Hendra virus using a chaotic micromixer was reported by Petkovic et al. (2017). A microfluidics chip for performing systematic evolution of ligands by exponential enrichment (SELEX) to select the single-stranded DNA aptamers for hemoglobin and HbA1c was demonstrated by Lin et al. (2015). The applications of micromixers for the synthesis of a wide range of nanoparticles have been explored. A T-shaped micromixer for the synthesis of nickel oxide nanoparticles using the hydrothermal method was developed by Kawasaki et al. (2010). Similar works has been done to synthesize TiO_2 (Kawasaki et al. 2009), ZnO (Li et al. 2010), Pt (Baumgard et al. 2013) nanoparticles, and CdSe quantum dots (Tian et al. 2016). Also, micromixers are crucial in optimizing various chemical reactors. A microfluidics device for DNA ligation using micromixers and microreactors was developed by Ko et al. (2011). The device could reduce the litigation time from 4 h to 5 min.

Micromixers are also widely used in polymerization processes to mix the initial reactants or the solutions in multi-stage block copolymerization processes to obtain polymers with narrow molecular weight distribution. They have been used for the polymerization of acrylic acid (Qiu et al. 2016), styrene radical (Ryu et al. 2017), anionic and cationic polymerizations (Nagaki et al. 2012). Micromixers have also been used for liquid–liquid extraction operations like the extraction of iodine in water to n-hexane in an emulsion (Sprogies et al. 2008), phenol from dodecane

(Mae et al. 2004), copper ions from sulphate solution containing copper, zinc, and iron (Jiang et al. 2018). They have been successfully employed in the purification process of DNA with high recovery efficiency (96 ± 11%) for polymerase chain reaction (Kastania et al. 2016). Micromixers have been a crucial component in developing lab-on-a-chip devices for a wide range of biological processes, including cell separation and lysis, cell disruption, protein quantifications, cancer biomarker detection, and different pathogen detection.

11.2 MICROVALVES FOR LAB-ON-A-CHIP APPLICATIONS

Valves are indispensable components that play a vital role in controlling and manipulating fluid flow. They are considered as traffic lights of flow in a microfluidics environment. They are indispensable components in the development of a standalone microfluidics device involving multiple fluids. The following sections describe some of the important aspects of a microvalve followed by its classifications and applications.

11.2.1 Principle of Microvalves

Depending on the initial position of the microvalves, they are classified as normally open or normally closed microvalves. A normally open microvalve permits the fluid flow in its initial conditions and obstructs the flow in response to an actuation, while a normally closed microvalve allows fluid flow only when actuated. Microvalves typically employ materials that can modify the fluid flow in response to an applied external field or by changing the flow environment and depend on the type of actuator employed. They are evaluated based on their leakage, capacity, temperature range, response time, power consumption, pressure range of operation, biocompatibility, chemical compatibility, and reliability (Nguyen and Wereley 2006).

11.2.2 Classification of Microvalves

Microvalves are classified as active microvalves and passive microvalves based on their energy requirements. They are further classified based on their actuation mechanism and their actuation force. The following section explains the various classifications of microvalves.

11.2.2.1 Active Microvalves

Active microvalves employ external energy sources for the control of fluid within the microfluidics device. They are basically classified into three types depending on their actuation principle – (i) mechanical active microvalves, which make use of moving mechanical components coupled to different actuation techniques, (ii) nonmechanical active microvalves that use functionalized smart materials or rely on physical or chemical properties of the working fluid to bring about the actuation, and (iii) external active microvalves which make use of external components like a pneumatic source to bring about the actuation.

11.2.2.1.1 Mechanical Active Microvalves

A mechanically active microvalve uses moving membranes that can be actuated by different sources like magnetic, electric, piezoelectric, and thermal energy. The following section briefly describes microvalves developed using each of these actuation techniques.

11.2.2.1.1.1 Magnetic Microvalves Magnetic active microvalves typically consist of a mechanical part that can respond to a magnetic field generated using an electromagnet or a permanent magnet. An electromagnetic valve consists of a permanent magnet and a coil pair. When an electric current is passed through the coil, magnetic fields are generated around it, which cause the actuation of the membrane of a microvalve. The first magnetic pinch microvalve was developed by Terry et al. which used a solenoid plunger on a nickel diaphragm (Terry et al. 1979). The movement of the plunger of the solenoid in response to the applied electric field served the valve action.

The most common method for the fabrication of deflectable magnetic membranes is by embedding magnetic particles in the PDMS membrane. Singh et al. demonstrated magnetic membranes made out of a nanocomposite of cobalt nanoparticles and PDMS for microvalve applications (Singh et al. 2014). Cobalt nanoparticles were synthesized by chemical methods and were mixed with PDMS to form the deflectable magnetic membrane. Magnetorheological fluids have higher magnetic permeability than magnetic particle-doped PDMS membranes, thereby allowing higher deformation. Gholizadeh and Javanmard demonstrated magnetorheological fluid patterned on elastomeric membranes to function as a microvalve that could be actuated using a permanent magnet (Gholizadeh and Javanmard 2016). Pradeep et al. demonstrated the use of mild steel metal pieces incorporated into PDMS membrane to function as the valve (Pradeep et al. 2018). The valves were actuated using in-built solenoids activated using a microcontroller (Figure 11.3). By using an array of such valves, automated actuation and controlled delivery of multiple fluids were demonstrated.

11.2.2.1.1.2 Electric Microvalves Electrostatic and electrokinetic microvalves belong to this class of microvalves. In electrostatic microvalves, membrane deflection is achieved using electrostatic forces. It typically consists of a valve-opening electrode, valve-closing electrode, and a deflectable membrane. By applying a

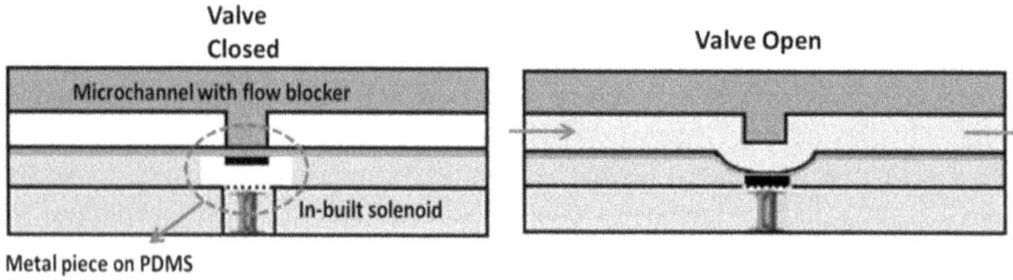

FIGURE 11.3 Magnetic microvalves using metal pieces on PDMS membrane (Pradeep et al. 2018).

voltage across an electrode in the deflecting membrane and a counter electrode, a charge is induced on the counter electrode that attracts the membrane and causes its deflection. These valves are mainly used to control the flow of gases rather than liquids, as the high operating voltage of the valves may cause electrolysis of liquids. Anjewierden et al. demonstrated an electrostatic valve for microfluidics applications using chrome–copper electrode pair. The copper foil was used as the deflecting membrane and chrome electroplated on PMMA served as the counter electrode (Anjewierden et al. 2012). The valves could withstand pressures up to 40 kPa and required a closing voltage of 680 V. Electrokinetic actuation principles used to manipulate liquids and particles are also employed to develop microvalves (Kirby et al. 2002, Li and Li 2018).

11.2.2.1.1.3 Piezoelectric Microvalves Piezoelectric microvalves are developed using the "inverse piezoelectric effect." The inverse piezoelectric effect refers to the mechanical stresses induced on a piezoelectric disc in response to an applied electric field. Membrane deflection is achieved by placing a piezoelectric disc on the membrane to be actuated. Piezoelectric actuation yields a large bending force but has small displacement. Since the extent of deflection obtained from a piezoelectric disc is low, it is often coupled with techniques like hydraulic displacement amplification mechanism (HDAM) using a PZT stack to increase the extent of deflection. Chen et al. reported a piezoelectric microvalve using a hydraulically displacement amplification ratio (Chen et al. 2016). A unimorph piezoelectric disc was used as the actuator, and a natural elastic rubber membrane was used as the valve stopper for sealing. The flexible valve stopper was made to shrink in response to an applied voltage to the piezoelectric disc, thus causing the valve action. A high amplification ratio of up to 14 was realized using this mechanism.

11.2.2.1.1.4 Thermal Microvalves Thermal microvalves include thermopneumatic valves and shape memory alloy valves. Thermopneumatic valves couple the volumetric thermal expansion to the membrane deflection. The actuator consists of a sealed chamber that has a deflectable membrane on one side. The chamber contains the liquid/gas system whose thermal expansion causes the valve action. The fluid is usually heated using resistors that are incorporated within the device. Zahra et al. reported a normally open thermal microvalve developed with PDMS microchannel and heater network on either side of a glass substrate (Zahra et al. 2015). Figure 11.4(A) shows the initial position of the microvalve. An air-filled chamber made of a deflectable PDMS membrane was positioned below the microchannel. The valve closure was achieved by applying a voltage to the heater unit, which resulted in increased pressure in the air in the chamber. This resulted in the deformation of the membrane in the upward direction, thus closing the microchannel (Figure 11.4(B)). Rich and Wise reported a thermopneumatically actuated microvalve using a volatile liquid like pentane to bring about the deflection of a corrugated silicon membrane (Rich and Wise 2003). The volatile fluid was heated using resistively heated grids, leading to increased pressure within the sealed chamber. This led to the deflection of the membrane, which functioned as the valve.

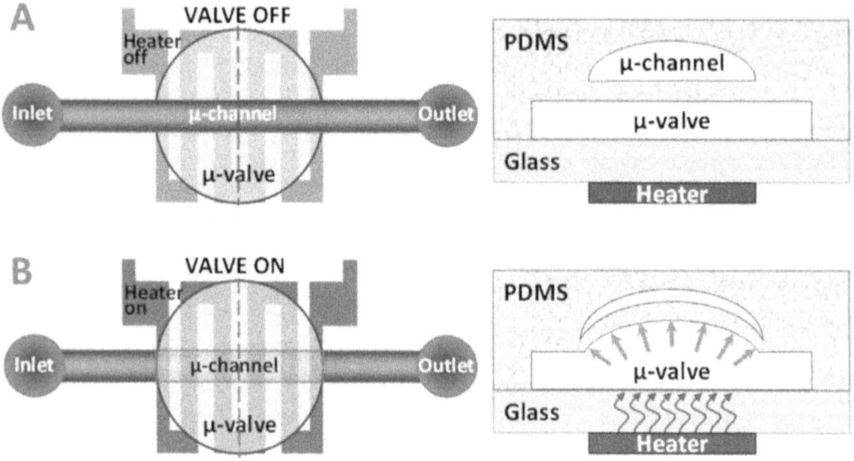

FIGURE 11.4 Working of the thermopneumatic microvalve. The valve in (A) OFF position (open) and (B) in ON position (closed) (Zahra et al. 2015).

Shape memory alloys (SMA) are materials that have two or more crystallographic phases and can undergo reversible transformations. Their unique property of returning to their initial state even after large deformations in response to a thermal stress field makes them well-suited for microvalve applications. Shape memory alloy-based multi-port valves for controlling the liquids and gases was demonstrated by Megnin and Kohl (2014). Ti-Ni sheet of thickness 20 μm was used as a bridge-microactuator that was loaded by a pressure-spring or a spherical plunger. The valve action was achieved by heating the SMA. Other SMA-based modifications like NiTiPd (Kohl et al. 1999) and NiTiCu (Ghadimi et al. 2013) have also been studied.

11.2.2.1.2 Nonmechanical Active Microvalves

In nonmechanical active microvalves, actuation is achieved in response to smart materials or properties of the fluidic system. Some of the common nonmechanical active valves include electrochemical microvalves, phase-change microvalves, and rheological microvalves.

11.2.2.1.2.1 Electrochemical Microvalves An electrochemical microvalve works by electrolyzing a solution using electrodes to produce gas bubbles which can deflect a membrane to perform the valve function. It typically consists of primary chambers containing solutions with a redox couple separated by a semipermeable membrane (Figure 11.5). On applying a potential across the chambers, ions pass from one chamber to another resulting in a volume change in the chamber. The solutions are separated from the fluidic channel using a deflectable elastomeric membrane, which can deflect in response to the change in volume of the primary chamber (Das and Payne 2016). An electrochemically actuated microvalve for standalone microfluidics applications was demonstrated by Watanabe et al. (2017). It was composed of a

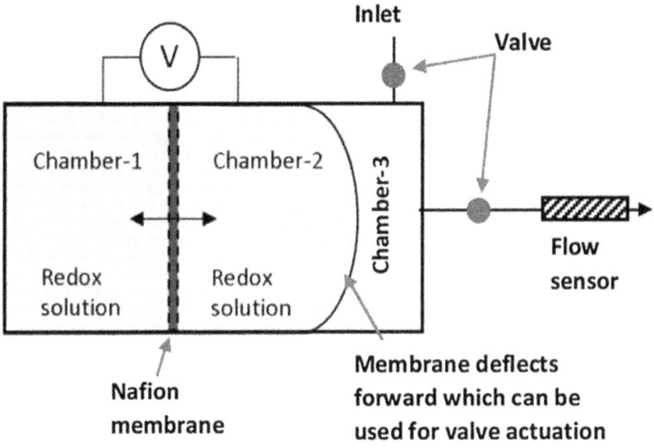

FIGURE 11.5 Working principle of an electrochemical actuation for microvalve applications (Das and Payne 2016).

single bimetallic Zn/Pt electrode, in which the Pt electrode surface was coated with hydrophobic self-assembled monolayers (SAM) and the Zn/Pt electrode was exposed to a control channel. The hydrophobic SAM initially stopped the fluid flow and functioned as the valve. The microvalve was actuated by passing an electrolyte through the control channel, which oxidized the Zn to produce a negative potential. This negative potential removed the SAM from the Pt electrode, thus allowing the fluid to flow.

11.2.2.1.2.2 Phase Change Microvalves In developing phase change microvalves, the phase change material is used either as a meltable plug or as a propellant for the membrane. The phase change nature of hydrogels, sol–gels, and paraffin wax are commonly explored to develop such microvalves.

A stimuli-responsive hydrogel can change its volume reversibly in response to small changes in environmental conditions. Lee et al. developed a thermally responsive hydrogel matrix using N, N-dimethyl acrylamide modified PolyN-isopropyl acrylamide (PNIPAm) hydrogels (Lee et al. 2014). Iron oxide nanoparticles were dispersed within the matrix that converted light energy into thermal energy to control the swelling of the hydrogels for valve action.

In microvalves based on paraffin wax, the low melting point of the wax is utilized to bring about the valve action. The solid-liquid phase transition of wax is coupled to a deformable membrane to achieve the opening and closing of the valve. Liu et al. demonstrated the phase change property of the paraffin wax to bring about the deflection of a PDMS membrane to act as a valve (Liu et al. 2019) (Figure 11.6). Heat energy was supplied to the wax by induction heating using three excitation coils. When a high-frequency alternating current was applied to the coil, the wax melted, resulting in the upward movement of the PDMS membrane to block the microchannel. When the power applied to the excitation coil was switched off, the wax solidified and returned to its initial position.

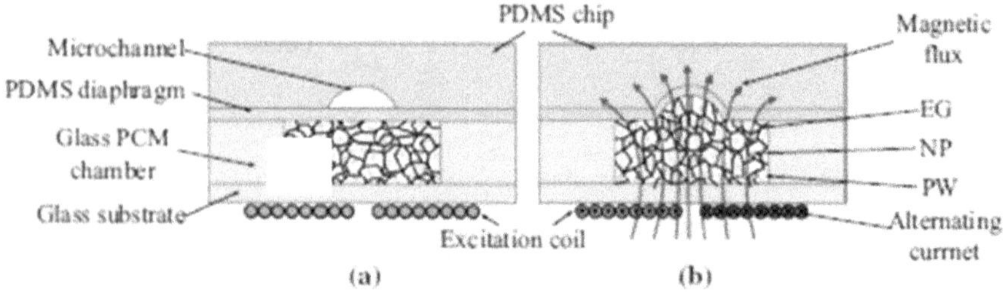

FIGURE 11.6 Phase change microvalves using paraffin wax (Liu et al. 2019).

11.2.2.1.2.3 Rheological Microvalves Rheological microvalves make use of the rheological properties of the fluid to bring about the valve action. Electrorheological fluids are widely used for developing lab-on-a-chip devices. These fluids are a type of colloidal suspension whose viscosity changes in response to an applied electrical field. They exhibit a solid-like behavior in an applied electric field of 1–2 kV/mm. Miyoshi et al. described the development of microvalves based on nematic liquid crystals whose viscosity is a function of the applied electric field (Miyoshi et al. 2016). Ferrofluids employ magnetic particles suspended in a carrier liquid to produce magnetic liquids that can respond to the localized magnetic field, thus providing easy actuation. Vinogradova et al. carried out numerical studies of a ferrofluid-based valve that responds to the magnetic field of an in-built current-carrying wire (Vinogradova et al. 2019).

11.2.2.1.3 External System-Based Microvalves

Actuation in this class of microvalves uses external systems like pneumatic or vacuum sources to control the fluid flow. One of the earliest and most popular pneumatic valves was developed by Quake et al. (Unger et al. 2000, Quake and Scherer 2000). It was developed by multilayer soft lithography with an upper pneumatic channel and a lower fluidic channel perpendicular to each other. When pressurized gas passed through the pneumatic chamber, it deflected downwards, thus pinching the flow in the lower fluidic channel (Figure 11.7). Grover et al. reported a similar microvalve using a vacuum source for lab-on-a-chip applications (Grover et al. 2003). Kim et al. reported programmable microfluidics platforms using pneumatically actuated microvalve circuits for chemical and biochemical analysis (Kim et al. 2016).

11.2.2.2 Passive Microvalves

Passive microvalves do not use external energy sources for actuation and prearranged fluid flow is achieved using microstructures or backpressure. Like active microvalves, they are also classified as mechanical and nonmechanical passive microvalves. The following section briefly explains each of these classes of microvalves.

11.2.2.2.1 Mechanical Passive Microvalves

Mechanical passive microvalves use deflecting membranes in which the actuation of the membrane is brought about by fluid motion. Hence, they are mostly used as

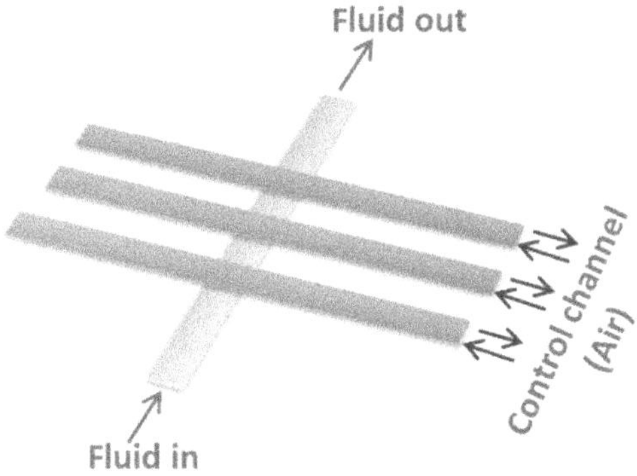

FIGURE 11.7 Illustration of the working of a pneumatic microvalve.

check valves, allowing fluid flow in one direction only. Some of the commonly used moving components in mechanical passive microvalves include membranes (Li and Chen 2007), flaps (Yoon 2014), and balls (Zhang et al. 2019). Bien et al. reported the fabrication of micromachined polycrystalline silicon membrane for unidirectional flow control in the microfluidic environment (Bien et al. 2003). It consisted of a 400 × 400 μm movable silicon membrane whose position determined the valve action.

11.2.2.2.2 Nonmechanical Passive Microvalves

In this class of microvalves, there are no moving components for valve actuation. Instead, the valve action is achieved using capillary forces or a nozzle-diffuser arrangement. Different capillary forces like electrocapillary, thermocapillary, or passive capillarity have also been used to develop microvalves. Zimmermann et al. demonstrated the application of valves for autonomous capillary systems by employing an abruptly changing geometry to control the flow rate (Zimmermann et al. 2008) (Figure 11.8).

11.2.3 Applications of Microvalves

Microvalves are an indispensable component of any standalone microfluidics device. They have a widespread application in controlling fluid flow in different domains, especially in chemistry and life sciences. Leak-proof microvalves capable of withstanding high input pressures are crucial in developing multiplexed biochemical assays. Zhu et al. demonstrated the application of a solenoid valve and a pneumatic valve for DNA extraction and PCR for genotyping human papillomavirus (HPV) in clinical samples (Zhu et al. 2019). Pneumatic microvalves have been used in synthetic biology for the automated design, construction, testing, and analysis of DNA by Linshiz et al. (2016). Hong et al. demonstrated the application of pneumatic microvalves for the automated staining of tumor cells (Hong et al. 2019). Hulme et al. demonstrated the application of prefabricated screws and pneumatic and solenoid valves to generate gradients within a microfluidics device (Hulme et al. 2009).

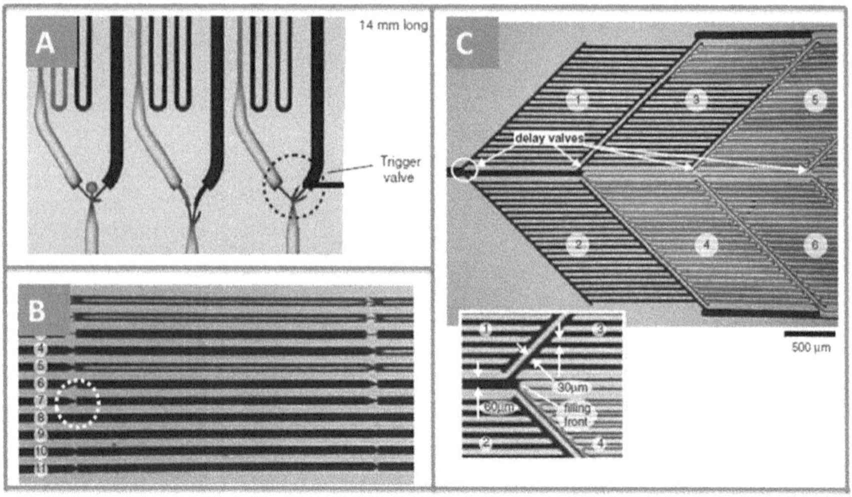

FIGURE 11.8 Illustration of the abrupt change in geometry for valve action. (A) Trigger valves, (B) stop valves, and (C) delay valves for capillary systems (Zimmermann et al. 2008).

Membrane microvalves for enzymatic immunoassays have been demonstrated by Ren et al. (2012). Microvalves also find numerous applications in sample preparations, synthesis of nanoparticles, and drug delivery applications.

11.3 CONCLUSION

Microfluidics technology enables the development of several miniaturized components capable of replicating laboratory functions on a microscale. This chapter provides an insight into the different micromixers and microvalves for lab-on-a-chip applications. With the recent advances in microfabrication techniques, low-cost micromixers and microvalves that can be easily integrated with other microfluidic components can be developed. This paves the way for developing an ideal PoCT device that is automated and has minimal user intervention capable of providing instantaneous results.

REFERENCES

Afzal, A., and K.Y. Kim. 2012. Passive Split and Recombination Micromixer with Convergent-Divergent Walls. *Chemical Engineering Journal*, 203: 182–192. http://dx.doi.org/10.1016/j.cej.2012.06.111.

An Le, N.H., H. Deng, C. Devendran, N. Akhtar, X. Ma, C. Pouton, H.K. Chan, A. Neild, and T. Alan. 2020. Ultrafast Star-Shaped Acoustic Micromixer for High Throughput Nanoparticle Synthesis. *Lab on a Chip*, 20: 582–591, https://doi.org/10.1039/c9lc01174a.

Anjewierden, D., G.A. Liddiard, and B.K. Gale. 2012. An Electrostatic Microvalve for Pneumatic Control of Microfluidic Systems. *Journal of Micromechanics and Microengineering*, 22(2): 025019. https://doi.org/10.1088/0960-1317/22/2/025019.

Balasubramaniam, L., R. Arayanarakool, S.D. Marshall, B. Li, P.S. Lee, and P.C.Y. Chen. 2017. Impact of Cross-Sectional Geometry on Mixing Performance of Spiral Microfluidic Channels Characterized by Swirling Strength of Dean-Vortices. *Journal of Micromechanics and Microengineering*, 27(9): 095016. https://doi.org/10.1088/1361-6439/aa7fc8.

Baumgard, J., A.M. Vogt, U. Kragl, K. Jähnisch, and N. Steinfeldt. 2013. Application of Microstructured Devices for Continuous Synthesis of Tailored Platinum Nanoparticles. *Chemical Engineering Journal*, 227: 137–144. https://doi.org/10.1016/j.cej.2012.08.066.

Bernassau, A.L., P. Glynne-Jones, F. Gesellchen, M. Riehle, M. Hill, and D.R.S. Cumming. 2014. Controlling Acoustic Streaming in an Ultrasonic Heptagonal Tweezers with Application to Cell Manipulation. *Ultrasonics,* 54: 268-–74, https://doi.org/10.1016/j.ultras.2013.04.019.

Bessoth, F.G., A.J. DeMello, and A. Manz. 1999. Microstructure for Efficient Continuous Flow Mixing. *Analytical Communications,* 36(6): 213–215. https://doi.org/10.1039/a902237f.

Bhagat, A.A.S., and I. Papautsky. 2008. Enhancing Particle Dispersion in a Passive Planar Micromixer Using Rectangular Obstacles. *Journal of Micromechanics and Microengineering* 18, 18(8). https://doi.org/10.1088/0960-1317/18/8/085005.

Bien, D.C.S., S.J.N. Mitchell, and H.S. Gamble. 2003. Fabrication and Characterization of a Micromachined Passive Valve. *Journal of Micromechanics and Microengineering*, 13(5): 557–562. https://doi.org/10.1088/0960-1317/13/5/305.

Cao, Q., X. Han, and L. Li. 2015. An Active Microfluidic Mixer Utilizing a Hybrid Gradient Magnetic Field. *International Journal of Applied Electromagnetics and Mechanics*, 47(3): 583–592. https://doi.org/10.3233/JAE-140057.

Chen, S., S. Lu, Y. Liu, J. Wang, X. Tian, G. Liu, and Z. Yang. 2016. A Normally-Closed Piezoelectric Micro-Valve with Flexible Stopper. *AIP Advances*, 6(4). https://doi.org/10.1063/1.4947301.

Cheng, K.C., R.C. Lin, and J.W. Ou. 1976. Fully Developed Laminar Flow in Curved Rectangular Channels. *Journal of Fluids Engineering, Transactions of the ASME*, 98(1): 41–48. https://doi.org/10.1115/1.3448205.

Chung, C.K., C.Y. Wu, T.R. Shih, C.F. Wu, and B.H. Wu. 2006. Design and Simulation of a Novel Micro-Mixer with Baffles and Side-Wall Injection into the Main Channel. In Proceedings of 1st IEEE International Conference on Nano Micro Engineered and Molecular Systems, 1st IEEE-NEMS, Zhuhai, China, 721–724, https://doi.org/10.1109/NEMS.2006.334881.

Collins, D.J., Z. Ma, and Y. Ai. 2016. Highly Localized Acoustic Streaming and Size-Selective Submicrometer Particle Concentration Using High Frequency Microscale Focused Acoustic Fields. *Analytical Chemistry*, 88(10): 5513–5522. https://doi.org/10.1021/acs.analchem.6b01069.

Das, C., and F. Payne. 2016. Design and Characterization of Low Power, Low Dead Volume Electrochemically-Driven Microvalve. *Sensors and Actuators A: Physical*, 241: 104–112. https://doi.org/10.1016/j.sna.2016.01.038.

Elizabeth Hulme, S., S.S. Shevkoplyas, and G.M. Whitesides. 2009. Incorporation of Prefabricated Screw, Pneumatic, and Solenoid Valves into Microfluidic Devices. *Lab on a Chip*, 9(1): 79–86. https://doi.org/10.1039/b809673b.

Fu, L.M., R.J. Yang, C.H. Lin, and Y.S. Chien. 2005. A Novel Microfluidic Mixer Utilizing Electrokinetic Driving Forces under Low Switching Frequency. *Electrophoresis*, 26(9): 1814–1824. https://doi.org/10.1002/elps.200410222.

Fu, L.M., W.J. Ju, C.H. Tsai, H.H. Hou, R.J. Yang, and Y.N. Wang. 2013. Chaotic Vortex Micromixer Utilizing Gas Pressure Driving Force. *Chemical Engineering Journal*, 214: 1–7. https://doi.org/10.1016/j.cej.2012.10.032.

Ghadimi, M., A. Shokuhfar, A. Zolriasatein, and H.R. Rostami. 2013. Morphological and Structural Evaluation of Nanocrystalline NiTiCu Shape Memory Alloy Prepared via Mechanical Alloying and Annealing. *Materials Letters*, 90: 30–33. https://doi.org/10.1016/j.matlet.2012.09.008.

Gholizadeh, A., and M. Javanmard. 2016. Magnetically Actuated Microfluidic Transistors: Miniaturized Micro-Valves Using Magnetorheological Fluids Integrated with Elastomeric Membranes. *Journal of Microelectromechanical Systems*, 25(5): 922–928. https://doi.org/10.1109/JMEMS.2016.2586420.

Grover, W.H., A.M. Skelley, C.N. Liu, E.T. Lagally, and R.A. Mathies. 2003. Monolithic Membrane Valves and Diaphragm Pumps for Practical Large-Scale Integration into Glass Microfluidic Devices. *Sensors and Actuators B: Chemical*, 89(3): 315–323. https://doi.org/10.1016/S0925-4005(02)00468-9.

Heshmatnezhad, F., H. Aghaei, and A.R. Solaimany Nazar. 2017. Parametric Study of Obstacle Geometry Effect on Mixing Performance in a Convergent-Divergent Micromixer with Sinusoidal Walls. *Chemical Product and Process Modeling*, 12(1). https://doi.org/10.1515/cppm-2016-0025.

Hong, S.L., M. Tang, Z. Chen, Z. Ai, F. Liu, S. Wang, N. Zhang, and K. Liu. 2019. High-Performance Multiplex Microvalves Fabrication and Using for Tumor Cells Staining on a Microfluidic Chip. *Biomedical Microdevices*, 21(4). https://doi.org/10.1007/s10544-019-0434-5.

Hossain, S., and K.Y. Kim. 2014. Mixing Analysis of Passive Micromixer with Unbalanced Three-Split Rhombic Sub-Channels. *Micromachines*, 5(4): 913–928. https://doi.org/10.3390/mi5040913.

Jang, L.S., S.H. Chao, M.R. Holl, and D.R. Meldrum. 2007. Resonant Mode-Hopping Micromixing. *Sensors and Actuators A: Physical*, 138(1): 179–186. https://doi.org/10.1016/j.sna.2007.04.052.

Jiang, F., S. Yin, L. Zhang, J. Peng, S. Ju, J.D. Miller, and X. Wang. 2018. Solvent Extraction of Cu(II) from Sulfate Solutions Containing Zn(II) and Fe(III) Using an Interdigital Micromixer. *Hydrometallurgy*, 177: 116–122, https://doi.org/10.1016/j.hydromet.2018.03.004.

Kastania, A.S., K. Tsougeni, G. Papadakis, E. Gizeli, G. Kokkoris, A. Tserepi, and E. Gogolides. 2016. Plasma Micro-Nanotextured Polymeric Micromixer for DNA Purification with High Efficiency and Dynamic Range. *Analytica Chimica Acta*, 942: 58–67. https://doi.org/10.1016/j.aca.2016.09.007.

Kawasaki, S.I., Y. Xiuyi, K. Sue, Y. Hakuta, A. Suzuki, and K. Arai. 2009. Continuous Supercritical Hydrothermal Synthesis of Controlled Size and Highly Crystalline Anatase TiO2 Nanoparticles. *Journal of Supercritical Fluids*, 54(1): 96–102, https://doi.org/10.1016/j.supflu.2010.03.001.

Kawasaki, S.I., K. Sue, R. Ookawara, Y. Wakashima, A. Suzuki, Y. Hakuta, and K. Arai. 2010. Engineering Study of Continuous Supercritical Hydrothermal Method Using a T-Shaped Mixer: Experimental Synthesis of NiO Nanoparticles and CFD Simulation. *Journal of Supercritical Fluids*, 50(3), 276–282. https://doi.org/10.1016/j.supflu.2009.06.009.

Kim, Jeongyun, D. Taylor, N. Agrawal, H. Wang, H. Kim, A. Han, K. Rege, and A. Jayaraman. 2012. A Programmable Microfluidic Cell Array for Combinatorial Drug Screening. *Lab on a Chip*, 12(10): 1813–1822. https://doi.org/10.1039/c2lc21202a.

Kim, Jungkyu, A.M. Stockton, E.C. Jensen, and R.A. Mathies. 2016. Pneumatically Actuated Microvalve Circuits for Programmable Automation of Chemical and Biochemical Analysis. *Lab on a Chip*, 16(5): 812–819. https://doi.org/10.1039/c5lc01397f.

Kirby, B.J., T.J. Shepodd, and E.F. Hasselbrink. 2002. Voltage-Addressable on/off Microvalves for High-Pressure Microchip Separations. In *Journal of Chromatography A*, 979: 147–154. https://doi.org/10.1016/S0021-9673(02)01453-X.

Ko, Y.J., J.H. Maeng, Y. Ahn, and S.Y. Hwang. 2011. DNA Ligation Using a Disposable Microfluidic Device Combined with a Micromixer and Microchannel Reactor. *Sensors and Actuators B: Chemical*, 157(2): 735–741. https://doi.org/10.1016/j.snb.2011.05.016.

Kohl, M., K.D. Skrobanek, and S. Miyazaki. 1999. Development of Stress-Optimised Shape Memory Microvalves. *Sensors and Actuators A: Physical*, 72(3): 243–250. https://doi.org/10.1016/S0924-4247(98)00221-0.

Lee, E., H. Lee, S. Il Yoo, and J. Yoon. 2014. Photothermally Triggered Fast Responding Hydrogels Incorporating a Hydrophobic Moiety for Light-Controlled Microvalves. *ACS Applied Materials and Interfaces* 6(19): 16949–16955. https://doi.org/10.1021/am504502y.

Lee, K.H., G. Yang, B.E. Wyslouzil, and J.O. Winter. 2019. Electrohydrodynamic Mixing-Mediated Nanoprecipitation for Polymer Nanoparticle Synthesis. *ACS Applied Polymer Materials* 1(4): 691–700. https://doi.org/10.1021/acsapm.8b00206.

Lee, Y.K., J. Deval, P. Tabeling, and C.M. Ho. 2001. Chaotic Mixing in Electrokinetically and Pressure Driven Micro Flows. In *Proceedings of the IEEE Micro Electro Mechanical Systems (MEMS)*, 483–486. https://doi.org/10.1007/978-3-642-56763-6_20.

Li, B., and Q. Chen. 2007. Development of Robust Microvavles for Compact Robust Pumps/Hydraulic Actuators. In *Solid State Phenomena*. Trans Tech Publications Ltd, 121–123, 1207–1210. https://doi.org/10.4028/www.scientific.net/ssp.121-123.1207.

Li, M., and D. Li. 2018. Microvalve Using Electrokinetic Motion of Electrically Induced Janus Droplet. *Analytica Chimica Acta*, 1021: 85–94. https://doi.org/10.1016/j.aca.2018.03.001.

Li, S., S. Meierott, and J.M. Köhler. 2010. Effect of Water Content on Growth and Optical Properties of ZnO Nanoparticles Generated in Binary Solvent Mixtures by Micro-Continuous Flow Synthesis. *Chemical Engineering Journal*, 165(3): 958–965. https://doi.org/10.1016/j.cej.2010.08.033.

Lin, H.I., C.C. Wu, C.H. Yang, K.W. Chang, G. Bin Lee, and S.C. Shiesh. 2015. Selection of Aptamers Specific for Glycated Hemoglobin and Total Hemoglobin Using On-Chip SELEX. *Lab on a Chip*, 15(2): 486–494. https://doi.org/10.1039/c4lc01124d.

Linshiz, G., E. Jensen, N. Stawski, C. Bi, N. Elsbree, H. Jiao, J. Kim, R. Mathies, J.D. Keasling, and N.J. Hillson. 2016. End-to-End Automated Microfluidic Platform for Synthetic Biology: From Design to Functional Analysis. *Journal of Biological Engineering*, 10(1). https://doi.org/10.1186/s13036-016-0024-5.

Liu, B., J. Yang, J. Yang, D. Li, G. Gao, and Y. Wang. 2019. A Thermally Actuated Microvalve Using Paraffin Composite by Induction Heating. *Microsystem Technologies*, 25(10): 3969–3975. https://doi.org/10.1007/s00542-019-04373-8.

Mae, K., T. Maki, I. Hasegawa, U. Eto, Y. Mizutani, and N. Honda. 2004. Development of a New Micromixer Based on Split/Recombination for Mass Production and Its Application to Soap Free Emulsifier. *Chemical Engineering Journal*, 101(1–3): 31–38. https://doi.org/10.1016/j.cej.2003.10.011.

Megnin, C., and M. Kohl. 2014. Shape Memory Alloy Microvalves for a Fluidic Control System. *Journal of Micromechanics and Microengineering*, 24(2). https://doi.org/10.1088/0960-1317/24/2/025001.

Mehrdel, P., S. Karimi, J. Farré-Lladós, and J. Casals-Terré. 2018. Novel Variable Radius Spiral-Shaped Micromixer: From Numerical Analysis to Experimental Validation. *Micromachines*, 9(11): https://doi.org/10.3390/mi9110552.

Meng, J., S. Li, J. Li, C. Yu, C. Wei, and S. Dai. 2018. AC Electrothermal Mixing for High Conductive Biofluids by Arc-Electrodes. *Journal of Micromechanics and Microengineering*, 28(6), https://doi.org/10.1088/1361-6439/aab39b .

Mensing, G.A., T.M. Pearce, M.D. Graham, and D.J. Beebe. 2004. An Externally Driven Magnetic Microstirrer. *Philosophical Transactions of the Royal Society A: Mathematical, Physical and Engineering Sciences*, 362(1818): 1059–1068. https://doi.org/10.1098/rsta.2003.1362.

Miyoshi, T., K. Yoshida, J.W. Kim, S.I. Eom, and S. Yokota. 2016. An MEMS-Based Multiple Electro-Rheological Bending Actuator System with an Alternating Pressure Source. *Sensors and Actuators, A: Physical*, 245: 68–75. https://doi.org/10.1016/j.sna.2016.04.041.

Nagaki, A., Y. Takahashi, K. Akahori, and J.I. Yoshida. 2012. Living Anionic Polymerization of Tert-Butyl Acrylate in a Flow Microreactor System and Its Applications to the Synthesis of Block Copolymers. *Macromolecular Reaction Engineering*, 6(11): 467–472. https://doi.org/10.1002/mren.201200051.

Nguyen, N.T., and S. Wereley. 2006. *Fundamentals and Applications of Microfluidics*. Artech House.

Ould El Moctar, A., N. Aubry, and J. Batton. 2003. Electro-Hydrodynamic Micro-Fluidic Mixer. *Lab on a Chip*, 3(4): 273–280. https://doi.org/10.1039/b306868b.

Petkovic, K., G. Metcalfe, H. Chen, Y. Gao, M. Best, D. Lester, and Y. Zhu. 2017. Rapid Detection of Hendra Virus Antibodies: An Integrated Device with Nanoparticle Assay and Chaotic Micromixing. *Lab on a Chip*, 17(1): 169–177. https://doi.org/10.1039/c6lc01263a.

Pradeep, A., J. Raveendran, T. Ramachandran, B.G. Nair, and S.B. Satheesh. 2016. Computational Simulation and Fabrication of Smooth Edged Passive Micromixers with Alternately Varying Diameter for Efficient Mixing. *Microelectronic Engineering*, 165: 32–40. http://dx.doi.org/10.1016/j.mee.2016.08.009.

Pradeep, A., S. Vineeth Raj, J. Stanley, B.G. Nair, and T.G.S. Babu. 2018. Automated and Programmable Electromagnetically Actuated Valves for Microfluidic Applications. *Sensors and Actuators, A: Physical*, 283: 79–86. https://doi.org/10.1016/j.sna.2018.09.024.

Qiu, L., K. Wang, S. Zhu, Y. Lu, and G. Luo. 2016. Kinetics Study of Acrylic Acid Polymerization with a Microreactor Platform. *Chemical Engineering Journal*, 284: 233–239. https://doi.org/10.1016/j.cej.2015.08.055.

Quake, S.R., and A. Scherer. 2000. From Micro- to Nanofabrication with Soft Materials. *Science*, 290: 1536–1540, https://doi.org/10.1126/science.290.5496.1536.

Rasouli, M.R., and M. Tabrizian. 2019. An Ultra-Rapid Acoustic Micromixer for Synthesis of Organic Nanoparticles. *Lab on a Chip*, 19(19): 3316–3325. https://doi.org/10.1039/c9lc00637k.

Ren, L., J.C. Wang, W. Liu, Q. Tu, R. Liu, X. Wang, J. Xu, et al. 2012. An Enzymatic Immunoassay Microfluidics Integrated with Membrane Valves for Microsphere Retention and Reagent Mixing. *Biosensors and Bioelectronics*, 35(1): 147–154. https://doi.org/10.1016/j.bios.2012.02.034.

Rich, C.A., and K.D. Wise. 2003. A High-Flow Thermopneumatic Microvalve with Improved Efficiency and Integrated State Sensing. *Journal of Microelectromechanical Systems*, 12(2): 201–208. https://doi.org/10.1109/JMEMS.2002.808459.

Rife, J.C., M.I. Bell, J.S. Horwitz, M.N. Kabler, R.C.Y. Auyeung, and W.J. Kim. 2000. Miniature Valveless Ultrasonic Pumps and Mixers. *Sensors and Actuators, A: Physical*, 86(1–2): 135–140. https://doi.org/10.1016/S0924-4247(00)00433-7.

Ryu, M., J.A. Kimber, T. Sato, R. Nakatani, T. Hayakawa, M. Romano, C. Pradere, A.A. Hovhannisyan, S.G. Kazarian, and J. Morikawa. 2017. Infrared Thermo-Spectroscopic Imaging of Styrene Radical Polymerization in Microfluidics. *Chemical Engineering Journal*, 324: 259–265. https://doi.org/10.1016/j.cej.2017.05.001.

Singh, A., L. Hirsinger, P. Delobelle, and C. Khan-Malek. 2014. Rapid Prototyping of Magnetic Valve Based on Nanocomposite Co/PDMS Membrane. *Microsystem Technologies*, 20(3), 427–436. https://doi.org/10.1007/s00542-013-1972-z.

Sprogies, T., J.M. Köhler, and G.A. Groß. 2008. Evaluation of Static Micromixers for Flow-through Extraction by Emulsification. *Chemical Engineering Journal*, 135(SUPPL. 1). https://doi.org/10.1016/j.cej.2007.07.032.

Stroock, A.D., S.K.W. Dertinger, A. Ajdari, I. Mezić, H.A. Stone, and G.M. Whitesides. 2002. Chaotic Mixer for Microchannels. *Science*, 295(5555): 647–651. https://doi.org/10.1126/science.1066238.

Tan, H. 2019. Numerical Study of a Bubble Driven Micromixer Based on Thermal Inkjet Technology. *Physics of Fluids*, 31(6). https://doi.org/10.1063/1.5098449.

Terry, S.C., J.H. Herman, and J.B. Angell. 1979. A Gas Chromatographic Air Analyzer Fabricated on a Silicon Wafer. *IEEE Transactions on Electron Devices*, 26(12): 1880–1886. https://doi.org/10.1109/T-ED.1979.19791.

Tian, Z.H., Y.J. Wang, J.H. Xu, and G.S. Luo. 2016. Intensification of Nucleation Stage for Synthesizing High Quality CdSe Quantum Dots by Using Preheated Precursors in Microfluidic Devices. *Chemical Engineering Journal*, 302: 498–502. https://doi.org/10.1016/j.cej.2016.05.070.

Unger, M.A., H.P. Chou, T. Thorsen, A. Scherer, and S.R. Quake. 2000. Monolithic Microfabricated Valves and Pumps by Multilayer Soft Lithography. *Science*, 288(5463): 113–116. https://doi.org/10.1126/science.288.5463.113.

Veldurthi, N., S. Chandel, T. Bhave, and D. Bodas. 2015. Computational Fluid Dynamic Analysis of Poly(Dimethyl Siloxane) Magnetic Actuator Based Micromixer. *Sensors and Actuators, B: Chemical*, 212: 419–424. https://doi.org/10.1016/j.snb.2015.02.048.

Vinogradova, A.S., V.A. Turkov, and V.A. Naletova. 2019. Modeling of Ferrofluid-Based Microvalves in the Magnetic Field Created by a Current-Carrying Wire. *Journal of Magnetism and Magnetic Materials*, 470: 18–21. https://doi.org/10.1016/j.jmmm.2017.11.090.

Watanabe, T., G.C. Biswas, E.T. Carlen, and H. Suzuki. 2017. An Autonomous Electrochemically-Actuated Microvalve for Controlled Transport in Stand-Alone Microfluidic Systems. *RSC Advances*, 7(62): 39018–39023. https://doi.org/10.1039/c7ra07335f.

Yoon, G.H. 2014. Compliant Topology Optimization for Planar Passive Flap Micro Valve. *Journal of Nanoscience and Nanotechnology*, 14(10): 7585–7591. https://doi.org/10.1166/jnn.2014.9552.

Zahn, J.D., and V. Reddy. 2006. Two Phase Micromixing and Analysis Using Electrohydrodynamic Instabilities. *Microfluidics and Nanofluidics*, 2(5): 399–415. https://doi.org/10.1007/s10404-006-0082-y.

Zahra, A., R. Scipinotti, D. Caputo, A. Nascetti, and G. De Cesare. 2015. Design and Fabrication of Microfluidics System Integrated with Temperature Actuated Microvalve. *Sensors and Actuators, A: Physical*, 236: 206–213. https://doi.org/10.1016/j.sna.2015.10.050.

Zhang, A.L., Y.W. Cai, Y. Xu, and W.Y. Liu. 2019. A New Ball Check Valve Based on Surface Acoustic Wave. *Ferroelectrics*, 540(1): 138–144. https://doi.org/10.1080/00150193.2019.1611114.

Zhu, C., A. Hu, J. Cui, K. Yang, X. Zhu, Y. Liu, G. Deng, and L. Zhu. 2019. A Lab-on-a-Chip Device Integrated DNA Extraction and Solid Phase PCR Array for the Genotyping of High-Risk HPV in Clinical Samples. *Micromachines*, 10(8). https://doi.org/10.3390/mi10080537.

Zimmermann, M., P. Hunziker, and E. Delamarche. 2008. Valves for Autonomous Capillary Systems. *Microfluidics and Nanofluidics*, 5(3): 395–402. https://doi.org/10.1007/s10404-007-0256-2.

12 Microfluidic Contact Lenses for Ocular Diagnostics

Antonysamy Dennyson Savariraj, Ammar Ahmed Khan, Mohamed Elsherif, Fahad Alam, Bader AlQattan, Aysha. A. S. J. Alghailani, Ali K. Yetisen, and Haider Butt

CONTENTS

12.1 INTRODUCTION

In recent times, public healthcare has become an important aspect of life, and quality of life demands healthy living, with early diagnosis, continuous monitoring, and rapid treatment of diseases. Continuous monitoring of diagnostic parameters is necessary to detect and treat several health disorders (Akram et al. 2015) which is often unmet by the conventional techniques. With the advancement of prototype devices

DOI: 10.1201/9781003033479-12

and rapid diagnostics, real-time monitoring of the diseases has been made possible. Until recently, contact lenses (CLs) were considered to be polymer-based gadgets worn for vision correction and were extended to meet stringent clinical requirements. The progress in CLs as a diagnostic tool led to improved health care by meeting the pharmaceutical and extending the medical prognosis. CLs stand out as a sophisticated platform that offer non/minimally invasive diagnostics and drug delivery (Farandos et al. 2015). The complications associated with severe health disorders such as diabetes and intraocular pressure (IOP) demand close monitoring and strict control of blood glucose concentration and IOP (Nathan et al. 1993). Inspired by a heads-up display, the incorporation of biosensors and semi-transparent displays in CLs for different functions promises ocular diagnostics (Badawy et al. 2018, Elsherif et al. 2021, Salih et al. 2020, Salih et al. 2021, Yetisen et al. 2020) and drug delivery (Ho et al. 2008).

CL-based embedded sensors for clinical purposes should be portable, compact, and discrete. Furthermore, CL-based devices should also have high oxygen and water permeability, in addition to transparency and flexibility to be incorporated in to CLs (Chen et al. 2013, Rim et al. 2015, Salvatore et al. 2014, Yao et al. 2011). Though hard lenses or rigid gas permeable (RGP) lenses are known for maintaining their shape on the eye and for high oxygen permeability, their susceptibility to scratches and displacement from the center of the eye and prolonged adoption time make soft and microfluidic contact lenses (MCLs) a better choice. On the contrary, MCLs with pronounced flexibility are more comfortable for the wearers than hard lenses, easy to adjust, and their capacity to take up water by soaking helps in enhancing oxygen to flow to the cornea. Microfluidics technology ensures control of small fluidic volumes up to picoliter level extending its application in medical diagnostics (Araci et al. 2014, Unger et al. 2000). Moreover, microfluidic materials facilitate the incorporation of electrically conductive elements for potential application in wearable technologies for healthcare, such as colorimetric sensing of sweat (Koh et al. 2016) to monitor physiochemical tissue properties (Mostafalu et al. 2016), tactile Sensing (Nie et al. 2014), etc. When CLs are fabricated out of such microfluidic materials like soft polymers, these lenses surpass the existing rigid/hard lenses in terms of flexibility and stretchability to have promising healthcare application in ophthalmology (An et al. 2018, Jiang et al. 2018).

Therefore, beyond vision correction, MCLs employed in ocular diagnostics can have added advantages for continuous monitoring of IOP (An et al. 2019, Araci et al. 2014) and glucose level in tear fluids (Elsherif et al. 2018b, Iguchi et al. 2007, Kim et al. 2017, Yang et al. 2020). Hence, regular monitoring of IOP can detail the cause for elevated IOP (Tseng et al. 2012). This chapter provides a detailed report on MCL sensors, device fabrication, and their use for continuous ocular diagnostics particularly estimating the concentration of tear glucose (Elsherif et al. 2018b, Elsherif et al. 2019), proteins, nitrite ion, chlorine ion, urea (Yang et al. 2020, Yetisen et al. 2020), IOP, pH (Riaz et al. 2019), and corneal temperature (Moreddu et al. 2019), and the challenges involved and suggestions for the further development of the devices the future.

12.2 SIGNIFICANCE OF MICROFLUIDIC CONTACT LENSES FOR OCULAR DIAGNOSTICS

MCLs are a subclass of soft CLs fabricated using hydrogels by incorporating microfluidics capabilities such as microcavities and microchannels (Jiang et al. 2018, Yang et al. 2020, Yetisen et al. 2019). They have the advantage of continuous fluid analysis (Karle et al. 2016, Yetisen et al. 2020), an essential element for accurate tear fluid collection, sample preparation, and to carry out in vivo and/or in vitro analysis. Several conventional methods including polymer casting (Becker and Gärtner 2008), thermoforming (An et al. 2018), microlithography (Jiang et al. 2018), injection molding (Mair et al. 2006), and imprinting (Schirhagl et al. 2012) have been adopted to develop microfluidics devices of desired parameters and geometries using templates. Irrespective of the beneficial outcomes, these techniques involve a tedious and time-consuming procedure besides taking in to account the cost factor involved. The shortcomings experienced using such traditional methods have been overcome, by coupling laser patterning and embedded templating to fabricate MCLs (Alqurashi et al. 2018). With tailor-made properties, MCLs complement soft CLs in terms of offering POC diagnostics and controlled drug release (Jiang et al. 2018).

Portable and self-contained point-of-care (POC) diagnostics were made possible by using miniature devices for biomedical analysis (Mair et al. 2006). In the context the evolution of wearable CLs turns out to be as minimally invasive diagnostic tool providing continuous measurements of physiological parameters. Flexible microfluidic sensors have already been demonstrated for tactile sensors (Kenry et al. 2016, Yeo et al. 2016b). Although the CL-based sensors use the advancements in biomaterials electronics and microfabrication, proving to be an efficient platform to investigate the biomarkers of interest in tear fluids, the sensor components increase the stiffness of the CLs and lack direct detection of biomarkers and tear fluid storage (Yang et al. 2020). Therefore, sensors capable of monitoring glucose and IOP without involving electronic components are quintessential since they are easy to interface with the cornea and the external environment as well (Agaoglu et al. 2018). Microfluidics technology offers soft and stretchable materials as a suitable platform for sensing and drug delivery applications. With the advancement in microfluidics fabrication, incorporation of sensing elements and/or liquid manipulation can be adopted to carry out several biosensing applications to offer complete sample-to-answer solutions. Microfluidics-based sensing devices use microstructures for storage and liquid handling which provides sensing accuracy and reliability, because precise amounts of liquids are used for manipulation. Since microstructured canals and cavities are used, only a small volume of tear fluid is required for sensing, benefiting the wearer because tear fluid secretion is limited in volume. The microstructures serve as physical containment for solutes storage and conduits for electronics, providing electrical leak-proof liquid storage in flexible substrates. Moreover, these devices make use of a lower volume of reagents and the analyzing process is rapid since the samples travel a very short distance to traverse (Mair et al. 2006). The advantage with MCLs is the microstructures/microchannels that are able to accurately store, handle, and dispense the liquids (Jiang et al. 2018, Whitesides 2006, Yeo et al. 2016a, Yetisen et al. 2019). Using a conductive liquid in soft elastomeric substrates holds promising application since the liquids have intrinsic mechanical deformability capable of withstanding a high degree of deformation which is advantageous over conventional solid-state devices provided the conducting liquid has sufficient physicochemical stability (Ota et al. 2014, Yeo et al. 2016b). These merits in MCL-based sensors offer pronounced flexibility and sensing accuracy and reliability to facilitate continuous healthcare monitoring applications. Electronics functionality and microstructure design are two key aspects in microfluidics-based sensing (Yeo et al. 2016a) and the merits of the devices are based on the material properties of liquids (An et al. 2018).

12.3 FIVE METHODS OF MANUFACTURING MICROFLUIDIC CONTACT LENSES

In developing MCLs, fabrication and usability tests are two important aspects involved. MCL fabrication can be carried out in different ways (An et al. 2018, Jiang et al. 2018). Since the conventional micro-fabrication process to obtain microfluidics devices is a planar process, employing it to fabricate 3D spherical devices to fit the surface of a cornea is very hard (Alam et al. 2020, 2021). Therefore, accurate calculations along with suitable processes are required for the fabrication of MCLs. There are several methods of fabricating MCLs.

12.3.1 Thermoforming

This method involves heat-assisted formation of irreversible bonding to fabricate MLCs. Polydimethylsiloxane (PDMS), being biocompatible, highly oxygen-permeable, soft, and readily converted to be hydrophilic (Leonardi et al. 2009, Yeo et al. 2016a) makes it a fitting material but its thermosetting nature poses practical difficulties because they cannot be softened after curing and lack stability beyond 200°C (Armani amd Liu 2000, Blaga 1978). So, polyethylene terephthalate (PET) – polydimethylsiloxane (PDMS) – assemblies were made and thermoformed to make a PDMS side where the channels can be integrated due to its soft nature. The bonding methods, surface modifications (Eddings et al. 2008), and chemical coupling treatments (Vlachopoulou et al. 2008) have to be engineered carefully, to ensure higher bond strength between PET-PDMS to retain intactness even after thermoforming.

As shown in Figure 12.1(A–C) using photolithography and replica molding techniques a series of microchannels (annularly/linearly distributed) were fabricated using PDMS. The microchannels fabricated with PDMS were used as substrates over which a PET cover plate has to be bonded after chemically modifying it. Firstly, PET was subjected to activation in an oxygen plasma chamber to bring hydrophilization on its surface (Figure 12.1(D–F)). Then PET membranes were separately surface-functionalized with the help of 3-aminopropyltriethoxysilane (APTES) to form silylation to form a bond with PDMS. The functionalized PET and PDMS substrates were kept in an oxygen plasma chamber to make a Si-OH linkage on the couple followed by bringing them into contact and heat-treated to establish irreversible bond (Figure 12.1(G)). Later the PET-PDMS couple was heat-treated on hot plate

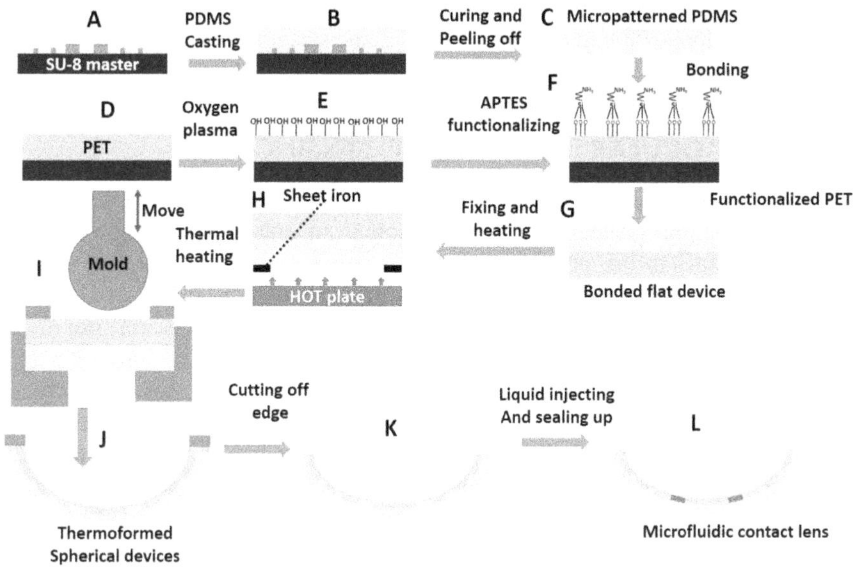

FIGURE 12.1 Schematic illustration of the fabrication of MCLs by thermoforming method. (A–C) The schematic of process flow for manufacturing micropatterned soft elastomer. (D–F) The schematic of the process flow for PET treatment. (G) Picture of bonded PET-PDMS assembly. (H–J) Schematic illustration of thermoforming progress to press the plate into the spherical cap. (K–L) Schematic illustration of liquid injecting.

to achieve soft form of PET and molded into spherical-shaped CL with the help of a concave die and mirrored spherical terrace (Figure 12.1(H–L)). The effect of thermoforming processes and the blockage and leak tests were conducted to evaluate the intactness of the fabricated lenses (An et al. 2018, 2019).

12.3.2 Microlithography

In this method microconcavities and microchannels were created in hydrogels using laser patterning and embedded templating respectively (Alqurashi et al. 2018). Fabrication of MCLs with embedded microchannels connected to microconcavities by this method is rapid (10 min) and easy compared to a conventional photolithography method (Isiksacan et al. 2016, Yetisen et al. 2019). Laser ablation used for inscribing 1D and 2D nanopatterns (AlQattan et al. 2018) is adopted to create microstructures on the polymer if the intensity of the laser beam is greater than the binding energy of the polymer and CO_2 laser-assisted fabrication used here provides precise control on the position and width of the laser beam (Liu et al. 2009, Prakash and Kumar 2016). Usage of a CO_2 laser reduces the risk of damage and ensures lens materials within the focal point of the beam, since it is operated in a mid-infrared wavelength of 10,600 nm (Omi and Numano 2014). On the top of it, the control over laser ablation power and beam speed dictates the depth and size of the microcavity. The microcavities of the CLs can be functionalized with individual assay (colorimetric and fluorescent assays) to suit sensing of specific analyte and have potential application in ocular diagnostics and sustained drug release.

12.4 MICROFLUIDIC CONTACT LENSES FOR INTRAOCULAR PRESSURE (IOP) SENSING

IOP is another health indicator obtained from human eyes and high IOP is an alarming sign of potential glaucoma. Severe glaucoma can cause loss of vision and IOP mainly contributes to this pathogenesis and therefore monitoring of IOP is mandatory for healthcare and diagnosis (Yao et al. 2011). It is both important and challenging to diagnose and treat glaucoma, since this intractable disease progresses slowly without any symptoms. In most cases patients measure IOP during daylight in the hospital that may not help to alert the patients, since IOP reaches its peak (ocular hypertension) only at night, which implies the requirement of round-the-clock monitoring of IOP, a must indicator for the diagnosis and treatment of glaucoma. IOP differs during the diurnal and nocturnal periods and based on the body posture (Liu 1998, 2003) and the peak IOP exhibited during the nocturnal period for glaucoma patients and other normal subjects was first reported by Liu et al. (2003a, b). The reason for this increase in IOP during a nocturnal period is attributed to the increase of episcleral venous pressure and the redistribution of body fluid (Friberg et al. 1987, Mansouri and Shaarawy 2011), which could even be biased, because the effect of awakening can influence IOP measurement during sleep (Weitzman et al. 1975, Weinreb and Liu 2006). Therefore, continuous monitoring of IOP is essential since elevation of IOP is the indicator of glaucoma.

While designing a CL for IOP sensing, one must keep in mind the forces acting on it. When a CL is worn on the eye, there are six forces in total acting on it. Among them there are five major forces and one minor force. The interplay of such forces in various positions of the eye and under different conditions dictate the fit of the lens. The five major forces are: The atmospheric pressure (P_1) which envelops the free anterior and peripheral surfaces of the lens-tear complex. Hydrostatic pressure (P_0) which refers to the pressure of the precorneal (postlens) tear film that acts against the lens similar to water acting against the wall of a dam. P_0 supports P_1 as its free surface. The force of gravity (P) is the weight of the lens, and the lid force (F_{lid}) represents the force developed by the lids during a blink which varies with the arc of excursion, duration, and lid tightness and the surface tension force ($F\sigma$) corresponds to the cohesive force existing between the molecules of tears at an interface like the interface between tear and the plastic the lens is made of, air and tears. The minor force is the force of viscosity (F_v) which is generally neglected, since the coefficient of tear viscosity is small and seldom assumed remarkable proportions (Miller 1963).

12.4.1 Microfluidic IOP Sensors

MCLs stand out to be a complement to soft CLs with added advantages including elimination of rigid electronic components, complex electronic circuits, radiofrequency, and associated risks to the eye besides having warpage-free CLs with good elastic deformation management capable of monitoring multiple biomarkers simultaneously (Yang et al. 2020). The change in the pressure results in the deformation of the sensing chamber that induces the displacement of the dyed liquid and correlating such displacement of the dyed liquid's interface in the sensing channel can provide information about IOP.

In general, PDMS and silicone hydrogel materials chosen to fabricate MCLs and thermoplastic polymers, do hold a promising future for this technology. PDMS and dyed glycerol-based wireless microfluidics pressure sensor using soft lithography working based on Laplace's principle have already been developed. PDMS was chosen for its biocompatibility, nontoxicity, optical transparency, chemically inertness, electrically isolative, flexibility, elasticity, and superior physical properties including a Young's modulus (E) of ~500 kPa. The pressure inside the hydraulic chamber (Pt) is dynamically estimated by loading pressure and the built-in microfluidics pressure sensor. The difference in pressure is indicated by the deflection of the sensing membrane and at each point the pressure difference (P_s) is measured through the individual sensing element. The exact device is presented in Figure 12.2(A), which has a large circular sensing chamber of network height H and radius r followed by a sensing channel with width w and height h acting as the sensing elements. Upon applying localized pressure on the sensing chamber network, it results in eventual strain induced on the elastomer unit by which the microfluidic network generates an internal pressure. When the sensing chamber undergoes compression fluid flows towards the sensing channel from the sensing chamber and the geometrical difference between the two, mechanical displacement amplification takes place as a result of the conservation of mass and a negative pressure created by the elastomer though the recovery properties of PDMS withdraws fluid from the channel. The incorporation of the sensor into the PDMS network as executed by bottom-up micromachining, replica molding, oxygen plasma, and fluidic injection as shown in Figure 12.2(B). Pressure is exerted on the sensing chamber with the aid of a force gauge connected to a step motor and the resulting fluidic displacement is recorded using a digital microscope, thus making it both wireless and noninvasive (Yan 2011).

A smartphone-assisted passive microfluidics pressure sensor was developed for self-monitoring of IOP (Araci et al. 2014). Passive visual readout is advantageous in terms of easy visual accessibility and the mechanical sensitivity of the cornea – the target organ (Agaoglu et al. 2018). The typical sensor setup is made of an airtight

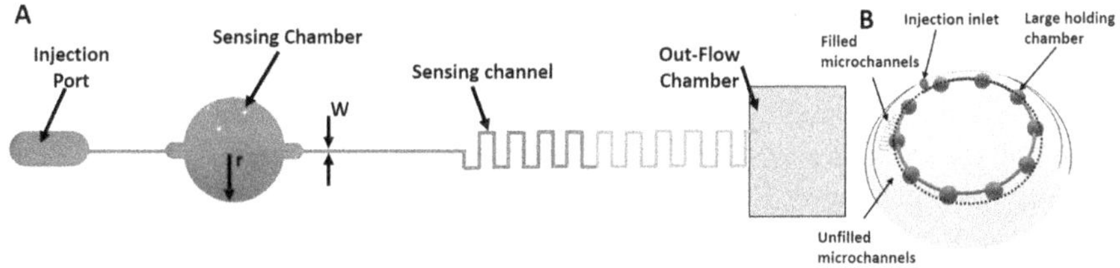

FIGURE 12.2 (A) Schematic illustration of calibration device. (B) Schematic design of a microfabricated PDMS-based device.

microfluidics channel with one end having access to the aqueous intraocular liquid and the other end is connected to a gas reservoir (Araci et al. 2014). The liquid inside the microchannel is displaced, combined by capillary force, and IOP until equilibrium is reached. The liquid–water interface is shifted towards the gas reservoir upon increase in IOP and the reverse occurred when IOP decreased. This setup was embedded into the CL and can be implanted into a patient's eyes during cataract surgery or as a standalone device. The shift in equilibrium pressure can be read out with image analysis software. The sensor tested both in a chamber and on porcine eyes exhibited high accuracy and reproducibility.

The sensitivity of the sensor is displacement of the air–liquid interface per 1 mmHg change in IOP and increasing the ratio of the reservoir volume to channel cross-section increased the sensitivity (Araci et al. 2014). Following the said rule of thumb, a MCL for IOP monitoring was developed with reduced channel width to enhance sensitivity and stiffness and the operating principle of this sensor is the volume expansion of the microfluidic channel network (Agaoglu et al. 2018). Mechanical deformation induces electrical resistance, inductance, or capacitance (Zhu et al. 2018). Based on a novel transduction mechanism a microfluidic which translates small strain changes to a large fluidic volume expansion microfluidic IOP sensor was fabricated. Polydimethylsiloxane (PDMS; RTV 615) and two polyurethane based elastomers Clearflex and NOA65 were the materials of choice for device fabrication. The sensor consists of a liquid reservoir, sensing channel, and air reservoir (Figure 12.3(A–C)) and the sensing channel is connected to both the liquid reservoir and air reservoir. The axial stretching-and-release results in a vacuum effect that causes displacement of the air–liquid interface. The displacement of the interface corresponding to strain (ε) can be defined as; $\text{Displacement}_{\varepsilon} = -\Delta V_{\varepsilon}/A$, where A is the cross-section area of the sensing channel. The choice of the sensor material is based on its compatibility with the guide liquid since hysteresis-free operation (Araci et al. 2014) of the sensor demands a guide liquid possessing low surface energy like oil. Clearflex and NOA65 are chosen for fabricating strain sensor based on their oleophobicity observed during a weeklong dyed oil absorption experiment, while PDMS is not picked out due to its high oleophilicity (Figure 12.3(D)) (Choi et al. 2011, Lei et al. 2011). The sensing readouts were obtained with a combination of smartphone-macrolens making it an easy operating device by the patients. This sensor was able detect as low as <0.06% for uniaxial and <0.004% for biaxial strain. The silicone-based CLs with the incorporation this sensor was tested for IOP in porcine eyes performed a continuous operation for more than 19 h and had a lasted for more than seven months. The good agreement between the simulation and experimental results indicated the reliability of the device (Agaoglu et al. 2018). This work throws light on the oleophilic nature of PDMS, through a dyed oil absorption test prompting the exploration of oleophobic substitutes.

A new combination of PDMS- and PET-based microfluidic soft CL IOP sensor was also demonstrated (An et al. 2018, 2019). The fabrication of the lens involved the following steps soft lithography, silanization of the substrate, plasma-treated bonding, thermoforming and liquid injection (Figure 12.4). The CL sensor consisted of a sensing layer made of micropatterned PDMS and a hard PET reference layer to

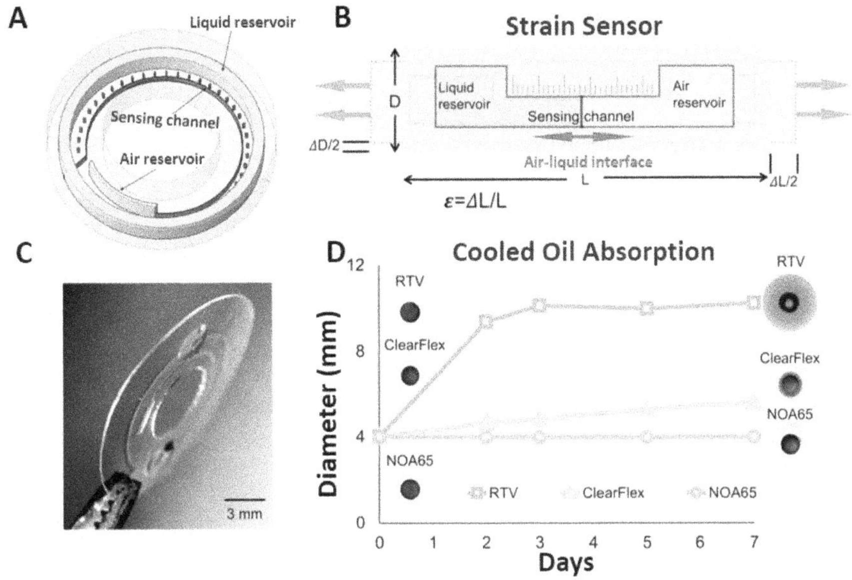

FIGURE 12.3 (A) 3D schematics of a strain sensor. The sensor is composed of a liquid reservoir, an air reservoir, and a sensing channel. (B) Cartoon sketch showing the strain sensor operation principle. (C) Digital image of a wearable microfluidic strain sensor (150 µm thickness). (D) Results of a dyed oil absorption experiment for different materials namely RTV (PDMS), Clearflex, and NOA65. The inset shows the microscope images of the wells fabricated from corresponding materials comparing the initial and the final states of dyed oil absorption. Reproduced with permission. Copyright 2018, The Royal Society of Chemistry (Agaoglu et al. 2018).

make the sensing chamber, sensing channel, and the buffer chamber A dyed liquid is filled in the sensing chamber while the sensing channel is filled partially with the red liquid and the dyed glycerol is accommodated in the rest of the chamber (Figure 12.4(A–F)). When IOP is increased it deforms the cornea causing a volume decrease in the sensing chamber and resulting in displacement change of the dyed liquid's interface in the sensing channel and the reverse flow of the liquid to the sensing chamber occurs if IOP is reduced which is attributed to vacuum force originating from sensing layer recovery. The displacement change was captured with the help of a smartphone camera and the displacement of liquid interface (Δl) and sensitivity (t) were calculated. Simultaneous finite element modeling (FEM) was used to study the influence of the variation of different parameters on the sensing mechanism and sensing IOP to the cross-sectional area of the sensing chambers with different distributions. The experimental observation exhibited liquid displacement response and linearly corresponded to the IOP, and in the case of the outer distributed chamber the liquid displacement nonlinearly corresponded to the IOP. The ex vivo clinical tests carried out on porcine eyes exhibited a sensitivity of 0.2832 mm/mmHg in a range of 8–32 mmHg with good reproducibility, reversibility (Figure 12.4(G)), and

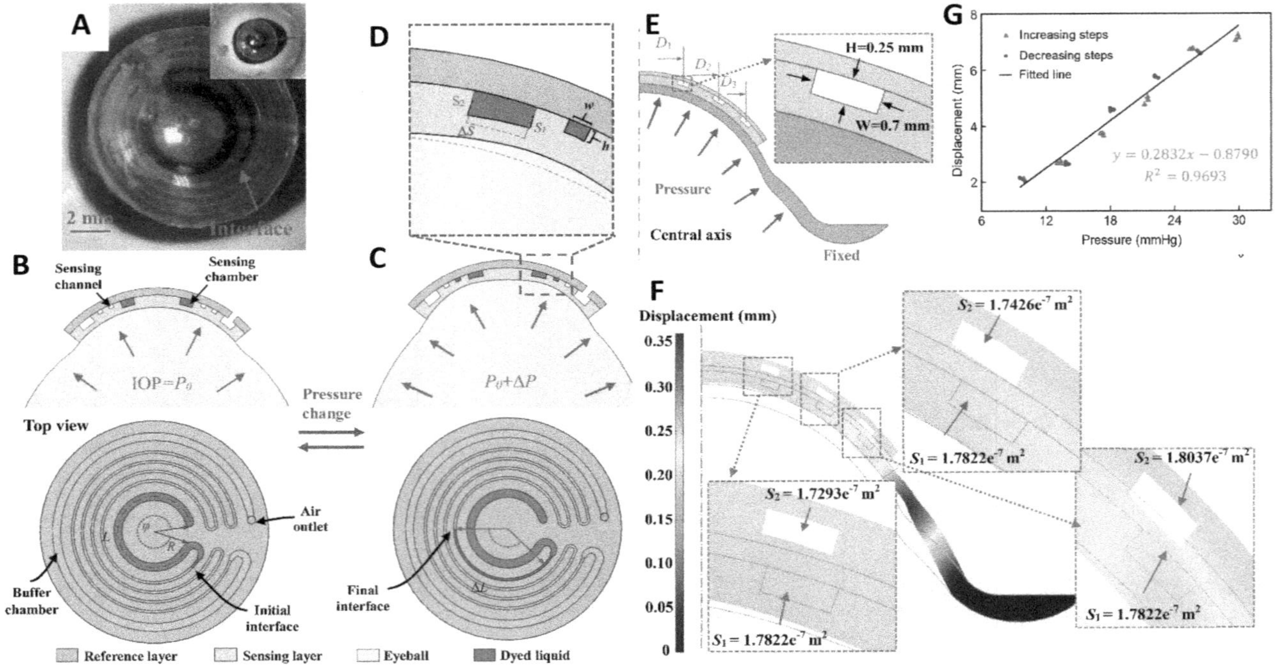

FIGURE 12.4 (A) Digital images of fabricated MCLs using PDMS and PET, worn on the porcine eye ex vivo. (B, C) Sectional view and top view of MCLs under different IOP of P_0 and $P_0 + \Delta P$, (D) Sectional area change of sensing chamber; FEM results of microfluidic CL. (E) Simulation illustration. (F) Sectional view of the sensor with a pressure of 40 mm Hg in relation to deformation. (G) Variation of displacement and IOP during three-cycle increasing and decreasing pressure. Reproduced with permission. Copyright 2019, Elsevier (An et al. 2019).

long-time stability; however the sensitivity decrease is experienced with the increase in the width of the channel. Microfluidic-based CL sensors are simple, unpowered, completely noninvasive, passive, reliable, and long standing which all make them a better alternative.

12.5 MICROFLUIDIC CONTACT LENSES FOR GLUCOSE SENSING

Blood glucose level is an essential biomarker for the early diagnosis as well as treatment and management of diabetes. Diabetes is a global pandemic found in 415 million people around the world as a result of either deficiency or resistance to insulin (Elsherif et al. 2018b, Zimmet et al. 2014), posing a threat to health and economies (Chen et al. 2012). Prompt diagnosis and continuous monitoring of glucose levels are inevitable for patients in order to stay safe from this chronic illness, as diabetes can cause blindness, neuropathy, heart disease, kidney failure, and limb amputation (Zhang et al. 2011). There have been several techniques used for glucose detection so far, including invasive methods such as subcutaneous sensor, microdialysis, and intravenous implantable, minimally invasive methods such as like, micropore/microneedle and noninvasive methods including Kromoscopy, photoacoustic spectroscopy, optical coherence tomography, scattering/occlusion spectroscopy, polarimetry, thermal infrared, fluorescence, Raman spectroscopy, mid-infrared (MIR) spectroscopy, near infrared (NIR) spectroscopy, impedance spectroscopy, optical rotation, fluorescence, colorimetric and Amperometric detection. Though each of them evidence promising results, they all pose certain challenges and limitations in one way or another (DiCesare and Lakowicz 2002, Oliver et al. 2009).

Although enzymatic assay-based methods are very reliable in estimating glucose concentration, the byproducts generated during the process such as hydrogen peroxide are highly reactive and can be both toxic and cause damage to biological composites (DiCesare and Lakowicz 2002) making it less preferred. Therefore, portable glucometer-assisted blood glucose monitoring is an alternative and most common self-measurement method adopted by the diabetes patients. Since this method is both inconvenient and painful (Bruen et al. 2017), it drastically brings down a patient's interest in frequent glucose monitoring. In addition to the abovementioned bottlenecks the lack of accuracy by such home glucose meters (±15%), the risk of bloodborne infection due to invasive sensors (Glasbey et al. 2005, Liu et al. 2010), drives the need for a real-time noninvasive, painless, and continuous monitoring strategy as a replacement to serve the purpose. Moreover, the finger prick method gives only a temporary glucose value but for accurate diagnosis continuous monitoring is necessary (Kimet et al. 2017), which cannot be achieved by this method.

Several new technologies both under noninvasive and minimally invasive strategies such as skin patches, tattoos, and implantable sensors combined with automatic insulin pumps claim to be less painful and complete, nonetheless they are neither truly noninvasive nor capable of continuous monitoring of glucose levels (Badugu et al. 2004). Therefore, a CL-based glucose sensor is the only reliable noninvasive technique that can offer painless continuous monitoring of glucose concentration and help the wearers. Recently, ample research has been performed on developing tiny and reliable sensors including non/minimally invasive gadgets which enable glucose monitoring in the patients (Yao et al. 2011). Among several body fluids such as saliva, urine, intestinal fluid, and tear fluid, useful for noninvasive diagnosis and diabetes monitoring, tear fluid has become an important body fluid to develop current sensor models (Ascaso and Huerva 2016, Zhang et al. 2011). Tear fluid is less vulnerable to dilution than urine and uninterruptedly procured and can be obtained more easily than interstitial fluid or blood (Chenet al. 2012, Leonardi et al. 2004, Mansouri and Shaarawy 2011, Mansourie et al. 2012, Zhang et al. 2011). Being a part of the optical system, tear fluid is responsible for lubricating the eye and nourishing the cornea. The study of tear glucose dates back to 1930 and through scientific research it is evident that tear glucose levels in diabetic patients are higher than those of healthy people. While designing a tear fluid-based glucose sensor, sensitivity enhancement and interference rejection are especially of paramount important since glucose concentration in the tear fluid (0.1–0.6 mM) is lower than in serum (4–6 mM). Therefore, a tear glucose sensor requires more rigorous detection than a blood glucose sensor (Whikehart 2003). Michail et al. were the pioneers in reporting elevation in tear glucose levels during hyperglycemia in 1937, and demonstrated that by determining the tear glucose level, the blood glucose can be estimated (Michail et al. 1937, Michail and Zolog 1937) and by making a correlation between blood and tear fluid, offering a wider window for envisaging noninvasive and continuous methods for glucose monitoring (Van Haeringen 1981). There have been correlations established between tear glucose level and blood glucose level using capillary electrophoresis too (Sen and Sarin 1980, Yetisen et al. 2019). Therefore, employing CL-based sensors to monitor tear glucose concentration can benefit as a noninvasive technique to offer POC diagnostics.

12.5.1 Microfluidic Contact Lens Sensors for Multiple Targets

MCLs can accommodate more than one sensing unit unlike hard and soft CLs which could monitor glucose (Elsherif et al. 2018a) and/or IOP (Kim et al. 2017). This added advantage of MCLs makes them superior in carrying out detection of more than one biomarker simultaneously (Moreddu et al. 2020b, Yang et al. 2020, Yetisen et al. 2020). This helps the wearer to customize the CL sensing ability based on individual health requirements. A multitasking MCL to detect the concentration of glucose, chlorine, and urea was developed using methacrylated poly (dodecanediol citrate) polymer (mPDC), a UV-curable biomaterial with good mechanical properties, hydrophilicity, and biocompatibility. The colorimetric method makes the wearers comfortable in passing by electronic structures, power supplies, and inductors. The MCL has three parts, (i) inner CL, in contact with the cornea with microchannels, (ii) colorimetric analysis module incorporated in the cavity of the inner lens which can show a color change upon reacting with the biomarkers in tears, and (iii) outer lens that is in contact with the eyelid which establishes an open-loop microchannel and reservoirs together with the inner lens. Three identical microchannels were placed in the inner lens and each of them have the inlet, the detection zone, the reservoir, and the outlet. Upon entry of the tears in to the microchannel the particular biomarkers react with the colorimetric analysis modules (commercial assay kit) and show a visible color change, and this red, green and blue (RGB) value can be read out using a smartphone camera to know the concentration of the biomarkers. The in vitro tests conducted on an artificial microfluidics hydrogel eyeball device proved the reliability of the device. When glucose is enzymatically oxidized to gluconic acid it yields hydrogen peroxide as a byproduct and when this hydrogen peroxide is decomposed by peroxidase it is not only converted into water and oxygen but also condenses to colorless 4-aminoantipyrine and phenol into a red-colored compound. Based on the glucose concentration the colorimetric reagent gives a color change that can be captured as images and whose RGB values can be analyzed with software. Similarly, the concentration of chlorine ions helps in maintaining the osmotic pressure and acid-base balance. Hyper/hypochloremia is caused when there is a variation in chlorine ion concentration. The assay kit consists of Fe^{2+}, Hg^{2+}, and 2,4,6-tripyridyl-s-triazineindicator. When Hg^{2+} reacts with Cl^- to give $HgCl_2$ the color changes from colorless to blue and the intensity depends on the concentration of chlorine ions. The higher concentration of urea in the tear fluid over blood due to its small molecular weight and high diffusivity, is advantageous in detecting its concentration noninvasively (Farkas et al. 2003). Urea, when complexed with a colorimetric agent, gives a yellow color and increased concentration of urea results in yellow/brown color.

Besides being noninvasive, this method was advantageous in terms of sensing three different analytes simultaneously (Yang et al. 2020), however, the shortage of tear fluids for all the three microchannels in the case of dry eyes and distinguishing colorimetric sensing units if the RGB values of two of them overlap, are the challenges to be addressed.

Moreddu et al developed a MCL sensor to detect four analytes including glucose, pH nitrite ions, and proteins (Moreddu et al. 2020b). Using CO_2 laser ablation, a ring-like microchannel with four branches was carved on a commercial CL and the colorimetric sensors were incorporated in the microcavities (Figure 12.5(A–F)). This semi-quantitative MCL sensor for glucose sensing was designed based on a two-step enzymatic method. Hydrogen peroxide, a byproduct formed during the oxidation of D-glucose to D-gluconolactone undergoes further oxidation via peroxide catalysis to yield 3,3′,5,5′-tetramethylbenzidin (TMB). The tear glucose sensor detects glucose level by showing a color change from yellow to different shades of green depending on the concentration of glucose (0–20 mmol L^{-1}) with a sensitivity of 1.4 nm/mmol L^{-1} of glucose, and a limit of detection (LOD) of 1.84 mmol L^{-1} (Figure 12.5(G)(ii), (H)(ii)). Sensing a yellow/green color is an indication of a healthy condition while a yellow color is ascribed to a down-regulated sugar level, and darker green shades indicate a high concentration of glucose in the tear. The results can be captured and analyzed with the help of a smartphone camera which is convenient for the wearer to do self-examination.

12.6 MICROFLUIDIC CONTACT LENSES FOR PH SENSING

Tear pH in a healthy person ranges between 6.5 to 7.6 (Abelson et al. 1981) and a deviation from this towards the alkaline window (pH $\approx$ 8) is an indication of the

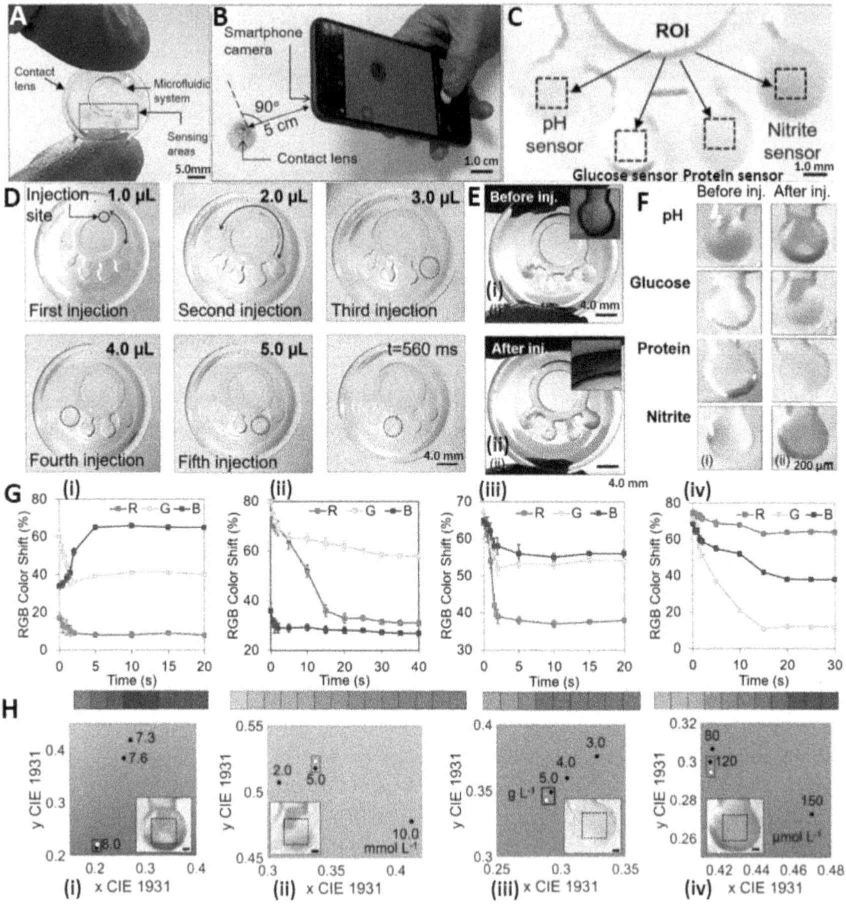

FIGURE 12.5 (A) Digital image of a CL sensing platform with multiple targets. (B) color change of the sensors is imaged using a smartphone camera. (C) Photographs of the sensors serve as inputs to the customized MATLAB algorithm, where the region of interest can be selected; characterization of MCLs. (D) Fluid flow characterization with fluorescein aqueous solution. Five consecutive injections amounting to 1 µL each were performed from the indicated injection site. Within 560 ms, the fluid reached all the sensing sites. (E) Characterization of CL sensors with artificial tear fluid. Photographs of a CL sensor before (i) and after (ii) artificial tear fluid injection. (F) Representation of smartphone readouts on CL sensors before (i) and after (ii) artificial tear fluid injection. (G) Red, green, and blue color shift over time for pH (i), glucose (ii), protein (iii), and nitrite (iv) biochemical sensors. (H) CIE 1931 chromaticity diagrams obtained with the algorithm after inputting the imaged sensors. The algorithm allowed selection of the region of interest, indicated with black dotted lines. The corresponding normalized color is plotted in the chromaticity space calibrated with the points of the sensor of interest (white dots) and compared to the calibration values (black dots). The nearest calibration point gives the concentration readout. Readouts refer to pH (i), glucose (ii), protein (iii), and nitrite (iv) sensors. Reproduced with permission. Copyright 2020, Elsevier. (Moreddu et al. 2020b).

early stage of rosacea, a chronic dermatosis causing corneal melting and stromal scarring (Abelson et al. 1980, An et al. 2005). Its occurrence in the ocular environment prior to skin will go unnoticed unless the variation in tear pH is detected. A deviation in pH is not only found in rosacea patients (8.0 ± 0.32) but also in cataract patients in pre- (7.26 ± 0.23) and post- (7.50 ± 0.23) operation stages (Thygesen and Jensen 1987). Also tear pH and buffering capacity are the important yardsticks to evaluate the extent of ocular penetration of drugs (Ahmed and Patton 1984) indicative of the pressing need to monitor tear pH. Most techniques used to date involve electrodes (Fischer and Wiederholt 1982), microelectrodes (Carney et al. 1990), glass probes (Abelson et al. 1981), or fluorescent probes (Chen and Maurice, 1990) to be in contact with the cornea which all could cause inconvenience to the patient and lack continuity in monitoring. Benzenedicarboxylic acid (Yetisen et al. 2020) and anthocyanin (Riaz et al. 2019) functionalized CL pH sensors were truly minimally/noninvasive with the ability to continuously monitor pH in a short range (pH 7.0–8.0) (Yetisen et al. 2020) and pH 6.5–7.5 (Riaz et al. 2019)) limiting them from detecting pH beyond this window. At this juncture MCLs with colorimetric sensing units comprising of a methyl red, bromothymol blue, and phenolphthalein to suit a range of pH both in acidic and alkaline windows (Figure 12.5(G)(i), (H)(i)) (Moreddu et al. 2020b) help in close monitoring of tear pH. Methyl red, a carboxylic acid group, and an amine functional group can work in both acid and base media to detect pH in the range of 4.3–6.2. On the contrary, phenolphthalein is colorless in acidic media, and shows a color shift in alkaline media from pink to fuchsia, due to the formation of quinoid-carboxylated structure with the ability to detect pH from 8.2 to 12.0. Bromothymol blue, being a weak acid, exhibits a color shift from yellow to blue in alkaline media due to the formation of triphenylmethane. This pH sensor assay with the combination of methyl red, bromothymol blue, and phenolphthalein show a green color in neutral environments and its color intensity varies based on the pH. The wearer can capture the color change with a smartphone and the RGB values can be analyzed to detect the pH. Green color indicates a healthy eye and yellow color represents mild acidic condition ascribed to changes in tear buffering capacity (Norn 1985) but an alkaline pH exhibited with blue color indicates rosacea disease (Abelson et al. 1980). This sensor has a quick response of 15 s and has a high sensitivity of 12.23 nm/pH unit.

12.7 MICROFLUIDIC CONTACT LENSES FOR PROTEIN SENSING

The protein concentration in tears is in the range of 3–7 μg μL^{-1}, and tear proteomics is of great interest, since it is useful in the diagnosis of several disorders such as aniridia (Ihnatko et al. 2013, Moreddu et al. 2020a), different dry eye diseases (Soria et al. 2013), diabetic retinopathy (Chiang et al. 2012, Csősz et al. 2012, Kim et al.2012), and keratoconus (Balasubramanian et al. 2012). The keratoconic tears have (3.86 mg mL^{-1}) nearly half the amount of protein present in the healthy control (7.00 mg mL^{-1}) and besides that, a decrease is observed in the amount of individual proteins such as lactoferrin and secretory IgA. So, it is necessary to monitor tear protein levels to diagnose and treat keratoconus in the early stage itself

(Balasubramanian et al. 2012, Yetisen et al. 2020). The protein sensor is constructed based on the reaction of 3′,3′,5′,5′-tetrachlorphenol-3,4,5,6-tetrabromsulfophthalein with hydrogen ions present on the amino acid to give an anode of the same compound reflecting a different color. The color changes from beige to light blue for the concentration of proteins ranging from 0.5 to 5.0 g L^{-1}. The device showed a high sensitivity of 0.49 nm/g L^{-1} for proteins, and a LOD of 0.63 g L^{-1} (Figure 12.5(G)(iii), (H)(iii)) (Moreddu et al. 2020b). The color displayed by the protein sensor is always blue in color irrespective of the amount of proteins present in the tears while intense blue color represents healthy protein corresponding to 5 g L^{-1} and above and reduced intensity is attributed to low protein level (3 g L^{-1}, even in early stages) due to keratoconus, which causes bulging of cornea into a conical shape (Balasubramanian et al. 2012).

12.8 MICROFLUIDIC CONTACT LENSES FOR NITRITE ION SENSING

Nitric oxide is a vital regulator of important homeostatic processes in the eye, including phototransduction, aqueous humor dynamics and retinal neurotransmission. A healthy control should have a nitrite level of 120 µmol $L^{-1}$and any deviation from this is an indication of an inflammatory state such as retinitis, uveitis, Behcet's syndrome, and degenerative diseases like glaucoma. The tear nitric oxide of Behcet's patients easily gets oxidized to peroxynitrite resulting in lower levels of nitrites and nitrates (82.29 µmol L^{-1}) and this can be detected by measuring tear nitrites and nitric oxide byproducts (Mirza et al. 2001). The principle used to sense nitrite ions is based on the nitrite ion's reaction with sulfanilamide to yield diazonium salt which, upon binding with N-(1-naphthyl)-ethylenediamine dihydrochloride, gives a pink azo dye with an absorption intensity of 528 nm and whose intensity increases along with the increase of the nitrite ion's concentration. Healthy tears (nitrite levels ≈120 µmol L^{-1}) with high nitrite ion concentration will yield an intense pink color and light pink color due to low concentration of nitrite ions (nitrite levels ≈80 µmol L^{-1}) is attributed to uveitis (Mirza et al. 2001). Therefore, smartphone-assisted nitrite ion sensing can detect the nitrite ion level in the tear. This sensor exhibited a sensitivity of 0.03 nm/µmol L^{-1} of nitrites, and a LOD of 24.4 µmol L^{-1} (Figure 12.5(G)(iv), (H)(iv)) (Moreddu et al. 2020b).

12.9 MICROFLUIDIC CONTACT LENS SENSOR FOR CORNEAL TEMPERATURE SENSING

Continuous monitoring of ocular surface temperature (OST) is helpful to review ocular conditions such as glaucoma, vascular neuritis, diabetic retinopathy, carotid artery stenosis, and dry eye disease (DED). The positive correlation observed between body temperature and ocular temperature (Purslow and Wolffsohn 2005) and between ocular temperature difference values (TDVs) and dry eye parameters suggest that OST monitoring is important to obtain POC ocular diagnostics. OST increase observed in patients with meibomian gland dysfunction (MGD) (Terada et al. 2004), phakic and pseudophakic (Sniegowski et al. 2015) psychiatric disorders (Monge-Roffarello et al. 2014, Tan et al. 2009) in post-corneal transplant undergoing inflammation (Sniegowski et al. 2018), and dogs with keratoconjunctivitis sicca (Biondi et al. 2015) further evidences for the necessity for the measurement of OST to monitor disease progression in personalized diagnostics (Moreddu et al. 2019).

Thermochromic liquid crystals (TLCs) show reversible color change to temperature and this principle was adopted to fabricate a CL-based temperature sensor. TLCs, with their reversibility, multiple color transitions, fast responsive times, and very high accuracy (0.1°C) can replace existing electronic sensors for temperature sensing. TLCs can be prepared by melting esters such as cholesteryl oleyl carbonate, cholesteryl nonanoate, and cholesteryl benzoate (COC/CN/CB wt%, 0.35:0.55:0.10) which share the same molecular skeleton, bonded to a different radical. Melting results in establishing weak inter- and intra-molecular bonding while retaining chemical properties, and the physical properties of TLCs can be tuned based on the composition. The temperature sensing principle by liquid crystals is shown in Figure 12.6(A, B) where the cholesteric phase formed from chiral, rigid, and rod-shaped molecules possess a layered structure where molecules lightly change their orientation over a range of temperature (Abdullah et al. 2010, Gevers and Smeulders 2006)

The light incident on the TLCs gets reflected obeying the following formula:

$\lambda = nP$ (Abdullah et al. 2010, Sage 2011), Where λ is the reflected wavelength, n is the refractive index of the material, P is the pitch distance between two equally oriented layers.

When there is an increase in the temperature the helical stack is shortened, and the pitch decreases from P_1 to P_2 causing the reflection peak to shift from λ_1 to λ_2 which is reflected in color change.

The CL-based temperature sensor is fabricated by creating micropatterns on scleral lenses using CO_2 laser ablation and embedding the microstructures with liquid crystals followed by sealing them with circular glass pieces by UV curing Figure 12.6(C). The optical characterization carried out with the help of reflection spectroscopy (Figure 12.6(F)) showed a shift in the wavelength of the light from 738 $\pm$ 4 nm to 474 $\pm$ 4 nm when the temperature was increased from 29.0°C to 40.0°C with steps of 0.5°C (Figure 12.6(G)). For the abovementioned condition the live images of color change were recorded using a smartphone with color comparator and color name application in order to individuate the RGB triplets and to observe the temperature-dependent color change. For ex vivo conditions the CL sensors were fitted on porcine eyes (Figure 12.6(M)), and tests were carried out for ambient and heated up (40.0°C) temperatures. The colorimetric readouts were carried out by smartphone and infrared (IR) thermal (Figure 12.6(N)) camera. With the accuracy of detecting a temperature as low as 0.1°C, this device can be accurate and offer minimally invasive cost-effective diagnosis of ocular infections on the doorstep.

12.10 CONCLUSION AND FUTURE PROSPECTS

This chapter presents an overview of MCLs for ocular diagnostics to provide a smart healthcare solution. We focused on various architectures for sensing glucose,

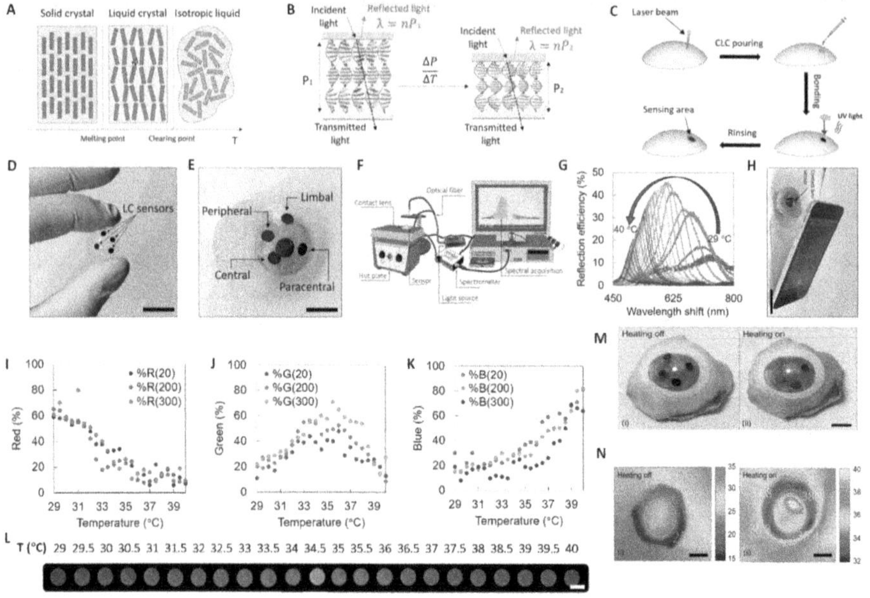

FIGURE 12.6 CL temperature sensor. (A, B) Sensing mechanism of thermochromic liquid crystals: (A) temperature-triggered state change, (B) reflection peak shift over temperature in the cholesteric phase, resulting from the preferentially reflected wavelength corresponding to the distance over which a 360°C rotation of the field director occurs, namely the pitch, P_1 and P_2. (C) Fabrication of CL sensor. (D, E) Digital photographs of a CL sensor with four active areas for continuous corneal temperature mapping. Scale bars: 2.0 cm and 1.0 cm; characterization of liquid crystal (COC/CN/CB wt%, 0.35:0.55:0.10) in the temperature range 29.0–40.0°C. (F) Reflection spectroscopy setup. (G) Wavelength shift over temperature; response of CL temperature sensors; smartphone readout method. (H) The smartphone is placed with the camera pointing at normal incidence to the sensor, at a distance of 5 cm; A CL sensor is fitted on an orbital prosthesis at ambient temperature. Scale bar: 2.5 cm; RGB percentages in the range 29.0–40.0°C at ambient light 200 lux; Individual (I) red, (J) green, and (K) blue contributions at ambient light levels of 20, 200, and 300 lux. (L) Color palette associated to the sensor under 200 lux in the range 29.0–40.0°C, captured with a smartphone camera. Scale bar: 2 mm; readout process tested on an ex vivo porcine eye. (M) Photograph of a CL sensor fitted on a porcine eye at ambient temperature (i) and under heating (ii). Scale bar: 50 mm. (N) Readout process tested on an ex vivo porcine eye at (i) ambient temperature and (ii) under heating. Reproduced with permission. Copyright 2019, The Royal Society of Chemistry (open Journal). (Moreddu et al. 2019).

chlorine ion, nitrite ion, urea, pH, corneal temperature, proteins, and IOP. Tear fluid is a rich bank of biomarkers that are secreted from blood vessels along the blood-tear barrier, and analysis of tear fluid can provide an excellent fingerprint to diagnose disorders such as diabetes and IOP, the indicator of glaucoma. The eye is an outstanding interface between an electronic gadget and the body, MCLs have transformative potential for healthcare applications and especially for personalized medicine. Smart

MCLs used for diagnostics are based on a change in volume or color of the sensing material without the need for an external power source. With the convergence in material and technology, suitable innovative modifications of MCLs such as proper manipulation of liquids, crosslinking of colorimetric and fluorescent optical sensors to execute simultaneous multiple sensing (Yetisen et al. 2017), and immobilization of chelating agents (Yetisen et al. 2014) can promise potential real-time healthcare.

The ardent research carried out on improving the MCLs as a platform for sensing in the recent past has resulted in present day wearable smart devices which offer POC diagnosis and envisage offering personalized medicine. Personalized medication is a combination of a therapeutic strategy with a common diagnostic tool to supply continuous real-time individual information at POC settings. Nonetheless, continuous monitoring of IOP and administration of the drug at night simultaneously can drastically affect oxygen permeability. Moreover, lens protein deposition due to extended wear, is another concern to be addressed by additional modification of the CL to arrest protein adsorption (Xu et al. 2018).

In spite of significant advancements in technology in ocular diagnostics, drug delivery developing theranostic CLs which can perform diagnostics and therapy simultaneously is still unmet. This can be realized by integrating biosensors, drug delivery, and data a communication system in the same CL to diagnose and treat the disease simultaneously. Developing such smart CLs can greatly help the wearer to have control of the disease with an automated supply of a controlled dosage of the drug, based on the diagnostic report. Besides this, introducing multitasking CLs that can perform more than one kind of diagnostics at the same time, can benefit patients with multiple disorders by providing custom-made diagnostic lenses depending on the individual's requirements. Such theranostic smart CLs capable of performing multiple diagnostics and supplying essential drug loads based on diagnostic readouts will be a significant step towards better healthcare, although their development and implementation still requires further research.

ACKNOWLEDGEMENTS

The authors acknowledge Khalifa University of Science and Technology (KUST) for the Faculty Startup Project (Project code: 8474000211-FSU-2019-04) and KU-KAIST Joint Research Center (Project code: 8474000220-KKJRC-2019-Health1) research funding in support on this research. They also acknowledge Sandooq Al Watan LLC and Aldar Properties for the research funding (SWARD Program – AWARD, Project code: 8434000391-EX2020-044).

REFERENCES

Abdullah, N., Abu Talib, A. R., Jaafar, A. A., Mohd Salleh, M. A. & Chong, W. T. (2010). The Basics and Issues of Thermochromic Liquid Crystal Calibrations. *Exp. Therm. Fluid Sci.*, *34*(8), 1089–1121. doi:https://doi.org/10.1016/j.expthermflusci.2010.03.011.

Abelson, M. B., Sadun, A. A., Udell, I. J. & Weston, J. H. (1980). Alkaline Tear pH in Ocular Rosacea. *Am. J. Ophthalmol.*, *90*(6), 866–869. doi:https://doi.org/10.1016/S0002-9394(14)75203-1.

Abelson, M. B., Udell, I. J. & Weston, J. H. (1981). Normal Human Tear pH by Direct Measurement. *Arch. Ophthalmol.*, *99*(2), 301–301. doi:10.1001/archopht.1981.03930010303017.

Agaoglu, S., Diep, P., Martini, M., KT, S., Baday, M. & Araci, I. E. (2018). Ultra-Sensitive Microfluidic wearable strain sensor for Intraocular Pressure Monitoring. *Lab Chip*, *18*(22), 3471–3483. doi:10.1039/C8LC00758F.

Ahmed, I. & Patton, T. F. (1984). Effect of pH and Buffer on the Precorneal Disposition and Ocular Penetration of Pilocarpine in Rabbits. *Int. J. Pharm.*, *19*(2), 215–227. doi:https://doi.org/10.1016/0378-5173(84)90164-9.

Akram, Muhammad Safwan, Ronan, Daly, da Cruz Vasconcellos, Fernando, Yetisen, Ali Kemal & Hutchings, Ian. (2015). Applications of Paper-Based Diagnostics. In: Castillo-León J., Svendsen W. (Eds.) *Lab-on-a-Chip Devices and Micro-Total Analysis Systems* (pp. 161–195). Berlin Heidelberg: Springer.

Alam, F., Elsherif, M., AlQattan, B., Ali, M., Ahmed, I. M. G., Salih, A.…Butt, H. (2021). Prospects for Additive Manufacturing in Contact Lens Devices. *Adv. Eng. Mater.*, *23*(1–13), 2000941. doi:https://doi.org/10.1002/adem.202000941.

Alam, F., Elsherif, M., AlQattan, B., Salih, A., Lee, S. M., Yetisen, A. K.…Butt, H. (2021). 3D Printed Contact Lenses. *ACS Biomater. Sci. Eng.*, *7*(2), 794–803. doi:10.1021/acsbiomaterials.0c01470.

AlQattan, B., Yetisen, A. K. & Butt, H. (2018). Direct Laser Writing of Nanophotonic Structures on Contact Lenses. *ACS Nano*, *12*(6), 5130–5140. doi:10.1021/acsnano.8b00222.

Alqurashi, Y., Bajgrowicz-Cieslak, M., Hassan, M. U., Yetisen, A. K. & Butt, H. (2018). Laser-Induced Surface Modification of Contact Lenses. *Adv. Eng. Mater.*, *20*(6), 1700963. doi:10.1002/adem.201700963.

An, H., Chen, L., Liu, X., Zhao, B., Ma, D. & Wu, Z. (2018). A Method of Manufacturing Microfluidic Contact Lenses by Using Irreversible Bonding and Thermoforming. *J. Micromech. Microeng.*, *28*(10), 105008. doi:10.1088/1361-6439/aaceb7.

An, H., Chen, L., Liu, X., Zhao, B., Zhang, H. & Wu, Z. (2019). Microfluidic Contact Lenses for Unpowered, Continuous and Non-Invasive Intraocular Pressure Monitoring. *Sens. Actuators A: Phys.*, *295*, 177–187. doi:https://doi.org/10.1016/j.sna.2019.04.050.

An, H. J., Ninonuevo, M., Aguilan, J., Liu, H., Lebrilla, C. B., Alvarenga, L. S. & Mannis, M. J. (2005). Glycomics Analyses of Tear Fluid for the Diagnostic Detection of Ocular Rosacea. *J. Proteome Res.*, *4*(6), 1981–1987. doi:10.1021/pr0501620.

Araci, I. E., Su, B., Quake, S. R. & Mandel, Y. (2014). An Implantable Microfluidic Device for Self-Monitoring of Intraocular Pressure. *Nat. Med.*, *20*(9), 1074–1078. doi:10.1038/nm.3621.

Armani, D. K. & Liu, C. (2000). Microfabrication Technology for Polycaprolactone, A Biodegradable Polymer. *J. Micromech. Microeng.*, *10*(1), 80–84. doi:10.1088/0960-1317/10/1/311.

Ascaso, F. J. & Huerva, V. (2016). Noninvasive Continuous Monitoring of Tear Glucose Using Glucose-Sensing Contact Lenses. *Optometry Vision Sci.*, *93*(4), 426–434.

Badawy, A., Hassan, M. U., Elsherif, M., Ahmed, Z., Yetisen, A. K. & Butt, H. (2018). Contact Lenses for Color Blindness. *Adv. Healthcare Mater.*, *7*(12), 1800152. doi:https://doi.org/10.1002/adhm.201800152.

Badugu, R., Lakowicz, J. R. & Geddes, C. D. (2004). Ophthalmic Glucose Monitoring Using Disposable Contact Lenses: A Review. *J. Fluoresc.*, *14*(5), 617–633. doi:10.1023/B:JOFL.0000039349.89929.da.

Balasubramanian, S. A., Pye, D. C. & Willcox, M. D. P. (2012). Levels of Lactoferrin, Secretory IgA and Serum Albumin in the Tear Film of People with Keratoconus. *Exp. Eye Res.*, *96*(1), 132–137. doi:https://doi.org/10.1016/j.exer.2011.12.010.

Becker, H. & Gärtner, C. (2008). Polymer Microfabrication Technologies for Microfluidic Systems. *Anal. Bioanal. Chem.*, *390*(1), 89–111. doi:10.1007/s00216-007-1692-2.

Biondi, F., Dornbusch, P. T., Sampaio, M. & Montiani-Ferreira, F. (2015). Infrared Ocular Thermography in Dogs with and without Keratoconjunctivitis Sicca. *Vet. Ophthalmol.*, *18*(1), 28–34. doi:10.1111/vop.12086.

Blaga, A. (1978). Use of Plastics in Solar Energy Applications. *Sol. Energy*, *21*(4), 331–338. doi:https://doi.org/10.1016/0038-092X(78)90010-5.

Bruen, D., Delaney, C., Florea, L. & Diamond, D. (2017). Glucose Sensing for Diabetes Monitoring: Recent Developments. *Sensors*, *17*(8), 1866. doi:10.3390/s17081866.

Carney, L. G., Mauger, T. F. & Hill, R. M. (1990). Tear buffering in contact lens wearers. *Acta Ophthalmol.*, *68*(1), 75–79. doi:10.1111/j.1755-3768.1990.tb01653.x.

Chen, F. S. & Maurice, D. M. (1990). The pH in the Precorneal Tear Film and Under a Contact Lens Measured with a Fluorescent Probe. *Exp. Eye Res.*, *50*(3), 251–259. doi:https://doi.org/10.1016/0014-4835(90)90209-D.

Chen, G., Chan, I. & Lam, D. C. C. (2013). Capacitive Contact Lens Sensor for Continuous Non-Invasive Intraocular Pressure Monitoring. *Sens. Actuators A: Phys.*, *203*, 112–118. doi:https://doi.org/10.1016/j.sna.2013.08.029.

Chen, L., Magliano, D. J. & Zimmet, P. Z. (2012). The Worldwide Epidemiology of Type 2 Diabetes Mellitus-Present and Future Perspectives. *Nat. Rev. Endocrinol.*, *8*(4), 228–236. doi:10.1038/nrendo.2011.183.

Chiang, S., Tsai, M., Wang, C., Chen, A., Chou, Y., Hsia, C.…Shui, H. (2012). Proteomic Analysis and Identification of Aqueous Humor Proteins with a Pathophysiological Role in Diabetic Retinopathy. J. *Proteomics*, *75*(10), 2950–2959. doi:https://doi.org/10.1016/j.jprot.2011.12.006.

Choi, S., Kwon, T., Im, H., Moon, D., Baek, D. J., Seol, M.…Choi, Y. (2011). A Polydimethylsiloxane (PDMS) Sponge for the Selective Absorption of Oil from Water. *ACS Appl. Mater. Interfaces*, *3*(12), 4552–4556. doi:10.1021/am201352w.

Csősz, É, Boross, P., Csutak, A., Berta, A., Tóth, F., Póliska, S.…Tőzsér, J. (2012). Quantitative Analysis of Proteins in the Tear Fluid of Patients with Diabetic Retinopathy. *J. Proteomics*, *75*(7), 2196–2204. doi:https://doi.org/10.1016/j.jprot.2012.01.019.

DiCesare, N. & Lakowicz, J. R. (2002). Charge Transfer Fluorescent Pprobes Using Boronic Acids for Monosaccharide Signaling. *J. Biomed. Opt.*, *7*(4), 538–545. doi:10.1117/1.1502263.

Eddings, M. A., Johnson, M. A. & Gale, B. K. (2008). Determining the Optimal PDMS–PDMS Bonding Technique for Microfluidic Devices. *J. Micromech. Microeng.*, *18*(6), 067001. doi:10.1088/0960-1317/18/6/067001.

Elsherif, M., Hassan, M. U., Yetisen, A. K. & Butt, H. (2018a). Glucose Sensing with Phenylboronic Acid Functionalized Hydrogel-Based Optical Diffusers. *ACS Nano*, *12*(3), 2283–2291. doi:10.1021/acsnano.7b07082.

Elsherif, M., Hassan, M. U., Yetisen, A. K. & Butt, H. (2018b). Wearable Contact Lens Biosensors for Continuous Glucose Monitoring Using Smartphones. *ACS Nano*, *12*(6), 5452–5462. doi:10.1021/acsnano.8b00829.

Elsherif, M., Hassan, M. U., Yetisen, A. K. & Butt, H. (2019). Hydrogel Optical fibers for Continuous Glucose Monitoring. *Biosens. Bioelectron.*, *137*, 25–32. doi:https://doi.org/10.1016/j.bios.2019.05.002.

Elsherif, M., Salih, A. E., Yetisen, A. K. & Butt, H. (2021). Contact Lenses for Color Vision Deficiency. *Adv. Mater. Technol.*, *6*(1), 2000797. doi:https://doi.org/10.1002/admt.202000797.

Farandos, N. M., Yetisen, A. K., Monteiro, M. J., Lowe, C. R. & Yun, S. H. (2015). Contact Lens Sensors in Ocular Diagnostics. *Adv. Healthcare Mater.*, *4*(6), 792–810. doi:10.1002/adhm.201400504.

Farkas, Á, Vámos, R., Bajor, T., Müllner, N., Lázár, Á & Hrabá, A. (2003). Utilization of Lacrimal Urea Assay in the Monitoring of Hemodialysis: Conditions, Limitations and Lacrimal Arginase Characterization. *Exp. Eye Res.*, *76*(2), 183–192. doi:https://doi.org/10.1016/S0014-4835(02)00276-2.

Fischer, F. H. & Wiederholt, M. (1982). Human Precorneal Tear Film pH Measured by Microelectrodes. *Graefe's Arch. Clin. Exp. Ophthalmol.*, *218*(3), 168–170. doi:10.1007/BF02215658.

Friberg, T. R., Sanborn, G. & Weinreb, R. N. (1987). Intraocular and Episcleral Venous Pessure Increase During Inverted Posture. *Am. J. Ophthalmol.*, *103*(4), 523–526.

Gevers, T. & Smeulders, A. W. M. (2006). Color Based Object Recognition. *Image Analysis and Processing*, Springer, Berlin Heidelberg, 32, 319–326. doi:10.1007/3-540-63507-6_217.

Glasbey, T. O., Newman, J. J., Newman, D. D., Sutton, H. S. & Tipton, W. M. (2005). WO 031400 A2.

Ho, H., Saeedi, E., Kim, S. S., Shen, T. T. & Parviz, B. A. (2008). *Contact Lens with Integrated Inorganic Semiconductor Devices.* Proceedings of the IEEE International Conference on Micro Electro Mechanical Systems (MEMS) 403–406, doi: 10.1109/MEMSYS.2008.4443678.

Iguchi, S., Kudo, H., Saito, T., Ogawa, M., Saito, H., Otsuka, K.…Mitsubayashi, K. (2007). A Flexible and Wearable Biosensor for Tear Glucose Measurement. *Biomed. Microdevices*, *9*(4), 603–609. doi:10.1007/s10544-007-9073-3.

Ihnatko, R., Edén, U., Lagali, N., Dellby, A. & Fagerholm, P. (2013). Analysis of Protein Composition and Protein Expression in the Tear Fluid of Patients with Congenital Aniridia. *J. Proteomics*, *94*, 78–88. doi:https://doi.org/10.1016/j.jprot.2013.09.003.

Isiksacan, Z., Guler, M. T., Aydogdu, B., Bilican, I. & Elbuken, C. (2016). Rapid Fabrication of Microfluidic PDMS Devices from Reusable PDMS Molds Using Laser Ablation. *J. Micromech. Microeng.*, *26*(3), 035008. doi:10.1088/0960-1317/26/3/035008.

Jiang, N., Montelongo, Y., Butt, H. & Yetisen, A. K. (2018). Microfluidic Contact Lenses. *Small*, *14*(15), 1704363. doi:10.1002/smll.201704363.

Karle, M., Vashist, S. K., Zengerle, R. & von Stetten, F. (2016). Microfluidic Solutions Enabling Continuous Processing and Monitoring of Biological Samples: A Review. *Anal.Chim.Acta*, *929*, 1–22. doi:https://doi.org/10.1016/j.aca.2016.04.055.

Kenry, Yeo, J. C., Yu, J., Shang, M., Loh, K. P. & Lim, C. T. (2016). Highly Flexible Graphene Oxide Nanosuspension Liquid-Based Microfluidic Tactile Sensor. *Small*, *12*(12), 1593–1604. doi:10.1002/smll.201502911.

Kim, H., Kim, P., Yoo, H. & Kim, C. (2012). Comparison of Tear Proteins Between Healthy and Early Diabetic Retinopathy Patients. *Clin. Biochem.*, *45*(1), 60–67. doi:https://doi.org/10.1016/j.clinbiochem.2011.10.006.

Kim, J., Kim, M., Lee, M., Kim, K., Ji, S., Kim, Y.…Park, J. (2017). Wearable Smart Sensor Systems Integrated on Soft Contact Lenses for Wireless Ocular Diagnostics. *Nat. Commun.*, *8*(1), 14997. doi:10.1038/ncomms14997.

Koh, A., Kang, D., Xue, Y., Lee, S., Pielak, R. M., Kim, J.…Rogers, J. A. (2016). A soft, wearable Microfluidic Device for the Capture, Storage, and Colorimetric Sensing of Sweat. *Sci. Transl. Med.*, *8*(366), 366ra165. doi:10.1126/scitranslmed.aaf2593.

Lei, Y., Liu, Y., Wang, W., Wu, W. & Li, Z. (2011). Studies on Parylene C-caulked PDMS (pcPDMS) for low permeability required microfluidics applications. *Lab Chip*, *11*(7), 1385–1388. doi:10.1039/C0LC00486C.

Leonardi, M., Leuenberger, P., Bertrand, D., Bertsch, A. & Renaud, P. (2004). First Steps toward Noninvasive Intraocular Pressure Monitoring with a Sensing Contact Lens. *Invest. Ophthalmol. Vis. Sci.*, *45*(9), 3113–3117. doi:10.1167/iovs.04-0015.

Leonardi, M., Pitchon, E. M., Bertsch, A., Renaud, P. & Mermoud, A. (2009). Wireless Contact Lens Sensor for Intraocular Pressure Monitoring: Assessment on Enucleated Pig Eyes. *Acta Ophthalmol.*, *87*(4), 433–437. doi:10.1111/j.1755-3768.2008.01404.x.

Liu, H. & Gong, H. (2009). Templateless Prototyping of Polydimethylsiloxane Microfluidic Structures Using a Pulsed CO2 Laser. *J. Micromech. Microeng.*, *19*(3), 037002. doi:10.1088/0960-1317/19/3/037002.

Liu, J., Shi, B., He, S., Yao, X., Willcox, M. D. P. & Zhao, Z. (2010). Changes to Tear Cytokines of Type 2 Diabetic Patients with or without Retinopathy. *Mol. Vision*, *16*, 2931–2938.

Liu, J. H. (1998). Circadian Rhythm of Intraocular Pressure. *J. Glaucoma*, *7*(2), 141–147.

Liu, J. H., Bouligny, R. P., Kripke, D. F., & Weinreb, R. N. (2003a). Nocturnal Elevation of Intraocular Pressure is Detectable in the Sitting Position. *Invest. Ophthalmol. Vis. Sci.*, *44*(10), 4439–4442.

Liu, J. H. K., Zhang, X., Kripke, D. F. & Weinreb, R. N. (2003b). Twenty-four-Hour Intraocular Pressure Pattern Associated with Early Glaucomatous Changes. *Invest. Ophthalmol. Vis. Sci.*, *44*(4), 1586–1590. doi:10.1167/iovs.02-0666.

Mair, D. A., Geiger, E., Pisano, A. P., Fréchet, J. M. J. & Svec, F. (2006). Injection Molded Microfluidic Chips Featuring Integrated Interconnects. *Lab Chip*, *6*(10), 1346–1354. doi:10.1039/B605911B.

Mansouri, K. & Shaarawy, T. (2011). Continuous Intraocular Pressure Monitoring with a Wireless Ocular Telemetry Sensor:Initial Clinical Experience in Patients with Open Angle Glaucoma. *Br. J. Ophthalmol.*, *95*(5), 627. doi:10.1136/bjo.2010.192922.

Mansouri, K., Medeiros, F. A., Tafreshi, A. & Weinreb, R. N. (2012). Continuous 24-Hour Monitoring of Intraocular Pressure Patterns with a Contact Lens Sensor: Safety, Tolerability, and Reproducibility in Patients with Glaucoma. *Arch. Ophthalmol.*, *130*(12), 1534–1539. doi:10.1001/archophthalmol.2012.2280.

Michail, D. & Zolog, N. (1937). Sur L'elimination Lacrymale Du Glucose Au Cours Dr L'hyperglycemie Alimentaire. *C. R. Soc. Biol. Paris*, *126*, 1042.

Michail, D., Vancea, P. & Zolog, N. (1937). Sur L'elimination Lacrymale Du Glucose Chez Les Diabetiques. *C. R. Soc. Biol.*, *125*, 1095.

Miller, D. (1963). An Analysis of the Physical Forces Applied to a Corneal Contact Lens. *Arch. Ophthalmol.*, *70*(6), 823–829. doi:10.1001/archopht.1963.00960050825018.

Mirza, G. E., Karaküçük, S., Er, M., Güngörmüş, N., Karaküçük, İ & Saraymen, R. (2001). Tear Nitrite and Nitrate Levels as Nitric Oxide End Products in Patients with Behçet's Disease and Non-Behcet's Uveitis. *Ophthal. Res.*, *33*(1), 48–51. doi:10.1159/000055641.

Monge-Roffarello, B., Labbe, S. M., Lenglos, C., Caron, A., Lanfray, D., Samson, P. & Richard, D. (2014). The Medial Preoptic Nucleus as a Site of the Thermogenic and Metabolic Actions of Melanotan II in Male Rats. *Am. J. Physiol. Regul. Integr. Comp. Physiol.*, *307*(2), R158–R166. doi:10.1152/ajpregu.00059.2014.

Moreddu, R., Elsherif, M., Butt, H., Vigolo, D. & Yetisen, A. K. (2019). Contact Lenses for Continuous Corneal Temperature Monitoring. *RSC Adv.*, *9*(20), 11433–11442. doi:10.1039/C9RA00601J.

Moreddu, R., Elsherif, M., Adams, H., Moschou, D., Cordeiro, M. F., Wolffsohn, J. S.… Yetisen, A. K. (2020a). Integration of Paper Microfluidic Sensors into Contact Lenses for Tear Fluid Analysis. *Lab Chip*, *20*(21), 3970–3979. doi:10.1039/D0LC00438C.

Moreddu, R., Wolffsohn, J. S., Vigolo, D. & Yetisen, A. K. (2020b). Laser-Inscribed Contact Lens Sensors for the Detection of Analytes in the Tear Fluid. *Sens. Actuators B: Chem.*, *317*, 128183. doi:https://doi.org/10.1016/j.snb.2020.128183.

Mostafalu, P., Akbari, M., Alberti, K. A., Xu, Q., Khademhosseini, A. & Sonkusale, S. R. (2016). A Toolkit of Thread-Based Microfluidics, Sensors, and Electronics for 3D Tissue Embedding for Medical Diagnostics. *Microsyst. Nanoeng.*, *2*(1), 16039. doi:10.1038/micronano.2016.39.

Nathan, D. M., Genuth, S., Lachin, J., Cleary, P., Crofford, O., Davis, M., ... & Siebert, C. (1993). The Effect of Intensive Treatment of Diabetes on the Development and Progression of Long-term Complications in Insulin-Dependent Diabetes Mellitus. *New England Journal of Medicine*, *329*, 977–986.

Nie, B., Li, R., Brandt, J. D. & Pan, T. (2014). Iontronic Microdroplet Array for Flexible Ultrasensitive Tactile Sensing. *Lab Chip*, *14*(6), 1107–1116. doi:10.1039/C3LC50994J.

Norn, M. (1985). Tear pH After Instillation of Buffer in Vivo. *Acta Ophthalmol.*, *63*, 32–34. doi:10.1111/j.1755-3768.1985.tb06834.x.

Oliver, N. S., Toumazou, C., Cass, A. E. G. & Johnston, D. G. (2009). Glucose sensors: A Review of Current and Emerging Technology. *Diabetic Med.*, *26*(3), 197–210. doi:10.1111/j.1464-5491.2008.02642.x.

Omi, T. & Numano, K. (2014). The Role of the CO_2 Laser and Fractional CO_2 Laser in Dermatology. *Laser Ther.*, *23*(1), 49–60. doi:10.5978/islsm.14-RE-01.

Ota, H., Chen, K., Lin, Y., Kiriya, D., Shiraki, H., Yu, Z.…Javey, A. (2014). Highly Deformable Liquid-State Heterojunction Sensors. *Nat. Commun.*, *5*(1), 5032. doi:10.1038/ncomms6032.

Prakash, S. & Kumar, S. (2016). Experimental and Theoretical Analysis of Defocused CO2 Laser Microchanneling on PMMA for Enhanced Surface Finish. *J. Micromech. Microeng.*, *27*(2), 025003. doi:10.1088/1361-6439/27/2/025003.

Purslow, C. & Wolffsohn, J. S. (2005). Ocular Surface Temperature: A Review. *Eye Contact Lens*, *31*(3), 117–123. doi: 10.1097/01.icl.0000141921.80061.17.

Riaz, R. S., Elsherif, M., Moreddu, R., Rashid, I., Hassan, M. U., Yetisen, A. K. & Butt, H. (2019). Anthocyanin-Functionalized Contact Lens Sensors for Ocular pH Monitoring. *ACS Omega*, *4*(26), 21792–21798. doi:10.1021/acsomega.9b02638.

Rim, Y. S., Bae, S., Chen, H., Yang, J. L., Kim, J., Andrews, A. M.…Tseng, H. (2015). Printable Ultrathin Metal Oxide Semiconductor-Based Conformal Biosensors. *ACS Nano*, *9*(12), 12174–12181. doi:10.1021/acsnano.5b05325.

Sage, I. (2011). Thermochromic Liquid Crystals. *Liq. Cryst.*, *38*(11–12), 1551–1561. doi:10.1080/02678292.2011.631302.

Salih, A. E., Elsherif, M., Ali, M., Vahdati, N., Yetisen, A. K. & Butt, H. (2020). Ophthalmic Wearable Devices for Color Blindness Management. *Adv. Mater. Technol.*, *5*(8), 1901134. doi:https://doi.org/10.1002/admt.201901134.

Salih, A. E., Elsherif, M., Alam, F., Yetisen, A. K. & Butt, H. (2021). Gold Nanocomposite Contact Lenses for Color Blindness Management. *ACS Nano.*, *15*(3), 4870–4880. doi:10.1021/acsnano.0c09657.

Salvatore, G. A., Munzenrieder, N., Kinkeldei, T., Petti, L., Zysset, C., Strebel, I.…Troster, G. (2014). Wafer-scale Design of Lightweight and Transparent Electronics that Wraps Around Hairs. *Nat. Commun.*, *5*(1), 2982. doi:10.1038/ncomms3982.

Schirhagl, R., Ren, K. & Zare, R. N. (2012). Surface-Imprinted Polymers in Microfluidic Devices. *Sci. China Chem.*, *55*(4), 469–483. doi:10.1007/s11426-012-4544-7.

Sen, D. K. & Sarin, G. S. (1980). Tear Glucose Levels in Normal People and in Diabetic Patients. *Br. J. Ophthalmol.*, *64*(9), 693. doi:10.1136/bjo.64.9.693.

Sniegowski, M., Erlanger, M., Velez-Montoya, R. & Olson, J. L. (2015). Difference in Ocular Surface Temperature by Infrared Thermography in Phakic and Pseudophakic Patients. *Clin. Ophthalmol.*, *9*, 461–466. doi:10.2147/OPTH.S69670.

Sniegowski, M. C., Erlanger, M. & Olson, J. (2018). Thermal Imaging of Corneal Transplant Rejection. *Int. Ophthalmol.*, *38*(6), 2335–2339. doi:10.1007/s10792-017-0731-z.

Soria, J., Durán, J. A., Etxebarria, J., Merayo, J., González, N., Reigada, R.…Suárez, T. (2013). Tear Proteome and Protein Network Analyses Reveal a Novel Pentamarker Panel for Tear Film Characterization in Dry Eye and Meibomian Gland Dysfunction. *J. Proteomics*, *78*, 94–112. doi:https://doi.org/10.1016/j.jprot.2012.11.017.

Tan, L., Cai, Z. & Lai, N. (2009). Accuracy and Sensitivity of the Dynamic Ocular Thermography and Inter-Subjects Ocular Surface Temperature (OST) in Chinese Young Adults. *Contact Lens Anterior Eye*, *32*(2), 78–83. doi:https://doi.org/10.1016/j.clae.2008.09.003.

Terada, O., Chiba, K., Senoo, T. & Obara, Y. (2004). Ocular Surface Temperature of Meibomia Gland Dysfunction Patients and the Melting Point of Meibomian Gland Secretions. *Nippon Ganka Gakkai zasshi, 108*(11), 690–693.

Thygesen, J. E. M. & Jensen, O. L. (1987). pH Changes of the Tear Fluid in the Conjunctival Sac During Postoperative Inflammation of the Human Eye. *Acta Ophthalmol., 65*(2), 134–136. doi:10.1111/j.1755-3768.1987.tb06990.x.

Tseng, C., Huang, Y., Tsai, S., Yeh, G., Chang, C. & Chiou, J. (2012). *Design and Fabricate a Contact Lens Sensor with a Micro-Inductor Embedded for Intraocular Pressure Monitoring.* SENSORS, IEEE, 2012, pp. 1–4. doi: 10.1109/ICSENS.2012.6411234.

Unger, M. A., Chou, H., Thorsen, T., Scherer, A. & Quake, S. R. (2000). Monolithic Microfabricated Valves and Pumps by Multilayer Soft Lithography. *Science, 288*(5463), 113. doi:10.1126/science.288.5463.113.

Van Haeringen, N. J. (1981). Clinical Biochemistry of Tears. *Surv.Ophthalmol., 26*(2), 84–96. doi:https://doi.org/10.1016/0039-6257(81)90145-4.

Vlachopoulou, M., Tserepi, A., Pavli, P., Argitis, P., Sanopoulou, M. & Misiakos, K. (2008). A Low Temperature Surface Modification Assisted Method for Bonding Plastic Substrates. *J. Micromech. Microeng., 19*(1), 015007. doi:10.1088/0960-1317/19/1/015007.

Weinreb, R. N. & Liu, J. H. K. (2006). Nocturnal Rhythms of Intraocular Pressure. *Arch. Ophthalmol., 124*(2), 269–270. doi:10.1001/archopht.124.2.269.

Weitzman, E. D., Henkind, P. A. U. L., Leitman, M. A. R. K. & Hellman, L. (1975). Correlative 24-Hour Relationships Between Intraocular Pressure and Plasma Cortisol in Normal Subjects and Patients with Glaucoma. *Br. J. Ophthalmol., 59*(10), 566–572.

Whikehart, D. R. (2003). Chapter 3 - Enzymes: Ocular Catalysts. In D. R. Whikehart (Ed.) *Biochemistry of the Eye* (Second Edition, pp. 55–84). Philadelphia, PA: Butterworth-Heinemann. doi:https://doi.org/10.1016/B978-0-7506-7152-1.50007-2.

Whitesides, G. M. (2006). The Origins and the Future of Microfluidics. *Nature, 442*(7101), 368–373. doi:10.1038/nature05058.

Xu, J., Xue, Y., Hu, G., Lin, T., Gou, J., Yin, T.…Tang, X. (2018). A Comprehensive Review on Contact lens for Ophthalmic Drug Delivery. *J. Controlled Release, 281*, 97–118. doi:https://doi.org/10.1016/j.jconrel.2018.05.020.

Yan, J. (2011). An Unpowered, Wireless Contact Lens Pressure Sensor for Point-of-Care Glaucoma Diagnosis. *Annu. Int. Conf. IEEE Eng. Med. Biol. Soc.*, 2522–2525. doi: 10.11 09/IEMBS.2011.6090698.

Yang, X., Yao, H., Zhao, G., Ameer, G. A., Sun, W., Yang, J. & Mi, S. (2020). Flexible, Wearable Microfluidic Contact Lens with Capillary Networks for Tear Diagnostics. *J. Mater. Sci., 55*(22), 9551–9561. doi:10.1007/s10853-020-04688-2.

Yao, H., Shum, A. J., Cowan, M., Lahdesmaki, I. & Parviz, B. A. (2011). A Contact Lens with Embedded Sensor for Monitoring Tear Glucose Level. *Biosens. Bioelectron., 26*(7), 3290–3296. doi:https://doi.org/10.1016/j.bios.2010.12.042.

Yeo, J. C., Kenry & Lim, C. T. (2016a). Emergence of Microfluidic Wearable Technologies. *Lab Chip, 16*(21), 4082–4090. doi:10.1039/C6LC00926C.

Yeo, J. C., Kenry, Yu, J., Loh, K. P., Wang, Z. & Lim, C. T. (2016b). Triple-State Liquid-Based Microfluidic Tactile Sensor with High Flexibility, Durability, and Sensitivity. *ACS Sens., 1*(5), 543–551. doi:10.1021/acssensors.6b00115.

Yetisen, A. K., Naydenova, I., da Cruz Vasconcellos, F., Blyth, J. & Lowe, C. R. (2014). Holographic Sensors: Three-Dimensional Analyte-Sensitive Nanostructures and Their Applications. *Chem. Rev., 114*(20), 10654–10696. doi:10.1021/cr500116a.

Yetisen, A. K., Jiang, N., Tamayol, A., Ruiz-Esparza, G., Zhang, Y. S., Medina-Pando, S.… Yun, S. (2017). Paper-Based Microfluidic System for Tear Electrolyte Analysis. *Lab Chip, 17*(6), 1137–1148. doi:10.1039/C6LC01450J.

Yetisen, A. K., Soylemezoglu, B., Dong, J., Montelongo, Y., Butt, H., Jakobi, M. & Koch, A. W. (2019). Capillary Flow in Microchannel Circuitry of Scleral Lenses. *RSC Adv.*, *9*(20), 11186–11193. doi:10.1039/C9RA01094G.

Yetisen, A. K., Jiang, N., Castaneda Gonzalez, C. M., Erenoglu, Z. I., Dong, J., Dong, X.… Koch, A. W. (2020). Scleral Lens Sensor for Ocular Electrolyte Analysis. *Adv. Mater.*, *32*(6), 1906762. doi:10.1002/adma.201906762.

Zhang, J., Hodge, W., Hutnick, C. & Wang, X. (2011). Noninvasive Diagnostic Devices for Diabetes through Measuring Tear Glucose. *J. Diabetes Sci. Technol.*, *5*(1), 166–172. doi:10.1177/193229681100500123.

Zhu, Z., Li, R. & Pan, T. (2018). Imperceptible Epidermal-Iontronic Interface for Wearable Sensing. *Adv. Mater.*, *30*(6), 1705122. doi:10.1002/adma.201705122.

Zimmet, P. Z., Magliano, D. J., Herman, W. H. & Shaw, J. E. (2014). Diabetes: A 21st Century Challenge. *Lancet Diabetes Endocrinol.*, *2*(1), 56–64. doi:10.1016/s2213-8587(13)70112-8.

13 Microfluidic Platforms for Wound Healing Analysis

Lynda Velutheril Thomas and Priyadarsini Sreenivasan

CONTENTS

13.1 INTRODUCTION

A structured therapeutic approach which involves regular feedback is the main prerequisite for proper wound management. This requires rigorous assessment of the wound environment on a timely basis and is central to an adequate treatment plan (Ubbink et al. 2014, Gupta et al. 2017). Unfortunately, in most cases, such diagnosis for wound assessment requires the swabbing of the wound surface and transporting the swab sample to a centralized equipped laboratory for testing, which becomes a time-consuming process and a bottleneck in resource-limited settings (Santy 2008,

DOI: 10.1201/9781003033479-13

Kiernan 1998, Healy and Freedman 2006). For people living in such areas where there is a dearth of necessary technical and economic resources, wound assessment thereby becomes practically impossible, hindering the clinician from providing accurate diagnosis on the healing of wounds and giving treatment at the right time. The development of simple and affordable devices capable of performing point-of-care (POC) testing rapidly on different parameters which indicate the status of wound healing becomes significant on these grounds (Brown et al.2018). It is in this realm that the use of microfluidics in such devices becomes well suited in meeting the requirements of POC diagnostic devices (Coltro et al. 2014). Microfluidics platforms have been made using several materials like silicon, polydimethylsiloxane (PDMS), poly (methyl methacrylate) (PMMA), and other polymers, with different modes of detection, each with their own advantages and disadvantages (Lazar 2015, Carrell et al. 2019). In this chapter, we discuss the several aspects of wound healing, the different parameters used for assessing the wounds, and how they can be integrated or multiplexed into a single detection platform.

13.2 WOUND FLUID ANALYSIS – CHALLENGES AND KEY CONSIDERATIONS

Wounds are a transitional tissue that occur due to any break in skin anatomy. The majority of wounds heal without any complications, however, in some cases healing is impeded due to many underlying reasons (Darlene et al.2005, Spear 2012, Percival 2002). Hence, it becomes mandatory to understand the underlying conditions that might impede the healing process. Several studies are carried out to assess the healing process of wounds. Wound swabbing and biopsies are used as they are a reservoir for several potential markers which provide quantitative information on the progress in the healing of wounds and the complications involved. These biochemical measurements are also important to understand the contribution of several factors, like the cellular products that constitute the wound fluid, the wound surface, and its milieu, which determines the rate at which tissue repair is initiated. This information would also play a major role in developing customized and evidence-based treatment protocols. Wound fluid is easily accessible and indicative of the status of wound healing and hence it will be ideal to probe these exudate specimens, apart from clinical examination and microbiological swabbing, as they contain a plethora of information. As reported by Cutting (2003), wound exudate has high protein content with a specific gravity greater than 1.020. It is basically composed of essential nutrients for epithelial cells, facilitates the ingress of white cells, and also contains electrolytes and a number of inflammatory components, such as leukocytes, fibrinogen, and fibrin. These exudates are mostly released in the inflammatory and proliferative phases of the wound healing process and the volume will vary depending not only on the extent of healing but also on the wound type, its origin, and location. The presentation of exudates is also different depending on the wound type.

There are four types of wound exudates – serous, sanguineous, serosanguinous, and purulent (Tickle et al. 2016) [Figure 13.1]. Serous exudates are clear, thin, and watery. These are mainly observed during the normal inflammatory healing stage.

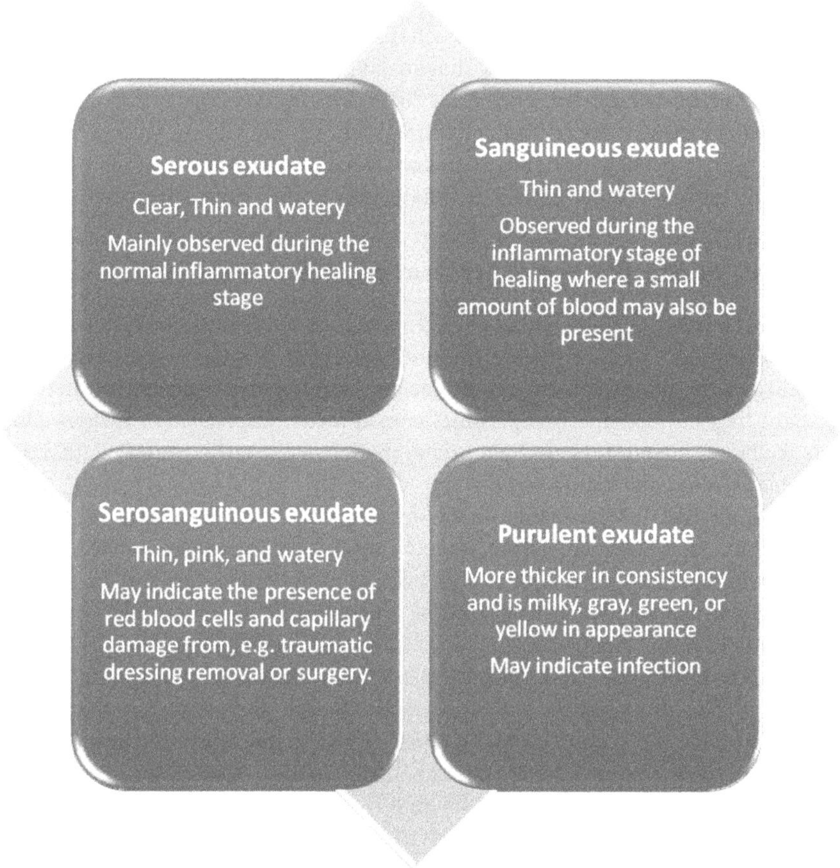
Serous exudate
Clear, Thin and watery
Mainly observed during the normal inflammatory healing stage
Sanguineous exudate
Thin and watery
Observed during the inflammatory stage of healing where a small amount of blood may also be present
Serosanguinous exudate
Thin, pink, and watery
May indicate the presence of red blood cells and capillary damage from, e.g. traumatic dressing removal or surgery.
Purulent exudate
More thicker in consistency and is milky, gray, green, or yellow in appearance
May indicate infection

FIGURE 13.1 Types of wound exudates, their presentation, and indication.

Sanguineous drainage is also seen during the inflammatory stage of healing, where a small amount of blood may also be present. If it is seen outside of the inflammatory phase, sanguineous drainage can be a result of trauma to the wound.

Serosanguinous exudate is thin, pink, and watery in presentation. Purulent drainage is thicker in consistency and is milky, gray, green, or yellow in appearance. A very thick fluid can be a sign of infection. Wound fluid analysis opens scope for examination of the wound microenvironment and hence development of platforms to test the wound exudate would be ideal. Moreover, microfluidics platforms integrated with different detection methods, which use very small quantities of exudate fluid to generate information regarding wound healing and infection, would be a breakthrough especially as POC devices in areas that do not have access to sophisticated diagnostic laboratories.

The analysis of such biochemical markers from wound exudates might be of great clinical value and helps in predicting the probability of wound healing, which will assist the clinicians in arriving at the right conclusions during treatment. This is vital, as infection is a major cause of much disease-related morbidity and mortality, like diabetes. In this scenario, novel diagnostic tools that support prompt decision-making are urgently required (Hsu et al. 2019). Ideally, development of a microfluidics-based diagnostic device that measures such wound healing parameters from minimal wound exudate volumes that can be collected noninvasively with little effort, will result in a better patient as well as greater physician acceptance.

13.3 MICROFLUIDICS-BASED DIAGNOSTIC DEVICES

The birth of microfluidics can be attributed to the ground breaking work of Ukrainian scientists Izmailov and his student Shraiber (in 1938) in which they analyzed the feasibility of the spot chromatographic technique for separating plant extracts. For the study, they utilized microscope slides coated with a suspension of various adsorbents (calcium, magnesium, and aluminum oxide) which were overlaid with a drop of sample solution and a drop of solvent.

The components of the plant extract solution were thereby separated into concentric rings on the slide and were detected by the fluorescence exhibited under a UV lamp (Shostenko et al. 2000). This age-old technique which utilizes droplets of sample and solvent can be regarded as a progenitor of present-day microfluidics. The microfluidic analytical devices of today are mainly employed in four main domains which are molecular analysis, biodefense, molecular biology, and microelectronics. The invention of analytical methods like gas phase chromatography (GPC), high pressure liquid chromatography (HPLC), and capillary electrophoresis (CE), which revolutionized the arena of chemical analysis also gave birth to microfluidic microanalytical methods with the first functional lab-on-a-chip being created for gas chromatography at Stanford University in 1979. The development of various novel microanalytical techniques which would require minimum sample volume and maximum miniaturization was nurtured by the adoption of laser-based optical detection techniques, which in addition, improved the accuracy of detection. The necessity of microfluidics platforms was further magnified in the 1980s with the advent of high-throughput genomic techniques which demanded enhanced sensitivity and resolution. The successful application of microfluidics operations scaled down the whole process to the level of a single cell for the first time. In the 1990s towards the end of the Cold War, the focus was on the development of field-deployable microfluidics sensors against chemical and biological weapons which posed major military and terrorist threats. The Defense Advanced Research Projects Agency (DARPA) of the US Department of Defense funded a series of programs to address this concern. This further catalyzed the growth of microfluidics technology.

The 1990s also saw the beginning of micro total analysis systems where plastics implemented instead of silicon and glass-based fluidic systems enhanced the integration of various chemical, biological, and biomedical protocols onto a single chip. Such platforms equipped with plastic-based microelectronic systems could handle the complete protocol beginning from sample collection to detection. Towards the beginning of the 21st century, soft-lithography techniques, which were rapid and cost effective, started gaining attention for the fabrication of microchannels using polymers like PDMS.POCKET, another microfluidic platform, was developed in 2004 by the Whitesides group and was used to carry out immunoassays with the same efficiency as a bench-top enzyme-linked immunosorbent assay (ELISA), but in less time and in a cost-effective manner (Figure 13.2).

The remarkable attributes of microfluidics systems, like smaller sampling volumes, lower test costs, and faster turnaround times adequately satisfies the WHO-defined requirements for analytical and diagnostic devices. To achieve this, different technologies like directed fluid flow, immobilized reagents, sequential mixing in miniature channels, etc. are utilized in microfluidics systems. Such "bed-side testing" systems are being employed in diagnosing a wide spectrum of conditions ranging from infectious diseases to cancer. Paper- and polymer-based microfluidics devices, having the additional advantages of being disposable and cost effective, are already finding applications in pregnancy testing, HIV diagnosis, and glucose biosensors, etc. The review by Eicher and Merten (2011) gives an overall sketch of the variety of the microfluidics devices currently used, ranging from pathogen-detecting non/minimally instrumented POC testing to fully instrumented, diagnostic leading-edge technologies like droplet-based microfluidics and next-generation sequencing.

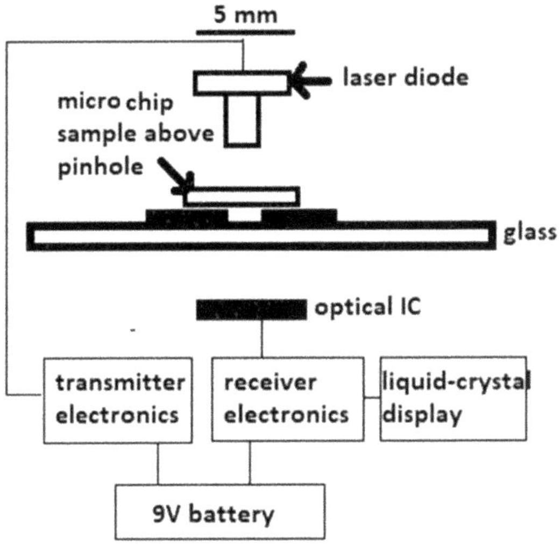

FIGURE 13.2 Schematic representation of the POCKET immunoassay, and performance of the optical detection device. A red light from the laser diode is passed through the silver-coated microwell containing the sample to the optical IC. A pinhole is used to block stray light that did not pass through the sample. The laser diode and the optical IC are driven by the same circuit, which also has an integrated liquid-crystal display that shows the measured transmittance value. (Adapted from Sia et al.2004.)

13.4 COST-EFFECTIVE PAPER-BASED MICROFLUIDICS: NEW TOOLS FOR POINT-OF-CARE DIAGNOSTICS

For resource-limited settings, the WHO defines the attributes of low-cost POC testing as "ASSURED" that is "affordable, sensitive, specific, user friendly, rapid, and robust, equipment free and delivered to those in need," all of which are befitting of microfluidic paper-based analytical devices (μPADs) (Bruzewicz et al. 2008). A paper-based microfluidics company, Diagnostics for All (DFA), has come up with a range of capillary flow-based, colorimetric test platforms for use in resource-limited countries. The concept of μPADs was first introduced by the Whitesides group at Harvard University in 2007. In the past, Müller and Clegg (1938) experimented with paper strips for pH determination by creating wax barriers on paper, which is similar to present-day μPADs. Today, the fabrication of a μPAD involves the creation of hydrophobic barriers on paper which can be patterned into hydrophilic channels that effectively wick the sample fluid to the different assay zones. Paper, with its inherent capillarity, directs the fluid flow by the cohesive and the adhesive forces that are dictated by the surface tension of the fluid and the surface chemistry of the cellulose fiber network, respectively. μPADs can be utilized for quantitative detection when coupled with smartphone cameras. The most widely used methods of fabrication are photolithography, plotting, inkjet printing, plasma etching, flexographic printing (Ispas et al. 2012), wax printing, and cutting patterns of channels from a sheet of paper (Nahavandi et al. 2014).The reagents are immobilized on μPADs using different methods like physical absorption, chemical coupling, carrier-mediated (via nanoparticles) deposition, sol–gel processing, dip casting, contact, and noncontact inkjet printing (Nath et al. 2015, Qi et al. 2015).

Several analytes like glucose, bovine serum albumin, nitrites, ketones, alkaline phosphatase, and cholesterol (Martinez et al. 2010) have been detected colorimetrically using μPADs. Other methods of detection include electrochemical (Martinez et al. 2010), transmittance (Ellerbee et al. 2009), chemiluminescence (Delaney et al. 2011) and fluorescence.

The area of microfluidics paper-based analytical devices in the diagnostic arena is showing a lot of promise in meeting the critical demands of simple and cost-effective testing in remote and resource-limited settings. In the last ten years, significant progress has been made in the use of μPADs in POC diagnostics. However more healthcare testing and diagnosis areas need to be explored using this technology where parameters to detect wound healing may be assessed through the sampled wound exudates. Furthermore, several challenges also need to be addressed which include improving reproducibility, obtaining higher levels of specificity, and higher limits of detection, improving the long-term storage of such devices and addressing the needs of multiplexing with more quantitative data interpretation.

13.5 PARAMETERS ASSESSED TO DETERMINE WOUND HEALING

Since wound presentation during different stages of the healing process is different, real-time or continuous monitoring is necessary to determine the healing process.

There are several markers which are thought of as possible diagnostic targets for assessing wounds. As per Dargaville et al. (2013) some of the potential markers which can be assessed for wound healing include bacterial load/specific microbial species/biofilms, cytokine release in response to specific microbial antigens, DNA – e.g., gene polymorphisms to indicate susceptibility to disease, poor healing, or infection; enzymes, e.g., matrix metalloproteinases and extracellular matrix; growth factors, e.g., platelet-derived growth factor (PDGF); immunohistochemical markers – e.g., integrins, chemokine receptors, and transforming growth factor beta II receptors to monitor healing status; inflammatory mediators – e.g., cytokines, nitric oxide; nutritional factors – e.g., zinc, glutamine, vitamins, pH of wound fluid, reactive oxygen species, transepidermal water loss from periwound skin, etc. Some of the parameters, along with the mode of measurements using microfluidics, are discussed below.

13.5.1 Microbial Load and Activity

The epidermis layer of the skin, being the first line of defense, is highly vulnerable and prone to bacterial colonization. The biofilms produced by these microorganisms gain resistance to common treatment methodologies through a phenomenon called quorum sensing. They produce toxins and virulence factors which are excreted to the bloodstream. A real-time bacterial growth-monitoring sensor which could be directly employed at the wound site was developed by Farrow et al. (2012). The presence or absence of bacteria was indicated by differences in the captured impedance profiles which was confirmed by plate-counting methods. Likewise, Miller et al. (2019) established the use of solution-blown spun poly (lactic acid)/multiwalled carbon nanotube nanofiber composites in vitro to detect the presence and quantify the concentration of *Pseudomonas putida* using changes in impedance. Simoska and Stevenson (2019) worked on the latest electrochemical-based approaches for pathogen sensing and obtained similar results to conventional techniques like cell culturing, mass spectrometry, and fluorescent-based approaches. In general, impedance microbiology is widely employed in areas like cancer detection, tuberculosis screening, and pregnancy tests. Many devices have been developed that use fluorescence detection of pathogens and bacterial proteases on wound surfaces giving real time information like MolecuLight i:X and WoundCheck which have been found to be very effective. The MolecuLight i:X device has obtained FDA clearance. The device uses the principle of fluorescence imaging to detect bacteria with a bacterial load of $>10^4$ cfu/g (colony-forming units per gram) which indicates them through red highlighting in the device. Gahlaut et al. (2019) developed a smartphone-based dual mode in situ detection of viability of bacteria using an Ag nanorods array. Wang et al. (2012) developed an immune-based microchip technology that can rapidly detect and quantify bacterial presence in various sources including physiologically relevant buffer solutions. Another group, Altintas et al. (2018) came out with a fully automated microfluidics-based electrochemical biosensor for pathogen detection and the quantification of *Escherichia coli* was performed with standard and nanomaterial-amplified immunoassays in the concentration ranges of

0.99×10^4–3.98×10^9 cfu mL^{-1} and 10–3.97×10^7 cfu mL^{-1} which resulted in detection limits of 1.99×10^4 cfu mL^{-1} and 50 cfu mL^{-1}, respectively. Furst and Francis (2019) report on the use of impedance-based POC detection of bacteria. Although the authors are skeptical about the commercialization of these sensors, they confirm that detecting pathogens from real-world environments using this detection method is promising.

13.5.2 Enzymes and Their Substrates

Matrix metalloproteinases and extracellular matrices are also potential markers for wound healing. Some of the most commonly used analytes include matrix metalloproteinase-9 (MMP-9), matrix metalloproteinase-2 (MMP-2), tissue inhibitor of metalloproteinase (TIMP), neutrophil elastase (NE), and albumin. Matrix metalloproteinases are commonly analyzed using immunoassays (e.g., antibodies labeled with fluorescent dyes) or by measuring the enzymatic activity (Lei et al. 2020). Song et al. (2013) developed a novel graphene oxide-based fluorescence resonance energy transfer (FRET) biosensor wherein fluorescein isothiocyanate-labeled peptide (Pep-FITC) bound to a graphene oxide surface which detected MMP-2 in complex serum samples. Even in the presence of other MMPs, this biosensor has displayed high selectivity and has recorded a detection limit of 2.5 ng/mL. Puchberger-Enengl et al. (2015) summarized that the MMP-9 levels for healing wounds are in the range of 1.18 ± 1.21 μg/mL and nonhealing wounds are in the range of 2.9 ± 1.64 μg/mL. A microfluidics platform produced by the in situ polymerization of polyethylene glycol diacrylate (PEG-DA) in a single step was used for the production of MMP-9 assaying biosensing hydrogels. A potentiometric sensor developed by Milne et al. (2014)was used for monitoring different aspects like wound moisture, the pH of the fluid, and wound matrix metalloproteinases (MMP) enzyme activity. The sensor required a very minimal volume of sample, as little as 10 μL, for accurate pH detection. The moisture sensor used in this device was fabricated by applying silver chloride ink onto a biocompatible polymer via screen printing and an electrochemical-based MMP-9 sensor was also developed.

13.5.3 Immunohistochemical Markers

Immunohistochemical markers like extracellular and intracellular proteins can also be utilized as cellular biological markers for wound assessment. This application of biosensing the intracellular and extracellular proteins was studied by Becker et al. (1986), Ouhayoun et al. (1990), and Carmichael et al. (1991). Some examples of intracellular markers in wounds are cytokeratins (CKs), vimentin, and vinculin, and extracellular markers are collagen IV, laminin, and fibronectin (Mai et al. 2009). CKs are a part of the cytoskeleton which are intermediate filament proteins found in the cytoplasm of epithelial cells. Moll et al. (1982)identified twenty different CK polypeptides in human tissues with molecular weights ranging from 40 to 68 kDa. There are basically two different types of CK, basically acid and neutral/basic CK, which can be distinguished using electrophoresis. The distribution patterns of CK are different but distinct in various epithelia and within stratified epithelia. Shabana et al. (1989) observed that characteristic distributions of CK are available in the stratum basale, stratum spinosum, stratum granulosum, and stratum corneum. Based on this, Boisnic et al. (1995) concluded that they can be utilized as specific markers for pathways of epithelial differentiation. Barui et al. (2011) investigated the impact of honey-based occlusive dressing on nonhealing (unresponsive to conventional antibiotics) traumatic lower limb wounds through clinicopathological and immunohistochemical (e.g., expression of p63, E-cadherin) and collagen I and III evaluations. Nevertheless, immunohistochemical markers are not yet implemented in microfluidics platforms and their capabilities are still being explored.

13.5.4 Nitric Oxide

An amperometric nitric oxide microfluidics sensor was developed by Hunter et al. (2013) using a standard photolithographic technique with a xerogel polymer as the ink. This sensor provided advantages like low background noise and minimal sample volume requirement (~250 μL). The sensor exhibited a nitric oxide sensitivity of 1.4 $pAnM^{-1}$, a limit of detection (LOD) of 840 pM, and excellent analytical performances in phosphate buffered saline.

13.5.5 Nutritional Factors

Chow and Barbul (2014) elicited the role of immunonutrition in the wound healing process. Certain amino acids like glutamine, arginine, etc. tend to play a crucial role in immune functioning, thereby impacting wound healing through direct and indirect means. For example, glutamine supplementation reduces the chances of infection and offers protection against inflammation by increasing the expression of heat shock proteins which provide protection against inflammation, injury, and stress. Moreover, glutamine also helps in modulating and retaining the gut function which is generally impaired during stress. Similarly, arginine, being a precursor for the formation of other amino acids like proline and ornithine, also accelerates the healing process by improving collagen accumulation since it is specifically recruited for collagen, polyamine, and nitric oxide synthesis. Arginine, which promotes peripheral blood lymphocytes activity and mitogenesis enhances post-traumatic intestinal epithelial reconstitution as well. However, during trials in patients or volunteers, it was observed that arginine has no effect on the re-epithelialization of skin graft donor sites, whether given orally or parenterally.

Another factor, vitamin A, has numerous positive effects on wound healing, for example, it increases collagen cross-linking and wound breaking strength. Vitamin A enhances lysosomal membrane lability, macrophage influx, and activates collagen synthesis, thereby improving the inflammatory response in wounds. Vitamin A increases the recruitment of monocytes and macrophages at the wound site early in the inflammatory phase which would enable epithelial cell differentiation. The most important capability is that it can act upon corticosteroid-induced inhibition of cutaneous wound healing and can facilitate the reversal of this condition. Hence its deficiency can become detrimental in the healing process.

Lin et al. (2017) have thrown light on the role of zinc in modulating the cellular and molecular mechanisms during the wound healing process. Zinc, being a cofactor in a number of intracellular enzymatic reactions, has an important role in the healing process. Apart from its antioxidant and anti-bacterial properties, it also offers resistance against epithelial apoptosis.

The role of selenium, which has antioxidant properties in wound healing for burn patients is currently being explored (Berger et al. 2007). The results of a randomized controlled trial which investigated the effect of high intravenous doses of trace elements (copper, selenium, and zinc) on patients with major burns suggested that with the increase in the cutaneous concentrations of copper, selenium, and zinc, the antioxidant status (as measured by normalization of plasma glutathione peroxidase level) was also improved, and wound healing was also accelerated which was indicated by a decreased graft requirement.

Iron, which is another cofactor required for collagen synthesis, moderates the immune response during inflammation. This is elucidated by the prolongation effect on inflammation as is seen in cases of iron deficiency. However, there are no proven studies about the effects of this factor, suggesting that iron supplementation alone would improve wound healing in the absence of a severe deficiency in the host. Lu et al. (2017) demonstrated a rapid diagnostic test and mobile-enabled platform for simultaneously quantifying iron (ferritin), vitamin A (retinol-binding protein), and inflammation (C-reactive protein) status. Yap et al. (2018) reviewed the potential microfluidics applications in the field of nutrition, specifically to diagnose iron deficiency anemia (IDA) detection. Li et al. (2013) discussed the potential chances for microfluidics in the field of nutrition in their review and highlighted some of the recent advances in microfluidics blood analysis systems that have the capacity to detect biomarkers of nutrition.

13.5.6 pH of Wound Fluid

Secretion of keratinocytes makes the pH of intact skin acidic in the range from 4 to 6. Acidic secretions like amino acids, lactic acid, and fatty acids from sebaceous and sweat glands render the skin pH acidic which also acts as a defense mechanism against pathogens. Moving inwards from the outer layer of skin, the natural pH of the underlying tissues is more in the neutral range, around 7.4. During the initial stages of healing, a temporary acidosis is reported due to production of organic acids which triggers the proliferation of fibroblasts, promoting epithelization and angiogenesis, controlling bacterial colonization (Jones et al. 2015), and facilitating the release of oxygen from oxyhemoglobin. Hence, an acidic pH in the initial stages favors wound healing. On the other hand, alkalinity can deprive the wound tissue of oxygen and can create an environment that accentuates bacterial growth, adversely affecting the healing process. An alkaline pH in the range of 7.15 to 8.93 is reported for delayed and nonhealing wounds (Gethin 2007). This environment can be detrimental to the healing process since the production of extracellular molecules can be halted (Jones et al. 2015). And as the healing progresses the pH slowly decreases from neutral to acidic (Schneider et al. 2007, Percival et al. 2014). pH is commonly measured using potentiometric methods and color-changing indicators. Santos et al. (2014) monitored pH using a simple and portable potentiometric method. Conventionally used indicators and dyes for physiological purposes are bromocresol purple (BCP), phenol red, and phenolphthalein. In such colorimetric approaches the indicator will be immobilized on a flexible substrate and the changes in the color of the indicator will be recorded against the changes in pH using various techniques (Morris et al. 2009) for example, bromocresol green (pH 3.8–5.4) and bromocresol purple (pH 5.2–6.8) being immobilized on tetraethoxysilane (TEOS) thin films. Curto et al. employed a chromic pH-sensitive dye (Curto et al. 2012) to monitor the pH of fresh sweat in a microfluidics platform constructed using PMMA. The operational lifetime was recorded as 135 min. Important advantages of the device include ease of fabrication and modification with reusability being the main limitation. Puchberger-Enengl et al. (2011) demonstrated the continuous measurements of wound pH using a miniaturized optical reflectance pH sensor based on organically modified silicate. Gethin et al. (2007) studied the relation between the mean pH level and mean wound size reduction and observed a significant correlation. During the study, the pH was reduced from 7.72 to 7.26 ($p < 0.001$) and the wound size decreased correspondingly from 10.1 cm^2 to 9.1 cm^2 ($p = 0.274$). A significant relationship could be drawn even though the reduction in wound size failed to have statistical significance. The study by Ono et al. (2015) found statistically significant decreases in wound pH measurement from the beginning of the study until re-epithelialization in wounds that healed without developing infection ($p < 0.05$). On the contrary, statistically significant increases in pH were noted for wounds that developed infection ($p < 0.01$). They could arrive at a positive relationship between wound healing and pH in which progression towards wound healing was seen with a decrease in pH from alkaline towards neutral. According to the previous studies a more acidic environment would aid the process of wound healing since an acidic environment promotes fibroblast proliferation, which facilitates epithelization and angiogenesis, in turn reducing bacterial colonization and improving the release of oxygen from oxyhemoglobin. As the wound progresses towards epithelialization the pH would decrease to reach par with the surface pH of intact skin which is naturally in the acidic pH range (4 to 6). Guinovart et al. (2014) proposed the development of a wearable potentiometric pH cell like a set of screen-printed silver-silver chloride electrodes, impregnated into an adhesive bandage for real-time continuous monitoring of pH changes in a wound. Farooqui and Shamim (2016) came up with a cost-effective continuous wireless monitoring system developed by inkjet printing on a standard bandage which is capable of transmitting signals corresponding to early warnings for conditions like irregular bleeding, variations in pH levels, and external pressure applied at wound sites. Changes in pH were measured using different acid and base solutions as the resistance reduces with increase in pH. The novel sensor developed has a very high sensitivity and specificity to the change in concentrations of $H^+$and OH^- ions and the reading therefore remains unaffected by the presence of other anions and cations. With changes in pH, the sensor recorded a maximum variation in the resistance of around ±2.6%.

13.5.7 Reactive Oxygen Species

For inflammatory disorders, reactive oxygen species (ROS) are important signaling molecules that play a vital role in the molecular mechanism of inflammatory disorders. They are associated with the delay in wound healing (Dunnill et al. 2017, Mittal et al 2013). Generally, radical and nonradical molecules formed from oxygen, such as superoxide anions, hydroxyl radicals, singlet oxygen, and hydrogen peroxide are collectively referred to as ROS. Neutrophils are actively involved in the defense against bacteria. As an initial response to a fresh wound, neutrophils release an oxidative burst of ROS via Nicotinamide Adenine Dinucleotide Phosphate Hydrogen (NADPH) oxidase complex localized in plasma membrane. Wlaschek and Scharffetter-Kochanek (2005) observed an extended inflammatory phase associated with increased number of neutrophils and macrophages as opposed to rapidly healing wounds since they resulted in high amounts of ROS. Some researchers have used the byproducts of the reaction of ROS with other stable molecules, like allantoin and uric acid, to assess ROS levels in chronic wounds since they have less stability and are difficult to measure. Chronic wounds showed higher levels of these byproducts (five-fold) than acute wounds. Similarly, another oxidized byproduct prostaglandin, 8-isoprostane, also revealed higher levels in chronic wound fluid than in acute wound fluid. Kwon et al. (2018) came up with fluorescent chemosensors for ROS and reactive nitrogen species (RNS) based on specific reactions. Hypochlorous acid (HOCl) which is a product from the reaction of hydrogen peroxide (H_2O_2) and $Cl^-$by myeloperoxidase (MPO), is a class of ROS. Some of the ROS which can be measured using fluorescent enhancements of the thiolactone and selenolactone rhodamine derivatives are H_2O_2, NO•, •OH, ROO•, $ONOO^-$HOCl, and $•O_2^-$, HOCl at a pH of 5.5. HOCl is a defensive mechanism of phagocytic cells that kill pathogens. Nevertheless, a high concentration of HOCl can lead to chronic inflammatory, cardiovascular, and kidney malfunctioning.

13.5.8 Temperature

One of the early indicators of wound infection is temperature, which is affected by both internal and external (environmental) factors. Body temperature plays a role in wound healing as it can affect local blood flow and lymphocyte extravasation. Nakagami et al. (2010) observed that wounds which exhibited an elevated wound temperature had an increased risk of delay in wound healing. The major wound characteristics, i.e., body temperature and body oxygenation, were the principal biomarkers in the biosensor developed by Sattar et al. (2019). They also tried to arrive at a correlation of these biomarkers with wound hydration level using a fuzzy interference system. There is also other evidence like the study done by Siah et al. (2019), which suggested that colder temperatures at the wound site indicated development of infection, and on the other hand, research by Robicsek et al. (1984), which suggested that persistently elevated temperatures at the wound site indicated the development of infection. Existing studies suggest that for normal acute post-incisional wound healing, temperature measurements are elevated during the first few days and within a two-week period returned to the normal measure. Dini et al. (2015) observed that as wound bed temperatures increased in the range of 33–35°C, there was progress in healing. Similar results were also obtained in animal research, where a lower wound bed temperature (below core body temperature) showed a delayed healing. This accounted for the inadequate collagen deposition and reduced recruitment of late-phase inflammatory cells and fibroblasts. Likewise, the results obtained using invitro models showed inadequate activity of neutrophil, fibroblast, and epithelial cells as the temperature was reduced below 33°C. Nevertheless, in the case of pressure ulcers, normal healing was observed when the wound bed temperature was lower than peri-wound skin and a delayed healing was observed in cases where the wound bed temperature is higher than peri-wound skin temperatures. The most commonly employed sensors for temperature measurement are resistance temperature detectors, thermocouples, thermistors, infrared sensors which are clinically used, and silicon-based sensors. Since most of the sensors are used directly on the wound surface, the use of microfluidic detection devices using wound fluid has not been extensively used for temperature detection.

13.5.9 Transepidermal Water Loss

The phenomenon of passive diffusion of water molecules through the skin is called transepidermal water loss (TEWL). It is measured in $g/m^2/h$ and for healthy human skin, TEWL values range from 4 to 8 $g/m^2/h$. In conditions where the skin is damaged or wounded, the TEWL value increases up to a few hundreds of $g/m^2/h$, and hence TEWL can be used as a biomarker to assess the wound healing status. TEWL measurement can also be useful in order to understand the evolution and recovery of wounds or the efficacy of treatments. However, the commercially available devices on the market (e.g., Vapometer by Delfin Technologies Ltd, Tewameter by Courage-Khazaka Electronic, and Dermalab by Cortex Technologies) are not suitable to incorporate into wound dressings.

13.5.10 C-Reactive Protein

Human C-reactive protein (CRP), which is an important biomarker in wound healing, is an annular ligand-binding plasma protein which is calcium dependent and is composed of five identical non-glycosylated polypeptide subunits with cyclic pentameric symmetry (Sproston and Ashworth2018). CRP is produced in the liver, kidneys, and atherosclerotic tissues during conditions of acute inflammation. CRP synthesis is triggered during the inflammatory phase of acute wound healing which leads to a sudden rise in its concentration from about 0.8 mg/L to 600–1000 mg/L, attaining it speak value at about 48 h. When the stimulus for the increased CRP production dies, the concentration rapidly diminishes to the normal values and hence has a half-life of 19 h. The conventional CRP detection methods in clinical settings are immune nephelometric and immune turbidimetric assays using a single polyclonal antibody, and ELISAs which are time consuming and need trained personnel. Tsai et al. (2018) developed a lab-on-a-chip system consisting of a microfluidics chip and a label-free

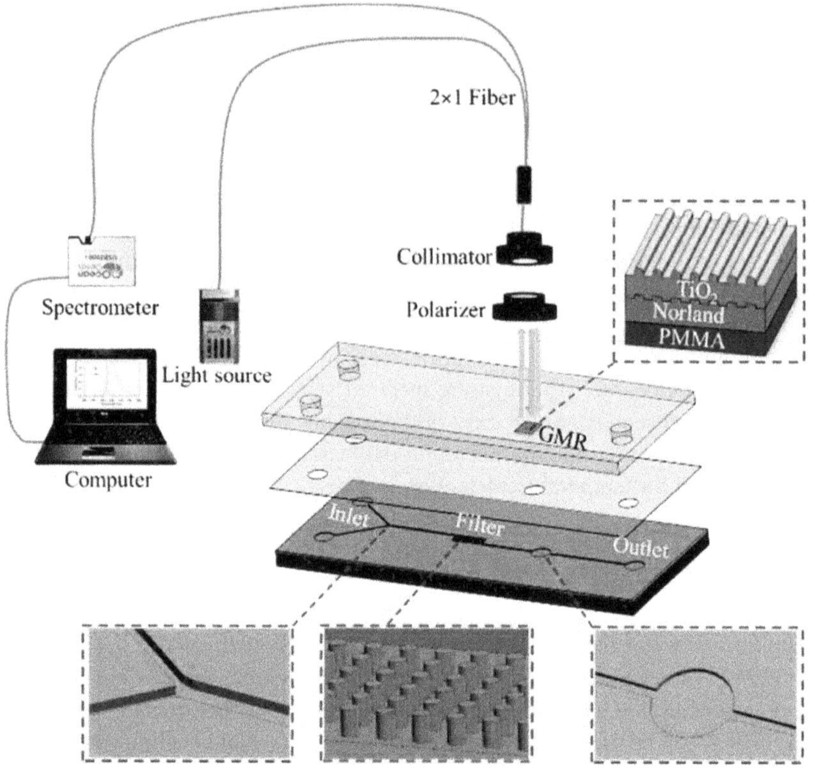

FIGURE 13.3 Schematic of three-layer lab-on-a-chip system and optical read-out setup (Tsai et al. 2018).

biosensor which can be used for the detection of CRP from whole blood samples (Figure 13.3). The advantage of this method is that it does not require sample pre-treatment. The developed guided-mode resonance sensor for the recombinant CRP spike in human serum records results comparable to ELISA with bulk sensitivity of 186 nm RIU-1, supporting a limit of detection of 3.2 ng mL^{-1}. Antibodies are in turn mapped with CRP concentration, a useful biomarker for inflammation and necrosis detection. This is implemented by binding CRP downstream which causes a measurable shift in wavelength of light passing through the device.

13.5.11 Interleukin-6

Mok et al. (2014) demonstrated the detection of interleukin-6 (IL-6) within microfluidics platforms employing anti-IL-6-conjugated microbeads. Impedance sensing mode with gold electrodes was used for this detection (Mok et al. 2014). Mandal et al. (2009) created a PDMS-based microfluidics device fabricated using the technique of soft lithography which was functionalized with biotinylated monoclonal antibodies. The sensing component for detecting IL-6 was a one-dimensional photonic crystal resonator array with planar photonic crystals (Mandal et al. 2009). However, the current sensors could achieve only a 50 pM which calls for further optimizations.

13.5.12 Uric Acid

Another marker which is indicative of bacterial colonization is a fall in uric acid concentration in exudate. Sharp et al. (2008) put forth the technique of incorporating carbon fiber-based electrodes in a bandage to monitor the uric acid concentration. The use of cellulose acetate barrier over the surface of electrodes helped to eliminate the fouling of extracellular components, e.g., proteins and fats. This sensor, with a linear response in the range of approximately 0–500 mM of uric acid, proved to be a robust technique in wound management.

13.5.13 Glucose

The glucose and bicarbonate levels are also important biomarkers which determine wound healing. During the process of wound healing glucose levels remain low probably due to dependency on neutrophils as an energy source. Generally, a delayed wound healing is observed in cases where blood glucose levels are relatively high. Glucose, being hygroscopic in nature, can make the cell walls firm and restrict the blood flow through small blood vessels located at the periphery of the wound, resisting the permeability of red blood cells, which are essential for the formation of dermal tissue. On the other hand, high glucose levels can impair the wound healing process by prohibiting the release of oxygen from hemoglobin. This leads to oxygen and nutrient deficits in the wound bed. Khan and Park (2015) employed glucose oxidase (GOx) immobilized to poly(acrylic acid-b-4-cynobiphenyl-4-oxyundecyl acrylate) (PAA-b-LCP) chains for detecting glucose in microdroplets consisting of 4-cyno-4-pentylbiphenyl (5CB) in an aqueous medium along with coating with PAA at the 5CB/water interface. This novel method, which utilized a functionalized liquid-crystal (LC) droplet, captured the radial to bipolar configurational change of glucose using polarized optical microscopy under crossed polarizers. This method also recorded glucose detection at a concentration as low as 0.03 mM, within a time interval of ~3 min (response time) and showed high selectivity for glucose compared to galactose.

13.6 FUTURE PERSPECTIVES

The use of microfluidics platforms for wound healing analysis is still in its infancy. Although all the parameters associated with wound healing have not been multiplexed into a single microfluidics device, a lot of work on assessing some parameters has been performed by various research groups. Most of these platforms are based on microfluidics paper-based analytical devices. Ratajczak and Stobiecka (2020) discuss the major advantages and disadvantages of using paper-based microfluidics analytical devices in their review paper. Even though paper-based devices are easy to construct and use without skilled personnel, much of their disadvantages are based on maintenance of the bioactivity and biospecificity of the analytes that are blotted or coated on the devices. The assessment of shelf life and maintenance of stability of the substrates are also a challenge. This is mainly because most of the reagents used consist of enzymes, antibodies, certain chromophores, and redox probe molecules which may be light sensitive, moisture sensitive, etc. Reactivity of the reagents with the cellulosic substrates is also a concern when the coated substrates are stored for longer periods. Much research is warranted in these areas. This also provides scope for research related to the storage of such devices to preserve their long-term stability and to provide consistent quantitative readouts. Another challenging aspect is the development of multiparametric detection devices, which can give a plethora of information on aspects of wound healing, enabling clinicians to decide on further therapeutic action. Quantitative assessment techniques may be varied for such multiparametric analysis and may require different readout mechanisms. Moreover, when considering the use of smartphones as readouts for such microfluidics platforms, consistency and accuracy in the data provided needs to be assessed. These challenges will pave the way for more modernized digital approaches wherein real-time assessment can also be done on the wound fluid exudates.

13.7 CONCLUSION

In this chapter we have tried to bring forward the use of microfluidics platforms for POC analysis of wound fluid. The chapter throws light on the parameters that are assessed to evaluate the healing of wounds using wound fluid exudate. The study of nonhealing or chronic wounds has been a major challenge for the medical community as it forms the basis for several therapeutic interventions. A cost-effective POC device in this respect will enable clinicians even in remote areas without access to medical laboratories to assess the condition of wounds. The different evaluation and imaging strategies that have been involved in detection of wound healing parameters are also discussed. The need for a multiparametric microfluidics paper-based analytical device with quantitative detection using devices like smartphones will result in more focused research in these fields.

REFERENCES

Altintas, Zeynep, Mete Akgun, Guzin Kokturk, and Yildiz Uludag 2018. A Fully Automated Microfluidic-Based Electrochemical Sensor for Real-Time Bacteria Detection. *Biosensors & Bioelectronics*100: 541–48.

Barui, A., P. Banerjee, R. K. Das, S. K. Basu, S. Dhara, and J. Chatterjee 2011. Immunohistochemical Evaluation of p63, E-cadherin, Collagen I and III Expression in Lower Limb Wound Healing Under Honey. *Evidence-Based Complementary and Alternative Medicine*, 2011.

Becker, J., D. Schuppan, E. G. Hahn, G. Albert, and P. Reichart 1986. The Immunohistochemical Distribution of Collagens Type IV, V, VI and of Laminin in the Human Oral Mucosa. *Archives of Oral Biology* 31(3): 179–86.

Berger, Mette M., Christophe Binnert, René L. Chiolero, William Taylor, Wassim Raffoul, Marie-Christine Cayeux, Messod Benathan, Alan Shenkin, and Luc Tappy 2007. Trace Element Supplementation after Major Burns Increases Burned Skin Trace Element Concentrations and Modulates Local Protein Metabolism but Not Whole-Body Substrate Metabolism. *The American Journal of Clinical Nutrition*85(5): 1301–6.

Boisnic, Sylvie, Jean-Pierre Ouhayoun, M. C. Branchet, C. Frances, J. Y. Béranger, Y. Le Charpentier, and H. Szpirglas 1995. Alteration of Cytokeratin Expression in Oral Lichen Planus. *Oral Surgery, Oral Medicine, Oral Pathology, Oral Radiology, and Endodontology* 79(2): 207–15.

Brown, Matthew S., Brandon Ashley, and Ahyeon Koh2018. Wearable Technology for Chronic Wound Monitoring: Current Dressings, Advancements, and Future Prospects. *Frontiers in Bioengineering and Biotechnology* 6: 47.

Bruzewicz, D. A., M. Reches, and G. M. Whitesides 2008. Low-cost Printing of Poly (Dimethylsiloxane) Barriers to Define Microchannels in Paper. *Analytical Chemistry* 80(9): 3387–92.

Carmichael, R. P., C. A. McCulloch, and G. A. Zarb 1991. Immunohistochemical Localization and Quantification of Desmoplakins I & II and Keratins 1 and 19 in Plastic-Embedded Sections of Human Gingiva. *Journal of Histochemistry & Cytochemistry* 39(4): 519–28.

Carrell, C., A. Kava, M. Nguyen, R. Menger, Z. Munshi, Z. Call, M. Nussbaum, and C. Henry 2019. Beyond the Lateral Flow Assay: A Review of Paper-based Microfluidics. *Microelectronic Engineering* 206: 45–54.

Chow, O., and A. Barbul 2014. Immunonutrition: Role in Wound Healing and Tissue Regeneration. *Advances in Wound Care* 3(1): 46–53.

Coltro, Wendell, Karlos Tomazelli, Chao-MinCheng, Emanuel Carrilho, and Dosil Pereira de Jesus 2014. Recent Advances in Low-Cost Microfluidic Platforms for Diagnostic Applications. *Electrophoresis* 35(16): 2309–24.

Curto, Vincenzo F., Cormac Fay, Shirley Coyle, Robert Byrne, Corinne O'Toole, Caroline Barry, Sarah Hughes, Niall Moyna, Dermot Diamond, and Fernando Benito-Lopez 2012. Real-Time Sweat PH Monitoring Based on a Wearable Chemical Barcode Micro-Fluidic Platform Incorporating Ionic Liquids. *Sensors and Actuators B: Chemical* 171–172: 1327–34.

Cutting, K. F. 2003. Wound Exudate: Composition and Functions. *British Journal of Community Nursing* 8(Sup3): S4–S9.

Dargaville, Tim R., Brooke L. Farrugia, James A. Broadbent, Stephanie Pace, Zee Upton, and Nicolas H.Voelcker 2013. Sensors and Imaging for Wound Healing: A Review. *Biosensors and Bioelectronics* 41: 30–42.

Delaney JL, Hogan CF, Tian J et al. 2011. Electrogenerated Chemiluminescence Detection in Paper-based Microfluidic Sensors. *Analytical Chemistry* 83: 1300–6.

Dini, Valentina, Pietro Salvo, Agata Janowska, Fabio Di Francesco, Alessandro Barbini, and Marco Romanelli 2015. Correlation Between Wound Temperatures Obtained with an Infrared Camera and Clinical Wound Bed Score in Venous Leg Ulcers. *Wounds: A Compendium of Clinical Research and Practice* 27: 274–8.

Dunnill, C., T. Patton, J. Brennan, J. Barrett, M. Dryden, J. Cooke, D. Leaper, and N. T. Georgopoulos 2017. Reactive Oxygen Species (ROS) and Wound Healing: The Functional Role of ROS and Emerging ROS-modulating Technologies for Augmentation of the Healing Process. *International Wound Journal* 14(1): 89–96.

Eicher, D., and C. A. Merten 2011. Microfluidic Devices for Diagnostic Applications. *Expert Review of Molecular Diagnostics* 11(5): 505–19.

Ellerbee AK, Phillips ST, Siegel AC et al. 2009. Quantifying Colorimetric Assays in Paper-based Microfluidic Devices by Measuring the Transmission of Light Through Paper. *Analytical Chemistry* 81: 8447–52.

Farooqui, M. F., and A. Shamim 2016. Low Cost Inkjet Printed Smart Bandage for Wireless Monitoring of Chronic Wounds. *Scientific Reports* 6(1): 1–13.

Farrow, Malcolm J., Iain S. Hunter, and Patricia Connolly 2012. Developing a Real Time Sensing System to Monitor Bacteria in Wound Dressings. *Biosensors* 2(2): 171–88.

Furst, Ariel L., and Matthew B. Francis 2019. Impedance-Based Detection of Bacteria. *Chemical Reviews* 119(1): 700–26.

Gahlaut, Shashank K., Neeti Kalyani, C. Sharan, Prashant Mishra, and J. P. Singh. Smartphone Based Dual Mode in Situ Detection of Viability of Bacteria Using Ag Nanorods Array 2019. *Biosensors & Bioelectronics* 126: 478–84.

Gethin, G. 2007. The Significance of Surface pH in Chronic Wounds. *Wounds uk* 3(3): 52.

Guinovart, Tomàs, Gabriela Valdés-Ramírez, Joshua R. Windmiller, Francisco J. Andrade, and Joseph Wang 2014. Bandage-Based Wearable Potentiometric Sensor for Monitoring Wound PH. *Electroanalysis* 26(6): 1345–53.

Gupta, Subhas, Charles Andersen, Joyce Black, Jeande Leon, Caroline Fife, John C. Lantis Ii, Jeffrey Niezgoda, et al. 2017. Management of Chronic Wounds: Diagnosis, Preparation, Treatment, and Follow-Up. *Wounds: A Compendium of Clinical Research and Practice* 29(9): S19–36.

Hanson, Darlene, Diane Langemo, Pat Thompson, Julie Anderson, and Susan Hunter 2005. Understanding Wound Fluid and the Phases of Healing. *Advances in Skin & Wound Care* 18(7): 360–362.

Healy, Brendan, and Andrew Freedman 2006. Infections. *British Medical Journal* 332(7545): 838–41.

Hsu, Jui-Tse, Yung-Wei Chen, Te-Wei Ho, Hao-Chih Tai, Jin-Ming Wu, Hsin-Yun Sun, Chi-Sheng Hung, Yi-Chong Zeng, Sy-Yen Kuo, and Feipei Lai 2019. Chronic Wound Assessment and Infection Detection Method. *BMC Medical Informatics and Decision Making*19(1): 99.

Hunter, Rebecca A., Benjamin J. Privett, W. Hampton Henley, Elise R. Breed, Zhe Liang, Rohit Mittal, Benyam P. Yoseph, et al. 2013. Microfluidic Amperometric Sensor for Analysis of Nitric Oxide in Whole Blood. *Analytical Chemistry* 85(12): 6066–72.

Ispas, C. R., G. Crivat and S. Andreescu 2012. Review: Recent Developments in Enzyme-Based Biosensors for Biomedical Analysis. *Analytical Letters* 45: 168–86.

Jones, Eleri M., Christine A. Cochrane, and Steven L. Percival 2015. The Effect of PH on the Extracellular Matrix and Biofilms. *Advances in Wound Care* 4(7): 431–9.

Kiernan, M 1998. Role of Swabbing in Wound Infection Management. *Community Nurse* 4(6): 45–46.

Khan, M., and S. Y. Park 2015. Liquid Crystal-based Glucose Biosensor Functionalized with Mixed PAA and QP4VP Brushes. *Biosensors and Bioelectronics* 68: 404–12.

Kwon, Nahyun, Ying Hu, and Juyoung Yoon 2018. Fluorescent Chemosensors for Various Analytes Including Reactive Oxygen Species, Biothiol, Metal Ions, and Toxic Gases. *ACS Omega* 3(10): 13731–51.

Lazar, Iulia M. 2015. Microfluidic Devices in Diagnostics: What Does the Future Hold? *Bioanalysis* 7(20): 2677–80.

Lei, Zhen, Minghong Jian, Xiaotong Li, Jia Wei, Xianying Meng, and Zhenxin Wang 2020. Biosensors and Bioassays for Determination of Matrix Metalloproteinases: State of the Art and Recent Advances. *Journal of Materials Chemistry B* 8(16): 3261–91.

Li, Sixing, Justin Kiehne, Lawrence I. Sinoway, Craig E. Cameron, and Tony Jun Huang 2013. Microfluidic Opportunities in the Field of Nutrition. *Lab on a Chip* 13(20): 3993–4003.

Lin, Pei-Hui, Matthew Sermersheim, Haichang Li, Peter H. U. Lee, Steven M. Steinberg, and Jianjie Ma 2017. Zinc in Wound Healing Modulation. *Nutrients* 10(1): 16.

Lu, Chuan-Pin, Bo-Xian Guo, Zi-Qing Fang, and Shu-Chiang Chung 2015. The Development of Image Base, Portable Microfluidic Paper-Based Analytical Device. In 2015 International Conference on Orange Technologies (ICOT), 144–8. https://doi.org/10.1109/ICOT.2015.7498497.

Mai, Ronald, Tomasz Gedrange, Henry Leonhardt, Nicole Sievers, and Günter Lauer 2009. Immunohistochemical Comparison of Markers for Wound Healing on Plastic-Embedded and Frozen Mucosal Tissue. *Cells Tissues Organs* 190(1): 34–41.

Mandal, Sudeep, Julie M. Goddard, and David Erickson 2009. A Multiplexed Optofluidic Biomolecular Sensor for Low Mass Detection. *Lab on a Chip* 9(20): 2924–32.

Martinez AW, Phillips ST, Whitesides GM et al. 2010. Diagnostics for the developing world: microfluidic paper-based analytical devices. *Analytical Chemisrty* 82: 3–10.
Miller, Craig, Madison Stiglich, Mark Livingstone, and Jordon Gilmore 2019. Impedance-Based Biosensing of Pseudomonas Putida via Solution Blow Spun PLA: MWCNT Composite Nanofibers. *Micromachines* 10(12): 879.
Milne, Stephen D., Patricia Connolly, Hanadi Al Hamad, and Ihab Seoudi 2014. Development of Wearable Sensors for Tailored Patient Wound Care. In 2014 36th Annual International Conference of the IEEE Engineering in Medicine and Biology Society, Chicago, IL, USA, 618–21.
Mittal, Manish, Mohammad Rizwan Siddiqui, Khiem Tran, Sekhar P. Reddy, and Asrar B. Malik 2013. Reactive Oxygen Species in Inflammation and Tissue Injury. *Antioxidants & Redox Signaling* 20(7): 1126–67.
Mok, J., M. N. Mindrinos, R. W. Davis, and M. Javanmard 2014. Digital Microfluidic Assay for Protein Detection. *Proceedings of the National Academy of Sciences* 111(6): 2110–5.
Moll, R., W. W. Franke, D. L. Schiller, B. Geiger, and R. Krepler 1982. The Catalog of Human Cytokeratins: Patterns of Expression in Normal Epithelia, Tumors and Cultured Cells. *Cell* 31(1): 11–24.
Morris, Deirdre, Shirley Coyle, Yanzhe Wu, King Tong Lau, Gordon Wallace, and Dermot Diamond 2009. Bio-Sensing Textile Based Patch with Integrated Optical Detection System for Sweat Monitoring. *Sensors and Actuators B: Chemical* 139(1): 231–36.
Müller, R. H., and D. L. Clegg 1949 Sep 1. Automatic Paper Chromatography. *Analytical Chemistry* 21(9): 1123–5.
Nakagami, G., H. Sanada, S. Iizaka, T. Kadono, T. Higashino, H. Koyanagi, and N. Haga 2010. Predicting Delayed Pressure Ulcer Healing Using Thermography: A Prospective Cohort Study. *Journal of Wound Care* 19(11): 465–72.
Nath, P., R. K. Arun, and N. Chanda 2015. Smart Gold Nanosensor for Easy Sensing of Lead and Copper Ions in Solution and Using Paper Strips. *RSC Advances* 5: 69024–31.
Ono, S., R. Imai, Y. Ida, D. Shibata, T. Komiya and H. Matsumura 2015. Increased Wound pH as an Indicator of Local Wound Infection in Second Degree Burns. *Burns* 41(4): 820–4.
Ouhayoun, J. P., J. C. Goffaux, M. H. Sawaf, A. H. M. Shabana, C. Collin, and N. Forest 1990. Changes in Cytokeratin Expression in Gingiva during Inflammation. *Journal of Periodontal Research* 25(5): 283–92.
Percival, Steven L., Sara McCarty, John A. Hunt, and Emma J.Woods 2014. The Effects of PH on Wound Healing, Biofilms, and Antimicrobial Efficacy. *Wound Repair and Regeneration* 22(2): 174–86.
Puchberger-Enengl, Dietmar, Christian Krutzler, and Michael J. Vellekoop 2011. Organically Modified Silicate Film PH Sensor for Continuous Wound Monitoring. In 2011 IEEE SENSORS, Limerick, Ireland, 679–82.
Puchberger-Enengl, Dietmar, Sandervan den Driesche, Christian Krutzler, Franz Keplinger, and Michael J. Vellekoop 2015. Hydrogel-Based Microfluidic Incubator for Microorganism Cultivation and Analyses. *Biomicrofluidics* 9(1): 014127.
Qi, Y. X., M. Zhang, A. Zhu, and G. Shi 2015. Terbium (III)/gold Nanocluster Conjugates: The Development of a Novel Ratiometric Fluorescent Probe for Mercury (II) and a Paper-based Visual Sensor. *Analyst* 40(16): 5656–61.
Ratajczak, Katarzyna, and Magdalena Stobiecka 2020. High-Performance Modified Cellulose Paper-Based Biosensors for Medical Diagnostics and Early Cancer Screening: A Concise Review. *Carbohydrate Polymers* 229: 115463.
Robicsek, F., T. N. Masters, H. K. Daugherty, J. W. Cook, J. G. Selle, P. J. Hess, and P.Vajtai 1984. The Value of Thermography in the Early Diagnosis of Postoperative Sternal Wound Infections. *The Thoracic and Cardiovascular Surgeon* 32(4): 260–65.

Santos, Lídia, Joana P. Neto, Ana Crespo, Daniela Nunes, Nuno Costa, Isabel M. Fonseca, Pedro Barquinha, et al. 2014. WO3 Nanoparticle-Based Conformable PH Sensor. *ACS Applied Materials & Interfaces* 6(15): 12226–34.

Santy, Julie2008. Recognising Infection in Wounds. *Nursing Standard* 23(7): 53–54.

Sattar, H., I. S. Bajwa, R. Ul-Amin, A. Mahmood, W. Anwar, B. Kasi, R. Kazmi, and U. Farooq 2019. An Intelligent and Smart Environment Monitoring System for Healthcare. *Applied Sciences* 9(19): 4172.

Schneider, Lars Alexander, Andreas Korber, Stephan Grabbe, and Joachim Dissemond 2007. Influence of PH on Wound-Healing: A New Perspective for Wound-Therapy? *Archives of Dermatological Research* 298(9): 413–20.

Shabana, A. H. M., J. P. Ouhayoun, M. H. Sawaf, and N. Forest 1989. A Comparative Biochemical and Immunological Analysis of Cytokeratin Patterns in the Oral Epithelium of the Miniature Pig and Man. *Archives of Oral Biology* 34(4): 249–59.

Sharp, Duncan, Stephen Forsythe, and James Davis 2008. Carbon Fibre Composites: Integrated Electrochemical Sensors for Wound Management. *Journal of Biochemistry* 144(1): 87–93.

Shostenkot, Y. V., V. P. Georgievskii, and M. G. Levin 2000 Sep 1. History of the Discovery of Thinlayer Chromatography. *Journal of Analytical Chemistry* 55(9): 904–5.

Sia, Samuel K., Vincent Linder, Babak Amir Parviz, Adam Siegel, and George M. Whitesides 2004. An Integrated Approach to a Portable and Low-Cost Immunoassay for Resource-Poor Settings. *Angewandte Chemie International Edition* 43(4): 498–502.

Siah, Chiew-Jiat Rosalind, Charmaine Childs, Chung King Chia, and Kin Fong Karis Cheng 2019. An Observational Study of Temperature and Thermal Images of Surgical Wounds for Detecting Delayed Wound Healing within Four Days after Surgery. *Journal of Clinical Nursing* 28(11–12): 2285–95.

Song, E., Cheng, D, Song, Y, Jiang, M, Yu, J, and Wang, Y 2013. A Graphene Oxide-Based FRET Sensor for Rapid and Sensitive Detection of Matrix Metalloproteinase 2 in Human Serum Sample. *Biosensors & Bioelectronics* 47: 445–50.

Spear, Marcia 2012. Wound Exudate: The Good, the Bad, and the Ugly. *Plastic Surgical Nursing* 32(2): 77–79.

Sproston, Nicola R., and Jason J. Ashworth 2018. Role of C-Reactive Protein at Sites of Inflammation and Infection. *Frontiers in Immunology* 9: 754.

Tickle, Joy 2016. Wound Exudate: A Survey of Current Understanding and Clinical Competency. *British Journal of Nursing* 25(2): 102–9.

Tsai, Meng-Zhe, Chan-Te Hsiung, Yang Chen, Cheng-Sheng Huang, Hsin-Yun Hsu, and Pei-Ying Hsieh 2018. Real-Time CRP Detection from Whole Blood Using Micropost-Embedded Microfluidic Chip Incorporated with Label-Free Biosensor. *The Analyst* 143(2): 503–10.

Ubbink, Dirk T., Fleur E. Brölmann, Peter M. N. Y. H. Go, and Hester Vermeulen 2014. Evidence-based care of acute wounds: A perspective. *Advances in Wound Care* 4(5): 286–94.

Wang, Shoumei, Lei Ge, Xianrang Song, Jinghua Yu, Shenguang Ge, Jiadong Huang, and Fang Zeng 2012. Based Chemiluminescence ELISA: Lab-on-paper Based on Chitosan Modified Paper Device and Wax-screen-Printing. *Biosensors and Bioelectronics* 31(1): 212–8.

Wlaschek, M., and K. Scharffetter-Kochanek 2005. Oxidative Stress in Chronic Venous Leg Ulcers. *Wound Repair and Regeneration* 13(5): 452–61.

World Union of Wound Healing Societies (WUWHS) 2007. *Principles of Best Practice: Wound Exudate and the Role of Dressings. A Consensus Document.* London: MEP Ltd.

Yap, B. K., M. Soair, S. Nur'Arifah, N. A. Talik, W.F. Lim, and I. L. Mei 2018. Potential Point-of-care Microfluidic Devices to Diagnose Iron Deficiency Anemia. *Sensors* 18(8): 2625.

14 Chromatographic Separation and Visual Detection on Wicking Microfluidics Devices

Keisham Radhapyari, Nirupama Guru Aribam, Suparna Datta, Snigdha Dutta, Rinkumoni Barman, and Raju Khan

CONTENTS

DOI: 10.1201/9781003033479-14

14.1 INTRODUCTION

The outbreak of pandemics and times of emergency pose serious global public health problems. Facing the emergence of such a pandemic, it is of the utmost importance to establish a simple, portable, sensitive, rapid, inexpensive, yet accurate and robust point-of-care (POC) diagnostic device capable of efficiently generating reliable qualitative and quantitative data for on-site monitoring processes. POC diagnostic devices have created a boom in the development of portable sensing devices for environmental monitoring, clinical testing, and food safety (Luppa et al. 2011) with diverse applicability such as affordability, accessibility, ease of use, speed, sensitivity, and accuracy (Tabak 2007) is compared with conventional analytical technologies (Gouveia et al. 2019). Furthermore, the conventional analytical techniques require extensive sample preparation, expensive equipment, highly trained operators, and therefore are only suitable for routine analysis in a centralized laboratory (Kung et al. 2019).

Microfluidic devices as an analytical tool have been a growing trend and have influenced many fields, viz., environmental monitoring, detection, and applications (Nabavi et al. 2016), medicine, and pharmaceutics (Whitesides 2006). Microfluidics has found applications in detecting viruses and disease, detecting protein and glucose, detecting contaminants, applying sensors, and material preparation (Chen et al. 2018). The distinctive small dimensions of the microfluidics device result in high surface-area-to-volume ratio, laminar flow, greater surface tension, and enhanced capillary effects, which help in miniaturizing and improving conventional methods for separation, detection, and analysis (Yager et al. 2006). Miniaturized analytical microfluidics systems offer several advantages over traditional instrumental approaches, such as reducing the quantity of reagents from microliters or nanoliters, rapid analysis, portability, low cost, disposability, and high capacity for multiplexing assay. They also have the advantages of less consumption of hazardous materials, high throughput sample detection, being handy and portable, and having assorted designs for different functional modules (Mehling and Tay 2014).

Wicking microfluidics analytical devices have found application in various sections and areas of analytical chemistry (Kumar et al. 2013). Wicking microfluidics devices can make rapid, inexpensive analytical techniques that can be applied to various areas of study. Membrane-based wicking microfluidics devices with simple, stable assay chemistries are ideal devices that facilitate portable analysis (Pengpumkiat et al. 2017) and are gaining much recognition in the analytical field. Such portable diagnostic devices based on paper, polymeric, glass microfluidic systems, or lateral-flow and dipstick assays were such techniques that are high in demand. The intrinsic wicking ability of the paper matrix facilitates an ideal membrane for a microfluidic device with the advantages of rapid sample elution, simple and portable platforms requiring only a small quantity of target sample for detecting various analytes. It has other advantageous features, such as it can transport target samples and reagents via capillary action without the need for mechanical components, external power sources (Hu et al. 2014), or controllers.

Their fabrication is done by forming hydrophobic barriers onto various paper platforms (Xia et al., 2016). Furthermore, paper is termed as a smart material for biosensors (Liana et al. 2012) owing to its attractive component for POC devices such as its abundance, biodegradability, disposability, and it can be chemically or physically modified, altering its properties in wettability, permeability, and reactivity. Another platform for microfluidics which is low cost and with suitable wicking property and flexibility, is cotton thread. Combining 1D cotton thread materials with other 2D paper materials makes it feasible to fabricate three-dimensional (3D) microfluidic devices (Lin et al. 2014). Furthermore, polymer inclusion membranes, glass fiber membranes, polyvinyl chloride, polycaprolactone-filled glass microfiber membranes, etc., are being used as a good platform for wicking microfluidics devices. However, in microfluidics devices, eliminating interferences to enable element-specific detection is one of the major challenges in adapting a simple chemical assay, making the process expensive by pre-treatment of the sample, making the process more complicated and challenging to adjust to a portable microfluidics device. Hence, there is a requirement for alternative methods that could gain specificity in a microfluidic assay without compromising the measurement quality.

Chromatographic separation techniques in conjunction with microfluidics devices are the important methods to separate closely related components of complex mixtures facilitating the rapid acquisition of analytical results immediately after testing or analysis. Various approaches were made that present an inexpensive and simple chromatographic separation approach on a lateral flow microfluidic channel resulting in minimization of interference. One such device incorporates a non-specific polymer inclusion membrane-based assay that enables simple visual observation to generate quantitative information coupled with a dot-counting quantification approach. Paper-based separation devices are a subclass of those devices in which a chromatographic separation takes place as the solvent moves up through the paper channel. The fundamentals of the devices are simple, but the achievements could be huge. Patterned paper substrates were used for microfluidic multivariate analysis. Biological reagents can flow spontaneously in the paper channel without additional pumps to drive the fluid. This approach not only reduces the sample volume but also achieves the goal of multivariate detection.

The choice of detection techniques is essential for any chemical assay. Visual detection is becoming popular among the detection techniques due to its simplicity, fast analysis, lack of complicated instrumentation, and ease of application. The visual detection system is adequate when a "*yes*" or "*no*" answer is sufficient for the diagnosis. Colorimetric detection is possible due to chemical or enzymatic reactions leading to a different color or optical density product. The availability of chromogenic reagents and relative ease of interpretation from colorimetric reactions make colorimetry suitable for analyzing a wide range of analytes (Morbioli et al. 2017). Detection sensitivity in colorimetric detection can be enhanced by signal amplification using enzyme-based immunoassays or nanoparticles, and selectivity can also be enhanced by combining multiple chromogenic reagents in the same device.

The global demand for point-of-care and point-of-need devices is growing at an accelerated pace. Simple, low-cost, paper-based devices for rapid analysis will soon be the technologies of choice for developing such robust and flexible devices. Nonetheless, new microfluidics technologies utilizing chromatographic separation and visual detection can establish widespread utilization and such a device for broad applications. This chapter presents the various inexpensive and simple visual detection quantification approaches on a wicking microfluidics device when coupled with a chromatographic separation approach that minimizes interference. The chapter discusses wicking microfluidics platforms, their applicability, the advantages of visual detection in wicking microfluidics devices, and their application. The chapter also examines any challenges to be tackled before this device, i.e., chromatographic separation and visual detection on wicking microfluidics can be more broadly accepted as a new diagnostic tool for various applications, particularly in health care, biomedical, and environmental monitoring.

14.2 OVERVIEW

Microfluidics is a well-known branch of science that studies the control and the behaviors of fluids and particles at tens to hundreds of micrometers inside micrometer-sized channels (Whitesides 2006). Much research has been done on studying the potential and role of micro fluids to modernize current emerging technologies. Microfluidics devices have distinctive small dimensions resulting in a high surface-area-to-volume ratio, laminar flow, greater surface tension, and enhanced capillary effects, which help in miniaturizing and improving conventional methods for separation, detection, and analysis (Yager et al. 2006). In recent years there has been a rapid development of microfluidic lab-on-a-chip technology, 3D cell culture, organs-on-a-chip, and droplet techniques, which provide innovative approaches for drug screening, active testing followed by the study of metabolism. Numerous studies showed noteworthy progress towards developing a multifunctional lab-on-a-chip package accomplished with the task of performing complete sets of processes required for biochemical assays using digital microfluidics (DMF) technology to perform complete sets of biochemical assays and possibly develop portable platforms for POC applications (Samiei et al. 2016).

There has been a growing trend of employing microfluidics technologies for on-site environmental monitoring, detection, and applications. Microfluidics has been incorporated for the separation and treatment of contaminated and wastewater (Dong et al. 2015), heavy metal detection and removal (Zhu et al. 2017), removal of oil from oil industry wastewater (Grem et al. 2013), carbon capture (Stolaroff et al. 2016), and carbon dioxide (Nabavi et al. 2016). A review highlighting the advances and focusing on potential advantages in portable trace arsenic detection by microfluidic and lab-on-a-chip technology has been done by Yogarajah and Tsai (2015). They have mentioned that the emerging methods may be practical alternatives for use in routine water monitoring efforts in arsenic-affected regions. When coupled to analytical techniques, microdevices have efficiently helped in the automation of the integrated analytical systems (Leung et al. 2012). Microfluidics has found applications in detecting viruses and disease, detecting protein and glucose, detecting contaminants, applying sensors, and material preparation (Chen et al. 2018).

14.3 FABRICATION

Fabrication methods commonly used for wicking microfluidics have been categorized as physical methods and chemical methods. Physical methods include (i) wax patterning (Lu et al. 2009), (ii) plotting (Bruzewicz et al. 2008), (iii) inkjet etching (Abe et al. 2010), (iv) flexographic printing (Olkkonen et al. 2010), laser patterning treatment (Chitnis et al. 2011), (v) paper cutting and shaping (Wang et al. 2010), (vi) screen printing (Dungchai et al. 2011), etc. and chemical methods include (i) photolithography (Klasner et al. 2012), (ii) plasma treatment (Li et al. 2010), (iii) chemical vapor-phase deposition, (iv) wet etching, (v) inkjet printing, etc. The essential principle for these fabrication techniques is to form hydrophilic-hydrophobic barriers on a chromatography or filter paper or membranes to craft a fluidic channel network. Liquids by capillary forces undergo hydrophilic wicking matrices that can be anticipated and examined by Lucas–Washburn and Darcy equations (Ahmed et al. 2016). Chemical modification techniques such as plasma treatment, inkjet printing, etc., result in proper solvent resistance owing to chemical agents being coupled with hydroxyl groups on cellulose fiber in chromatography paper covalently (Li et al. 2012). Some fabrication techniques employed using chromatographic separation and visual detection approaches are discussed below.

14.3.1 Plasma Treatment and Dot Counting

Bandara et al. (2018a) presented a simple, unique methodology to fabricate wicking microfluidics devices, which allowed the microfluidic channel to be fabricated after the assay had been dispensed. Fabrication is done using a hydrophobic polycaprolactone (PCL) -filled glass microfiber (GMF) substrate on which the polymer inclusion membrane (PIM) is evenly dispensed as a dot array. In another study, Bandara et al. (2018b) fabricated devices by utilizing the ability to modify hydrophilicity and hydrophobicity of the device in which hydrophilic GMF membranes become hydrophobic by filling them with PCL, masking the top and bottom surfaces of the PCL-filled GMF substrate with pattern masks cut from inexpensive masking tape. The substrate was then exposed to oxygen radicals in a homebuilt exposure chamber. The patterned mask alters the hydrophobic substrate-polymer surface chemistry to generate hydrophilic fluid flow pathways.

14.3.2 Chemical Vapor-Phase Deposition (CVD) Technique

Lam et al. (2017) fabricated a chemically patterned microfluidic paper-based analytical device (C-µPAD) by depositing vaporized trichlorosilane (TCS) on a heating block using a pre-vacuum fluidic-treated pattern piece of polyvinyl tape affixed to a chromatographic paper in a vacuum chamber. The paper area covered by the tape would be hydrophilic, while all other areas of the paper would be hydrophobic. After fabrication, silanized hydrophobic patterns were invisible to the eyes, and the modified area retained its original flexibility patterns on chromatography paper with food dyes. The resulting patterns were of both single-layered C-µPAD with multiple color depositions and double-layered C-µPAD. This chemically patterned chromatography paper was further evaluated using color dyes and successfully applied in the detection of glucose assay, immunoassay, and heavy metal detection on well-spot C-µPAD and lateral flow C-µPAD. Devadhasan et al. (2018) further exploited these C-µPADs in their study using chromatography paper. Functional-group-terminated silane compounds and colorimetric reagents were pre-immobilized in the appropriate order detection zones on a patterned chromatographic paper using condensation chemistry before heavy metal detection. It has properties of high portability and a rapid on-site monitoring platform for multiple heavy metal ion detection with extremely high repeatability, which is useful for resource-limited areas and developing countries.

14.3.3 Wax Patterning and Plotting

Schonhorn et al. (2014) developed a three-dimensional device using a combination of cellulose that readily wicks aqueous fluids and provides a suitable matrix for the storage of dried reagents for immunoassay nylon-based, porous materials. They used the wax printing method to pattern hydrophobic barriers into each layer. The hydrophilic zones produce a network of lateral and vertical conduits to control the wicking of fluids within the device. Furthermore, Lewis et al. (2012) had also developed a 3D paper-based microfluidics device by stacking alternating layers of wax-patterned paper and patterned double-sided adhesive tape to craft a device that contains four paper layers and three tape layers with a single hydrophilic conduit that extends in the z-direction from one end of the device to the other. Using this paper-based microfluidics device, they showed that the level of an analyte can be quantified by simply measuring time, and no external electronic reader is required for the quantitative measurement.

14.3.4 Photolithography

Busa et al. (2016) had fabricated μPADs by photolithography in which paper substrate was impregnated with SU-8 2010 photoresist. In a microfluidic paper-based assay, the wicking rate of the μPAD is an essential property to influence rapid flow detection. Properties of the paper substrates suggest that the degree of compactness of the cellulose fibers in the fiber network composition possibly affects the capabilities of the paper substrates to liberate the unpolymerized SU-8 photoresist during the washing procedure, therefore, resulting in the reduced wicking rates of the μPADs. The developed FP41 μPAD was utilized for the measurement of HRP assay by using the HRP-TMB-H_2O_2 reaction.

14.3.5 Laser Patterning Treatment

Nath et al. (2014) demonstrated a simple paper-based microfluidic device that can selectively detect arsenic at a very low concentration level (1 ppb) using a gold nanosensor (Au-TA-TG) prepared by chemical conjugation of gold nanoparticles (AuNPs) with thioctic acid (TA) followed by thioguanine (TG) molecules in the presence of EDC/NHS. A visual color changes from red to blue is observed because of interparticle coupled plasmon resonance from aggregated AuNPs. The change in coloration forms the root of the low cost, portable, efficient paper-based device for very low level (<10 ppb) arsenic detection.

14.3.6 Plotting, Cutter, and Shaper

Fang et al. (2014) demonstrated a highly reproducible fabrication method of microfluidics systems cut on a glass fiber membrane by a common cutter without using any other sophisticated equipment or organic solvents involving adhering, cutting, and weeding steps. The glass fiber membrane was found to be much more wettable and hydrophilic than the cellulose paper and was found to have good water-retaining capacity. A star-shaped microfluidic pattern was used in bioassays to analyze protein, pH, glucose, ketone bodies, and nitrite in human urine samples.

14.4 APPLICATIONS

14.4.1 Detection of Heavy Metal

Pollution caused by heavy metals is becoming an alarming problem for human health. They have an affinity for and bind to various essential cellular components and then accumulate in the body, which later becomes the cause of diseases, leading to disorders and organ failures (Li et al. 2015). Techniques for detecting heavy metals require expensive instrumentation and laborious operation, which can only be accomplished in centralized laboratories (Hossain et al. 2011). Owing to the numerous properties of microfluidics, many researchers developed simple, user-friendly, cheap, specific, sensitive, accurate, and environment-friendly detection devices based on wicking microfluidics using chromatographic separation and visual detection approaches. A detailed finding such as method, the substrate used, target molecule, sample type, linear range, detection limit, and various applications of chromatographic separation and visual detection on wicking microfluidic devices are discussed in Table 14.1.

14.4.1.1 Copper

For the first time, Bandara et al. (2018a) present a simple, visual detection method based on a chromatographic separation approach to quantify Cu^{2+} using a lateral flow microfluidic channel. Such a lateral flow microfluidic channel has shown remarkable performance to overcome interferences in the detection of Cu^{2+}. The fabrication approached has been discussed briefly in Section 14.3.1. ATR-FTIR spectra (Figure 14.1(II)) identified any significant chemistry change in the PIM assay. The finding suggests that exposure of the PIM-to-oxygen radicals did not measurably change the chemistry of the bulk PIM. The Si-O stretching signal (*e*) was noticed in the fingerprint region of every spectrum due to the overlap of the background signal produced by the GMF substrate. The wicking speed and distance traveled on the lateral flow channel is proportional to the concentration of Cu^{2+} in the sample. The Cu^{2+} traveled on the lateral flow channel until it was fully depleted from the sample by complexation with 1-(2-pyridylazo)-2-naphthol (PAN). Figure 14.1(I)(a) depicts the migration of the PAN-Cu^{2+} complex on the trimethylchlorosilane (TMCS)-modified PCL-filled GMF platform where higher concentrations of Cu^{2+} traveled further into the linear array of PIM dots, which increases the number of red zones which are directly proportional to the Cu^{2+} concentration. Figure 14.1(I) (b) shows

TABLE 14.1
Applications of Chromatographic Separation and Visual Detection on Wicking Microfluidics Devices

Method/Principle	Substrate Used	Target Molecule	Sample Type	Linear Range	Detection Limit	Application/POC	References
Lateral flow microfluidics devices (PIM), dot array	Polymer inclusion membrane (PIM)	Cu^{2+}	Surface, ground and drinking water	1.00 to 20.00 ppm	0.5 ppm Cu^{2+}	Environmental monitoring/yes	Bandara et al. (2018a)
C-µPADs	Masked chromatographic paper	Ni (II), Cr (VI), and Hg (II)	Lake samples	-	0.24 ppm for Ni(II), 0.18 ppm for Cr(VI) and 0.19 ppm for Hg (II)	Environmental monitoring targets and for point-of-care diagnostics/yes	Devadhasan and Kim (2018)
C-µPAD	Chemical vapor deposition (CVD) of trichlorosilane (TCS) on a chromatography paper	Glucose Ni TNFα immunoassay	Human blood for glucose human anti-TNFα capture antibody	0 to 160 mg/dL glucose 1~1000 ng/mL for TNFα immunoassay	13 mg/dL of LOD for well spot C-µPAD 23 mg/dL of LOD for lateral flow C-µPAD 150 µg/L of Ni 3 ng/mL for TNFα immunoassay	Demonstrating glucose assay, immunoassay and heavy metal detection/yes	Lam et al. (2017)
3D-paper-based microfluidics devices	Whatman Grade 4 qualitative filter paper	Human chorionic gonadotropin (hCG)	Human serum	0–25 µg/mL	1.2 µg/mL	Assays for HIV antibodies/yes	Schonhorn et al. (2014)
2D and 3D microfluidics devices	PCL/GMF substrates	Total protein content	Human blood serum	0.5–60 mg/mL	-	Protein concentrations analysis in real samples/yes	Bandara et al. (2018b)
Microfluidics paper-based devices	Cellulose-based paper	HRP-conjugated molecules	-	3 to 1000 ng/mL	0.69 fmol or 5.58 ng/mL	Horseradish peroxidase/yes	Basu et al. (2016)
3D paper-based microfluidics devices	Whatman Chromatography Paper No. 1.	Hydrogen peroxide	-	-	0.7 mm	POC diagnostic assays/yes	Lewis et al. (2012)
Paper-based microfluidics devices	Whatman filter paper	As^{3+}	Arsenic solution sample	10 ppm to 0.001 ppm	1.0 ppb	Arsenic detection/yes	Nath et al. (2014)

(Continued)

TABLE 14.1 (CONTINUED)
Applications of Chromatographic Separation and Visual Detection on Wicking Microfluidics Devices

Method/Principle	Substrate Used	Target Molecule	Sample Type	Linear Range	Detection Limit	Application/POC	References
Lateral flow immunoassay/ paper-based microfluidics	IgM-IgG combined antibody test/ chromatography	SARS-CoV-2	Blood sample	-	-	Biomedical/yes	Li et al. (2020)
Paper-based quantification method/wicking	Chromatography paper	Sickle cell disease	Blood	-	-	Biomedical/yes	Piety et al. (2016)
Paper-based quantification method/wicking/ visual Interpretation	Chromatography paper	Sickle cell disease	Blood	-	-	Biomedical/yes	Piety et al. (2017)
Paper-based microfluidics device/wicking	Chromatography paper	Hematocrit	Blood	-	-	Biomedical/yes	Berry et al. (2016)
Microfluidics paper-based/color change can be observed by the naked eye	Molecularly imprinted polymer	Bisphenol	Blood samples	10–1000 nM	6.18 nM	Environmental monitoring/yes	Kong et al. (2017)
Paper-based microfluidics	Glass fiber membranes and polyvinyl chloride	Protein, pH, glucose, ketone bodies and nitrite	Urine samples	-	0.05 mg/mL for glucose, 0.25 mg/mL for protein, 0.5 mg/ mL for ketone bodies and 0.25 mg/ mL for nitrite	POCT bioassays. ELISA or PCR/yes	Fang et al. (2014)
3D paper-based microfluidics devices	Cotton thread with polymer film	Nitrite ion (NO_{2-}) and uric acid	-	0–1000 mM/L	-	Health diagnostics, environment and food safety/yes	Li et al. (2010)

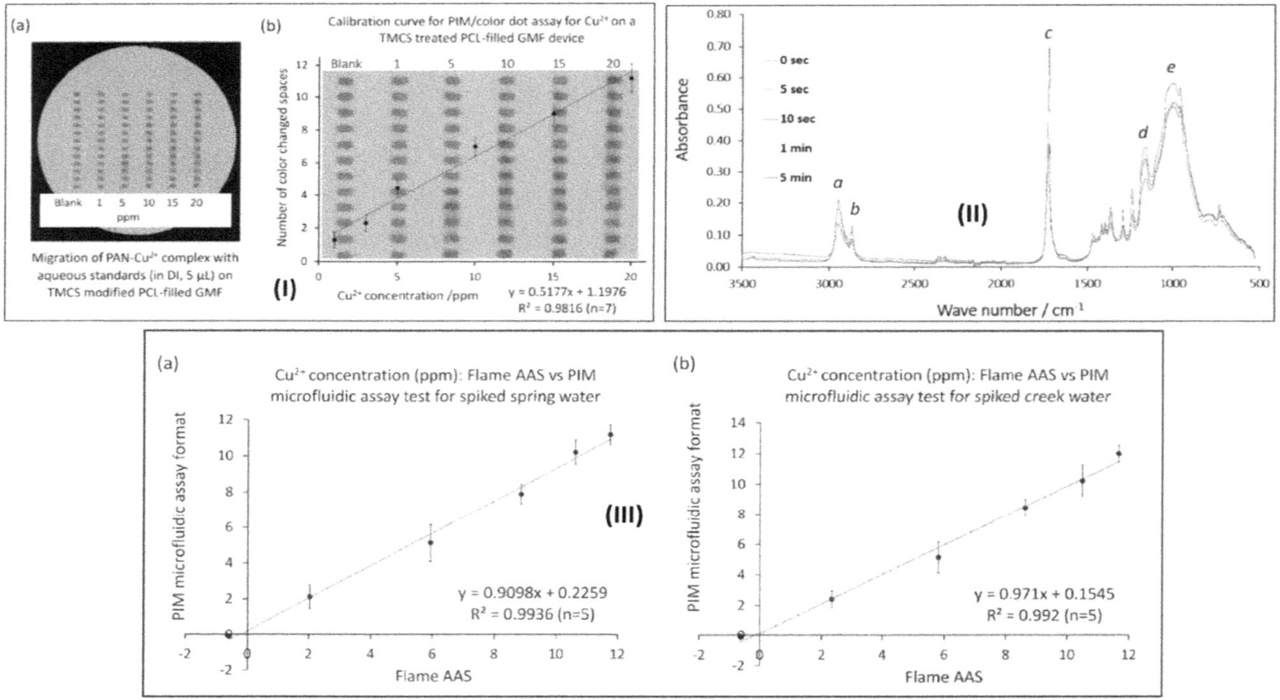

FIGURE 14.1 (I) Device-resident calibration curves for the PIM/dot assay for Cu^{2+}. (II) ATR-FTIR spectra of PCL-filled GMF membranes at different oxygen radical exposure time intervals. (III) A direct comparison between the flame-AAS and the microfluidic devices was performed by plotting Cu^{2+}concentration values for each method against each other. (Reproduced with permission from Bandara et al. (2018a).)

the calibration curve yielded a linear relationship over the concentration range of 1.0–20.0 ppm with a sample volume of the only 5 µL. The signal-detection limit was 1.5 ppm (n = 12) and the visual detection limit was 1 ppm. This dynamic study range is within the maximum allowable concentration of Cu^{2+} in drinking water, as stated by the U.S. Environmental Protection Agency (US EPA). In Figure 14.1(III)(a, b), a linear relationship between the new wicking microfluidic method and the flame atomic absorption spectrometry (AAS) method was observed for both drinking water and creek water samples with R^2 of 0.99, concluding and demonstrating that the new wicking microfluidics approach is capable of producing accurate quantitative data for an unknown test solution. Hence, this approach is highly acceptable for detecting Cu^{2+} in drinking water and real water samples.

14.4.1.2 Nickel, Chromium, and Mercury

A chemically functionalized microfluidic paper-based analytical device (C-µPAD) for multiplex heavy metal detection was developed by Devadhasan and Kim (2018). This novel method demonstrated silane coupling to immobilize three functional groups – amine (NH2), carboxyl (COOH), and thiol (SH)-on (C-µPAD). The fabrication and immobilization methods were briefly discussed in Section 14.3.2. The performance of the functionalized C-µPAD was investigated by using various concentrations of heavy metals. The metal ions reacted with their respective colorimetric reagents, forming colored metal complexes on the well-spots within 1 min. Figure 14.2(I)(a–c) shows the color intensities of the metal complexes formed by the different concentrations of metal ions. It has a negative control in the center spot. Ni(II) reacts with dimethylglyoxime giving a deep pink color complex on the well-spot. Cr(VI) reacts with 1,5-diphenylcarbazide to create a purple color complex on the well-spot. And Hg(II) solution produces a brown colored Michler's thioketone–Hg(II) complex. Color intensity decreases with metal concentration, and no color was developed in the negative-control zone. Figure 14.2(I)(d–f) shows the Euclidean distances for the metal ion concentrations. The colorimetric analysis exhibited a high detection sensitivity and detection limits as low as 0.24 ppm for Ni(II), 0.18 ppm for Cr(VI), and 0.19 ppm for Hg(II), respectively.

For specificity measurement, colorimetric reagents were applied to each column of three separate 3 × 3 well-spot C-µPADs. A pink dimethylglyoxime-Ni(II) complex was generated on the first column of the first device, and no color appeared on the other two columns. A purple-colored 1,5-diphenylcarbazide–Cr (VI) complex was formed on the second column of the device, and no color appeared on the other two columns. A brown Michler's thioketone–Hg(II) complex was seen in the third column of the third device, and no color appeared on the other two columns Figure 14.2(II)(b–d). As shown in Figure 14.2(II)(e), the Euclidean distances confirm the metal complexes formation in their intended zones and that there was no cross-reaction between the metals. Interferences were evaluated using high concentrations of NaCl and KCl, which are components found in drinking water. Additionally, they have used metal-selective chromogenic reagents and EDTA to eliminate the interferences.

The applicability of the functionalized C-µPAD was verified by detecting heavy metal ions in the lake water sample. The multiplexing capability was demonstrated

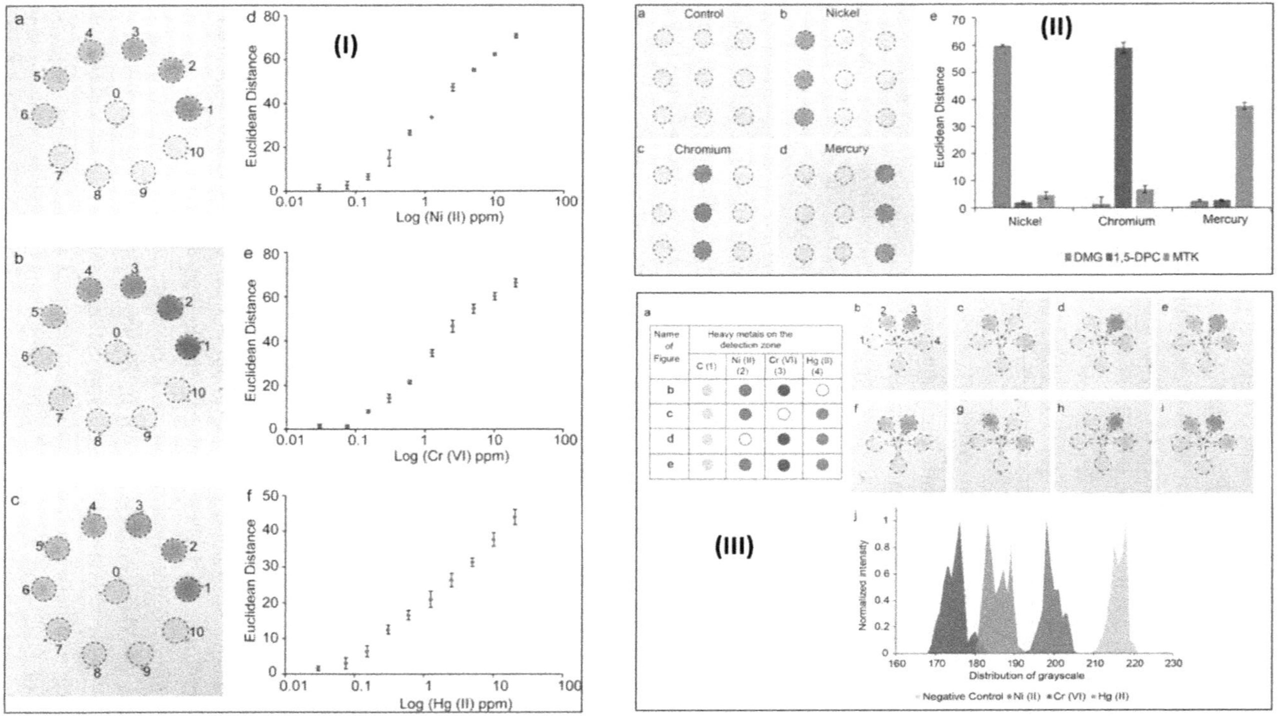

FIGURE 14.2 (I) Demonstration of the quantitative analysis conducted during heavy metal detection using the surface-modified well-spot C-μPAD, (II) Specificity analysis, (III) Heavy metal detection for standard and spiked lake-water samples using the multiplex C-μPAD pattern. (Reproduced with permission from Devadhasan er al. (2018).)

using the standard and the spiked lake water samples shown in Figure 14.2(III)(a). Figure 14.2(III)(j) shows histogram graphs, showing the color distribution of multiplex heavy metal detection for the standard and spiked lake samples. The histogram for Ni(II) shows better uniformity than the other two analytes. Quantitative analysis was the main challenge in the µPAD-based heavy metal detection since the metal complexes get flushed away or migrate towards the edges and heterogeneous color formation on the detection zones.

Lam et al. (2017) further used the chemical vapor deposition (CVD) method to create a thermally and chemically stable hydrophobic barrier for µPADs, which is called a chemically patterned microfluidic paper-based analytical device (C-µPAD), by utilizing TCS as a hydrophobic agent, a chromatography paper, and a vacuum chamber to create hydrophobic patterns for the detection of nickel. The fabrication and method are discussed in subsection 14.3.2. Using this unique property of C-µPAD, Ni detection was demonstrated on amine-functionalized well-spot C-µPAD by coupling dimethylglyoxime (DMG). This colorimetric reagent can be covalently attached to the amine-functionalized C-µPAD and reacts with nickel to form DMG-Ni complex. On the addition of various concentrations of Ni solution, color intensity is proportional to the concentration of Ni with limit of detection (LOD) as low as 150 µg/L. This heavy metal assay exhibits and proves that the C-µPAD enables us to use silane functionalization, which requires thermal condensation to form covalent bonds. The hydrophobicity on the C-µPAD was highly stable since the TCS molecules altered the wetting property of cellulose fibers in the chromatography paper.

14.4.1.3 Detection of Arsenic

As discussed in Section 14.3.1.5, a sensitive and selective paper-based microfluidics device was fabricated for rapid visual detection of very low concentration with a detection limit of 1.0 ppb of As^{3+} ions using gold nanosensor Au-TA-TG (Nath et al. 2014). A low flow rate or wicking speed through fibers and pores, the paper substrate, allows a very low concentration of arsenic to remain in a microchannel for ample interaction with nanosensors that can produce an intense signal term of the visible band for analysis. Au-TA-TG interacts with As^{3+} ions to develop a visible dark bluish-black precipitate at the interfacial zone. Due to the unique characteristics of gold nanosensors, the detection process becomes highly sensitive and rapid as it relies on the interparticle aggregation behavior of nanoparticles. The developed technique has many advantages, such as the overall system becomes power free, portable, cost effective, and safe for arsenic detection. The microfluidics device that prevents wicking with simple visual detection proves its potential as a completely miniaturized sensing device, removing numerous challenges which we commonly encounter with the existing state-of-the-art sensors for arsenic detection. The developed device is highly selective for arsenic with LOD down to 1 ppb, lower than the 10 ppb WHO reference limit for drinking water.

14.4.2 Detection of Glucose

The wicking microfluidics device developed by Lam et al. (2017) was further applied for detection of glucose assay using D-(+)-glucose dissolved in deionized water and successively diluted to prepare 0 ~ 160 mg/dL concentration of a standard glucose solution. The bioassay capabilities of C-μPAD were determined using glucose assays on well-spot C-μPAD and lateral flow C-μPAD platforms using standard glucose samples shows that color intensity increases as glucose concentration increases from 0 to 160 mg/dL in glucose well-spot C-μPADas shown in (-I(a). The LOD of 13 mg/dL was achieved in well-spot C-μPAD, which is equivalent to the LOD of the commercially available glucose meter. A lateral flow glucose assay was developed by using a dumbbell-shaped channel. Here, 2.5 μL of glucose sample was then applied to the sample inlet and allowed to flow and react in the detection zone. The gradient of color intensity depending on the glucose concentration (Figure 14.3(I)(b)) was observed. At the same time, Figure 14.3(I)(c) indicates the differences between the color intensity of each glucose concentration in lateral flow C-μPAD. Using the same glucose concentration range, lateral flow assay was successfully demonstrated at the well-spot C-μPAD glucose assay, and LOD of 23 mg/dL was achieved. The C-μPAD ability for POC diagnostics was demonstrated using a human blood glucose assay on the well-spot C-μPAD. A plasma separation membrane (Pall Corporation) was fixed on the frontal side of the well-spot C-μPAD to extract plasma from the blood (Figure 14.3(I)(d)). Figure 14.3(I)(e) shows the results on the reverse of the well-spot C-μPAD from glucose-spiked blood samples. Overall color intensity increases as the total amount of glucose increase from glucose-spiked blood samples. Compared to the standard well-spot glucose assay in Figure 14.3(I)(f), the color intensities from the blood samples are less than those of the standard glucose samples due to differences in fluidic properties and mild reaction inhibitors in the extracted plasma. Also, the overall intensity is shifted down a bit, sensitivity from the blood samples is almost identical to that from the standard glucose samples.

14.4.3 Detection of Horseradish Peroxidase

A simple, rapid, portable, and highly sensitive horseradish peroxidase (HRP) assay using a wicking microfluidics device was developed by Basu et al. (2016) that can be visually detected by the naked eye using blue color intensity within a short reaction time of 10 min. They have used photolithography as a fabrication method which is discussed in Section 14.3.1.4 where hydrophobic barriers separated hydrophilic test regions. 10 mM of 3,3′,5,5′-tetramethylbenzidine was used to immobilize these test regions for the HRP assay. The study also described the type of paper substrate that is most suitable to undergo photolithographic fabrication. Therefore, there is no need for further treatments to increase its hydrophilicity while maintaining the detection performance of the microfluidic device. The detection range for the developed method ranges from 0.37 to 124 fmol (or 3 to 1000 ng/mL) with detection limit 0.69 fmol (or 5.58 ng/mL). The developed portable microfluidics paper-based analytical devices for simple HRP assay can be applied to various target molecules for point-of-need testing.

14.4.4 Immunoassay

An array pattern well-spot C-μPAD with 4 mm diameter was developed for the immunoassay by Lam et al. (2017). First, 1 μg/mL TNFα antibody was incubated

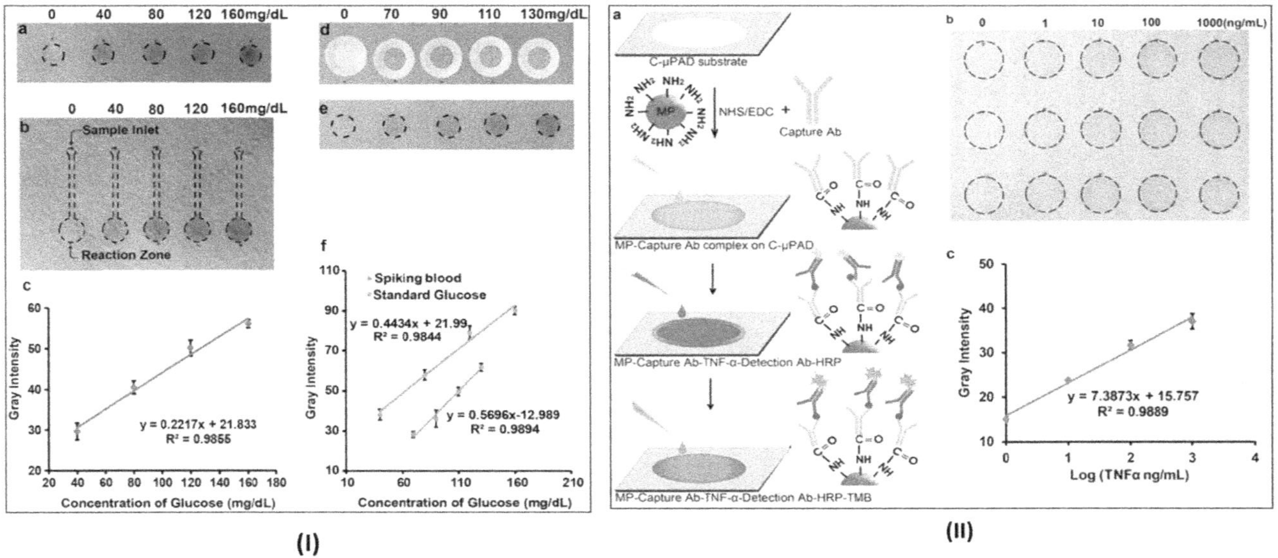

FIGURE 14.3 (I) Demonstration of glucose assay on well-spot C-μPAD and lateral flow C-μPAD. (II) (a) Procedure of immunoassay on C-μPAD. (b) C-μPAD shows a clear concentration gradient after the assay with different concentration of TNFα(0 ng/mL~1000 ng/mL). (c) Immunoassay results acquired by a smartphone camera and analyzed by image J software. These results show log-linear relationship between TNFαconcentrations and differential gray value. (Reproduced with permission from Lam et al. (2017).)

with 1 μg/mL of EDC/NHS for 30 min and applied on the 1 μm size of amine-functionalized magnetic particles. The particles were physically immobilized on the well-spot C-μPAD. Next, 1 μL concentrations ranging from 1, 10, 100, and 1000 ng/mL of human TNFα was then applied on each well-spot and incubated for 10 min. Then, 1 μg/mL of human TNFα biotinylated antibody, followed by 2 μL of streptavidin-HRP reagents, was applied on the substrate, which was incubated for 10 min at room temperature. In addition, a sandwich immunoassay for TNFα quantification was demonstrated on well-spot C-μPAD by following the procedure shown in Figure 14.3(II)(a). Since the pore size of the paper is in the range of ~1 μm and can physically trap the beads, 1 μm sized amine-terminated magnetic particles were applied on the well-spot C-μPAD. On a positive well-spot, the appearance of blue coloration confirms the formation of the immune complex. TMB stop solution interrupts the reaction, which leads to a color change from blue to yellow, as shown in Figure 14.3(II)(b). The yellow color intensity formed in each well-spot was proportional to the amount of TNFα formed on the well-spot C-μPAD. Figure 14.3(II)(c) shows a log-linear correlation of the concentrations of TNFα taken, demonstrating that the color intensity increased to the concentration ranges of 1~1000 ng/mL. The resulting LOD from the immunoassay demonstration was achieved to be 3 ng/mL under well-spot C-μPAD. Compared to the traditional ELISA, C-μPAD decreased the reaction time significantly since this C-μPAD offered a 3D matrix to create the immune complex by transporting all liquid samples with wicking force. Also, less than 15 μL of total reagents was sufficient to carry out this sandwich immunoassay on well-spot C-μPAD.

Li et al. (2020) developed a POC lateral flow immunoassay (LFIA) test product, which can visually detect IgM and IgG simultaneously in human blood within 15 min. The test kit consists of the pre-treated plastic backing, sample pad, conjugate pad, absorbent pad, and NC membrane. Twenty μL of whole blood sample (or 10 μL of serum/plasma samples) was pipetted into the sample port followed by adding two to three drops (70–100 μL) of dilution buffer (10 mM PBS buffer) to drive capillary action along the strip. A red/pink test line in the M and G region indicates anti-SARS-CoV-2 IgM and anti-SARS-CoV-2 IgG, and if only the control line (C) showed red, the sample is negative, or the control line does not appear red, the test is invalid. If the M or G line or both lines turn red, it indicates the presence of anti-SARS-CoV-2-IgM or anti-SARS-CoV-2-IgG or both antibodies in the specimen (Figure 14.4(I)). Testing on blood samples collected from 397 PCR-confirmed COVID-19 patients and 128 negative patients at eight different clinical sites was performed (Figure 14.4(II)). The quantification demonstrated that the paper-based assay had overall testing sensitivity of 88.66% and a specificity of 90.63%. Detection consistency was indicated on evaluation of clinical diagnosis results obtained from different types of fingerstick blood, serum, and plasma of venous blood.

14.4.5 Detection of Hematocrit of Whole Blood

Berry et al. (2016) describe a paper-based microfluidics device that is simple and low cost for measuring hematocrit (Hct) of whole blood using wax-patterned

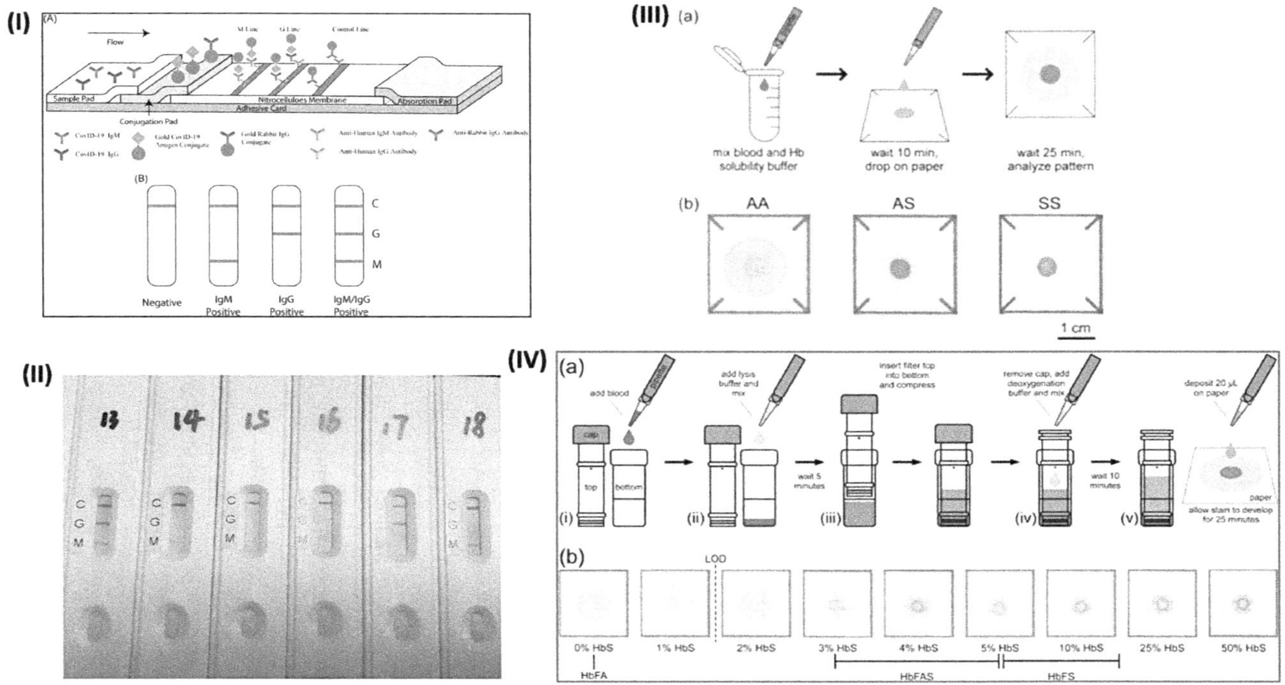

FIGURE 14.4 (I) Schematic illustration of rapid SARS-CoV-2 IgM-IgG combined antibody test. (II) Representative photo for different patient blood testing results. (Reproduced with permission from Li et al. (2020). (III) Overview of the paper-based SCA diagnostic test. (Reproduced with permission from Piety et al. (2016)). (IV) Paper-based newborn SCD screening test. (Reproduced with permission from Piety et al. (2017).)

chromatography paper. The device permits the transport of RBCs with the wicking front of plasma. They designed the device consisting of a top lamination sheet, a sample adding layer (made of chromatography paper) which prevents WBCs from entering the channel, a cutter-patterned double-sided adhesive, a wax-patterned microfluidic channel (made of chromatography paper) pretreated with 4.5 mM EDTA and 5 mM NaCl solutions for blood flow improvement, and a bottom lamination sheet. On administration of a whole blood sample, the WBCs get trapped in the sample adding layer, while the RBCs and plasma travel through it and go into the microfluidic channel. After a fixed period, separation occurs in which blood with lower (higher) hematocrit traveled a longer (shorter) distance. After 30 min into sample application Hct was detected at a range from ~30 to 55% with high linearity between blood travel distance and Hct value, having $R^2 = 0.991$ with good reproducibility (mean ± SD = 20.1 ± 3.2 mm and CV = 16%, when Hct = 42% and n = 20).

14.4.6 Detection of Sickle Cell Disease

Yang et al. (2013) achieved a major breakthrough towards a simple, rapid, and low-cost paper device for a POC diagnostic test for sickle cell disease (SCD) that can conclusively differentiate between blood samples from normal healthy individuals, sickle cell trait carriers and SCD patients using the characteristic blood stain using wax-patterned chromatography paper. They mixed whole blood with the standard hemoglobin solubility assay (SickleDex) at a ratio of 1:20 (volume ratio), and 20 µL of the blood mixture was applied after 5 min of incubation at room temperature onto the center of the paper device. The sickle hemoglobin (HbS), upon contact with SickleDex, polymerizes and becomes intertwined into the fibrous network of paper and forms a dark-red center spot. At the same time, the normal hemoglobin remains soluble and gets wicked through the paper upon contact and gives a uniform light-red circular pattern. The devices use Hb's natural color, giving a visual indication and differentiate between blood samples with or without sickle hemoglobin. The fully developed bloodstain on the paper device was digitally scanned, and image analysis was performed to capture the red color intensity change over the distance from the center of the bloodstain.

Later on, Piety et al. (2016) upgraded and utilized this diagnostic method for quantification of HbS. They performed sickle hemoglobin quantification using 88 human blood samples. The paper-based assay proves to be low cost, easy to use and is rapid with accuracy and reproducibility relative to conventional hemoglobin electrophoresis. Piety and co-workers further studied and performed a test on the same paper-based assay by identifying individuals whose blood contained any HbS (HbAS and HbSS). The assay was performed on 226 post-partum women at a primary obstetric hospital in Cabinda, Angola. It detected sickle cell anemia (SCA) (HbSS) with 93% sensitivity and 94% specificity for visual evaluation and 100% sensitivity and 97% specificity for automated analysis. Figure 14.4(III) gives an overview of the paper-based SCA diagnostic test. Notably, it permits instrument and electricity-free visual diagnostics within 30 min. The assay was also optimized by Piety et al. (2017) to detect the low levels of sickle hemoglobin (HbS) present in the blood among 159 newborns, allowing the direct screening for SCD and sickle cell trait (SCT; HbAS) in the immediate postnatal period. (Figure 14.4(IV)) illustrates the steps and procedure along with the LOD for visually detecting the presence of any HbS in a blood sample. The detection limit was 2% HbS, and it identified the presence of HbS with 81.8% sensitivity and 83.3% specificity and diagnosed SCD from newborns with 100.0% sensitivity and 70.7% specificity.

Kanter et al. (2015) developed a paper-based device, Sickle SCAN, based on colorimetric diagnostics for a novel POC test for SCD. It is a qualitative sandwich-type lateral flow immunoassay that detects HbA, HbS, and HbC. The test strip comprises a cellulose sample pad, fiberglass conjugate pad, cellulose wicking pad, and nitrocellulose analytical membrane with HbA, HbS, and HbC test lines and a control line. The device's corresponding test lines with HbA, HbS, and/or HbC become visible and serve as disease indicators. The overall accuracy of 99% was achieved on testing 71 human subjects using capillary blood; the device was able to correctly detect the presence of HbA, HbS, and HbC. A similar paper-based sandwich-type lateral flow immunoassay for SCD diagnostics was developed (Bond et al. 2017), consisting of a cellulose sample pad, fiberglass conjugate pad, and nitrocellulose analytical membrane consisting of HbA, HbS test lines, and a control line. The device detected SCD with 90% sensitivity and 100% specificity with either naked-eye or computer-aided analysis.

14.4.7 Detection of Nitrite Ion and Uric Acid

Wicking microfluidics devices based on thread and thread paper have potential applications in environmental monitoring, human health diagnostics, and food safety analysis and could be of great importance for the developing world because of their relatively low fabrication costs. Li et al. (2010) described a new and simple concept for fabricating low-cost, low-volume, easy-to-use microfluidics devices using cotton thread with a polymer film. The fabrication steps are simple and relatively low cost due to the requirement of only sewing needles or household sewing machines, which are affordable and commonly used. In the study, a hydrophilic line was immersed in the NO_2^- indicator solution and was then dried in an oven at 60°C for approximately 5 min. The indicator-treated thread (five short threads cut from it) was sewed onto an opaque polymer sheet with one stitch (~3 mm) exposed at the sheet's top side as the detection zone. Five serially diluted NO_2^- standard solution samples were deposited separately with a micro pipet onto the five sewed threads. Colorimetric alteration on each of the stitches can be seen with the naked eye. The detection range for the developed method ranges from 0 to 1000 mM/L. They developed thread paper-based microfluidic sensors by incorporating colorimetric indicators to detect two essential biomarkers, nitrite ion and uric acid.

14.4.8 Detection of Endocrine Disruptors

Kong et al. (2017) combined the adsorption capacity of molecularly imprinted polymer (MIP) membranes with $ZnFe_2O_4$ (Figure 14.5(I)) as peroxidase mimetics on

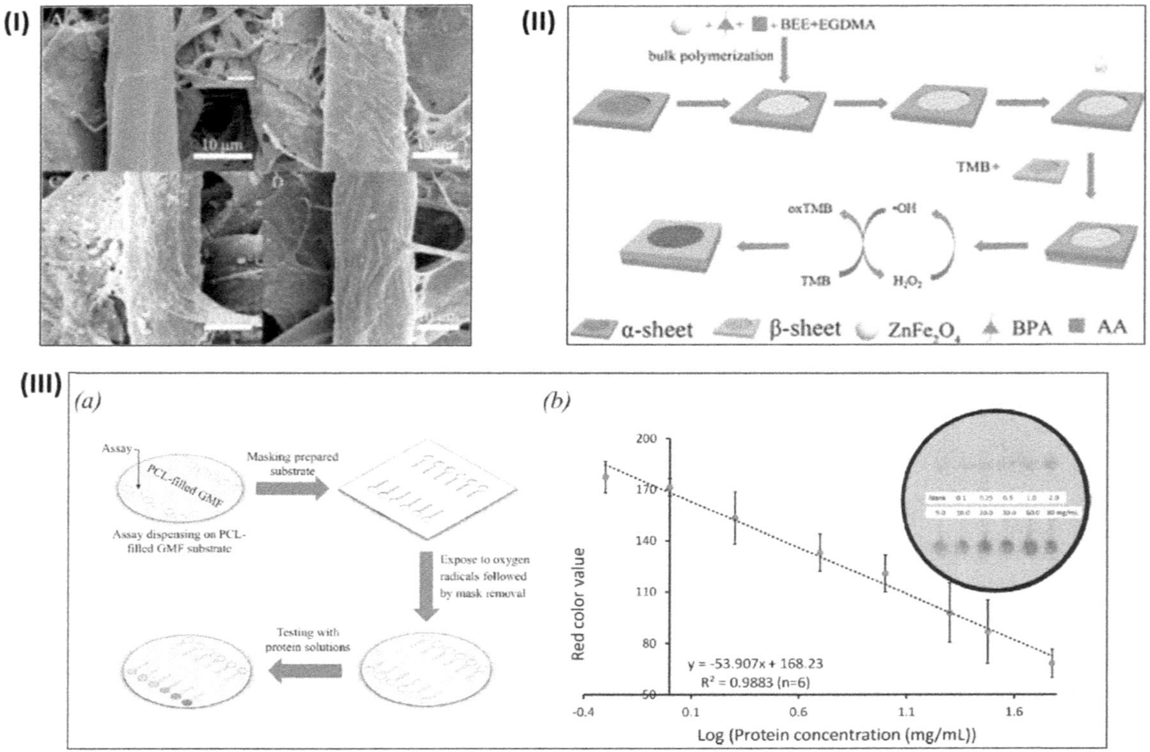

FIGURE 14.5 (I) SEM image showing fabrication process. (II) Schematic representation of the fabrication procedure for the paper-based colorimetric sensor. (Reproduced with permission from Kong et al. (2017).) (III) Fabrication approach for production of a total protein analysis device on PCL-filled GMF membrane media. (Reproduced with permission from Bandara et al. (2018b).)

one paper microzone through MIP membrane wrapping paper as well as its colorimetric potential to fabricate functional paper for the detection of bisphenol. The colorimetric sensor (Figure 14.5(II)) comprises two layers of patterned rectangular papers called an α-sheet comprising one working zone and a β-sheet with a circular contacting zone respectively from up to down. A wax screen-printed paper is oven treated at 130°C for 150 s so that the wax melts and penetrates the thickness of the paper creating hydrophobic barriers to form the microzone. The modification resulted in high performance and subsequent color development by H_2O_2 and 3,3',5,5' Tetramethylbenzidine (TMB), which can be observed with the naked eye and quantified by software. The colorimetric sensor for detecting Bisphenol A (BPA) at pH 4.0 HAc-NaAc buffer using $ZnFe_2O_4$ as a catalyst and H_2O_2 as an enzyme-substrate gave a correlation coefficient of 0.9945, with a LOD of 6.18 nM with high selectivity, sensitivity, and regeneration in a complicated matrix.

14.4.9 Detection of Hydrogen Peroxide

The WHO identified external readers as a challenge, and this issue needs to be addressed when creating ultimate POC diagnostic assays. The organization also listed "equipment-free" as one of seven necessary attributes for diagnostic tests in the developing world. To address this issue, Lewis et al. (2012) had described two complementary assay strategies using paper-based microfluidics devices. They showed that the level of an analyte could be quantified by simply measuring time devoid of any external electronic reading device for the quantitative measurement. The methods involve a digital assay where the time required for a sample to react is tracked. In due course, it was allowed to pass through a hydrophobic detection reagent in a single conduit within a 3D paper-based microfluidics device or an analog assay. Second, the study counted the number of colored bars after a fixed assay period in a related paper-based microfluidics device. In the method they described, only a timer, the ability to see color, and/or the ability to count to measure the quantity of a specific analyte is required. Furthermore, these two assays are based on selective changes in the wetting properties of paper. Hydrogen peroxide oxidatively cleaves a hydrophobic compound and is deposited into defined regions of the microfluidic conduits before assembling the devices. The reaction of hydrogen peroxide with the previous compound initiates an elimination reaction converting into hydrophilic byproducts. The change from hydrophobic to hydrophilic allows the sample to wick through the device. During this, it wets a detection region, where the time required to soak the detection region depends on the concentration of the analyte. The fabrication of the device is briefly described in Section 14.3.3. These assays do not require external instruments other than a timer, and this unique quantitative assay is operated by counting the number of colored bars at a fixed assay or by measuring flow-through time, making a handy platform for use in the developing world.

14.4.10 Detection of Protein, Ketone Bodies, and Nitrite

Fang et al. (2014) reported a simple, straightforward, low-cost, and highly reproducible fabrication method of microfluidics systems based on a glass fiber membrane.

The unique fabrication method has already been discussed briefly in Section 14.3.6. The device represents a novel type of wicking microfluidics paper-based device with a high resolution of the microchannel down to ~137 μm, similar to those made by conventional photolithography. The authors successfully applied the new fabrication method to microfluidics to create a star micro-array format of multiplexed urine tests in this study. The developed method was used to detect protein, glucose, ketone bodies, nitrite, and pH in urine.

The wicking microfluidics devices developed by Bandara et al. (2018b) using the fabrication approach as in Section 14.3.1 was used to conduct total protein assay by visual detection. The device was fabricated using the same approach as above and is illustrated in Figure 14.5(III)(a). In a quantitative detection approach, image-based detection and digital red (R), green (G), and blue (B) color analysis was used. The intensity of the RGB color channels over the assay zone was measured (Figure 14.5(III)(b), inset). Both red and green color channels showed a good linear trend over the range of 0.5–60 mg/mL with respective R^2 of 0.988 and 0.989. The red color channel was the best-fit color channel for quantitative detection as it exhibited higher sensitivity than the green channel. Normal human blood serum was tested in both undiluted and diluted forms to explore the reliability of the assay. The method developed is an efficient and unique process. The geometry of fabrication decides the unique properties of the channels and enables various unit operations such as microfluidic mixing, separations, delay, and timing devices, etc. A protein assay's practicability and performance on this novel microfluidic platform demonstrated its applicability with different real-world assay chemistries. Considering the fabrication cost, feasibility, and assay results, the performance of this device is found to be superior to the implementation of previously reported microfluidic devices using the same assay chemistry.

14.5 CONCLUSION

This chapter discusses various research findings and the recent development of wicking microfluidic devices based on multiple membranes, including modified polymer, paper, thread, etc. The various fabrication techniques involved in the preparation of wicking microfluidics devices based on visual detection and chromatographic separation approach for quantification is studied in the chapter. The different unique methodology that involved a new approach to fabricating wicking microfluidics devices is also discussed. The microfluidic channel could be manufactured after the assay has been dispensed. Such a new lateral flow microfluidic channel has shown remarkable performance in overcoming interferences in detecting various types of matrixes. A few interesting paper-based techniques based on patterned chromatography paper, visual interpretation, and wicking properties are also discussed, particularly in blood-based POC analysis. The device discussed in the chapter is shown as a powerful research tool for a portable analytical approach capable of efficiently generating reliable qualitative and quantitative data for on-site monitoring. The chapter stimulates and provide essential information for further study on the different backgrounds to solve more problems. They contribute the potential wicking microfluidics devices based on chromatographic and visual detection as a widely reliable, sensitive, rapid, portable, and genuinely affordable emerging device for various fields, viz., healthcare as a diagnostic device, environmental monitoring, etc.

REFERENCES

Abe, K., Kotera, K., Suzuki, K., & Citterio, D. 2010. Inkjet-printed paperfluidic immunochemical sensing device. *Analytical and Bioanalytical Chemistry.* 398(2): 885–893.

Ahmed, S., Bui, M.-P. N., & Abbas, A. 2016. Paper-based chemical and biological sensors: Engineering aspects. *Biosensors and Bioelectronics.* 77: 249–263.

Bandara, G. C., Heist, C. A., & Remcho, V. T. 2018a. Chromatographic separation and visual detection on wicking microfluidic devices: Quantitation of Cu2+ in surface, ground, and drinking water. *Analytical Chemistry.* 90(4): 2594–2600.

Bandara, G. C., Heist, C. A., & Remcho, V. T. 2018b. Patterned polycaprolactone-filled glass microfiber microfluidic devices for total protein content analysis. *Talanta.* 176: 589–594.

Berry, S. B., Fernandes, S. C., Rajaratnam, A., DeChiara, N. S., & Mace, C. R. 2016. Measurement of the hematocrit using paper-based microfluidic devices. *Lab on a Chip.* 16(19): 3689–3694.

Bond, M., Hunt, B., Flynn, B., Huhtinen, P., Ware, R., & Richards-Kortum, R. 2017. Towards a point-of-care strip test to diagnose sickle cell anemia. *Plos One.* 12(5): e0177732.

Bruzewicz, D. A., Reches, M., & Whitesides, G. M. 2008. Low-cost printing of poly(dimethylsiloxane) barriers to define microchannels in paper. *Analytical Chemistry.* 80(9): 3387–3392.

Busa, L. S. A., Maeki, M., Ishida, A., Tani, H., & Tokeshi, M. 2016. Simple and sensitive colorimetric assay system for horseradish peroxidase using microfluidic paper-based devices. *Sensors and Actuators B: Chemical.* 236: 433–441.

Chen, X., Zhang, S., Han, W., Wu, Z., Chen, Y., & Wang, S. 2018. A review on application of graphene-based microfluidics. *Journal of Chemical Technology & Biotechnology.* 93(12): 3353–3363.

Chitnis, G., Ding, Z., Chang, C.-L., Savran, C. A., & Ziaie, B. 2011. Laser-treated hydrophobic paper: An inexpensive microfluidic platform. *Lab on a Chip.* 11(6): 1161.

Devadhasan, J. P., & Kim, J. 2018. A chemically functionalized paper-based microfluidic platform for multiplex heavy metal detection. *Sensors and Actuators B: Chemical.* 273: 18–24.

Dong, Z., Xu, H., Bai, Z., Wang, H., Zhang, L., Luo, X., Tang, Z., Luque, R., & Xuan, J. 2015. Microfluidic synthesis of high-performance monodispersed chitosan microparticles for methyl orange adsorption. *RSC Advances.* 5(95): 78352–78360.

Dungchai, W., Chailapakul, O., & Henry, C. S. 2011. A low-cost, simple, and rapid fabrication method for paper-based microfluidics using wax screen-printing. *The Analyst.* 136(1): 77–82.

Fang, X., Wei, S., & Kong, J. 2014. Paper-based microfluidics with high resolution, cut on a glass fiber membrane for bioassays. *Lab on a Chip.* 14(5): 911.

Gouveia, F., Bicker, J., Goncalves, J., Alves, G., Falcão, A., & Fortuna, A. 2019. Liquid chromatographic methods for the determination of direct oral anticoagulant drugs in biological samples: A critical review. *Analytica Chimica Acta.* 1076: 18–31.

Grem, I. C. da S., Lima, B. N. B., Carneiro, W. F., Queirós, Y. G. de C., & Mansur, C. R. E. 2013. Chitosan microspheres applied for removal of oil from produced water in the oil industry. *Polímeros Ciência e Tecnologia.* 23(6): 705–711.

Hossain, S. M. Z., & Brennan, J. D. 2011. β-galactosidase-based colorimetric paper sensor for determination of heavy metals. *Analytical Chemistry.* 83(22): 8772–8778.

Hu, J., Wang, S., Wang, L., Li, F., Pingguan-Murphy, B., Lu, T. J., & Xu, F. 2014. Advances in paper-based point-of-care diagnostics. *Biosensors and Bioelectronics*. 54: 585–597.

Kanter, J., Telen, M. J., Hoppe, C., Roberts, C. L., Kim, J. S., & Yang, X. 2015. Validation of a novel point of care testing device for sickle cell disease. *BMC Medicine*. 13(1): 225.

Klasner, S. A., Price, A. K., Hoeman, K. W., Wilson, R. S., Bell, K. J., & Culbertson, C. T. 2012. Paper-based microfluidic devices for analysis of clinically relevant analytes present in urine and saliva. *Analytical and Bioanalytical Chemistry*. 397(5): 1821–1829.

Kong, Q., Wang, Y., Zhang, L., Ge, S., & Yu, J. 2017. A novel microfluidic paper-based colorimetric sensor based on molecularly imprinted polymer membranes for highly selective and sensitive detection of bisphenol A. *Sensors and Actuators B: Chemical*. 243: 130–136.

Kumar, S., Kumar, S., Ali, M., Anand, P., Agrawal, V. V., John, R., Maji, S., & Malhotra, B. D. 2013. Microfluidic-integrated biosensors: Prospects for point-of-care diagnostics. *Biotechnology Journal*. 8: 1267–1279.

Kung, C. T., Hou, C. Y., Wang, Y. N., & Fu, L. M. 2019. Microfluidic paper-based analytical devices for environmental analysis of soil, air, ecology and river water. *Sensors and Actuators B: Chemical*. 301: 126855.

Lam, T., Devadhasan, J. P., Howse, R., & Kim, J. 2017. A chemically patterned microfluidic paper-based analytical device (C-μPAD) for point-of-care diagnostics. *Scientific Reports*. 7(1): 1–10.

Leung, K., Zahn, H., Leaver, T., Konwar, K. M., Hanson, N. W., Page, A. P., … Hansen, C. L. 2012. A programmable droplet-based microfluidic device applied to multiparameter analysis of single microbes and microbial communities. *Proceedings of the National Academy of Sciences*. 109(20): 7665–7670.

Lewis, G. G., DiTucci, M. J., & Phillips, S. T. 2012. Quantifying analytes in paper-based microfluidic devices without using external electronic readers. *Angewandte Chemie International Edition*. 51(51): 12707–12710.

Li, M., Cao, R., Nilghaz, A., Guan, L., Zhang, X., & Shen, W. 2015. "Periodic-Table-Style" paper device for monitoring heavy metals in water. *Analytical Chemistry*. 87(5): 2555–2559.

Li, X., Ballerini, D. R., & Shen, W. 2012. A perspective on paper-based microfluidics: Current status and future trends. *Biomicrofluidics*. 6(1): 011301.

Li, X., Tian, J., & Shen, W. 2010. Progress in patterned paper sizing for fabrication of paper-based microfluidic sensors. *Cellulose*. 17(3): 649–659.

Li, Z., Yi, Y., Luo, X., Xiong, N., Liu, Y., Li, S., … Ye, F. 2020. Development and clinical application of a rapid IgM-IgG combined antibody test for SARS-CoV-2 infection diagnosis. *Journal of Medical Virology*. 92(9): 1518–1524.

Liana, D. D., Raguse, B., Gooding, J. J., & Chow, E. 2012. Recent advances in paper-based sensors. *Sensors*. 12(9): 11505–11526.

Lin, S.-C., Hsu, M.-Y., Kuan, C.-M., Wang, H.-K., Chang, C.-L., Tseng, F.-G., & Cheng, C.-M. 2014. Cotton-based diagnostic devices. *Scientific Reports*. 4(1): 1–12.

Lu, Y., Shi, W., Jiang, L., Qin, J., & Lin, B. 2009. Rapid prototyping of paper-based microfluidics with wax for low-cost, portable bioassay. *Electrophoresis*. 30(9): 1497–1500.

Luppa, P. B., Müller, C., Schlichtiger, A., & Schlebusch, H. 2011. Point-of-care testing (POCT): Current techniques and future perspectives. *TrAC Trends in Analytical Chemistry*. 30(6): 887–898.

Mehling, M., & Tay, S. 2014. Microfluidic cell culture. *Current Opinion in Biotechnology*. 25: 95–102.

Morbioli, G. G., Mazzu-Nascimento, T., Stockton, A. M., & Carrilho, E. 2017. Technical aspects and challenges of colorimetric detection with microfluidic paper-based analytical devices (μPADs): A review. *Analytica Chimica Acta*. 970: 1–22.

Nabavi, S. A., Vladisavljević, G. T., Eguagie, E. M., Li, B., Georgiadou, S., & Manović, V. 2016. Production of spherical mesoporous molecularly imprinted polymer particles containing tunable amine decorated nanocavities with CO_2 molecule recognition properties. *Chemical Engineering Journal*. 306: 214–225.

Nath, P., Arun, R. K., & Chanda, N. 2014. A paper based microfluidic device for the detection of arsenic using a gold nano sensor. *RSC Advances*. 4(103): 59558–59561.

Olkkonen, J., Lehtinen, K., & Erho, T. 2010. Flexographically printed fluidic structures in paper. *Analytical Chemistry*. 82(24): 10246–10250.

Pengpumkiat, S., Wu, Y., Boonloed, A., Bandara, G. C., & Remcho, V. T. 2017. A microfluidic detection system for quantitation of copper incorporating a wavelength-ratiometric fluorescent quantum dot pair. *Analytical Methods*. 9(7): 1125–1132.

Piety, N. Z., George, A., Serrano, S., Lanzi, M. R., Patel, P. R., Noli, M. P., Kahan, S., Nirenberg, D., Camanda, J. F., Airewele, G., & Shevkoplyas, S. S. 2017. A paper-based test for screening newborns for sickle cell disease. *Scientific Reports*. 7(1): 1–8.

Piety, N. Z., Yang, X., Kanter, J., Vignes, S. M., George, A., & Shevkoplyas, S. S. 2016. Validation of a low-cost paper-based screening test for sickle cell anemia. *Plos One*. 11(1): e0144901.

Samiei, E., Tabrizian, M., & Hoorfar, M. 2016. A review of digital microfluidics as portable platforms for lab-on a-chip applications. *Lab on a Chip*. 16(13): 2376–2396.

Schonhorn, J. E., Fernandes, S. C., Rajaratnam, A., Deraney, R. N., Rolland, J. P., & Mace, C. R. 2014. A device architecture for three-dimensional, patterned paper immunoassays. *Lab on a Chip*. 14(24): 4653–4658.

Stolaroff, J. K., Ye, C., Oakdale, J. S., Baker, S. E., Smith, W. L., Nguyen, D. T., Spadaccini, C. M., & Aines, R. D. 2016. Microencapsulation of advanced solvents for carbon capture. *Faraday Discussions*. 192: 271–281.

Tabak, L. A. 2007. Point-of-care diagnostics enter the mouth. *Annals of the New York Academy of Sciences*. 1098(1): 7–14.

Wang, W., Wu, W.-Y., Wang, W., & Zhu, J.-J. 2010. Tree-shaped paper strip for semiquantitative colorimetric detection of protein with self-calibration. *Journal of Chromatography A*. 1217(24): 3896–3899.

Whitesides, G. M. 2006. The origins and the future of microfluidics. *Nature*. 442(7101): 368–373.

Xia, Y., Si, J., & Li, Z. 2016. Fabrication techniques for microfluidic paper-based analytical devices and their applications for biological testing: A review. *Biosensors and Bioelectronics*. 77: 774–789.

Yager, P., Edwards, T., Fu, E., Helton, K., Nelson, K., Tam, M. R., & Weigl, B. H. 2006. Microfluidic diagnostic technologies for global public health. *Nature*. 442(7101): 412–418.

Yang, X., Kanter, J., Piety, N. Z., Benton, M. S., Vignes, S. M., & Shevkoplyas, S. S., 2013. A simple, rapid, low-cost diagnostic test for sickle cell disease. *Lab on a Chip*. 13(8): 1464–1467.

Yogarajah, N., & Tsai, S. S. H. 2015. Detection of trace arsenic in drinking water: Challenges and opportunities for microfluidics. *Environmental Science: Water Research & Technology*. 1(4): 426–447.

Zhu, Y., Bai, Z.-S., & Wang, H.-L. 2017. Microfluidic synthesis of thiourea modified chitosan microsphere of high specific surface area for heavy metal wastewater treatment. *Chinese Chemical Letters*. 28(3): 633–641.

15 Microfluidic Electrochemical Sensor System for Simultaneous Multi Biomarker Analyses

Mayank Garg, Reetu Rani, Amit L. Sharma, and Suman Singh

CONTENTS

15.1 INTRODUCTION

Diagnosis of various diseases worldwide is now possible through advancements in the medical field accompanied by progressions in the field of chemistry, engineering, and other associated fields. However, for the detection of most diseases, a visit to a local clinic or a hospital is still required. With the ever-growing human population, along with the rise of novel diseases, the world needs portable and easy-to-use biosensors for the detection of such diseases. In this regard, the use of microfluidics has been a key factor for the development of portable biosensors. The integration of microfluidics for the diagnostic applications of the biosensor has been fruitful

DOI: 10.1201/9781003033479-15

(Bhatt and Bhattacharya 2019). Microfluidics deals with handling very small volumes of liquid (usually in micro or nano liters) inside micron-sized channels (Tarn and Pamme 2014, Whitesides 2006). Many configurations of microfluidics for biosensors have been reported, such as droplet microfluidics, paper microfluidics, and digital microfluidics, etc. (Srinivasan and Tung 2015, Nikoleli et al. 2018, Noh et al. 2011). All these configurations have been used to develop various types of biosensors (Luka et al. 2015). The incorporation of microfluidics with biosensors ensures continuous flow of the liquid (having analyte or electrolytes) over the biorecognition elements, thereby providing continuous monitoring of the analytes, which is not possible in the case of regular biosensing systems which are usually static in nature. The advantages of microfluidic biosensors are enhanced by multiplexing. This is the detection of multiple analytes at the same time while saving on time, reagents, and possible automation of the sensing systems. Multiplexing provides endless opportunities on how many analytes can be sensed at once. A review article by Araz et al. has shown that researchers have reported multiplexing of up to 96 samples at a time (Araz et al. 2013). This is of high significance in the point-of-care testing devices wherein multiple analytes are detected at once. This ensures less patient time along with low sample consumption (Dincer et al. 2017). The microfluidic biosensors can be broadly classified into three main categories, namely electrochemical, optical, and mechanical sensors (Han et al. 2013). Of these, electrochemistry-based sensors are the most well-known (Sanjay et al. 2015). The integration of electrochemistry with the microfluidic biosensors provides a platform which is highly sensitive, robust, and stable towards analyte detection and has the potential to be developed as practical devices for biomarker detection (Rackus et al. 2015, Nesakumar et al. 2019). Microfluidic biosensors have been used for the detection of analytes such as nucleic acids, proteins, DNA, ingredients in beverages, and pharmaceutical residues, to name a few (Choi et al. 2011, Dutta et al. 2018, Prasad et al. 2019, Campaña et al. 2019). The portability, stability, and robustness of the electrochemical systems make them highly appealing and a favored candidate for practical device fabrications.

A detailed account of electrochemistry-based microfluidic biosensors is given in the coming sections.

15.2 PLATFORMS FOR MICROFLUIDIC ELECTROCHEMICAL SENSOR SYSTEMS AND APPLICATIONS

Electrochemistry incorporated with microfluidics provides a new perspective to the biosensing approach. Very fine electrical changes in a reaction can be easily monitored, thereby also giving very sensitive detection of rare analytes. The biorecognition elements such as antibodies, aptamers, nucleic acids, or enzymes are immobilized on the working electrode making them specific towards the analyte of interest.

In general, the electrochemical detection system requires set of electrodes. If detection of a single analyte is considered, there is only one working electrode modified with the biorecognition element for a single analyte. For the multiplexed approach, usually an array of electrodes are used, wherein multiple sets of electrodes (working electrode, reference electrode, and counter electrode) are joined together in a microfluidics device. An alternative approach is the use of multiple working electrodes with a common reference and counter electrode. The selection of the approach for the detection is usually application based. However, in all these cases, all the working electrodes are modified with different biorecognition elements allowing detection of various analytes of interest. These electrodes can be directly printed on the microfluidic channel. This chapter highlights the most common microfluidics approaches used for electrochemical detection systems. Researchers have used either plastic or paper-based substrates for electrode printing. These are discussed in the coming sections.

15.3 NON-PAPER-BASED DEVICES

This is a less popular and not a widely used approach for multiplexed electrochemical detection systems. This is due to the difficulties faced in microfluidic channel fabrication and the integration of the electrodes in the fabricated microfluidics devices. The cost and disposal of these devices also limits their use. However, despite these limitations, non-paper-based devices have a significant advantage over paper-based devices. These devices tend to be more sturdy and rugged and can be easily developed for daily practical applications. These devices have been used for the sensing of the biomarkers of life-threatening diseases such as cancer, hypertension, and cardiac diseases. They also have the potential to be used for the detection of DNA sequences and viruses.

The general approach for the fabrication of the non-paper-based microfluidics devices is the use of poly(dimethylsiloxane) (PDMS). This is a soft polymer providing flexibility to the microfluidic device. Due to its soft nature, the fabrication of the channels in this is also convenient. In some cases, the PDMS layer is sandwiched between layers of polymethyl methacrylate (PMMA) to provide strength to the device. The electrode array is placed underneath the PDMS channels for the purpose of electrochemical sensing. A pumping system is accompanied with the microfluidic channel for the flow of the electrolytes and other reagents. This increases the complexity of the overall microfluidics device. The bulkiness of the non-paper-based method is another limitation compared to the paper-based method.

Here are some applications of non-paper-based microfluidics devices for the electrochemical sensing of multiple biomarkers:

15.3.1 CANCER

For cancer diagnostics, the use of microfluidics can be of immense help and if it is interfaced with electrochemistry, it opens the door for highly selective and sensitive multiplexing. The joint use of microfluidics and electrochemistry is attempted by Chikkaveeraiah et al. who used eight electrode chips as shown in Figure 15.1 (Chikkaveeraiah et al. 2011). The chip was modified with superparamagnetic particle antibody conjugates for the simultaneous detection of cancer biomarkers, namely prostate specific antigen and iterleukin-6. The total assay time for this device was

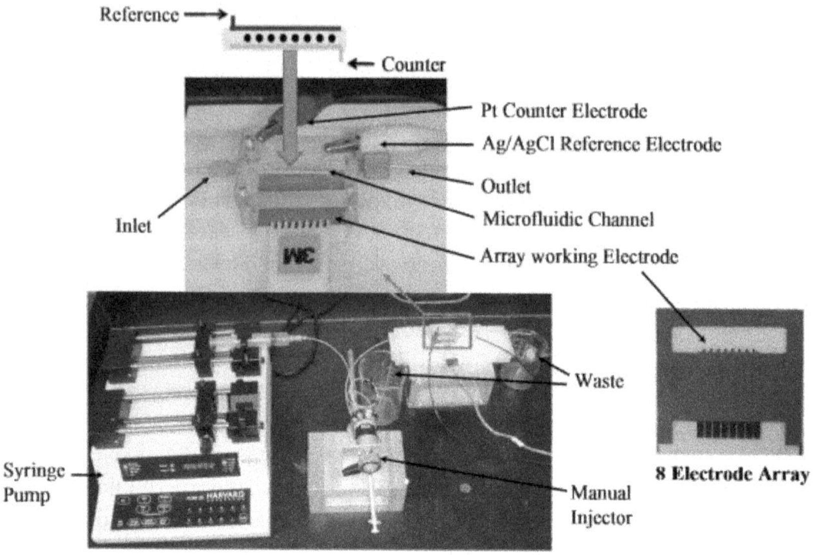

FIGURE 15.1 A microfluidics device with an eight-electrode array for the detection of cancer biomarkers. (Reproduced with permission from ref Chikkaveeraiah, Bhaskara V., Vigneshwaran Mani, Vyomesh Patel, J. Silvio Gutkind, and James F. Rusling. 2011. Microfluidic electrochemical immunoarray for ultrasensitive detection of two cancer biomarker proteins in serum. *Biosensors and Bioelectronics* 26 (11):4477-4483 (Chikkaveeraiah et al. 2011).) The setup shows a network of tubing and wires required to operate this microfluidic device.

1.15 h and had a detection limit in the pg/mL range. The developed sensor was able to detect these antigens even in the serum.

The same group demonstrated the detection of prostate specific antigen (PSA) and prostate specific membrane antigen by employing a composite of Fe_3O_4 and graphene oxide with the same setup. This system exhibited an enhanced PSA detection with the limit of detection being in the fg/mL range. In yet another effort by the same research group, an electrode array with 256 sensors was developed for the detection prostate specific antigen, prostate specific membrane antigen, interleukin-6, and platelet factor-4 in the serum wherein 32 electrodes were connected to an eight-port manifold (Tang et al. 2016). This method provided high throughput for the simultaneous detection of various cancer biomarkers. Generally, for microfluidics devices, pumps are required to disperse the liquid in continuous flow manner, but the use of a pumping system increases the complexity of these devices. To overcome this, Kallempudi et al. demonstrated the use of an on-chip actuator system for the flow of the liquid on a lab-on-a-chip developed by them for the detection of human-epidermal growth factor receptor and interleukin-6. The detection limit for both the analytes was in the ng/mL range (Kallempudi et al. 2012). Six biomarkers, potentially involved in the gastric cancer, namely carcinoembryonic antigen, carbohydrate antigen 19-9, *Helicobacter pylori* CagA protein, P53 oncoprotein, pepsinogen I, and pepsinogen II were simultaneously detected by Xie et al. in their work (Xie et al. 2015). An array of six electrodes, with each working area modified with an antibody selective towards an analyte was used. The study claimed to achieve better sensitivity as compared to the ELISA-based method for the cancer sera that they tested. Figure 15.2 shows the microfluidic device fabricated by them. A DNA-based sensor for the detection of DNA biomarkers for bladder cancer was also reported. A sensor array of 20 electrodes was fabricated by a series of steps involving physical vapor deposition, lithography, and laser cutting. These were shown to detect ~~the~~ three DNA biomarkers chosen by the researcher. The limit of detection for all the biomarkers was 250 fM which is well below the amount one would find in a urine sample. The device was shown to give a reading within 20 minutes (Pursey et al. 2017).

Zhu and co-workers employed a microfluidics device which was able to separate various analytes using in-built channels. These were conjugated with a screen-printed electrode which was used for the electrochemistry-based detection. The platform was demonstrated for the detection of various markers of leukemia such

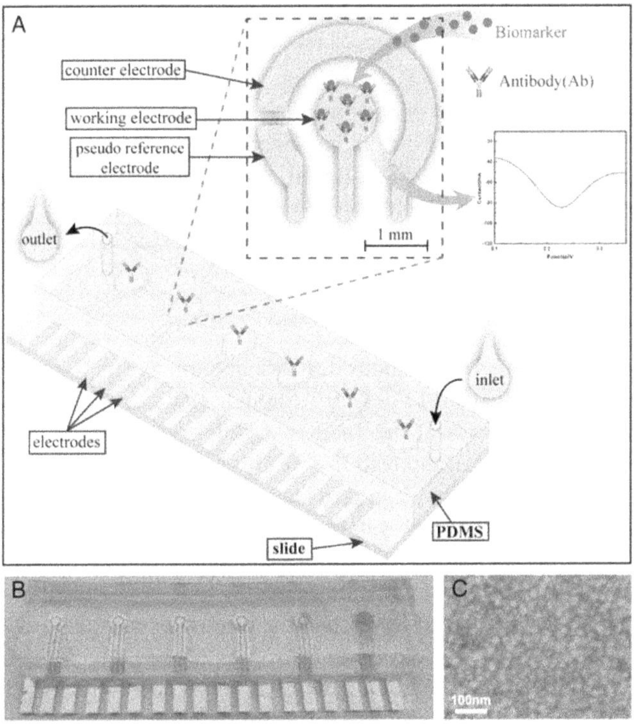

FIGURE 15.2 The figure shows the schematic of the fabricated microfluidic device (A); the picture of the electrode array (B); the SEM micrograph of the working electrode (C). (Reproduced with permission from ref Xie, Yao, Xiao Zhi, Haichuan Su, Kan Wang, Zhen Yan, Nongyue He, Jingpu Zhang, Di Chen, and Daxiang Cui. 2015. A Novel Electrochemical Microfluidic Chip Combined with Multiple Biomarkers for Early Diagnosis of Gastric Cancer. *Nanoscale Research Letters* 10 (1):477 (Xie et al. 2015).)

as methotrexate, lactate dehydrogenase, uric acid, and urea. The Limit of Detection (LOD) values obtained were much less than generally found in the human serum (Zhu et al. 2020).

15.3.2 Cardiac Disease and Hypertension

For the early detection of cardiac-related diseases, electrochemistry can be employed. Zhou et al. used a gold nanoparticle composite modified microfluidic chip for the simultaneous detection of two major cardiac disease biomarkers, cardiac troponin I and C-reactive protein. The antibodies for the specific analytes were bound to the gold nanoparticle composite-PDMS cells. A sandwich ELISA configuration was used for the detection of the analytes. The secondary antibodies were labeled with cadmium telluride (CdTe) and zinc selenide (ZnSe) QDs. The bound QDs on the antigen-antibody complex were dissolved using nitric acid. The remaining Cd^{2+} and Zn^{2+} were detected using square wave anodic stripping voltammetry. The detection limit achieved by this method was in the attomolar (aM) range and a correlation co-efficient of over 0.99 was attained (Zhou et al. 2010). Pulmonary hypertension biomarkers such as fibrinogen, adiponectin, LDL, and 8-isoprostane were shown to be detected in a microfluidic fashion using sensor arrays. PDMS channels were fabricated using a standard soft photolithographic approach. Pneumatic microvalves were fabricated on the chip working on the principle of the difference in the pressure. All the biomarkers were detected at 1 µg/mL concentration. Although this method reports a detection of a fairly high concentration of the biomarkers, this approach provided a rapid, and miniaturized sensor system (Lee et al. 2017).

15.3.3 Virus, DNA/RNA Sequences and Others

Influenza is one of the most contagious diseases in humans and unfortunately influenza viruses are very difficult to detect. Although methods like PCR and ELISA have been used for a fairly long time, their cost, time consumption, and labor intensiveness limit their practical applicability. In this regard, the use of microfluidics in tandem with electrochemistry can provide a solution to this problem. Han et al. developed a multi-virus detection system for the simultaneous amperometric-based detection of H1N1, H5N1, and H7N9 viruses. Zinc Oxide (ZnO) nanorods were used for the enhancement of sensitivity of the sensor system. Soft lithography was used to prepare channels in the PDMS layer, and the electrodes were deposited using the e-beam evaporator. This triple electrode array was able to detect these viruses as low as 1 pg/mL (Han et al. 2016). Recently it has been shown that non-coding small RNAs play a key role as biomarkers for various diseases. However, the detection of these RNA sequences is very difficult and cumbersome. Bruch and co-workers were the first to attempt a CRISPR/Cas13a-powered electrochemical biosensor for the nucleic acid amplification of the free mi-RNAs. The potential of the sensor was demonstrated by the detection of miR-19b and miR-20a, potential tumor markers. A limit of detection of 10 pM was achieved with the entire process taking less than 4 hours (Bruch et al. 2019).

Zupančič and co-workers demonstrated the use of bovine serum albumin and reduced graphene oxide nanoparticles for the multiplexed-based detection of various sepsis markers in whole blood. The markers targeted in the study were procalcitonin, C-reactive protein, and pathogen-associated molecular patterns. The developed sensor displayed good correlation with an ELISA-based method even for undiluted serum samples. Not only did the microfluidic component of this sensor allow for less sample consumption, but it also reduced the time needed for incubation of samples and labels (Zupančič et al. 2021).

15.4 PAPER-BASED DEVICES

Nowadays, most of the available diagnostic techniques are quite costly. Therefore, microfluidics paper-based analytical devices (µPADs) have emerged as great platforms for diagnostic purposes because these are based on the use of a common, affordable, and environment-friendly substrate, have the capability of multiplexed detection, and possess point-of-care testing characteristics (Adkins et al. 2015). But most of the µPADs are based on colorimetric response. Although colorimetric and visualization responses are helpful in some applications, their main limitation is their incapability to achieve low detection limits. Compared to these methods, electrochemistry-based detection is more sensitive, selective, and helps to achieve low detection limits (Liu et al. 2014). Therefore, electrochemistry-based detection coupled with µPADs offers a great solution to these problems. The very first microfluidic paper-based electrochemical devices (µPEDs) were reported by Dungchai et al. in 2009 for simultaneous detection of glucose, lactate, and uric acid in serum samples (Dungchai et al. 2009). Exponential growth in the fabrication and utilization of µPEDs has been observed since then for various applications. µPEDs are comprised of microfluidic channels on patterned sheets of paper (chromatography paper, polyester, or cellulose, etc.) and printed electrodes (Nie et al. 2010). The response of µPED is greatly dependent upon its fabrication and integration of electrodes. Two main strategies are usually employed to integrate different components of a µPED, either the device is fabricated on a single paper substrate with both an electrochemical transducer part and fluidic structure or electrode fabrication performed on a separate substrate and then assembled with paper fluidic component. Every paper-based device comprises of reaction zones and detection zones, irrespective of the detection method used (Gutiérrez-Capitán et al. 2020).

15.4.1 Hydrophobic Barrier Fabrication

Channels on paper-based microfluidic devices are fabricated using two methods. In the first method, hydrophobic barriers are constructed on paper and in the second, papers are cut directly into channels. Fabrication of hydrophobic barriers is a common practice to outline flow channels in paper-based devices. Photolithography is the most frequently used technique to make these hydrophobic barriers on paper. Photoresist materials (ultraviolet resins, octadecyltrichlorosilane etc.) are exposed to light through a photo mask to create barriers on paper.

Although photolithography benefits from good resolution, due to its high cost it is replaced by various printing techniques (Akyazi et al. 2018). Numerous materials are used for printing the barriers on papers such as wax, indelible ink, ultraviolet-curable inks, polystyrene, polyacrylate, silicone, etc. Among these, wax is the most frequently used barrier material. The printing of wax on paper is performed by different methods, viz., wax printing, screen printing, the dipping method, and inkjet printing (Jiang and Fan 2016, Yamada et al. 2015). Wax printing and inkjet printing are the representative methods of this category due to their ease of use, low-cost, and commercial availability (Dungchai et al. 2011). The second approach to channel fabrication, by direct cutting, is user friendly and can be performed with a simple cutting machine or equipment such as scissors, craft knives, hole punches, etc. There is no need of materials such as wax, polymers, or solvents to change the chemical properties of paper in this method. Laser cutter printers are also employed by some researchers to cut channels. Capillary action in paper is responsible for the flow of liquid. Vertical and horizontal flow are utilized for multiplexed sensing in μPEDs. The good stackability of paper allows it to make 3D platforms for complex assays.

15.4.2 Electrode Fabrication

Fabrication of conductive tracks (electrodes) on μPADs is the most crucial step affecting its performance. Numerous methods including screen printing, stencil printing, inkjet printing, and pencils are explored by different researchers to fabricate electrodes on paper-based devices depending upon the type of electrode. Carbon and noble metals are the most attractive choices to fabricate electrode on paper due to low-cost, ease of fabrication, and a wide potential window. Carbon was the first material incorporated in μPED by Dungchai et al. in 2009 (Dungchai et al. 2009). The screen printing method was employed to print carbon electrodes on paper substrate. In this method, customized patterned screens are placed on the paper and then carbon inks are pressed through open regions of the screen to create electrodes (Sekar et al. 2014). In stencil printing, instead of screens, adhesive tapes and transparencies are used as mask to create electrodes. Ink is pressed through open holes; therefore, it requires a thick ink consistency compared to the screen-printing method (Godino et al. 2012). Conductive inks prepared using binders and different forms of carbon, viz., graphite, carbon nanotubes, etc. are commonly employed as electrode fabrication materials. Along with these, electrode modifiers are also used to enhance electro-catalytical activity and current response of electrodes. The drop-casting method is usually adopted for electrode modification (Noviana et al. 2019). Performance of both screen and stencil printing greatly depends upon ink consistency and the features of mask and mesh pores (Adkins et al. 2015). Another simple and cost-effective method of electrode fabrication is the use of graphitic pencil lead to draw electrodes on paper. Additives can be added to pencil lead to enhance sensitivity and selectivity of electrochemistry-based detection. Although pencil drawing is a simple method, the reproducibility issue in this method limits its application of electrode fabrication (Santhiago and Kubota 2013). The painting of carbon ink on paper is an alternative way of electrode fabrication that does not involve equipment and masks. A simple paintbrush can be used as a tool to paint electrodes on paper substrate. However, electrode stability and adherence of the ink to the paper are some issues that should be taken care of during this method of electrode preparation (Adkins et al. 2015). Inkjet printing is also employed by some researchers to create electrodes on paper substrates as this method does not need templates, show high patterning capacity, and ability to print multiple materials simultaneously. However, the requirement for printers with high precision and control make this method expensive (Cinti et al. 2018). For papers with 3D structures, vacuum filtration is the preferred choice to create electrodes. It is one of the fast and scalable methods for this purpose. Filling density is a most crucial factor here, affecting the electroactive surface area of fabricated electrodes. Apart from carbon electrodes, metallic electrodes are also explored by various researchers due to the inherent properties of electron transfer in metals. Deposition of metal films using spraying or sputtering, evaporation through a mask are traditional means used for incorporation of metallic electrodes in paper substrates. An alternative technique used for this is the incorporation of microwires in μPEDs. Compared to carbon electrodes, microwires possess the advantage of low resistance (Ataide et al. 2020).

15.5 MULTIPLEXED DETECTION OF BIOMARKERS IN μPEDs

Various strategies are employed for simultaneous detection of biomarkers in μPED (Wu et al. 2013, Nontawong et al. 2018). Most of the biomarkers studied to date using μPEDs are redox inactive, therefore either application of different redox labels is employed or detection at separate zones with enzyme or immunoassay is performed (Noviana and Henry 2020). Some reports also show direct detection of biomarkers by monitoring their oxidation or reduction peaks. In multiplexed detection using direct methods, the common problem encountered is the overlapping of the oxidation/reduction peak of different analytes. To meet these challenges, modification of working electrodes with nanomaterials and electrocatalysts is usually employed to enhance selectivity and sensitivity of the device (Ataide et al. 2020, Nontawong et al. 2018). In enzymatic detection, biomarkers are selectively diagnosed based on the activity of the specific enzyme in the presence of molecules of interest. For this purpose, enzymes are immobilized over the working zone of paper using different immobilization strategies such as drop-casting, use of glutaraldehyde crosslinker and nafion membranes (Noviana and Henry 2020). Redox mediators or electroactive enzymatic byproducts are usually employed to monitor electrochemistry-based changes. Nicotinamide adenine dinucleotide, hydrogen peroxide (H_2O_2), and thiocholine are most frequently used electroactive enzymatic byproducts. Redox mediators such as ferrocyanide and hexamine-ruthenium (III) chloride are used to maintain electron flow in reaction zone (Ataide et al. 2020). Affinity or immunoassay-based detection relies on change in the electrochemical signal due to the binding of antibody or nucleic acid biorecognition elements to target biomarkers. The most crucial step in working with immunoassay-based μPED is the immobilization of biorecognition elements on working electrodes. In this regard, electrode surfaces are typically terminated with functional groups with crosslinker chemistries that can simply conjugate to biorecognition elements (Noviana et al. 2019).

Based on these strategies, various efforts made by different researchers to fabricate paper-based microfluidics devices for multiplexed electrochemistry-based detection of numerous biomarkers are discussed in next sections.

15.5.1 Cancer

To date immunoassay-based μPEDs have been prominently studied for the multiplexed detection of cancer biomarkers. A 3D microfluidics paper-based device combined with electrochemical immunoassay was reported by Wang et al. for the multiplexed detection of two tumor biomarkers. The 3D-μPED used for detection purposes was comprised of two layers, one wax-patterned paper layer and second, screen-printed electrode layer. The fabricated paper sensor contained an auxiliary zone of 8 mm diameter surrounded with two working zones of 4 mm diameter (carbon electrodes) connected through paper channels. Horseradish peroxidase (HRP)-labeled capture antibodies were immobilized on a working zone already modified with multi-walled carbon nanotubes (MWCNTs) and chitosan. The developed sensor was based on a typical HRP-O-phenylenediamine-H_2O_2 electrochemistry system. The electrochemistry-based immunosensor gave a linear response for carcinoma antigen 125 (CA125) and carcinoembryonic antigen (CEA) in the concentration range of 0.001–75 U/mL and 0.05–50 ng/mL respectively (Wang et al. 2012). Li et al. designed a 3D multiplex origami electrochemistry-based immunodevice (μ-OMEI) for sensitive and selective diagnosis of CEA and AFP cancer biomarkers. Metal ion labeled immunoassays (Cu^{2+} and Pb^{2+}) were used for simultaneous detection of non-redox active CEA and AFP biomarkers. Both labels being oxidized at distinct redox potentials, the voltage sweep mode was employed for detection purposes. The device was fabricated on a cellulose paper by wax printing and comprised of two tabs (sheets) including auxiliary and sample tabs. Screen-printed carbon on the sample tab was used as a working electrode and screen-printed carbon and Ag/AgCl on the auxiliary tab were used as counter and reference electrodes, respectively. In order to enhance the surface area and conductivity of working zone, a nanoporous silver (NPS) layer was grown over a paper working electrode. The NPS layer was used to immobilize and capture antibodies for target analytes. Apart from this, metal ion-labeled nanoporous gold was employed as signal amplifier and its hybrid with chitosan was used to modify electrode. Fabricated μ-OMEI was able to detect CEA and AFP up to 0.06 pg/mL and 0.08 pg/mL, respectively (Li et al. 2013). The work was further extended by the same research group to diagnose two other biomarkers, namely CA 125 and CA 199, by labeling with different metal ion-coated (Ag^+ and Cu^{2+}) nanoporous Ag/chitosan. Low detection limits up to 0.08 and 0.10 mU/mL were achieved with a 3D-origami-based immunosensor, respectively (Li et al. 2014). The same strategy was employed to detect a new combination of biomarkers, namely CEA and CA125. This time Cd^{2+} and Pb^{2+} metal ion-coated Au/bovine serum albumin nanospheres were explored as tracing tags (Ma et al. 2015). Simultaneous diagnosis of CEA and AFP was also explored with redox tracer molecules, viz., carboxyl ferrocene, and methylene blue. Paper working electrodes were modified with polyaniline-gold nanoparticles. The developed μ-OMEI showed LODs 0.5 and 0.8 pg/mL, respectively (Li et al. 2014). Based on similar eight-electrode architecture but a different electrode functionalization strategy, multiplexed detection of four different cancer biomarkers, namely CEA, AFP, CA125, and carbohydrate antigen 153 (CA153), was reported by Wu et al. The authors demonstrated the first ever integration of signal amplification strategy with μPEDs. The authors used graphene and SiO_2 nanoparticles for dual signal amplification as incorporation of graphene into the electrode surface facilitated enhanced electron transfer kinetics and SiO_2 nanoparticles increased the upload of the electrochemical tag. As designed, the immunosensor achieved a limit of detection as low as pg/mL (Wu et al. 2013). A new signal amplification strategy based on polymerization was also attempted by same group for ultrasensitive detection of cancer biomarkers. Glycidyl methacrylate (GMA) (monomer) with excess epoxy groups was used to immobilize HRP. Further amplification of the signal was performed by modification of the immuno-device surface with graphene. The device showed a limit of detection of 0.01, 0.01, 0.05, and 0.05 ng/mL CEA, AFP, CA125, and CA153, respectively (Wu et al. 2014). An aptamer-based μPED for multiplexed analysis of lung cancer biomarkers namely, neuron-specific enolase (NSE) and CEA, is recently reported. Wax printing and screen printing methods were employed to fabricate the device with layered configuration. Separate paper substrates were used to fabricate electrochemistry-based cell and microfluidic channels. For immobilization of biomarker-specific aptamer and enhanced detection sensitivity, Prussian blue-poly (3,4-ethylenedioxythiopene)-gold nanoparticle composite and amino functional graphene-thionin-gold nanoparticles composite were used, respectively (Wang et al. 2019). A novel μPED platform was developed with hairpin DNA nanostructures for multiplexed sensing of microRNA-based cancer biomarkers. Hierarchically assembled nanomaterials and metal organic framework-conjugated bio-probes resulted in selective and sensitive detection of microRNA-141 and microRNA-21 in serum samples (Tian et al. 2019). Recently, Kraatz and co-workers developed a dual readout platform that combines electrochemical and SERS signals for simultaneous detection of tumor biomarkers, namely AFP and CEA. A gold microelectrode array (GMA) electrode system was used for sensor fabrication which was prepared using numerous silica cavities and electrodeposited gold. The fabricated sensor was able to detect wide concentration range of both proteins with low detection limit of 0.01 ng/mL (CEA) and 8.0 ng/mL (AFP) (Gu et al. 2021).

15.5.2 Clinical Biomarkers

The most common clinical biomarkers diagnosed using μPED are glucose, lactate, ascorbic acid (AA), dopamine (DP), and uric acid (UA). Among these, redox active types are detected directly by oxidation or reduction. A μPED based on an array of eight electrochemical biosensors was used for simultaneous detection of metabolic biomarkers including glucose, lactate, and UA. Eight biosensing modules were fabricated on chromatographic paper where each module consisted of a hydrophobic wax channel and three screen-printed carbon electrodes. Chrono-amperometry technique was employed to monitor enzyme-catalyzed reactions specific to metabolic biomarkers. The fabricated device gave a linear response in the concentration range appropriate for application in clinical assays for all the three biomarkers (Zhao et al. 2015). Using an active paper-based hybridized chip (APHC) strategy, a programmable microfluidic sensor was introduced by Ruecha et al. to quantify glucose, DA, and UA metabolites in serum samples. In APHC methods, an active electro-wetting technique is employed as a substituent of the capillary action of paper to actuate quantified drops on substrate. The portable device was comprised of three main parts: The AHPC platform, driving part, and control part. An android smartphone was used as a wireless control system. Two working electrodes with single counter and reference electrodes were screen-printed on the paper. The working electrode was modified with graphene oxide and gold nanoparticles to enhance detection sensitivity of the device. Different coupled and de-coupled modules were employed for sensitive detection of three biomarkers (Ruecha et al. 2017). Based on a paper-folding approach, a disposable μPED comprised of two electrochemistry-based sensors were introduced by Fava et al. for simultaneous analysis of UA and creatinine in urine samples. The filter paper was cut using a home cutter printer for the sample injection spot and detection area. A single working electrode was cut into two spots so that a single spot could be utilized for detection of one analyte. Different detection strategies were employed for determination of two analytes. Direct oxidation of UA was performed at one of the spots by modifying it with graphene quantum dots (GQDs) and the enzymatic reaction was monitored for creatinine detection on the second spot which, was modified with GQDs, creatininase enzyme, and a ruthenium redox mediator, viz., hexaammineruthenium (III) chloride. A square wave voltammetry technique was employed for sensitive and selective detection of both biological markers simultaneously in real human urine samples (Cincotto et al. 2019). The same research group developed a similar disposable μPED with 16 independent microfluidic channels. A stencil printing approach was used to fabricate electrodes on filter paper. The fabricated μPED was comprised of 16 microfluidic channels radially distributed around a sample hole and 16 working electrodes. Working and reference electrodes were printed on the same sheet while a counter electrode was printed on a separate sheet. A four working electrode set was compiled with one reference electrode; resulting in overall four reference electrodes in the fabricated μPED. The device can record the electrochemical response of multiple replicates of a sample simultaneously. The feasibility of the device was checked running glucose as a model analyte in 16 replicates. The working electrodes were modified with chitosan, carbon black, glucose oxidase, and ferrocene-based redox mediator (Fava et al. 2019). The group further extended its research and used a similar 16-channel device to simultaneously detect glucose, creatinine, and UA in biological samples. Different electrochemistry-based detection principles were followed for all analytes. The electrical signal generated by electrochemical reaction between glucose oxidase and ferrocene carboxylic acid was monitored for estimation of glucose. For creatinine and UA detection, the signals generated due to Fe^{3+} ions and carbon black nanoparticles were monitored. The fabricated sensor was able to detect all three metabolites in real urine samples (Fava et al. 2020).

The application of a paper-based label-free immunosensor for simultaneous detection of two diabetes biomarkers, namely total hemoglobin (total Hb) and glycated hemoglobin (HbA1c), was introduced by Boonyasit et al. The developed 3D immunosensing device comprised of twin screen-printed conductive pads on wax-patterned paper combined with magnetic paper. Haptoglobin (Hp) and 3-aminophenylboronic acid (APBA) were used as recognition elements for biomarkers, respectively. Eggshell membranes (ESMs) were used to modify the screen-printed electrode and act as potential immobilizing platforms for Hp and APBA. Fe $(CN)_6^{3-/4-}$ solution was used as a working buffer and electrochemical impedance spectroscopy (EIS) at a single frequency was employed to monitor target analytes (Boonyasit et al. 2016). Apart from cancer and clinical biomarkers, paper-based microfluidics devices are also explored for multiplexed electrochemistry-based detection of human immunodeficiency virus (HIV) and hepatitis C virus (HCV) in serum samples. For detection purposes, an immunosensor array was coupled with a handheld potentiostat and wireless data transmission via universal serial bus (USB). Eight serum samples can run simultaneously to provide assay results in 20 minutes. The immunoassay-based sensor was able to achieve limit of detection of 300 pg/mL and 750 mg/mL respectively (Zhao and Liu 2016). Recently, an interesting work was reported by Liu et al. for simultaneous detection of biomarkers related to different diseases on a paper-based electrochemical device. The research group studied miR-21, alkaline phosphatase (ALP), and CEA biomarkers as model analytes. In this work, commercialized screen-printed electrodes were assembled with a ferrocene-labeled DNA (Fc-DNA) modified paper substrate using double-sided adhesive tape. Carbon nanotubes (CNTs) were used to modify working electrodes in order to amplify the detection signal. Specific target recognition probes, namely a microRNA probe (consisting of a ssDNA probe (P1), KF polymerase, and nicking endonuclease Nt.BbvCI), a phosphorylated hairpin probe, DNA aptamer probe, respectively, were used in the work. The detection assays depend on analyte-induced synthesis of Mg^{2+}-dependent DNAzyme for catalyzing the cleavage of Fc-DNA from paper (Liu et al. 2019). Recently, Boonkaew et al. reported multiplexed μPED for simultaneous detection of three cardiovascular disease (CVD) biomarkers including troponin I (cTnI), C-reactive protein (CRP), and procalcitonin (PCT). Prepared paper-based device was coupled with a label-free immunoassay for multiplexed detection of CVD biomarkers. A graphene oxide (GO)-modified carbon electrode was used to immobilize antibodies against target biomarkers. The fabricated μPED demonstrated good linearity with low detection limits for all biomarkers. Furthermore, the detection potential of the device was evaluated in serum samples and the results obtained were satisfactory (Boonkaew et al. 2021).

15.6 CONCLUSION AND FUTURE SCOPE

Recent advances in microfluidics devices make them appealing candidates for the diagnosis of various diseases at an early stage. Integration of electrochemistry with microfluidic detection results in improved detection limits and selectivity of the sensing system. A range of materials from polymers to papers are explored, to fabricate cost-effective and portable microfluidics devices. Various efforts have been made so far for the multiplexed sensing of clinically important biomarkers to improve the efficiency of analysis with various detection strategies. More programmable paper-based devices should be explored for high-throughput multiplexed sensing of different biomarkers in a cost-effective manner. These future developments in microfluidics devices will help to achieve comprehensive results for the early detection of disease biomarkers.

ACKNOWLEDGMENT

The authors are grateful to the Director, CSIR-Central Scientific Instruments Organisation (CSIR-CSIO), Chandigarh, India, for his constant encouragement and support. MG and RR would like to thank the Council of Scientific and Industrial Research (CSIR-HRDG), New Delhi, India, for their Senior Research Fellowship-GATE and Senior Research Fellowship respectively.

Conflict of interest: The authors declare no conflict of interest.

REFERENCES

Adkins, Jaclyn, Katherine Boehle, and Charles Henry. 2015. "Electrochemical paper-based microfluidic devices." *Electrophoresis* no. 36 (16):1811–1824.

Akyazi, Tugce, Lourdes Basabe-Desmonts, and Fernando Benito-Lopez. 2018. "Review on microfluidic paper-based analytical devices towards commercialization." *Analytica Chimica Acta* no. 1001:1–17.

Araz, M. Kursad, Augusto M. Tentori, and Amy E. Herr. 2013. "Microfluidic multiplexing in bioanalyses." *Journal of Laboratory Automation* no. 18 (5):350–366. doi: 10.1177/2211068213491408.

Ataide, Vanessa N, Letícia F Mendes, Lillia ILM Gama, William R de Araujo, and Thiago RLC Paixão. 2020. "Electrochemical paper-based analytical devices: Ten years of development." *Analytical Methods* no. 12 (8):1030–1054.

Bhatt, Geeta, and Shantanu Bhattacharya. 2019. "Biosensors on chip: A critical review from an aspect of micro/nanoscales." *Journal of Micromanufacturing* no. 2 (2):198–219. doi: 10.1177/2516598419847913.

Boonkaew, Suchanat, Ilhoon Jang, Eka Noviana, Weena Siangproh, Orawon Chailapakul, and Charles S. Henry. 2021. "Electrochemical paper-based analytical device for multiplexed, point-of-care detection of cardiovascular disease biomarkers." *Sensors and Actuators B: Chemical* no. 330:129336. doi: 10.1016/j.snb.2020.129336.

Boonyasit, Yuwadee, Orawon Chailapakul, and Wanida Laiwattanapaisal. 2016. "A multiplexed three-dimensional paper-based electrochemical impedance device for simultaneous label-free affinity sensing of total and glycated haemoglobin: The potential of using a specific single-frequency value for analysis." *Analytica Chimica Acta* no. 936:1–11.

Bruch, Richard, Julia Baaske, Claire Chatelle, Mailin Meirich, Sibylle Madlener, Wilfried Weber, Can Dincer, and Gerald Anton Urban. 2019. "CRISPR/Cas13a-powered electrochemical microfluidic biosensor for nucleic acid amplification-free miRNA diagnostics." *Advanced Materials* no. 31 (51):1905311. doi: 10.1002/adma.201905311.

Campaña, Ana Lucia, Sergio Leonardo Florez, Mabel Juliana Noguera, Olga P. Fuentes, Paola Ruiz Puentes, Juan C. Cruz, and Johann F. Osma. 2019. "Enzyme-based electrochemical biosensors for microfluidic platforms to detect pharmaceutical residues in wastewater." *Biosensors* no. 9 (1):41.

Chikkaveeraiah, Bhaskara V., Vigneshwaran Mani, Vyomesh Patel, J. Silvio Gutkind, and James F. Rusling. 2011. "Microfluidic electrochemical immunoarray for ultrasensitive detection of two cancer biomarker proteins in serum." *Biosensors and Bioelectronics* no. 26 (11):4477–4483. doi: 10.1016/j.bios.2011.05.005.

Choi, Seokheun, Michael Goryll, Lai Yi Mandy Sin, Pak Kin Wong, and Junseok Chae. 2011. "Microfluidic-based biosensors toward point-of-care detection of nucleic acids and proteins." *Microfluidics and Nanofluidics* no. 10 (2):231–247. doi: 10.1007/s10404-010-0638-8.

Cincotto, Fernando H, Elson L Fava, Fernando C Moraes, Orlando Fatibello-Filho, and Ronaldo C Faria. 2019. "A new disposable microfluidic electrochemical paper-based device for the simultaneous determination of clinical biomarkers." *Talanta* no. 195:62–68.

Cinti, Stefano, Noemi Colozza, Ilaria Cacciotti, Danila Moscone, Maxim Polomoshnov, Enrico Sowade, Reinhard R Baumann, and Fabiana Arduini. 2018. "Electroanalysis moves towards paper-based printed electronics: Carbon black nanomodified inkjet-printed sensor for ascorbic acid detection as a case study." *Sensors and Actuators B: Chemical* no. 265:155–160.

Dincer, Can, Richard Bruch, André Kling, Petra S. Dittrich, and Gerald A. Urban. 2017. "Multiplexed point-of-care testing – xPOCT." *Trends in Biotechnology* no. 35 (8):728–742. doi: 10.1016/j.tibtech.2017.03.013.

Dungchai, Wijitar, Orawon Chailapakul, and Charles S Henry. 2009. "Electrochemical detection for paper-based microfluidics." *Analytical Chemistry* no. 81 (14):5821–5826.

Dungchai, Wijitar, Orawon Chailapakul, and Charles S Henry. 2011. "A low-cost, simple, and rapid fabrication method for paper-based microfluidics using wax screen-printing." *Analyst* no. 136 (1):77–82.

Dutta, Gorachand, Joshua Rainbow, Uros Zupancic, Sotirios Papamatthaiou, Pedro Estrela, and Despina Moschou. 2018. "Microfluidic devices for label-free DNA detection." *Chemosensors* no. 6 (4):43.

Fava, Elson Luiz, Thiago Martimiano do Prado, Tiago Almeida Silva, Fernando Cruz de Moraes, Ronaldo Censi Faria, and Orlando Fatibello-Filho. 2020. "New disposable electrochemical paper-based microfluidic device with multiplexed electrodes for biomarkers determination in urine sample." *Electroanalysis* no. 32 (5):1075–1083.

Fava, Elson Luiz, Tiago Almeida Silva, Thiago Martimiano do Prado, Fernando Cruz de Moraes, Ronaldo Censi Faria, and Orlando Fatibello-Filho. 2019. "Electrochemical paper-based microfluidic device for high throughput multiplexed analysis." *Talanta* no. 203:280–286.

Godino, Neus, Robert Gorkin, Ken Bourke, and Jens Ducree. 2012. "Fabricating electrodes for amperometric detection in hybrid paper/polymer lab-on-a-chip devices." *Lab on a Chip* no. 12 (18):3281–3284.

Gu, Xuefang, Kaiyue Wang, Jiawei Qiu, Yajie Wang, Shu Tian, Zhenkuan He, Ran Zong, and Heinz-Bernhard Kraatz. 2021. "Enhanced electrochemical and SERS signals by self-assembled gold microelectrode arrays: A dual readout platform for multiplex immumoassay of tumor biomarkers." *Sensors and Actuators B: Chemical* no. 334:129674. doi: 10.1016/j.snb.2021.129674.

Gutiérrez-Capitán, Manuel, Antonio Baldi, and César Fernández-Sánchez. 2020. "Electrochemical paper-based biosensor devices for rapid detection of biomarkers." *Sensors* no. 20 (4):967.

Han, Ji-Hoon, Dongyoung Lee, Charleson Hong Chuang Chew, Taeheon Kim, and James Jungho Pak. 2016. "A multi-virus detectable microfluidic electrochemical immunosensor for simultaneous detection of H1N1, H5N1, and H7N9 virus using ZnO nanorods for sensitivity enhancement." *Sensors and Actuators B: Chemical* no. 228:36–42. doi: 10.1016/j.snb.2015.07.068.

Han, K. N., C. A. Li, and G. H. Seong. 2013. "Microfluidic chips for immunoassays." *Annual Review of Analytical Chemistry* no. 6:119–41. doi: 10.1146/annurev-anchem-062012-092616.

Jiang, Xiao, and Z Hugh Fan. 2016. "Fabrication and operation of paper-based analytical devices." *Annual Review of Analytical Chemistry* no. 9:203–222.

Kallempudi, Sreenivasa Saravan, Zeynep Altintas, Javed H. Niazi, and Yasar Gurbuz. 2012. "A new microfluidics system with a hand-operated, on-chip actuator for immunosensor applications." *Sensors and Actuators B: Chemical* no. 163 (1):194–201. doi: 10.1016/j.snb.2012.01.034.

Lee, GeonHui, JuKyung Lee, JeongHoon Kim, Hak Soo Choi, Jonghan Kim, SangHoon Lee, and HeaYeon Lee. 2017. "Single microfluidic electrochemical sensor system for simultaneous multi-pulmonary hypertension biomarker analyses." *Scientific Reports* no. 7 (1):7545. doi: 10.1038/s41598-017-06144-9.

Li, Long, Weiping Li, Hongmei Yang, Chao Ma, Jinghua Yu, Mei Yan, and Xianrang Song. 2014. "Sensitive origami dual-analyte electrochemical immunodevice based on polyaniline/Au-paper electrode and multi-labeled 3D graphene sheets." *Electrochimica Acta* no. 120:102–109.

Li, Weiping, Li Li, Shenguang Ge, Xianrang Song, Lei Ge, Mei Yan, and Jinghua Yu. 2014. "Multiplex electrochemical origami immunodevice based on cuboid silver-paper electrode and metal ions tagged nanoporous silver–chitosan." *Biosensors and Bioelectronics* no. 56:167–173.

Li, Weiping, Long Li, Meng Li, Jinghua Yu, Shenguang Ge, Mei Yan, and Xianrang Song. 2013. "Development of a 3D origami multiplex electrochemical immunodevice using a nanoporous silver-paper electrode and metal ion functionalized nanoporous gold–chitosan." *Chemical Communications* no. 49 (83):9540–9542.

Liu, Bingwen, Dan Du, Xin Hua, Xiao-Ying Yu, and Yuehe Lin. 2014. "Paper-based electrochemical biosensors: From test strips to paper-based microfluidics." *Electroanalysis* no. 26 (6):1214–1223.

Liu, Xiaojuan, Xiuyuan Li, Xin Gao, Lei Ge, Xinzhi Sun, and Feng Li. 2019. "A universal paper-based electrochemical sensor for zero-background assay of diverse biomarkers." *ACS Applied Materials & Interfaces* no. 11 (17):15381–15388.

Luka, George, Ali Ahmadi, Homayoun Najjaran, Evangelyn Alocilja, Maria DeRosa, Kirsten Wolthers, Ahmed Malki, Hassan Aziz, Asmaa Althani, and Mina Hoorfar. 2015. "Microfluidics integrated biosensors: A leading technology towards lab-on-a-chip and sensing applications." *Sensors* no. 15 (12):30011–30031.

Ma, Chao, Weiping Li, Qingkun Kong, Hongmei Yang, Zhaoquan Bian, Xianrang Song, Jinghua Yu, and Mei Yan. 2015. "3D origami electrochemical immunodevice for sensitive point-of-care testing based on dual-signal amplification strategy." *Biosensors and Bioelectronics* no. 63:7–13.

Nesakumar, Noel, Srinivasan Kesavan, Chen-Zhong Li, and Subbiah Alwarappan. 2019. "Microfluidic electrochemical devices for biosensing." *Journal of Analysis and Testing* no. 3 (1):3–18. doi: 10.1007/s41664-019-0083-y.

Nie, Zhihong, Christian A Nijhuis, Jinlong Gong, Xin Chen, Alexander Kumachev, Andres W Martinez, Max Narovlyansky, and George M Whitesides. 2010. "Electrochemical sensing in paper-based microfluidic devices." *Lab on a Chip* no. 10 (4):477–483.

Nikoleli, Georgia-Paraskevi, Christina G. Siontorou, Dimitrios P. Nikolelis, Spyridoula Bratakou, Stephanos Karapetis, and Nikolaos Tzamtzis. 2018. "Chapter 13: Biosensors based on microfluidic devices lab-on-a-chip and microfluidic technology." In *Nanotechnology and Biosensors*, edited by Dimitrios P. Nikolelis and Georgia-Paraskevi Nikoleli, 375–394. Elsevier.

Noh, Jongmin, Hee Chan Kim, and Taek Dong Chung. 2011. "Biosensors in microfluidic chips." In *Microfluidics: Technologies and Applications*, edited by Bingcheng Lin, 117–152. Springer.

Nontawong, Nongyao, Maliwan Amatatongchai, Wanchai Wuepchaiyaphum, Sanoe Chairam, Saichol Pimmongkol, Sirirat Panich, Suparb Tamuang, and Purim Jarujamrus. 2018. "Fabrication of a three-dimensional electrochemical paper-based device (3D-ePAD) for individual and simultaneous detection of ascorbic acid, dopamine and uric acid." *International Journal of Electrochemical Science* no. 13:6940–6957.

Noviana, Eka, and Charles S Henry. 2020. "Simultaneous electrochemical detection in paper-based analytical devices." *Current Opinion in Electrochemistry* no. 23:1–6.

Noviana, Eka, Cynthia P McCord, Kaylee M Clark, Ilhoon Jang, and Charles S Henry. 2019. "Electrochemical paper-based devices: Sensing approaches and progress toward practical applications." *Lab on a Chip* no. 20 (1):9–34.

Prasad, Alisha, Tiffany Tran, and Manas Ranjan Gartia. 2019. "Multiplexed paper microfluidics for titration and detection of ingredients in beverages." *Sensors* no. 19 (6): 1286.

Pursey, Joanna P., Yu Chen, Eugen Stulz, Mi Kyoung Park, and Patthara Kongsuphol. 2017. "Microfluidic electrochemical multiplex detection of bladder cancer DNA markers." *Sensors and Actuators B: Chemical* no. 251:34–39. doi: 10.1016/j.snb.2017.05.006.

Rackus, Darius G., Mohtashim H. Shamsi, and Aaron R. Wheeler. 2015. "Electrochemistry, biosensors and microfluidics: A convergence of fields." *Chemical Society Reviews* no. 44 (15):5320–5340. doi: 10.1039/C4CS00369A.

Ruecha, Nipapan, Jumi Lee, Heedo Chae, Haena Cheong, Veasna Soum, Pattarachaya Preechakasedkit, Orawon Chailapakul, Georgi Tanev, Jan Madsen, and Nadnudda Rodthongkum. 2017. "Paper-based digital microfluidic chip for multiple electrochemical assay operated by a wireless portable control system." *Advanced Materials Technologies* no. 2 (3):1600267.

Sanjay, Sharma T., Guanglei Fu, Maowei Dou, Feng Xu, Rutao Liu, Hao Qi, and XiuJun Li. 2015. "Biomarker detection for disease diagnosis using cost-effective microfluidic platforms." *Analyst* no. 140 (21):7062–7081. doi: 10.1039/C5AN00780A.

Santhiago, Murilo, and Lauro T Kubota. 2013. "A new approach for paper-based analytical devices with electrochemical detection based on graphite pencil electrodes." *Sensors and Actuators B: Chemical* no. 177:224–230.

Sekar, Nadia Chandra, Seyed Ali Mousavi Shaegh, Sum Huan Ng, Liya Ge, and Swee Ngin Tan. 2014. "A paper-based amperometric glucose biosensor developed with Prussian Blue-modified screen-printed electrodes." *Sensors and Actuators B: Chemical* no. 204: 414–420.

Srinivasan, Balaji, and Steve Tung. 2015. "Development and applications of portable biosensors." *Journal of Laboratory Automation* no. 20 (4):365–389. doi: 10.1177/2211068215581349.

Tang, Chi K., Abhay Vaze, Min Shen, and James F. Rusling. 2016. "High-throughput electrochemical microfluidic immunoarray for multiplexed detection of cancer biomarker proteins." *ACS Sensors* no. 1 (8):1036–1043. doi: 10.1021/acssensors.6b00256.

Tarn, M. D., and N. Pamme. 2014. "Microfluidics." In *Reference Module in Chemistry, Molecular Sciences and Chemical Engineering*. Elsevier.

Tian, Rong, Yujing Li, and Jingwei Bai. 2019. "Hierarchical assembled nanomaterial paper based analytical devices for simultaneously electrochemical detection of microRNAs." *Analytica Chimica Acta* no. 1058:89–96.

Wang, Panpan, Lei Ge, Mei Yan, Xianrang Song, Shenguang Ge, and Jinghua Yu. 2012. "Based three-dimensional electrochemical immunodevice based on multi-walled carbon nanotubes functionalized paper for sensitive point-of-care testing." *Biosensors and Bioelectronics* no. 32 (1):238–243.

Wang, Yang, Jinping Luo, Juntao Liu, Shuai Sun, Ying Xiong, Yuanyuan Ma, Shi Yan, Yue Yang, Huabing Yin, and Xinxia Cai. 2019. "Label-free microfluidic paper-based electrochemical aptasensor for ultrasensitive and simultaneous multiplexed detection of cancer biomarkers." *Biosensors and Bioelectronics* no. 136:84–90.

Whitesides, George M. 2006. "The origins and the future of microfluidics." *Nature* no. 442 (7101):368–373. doi: 10.1038/nature05058.

Wu, Yafeng, Peng Xue, Kam M Hui, and Yuejun Kang. 2014. "A paper-based microfluidic electrochemical immunodevice integrated with amplification-by-polymerization for the ultrasensitive multiplexed detection of cancer biomarkers." *Biosensors and Bioelectronics* no. 52:180–187.

Wu, Yafeng, Peng Xue, Yuejun Kang, and Kam M. Hui. 2013. "Paper-based microfluidic electrochemical immunodevice integrated with nanobioprobes onto graphene film for ultrasensitive multiplexed detection of cancer biomarkers." *Analytical Chemistry* no. 85 (18):8661–8668. doi: 10.1021/ac401445a.

Xie, Yao, Xiao Zhi, Haichuan Su, Kan Wang, Zhen Yan, Nongyue He, Jingpu Zhang, Di Chen, and Daxiang Cui. 2015. "A novel electrochemical microfluidic chip combined with multiple biomarkers for early diagnosis of gastric cancer." *Nanoscale Research Letters* no. 10 (1):477. doi: 10.1186/s11671-015-1153-3.

Yamada, Kentaro, Terence G Henares, Koji Suzuki, and Daniel Citterio. 2015. "Paper-based inkjet-printed microfluidic analytical devices." *Angewandte Chemie International Edition* no. 54 (18):5294–5310.

Zhao, Chen, and Xinyu Liu. 2016. "A portable paper-based microfluidic platform for multiplexed electrochemical detection of human immunodeficiency virus and hepatitis C virus antibodies in serum." *Biomicrofluidics* no. 10 (2):024119.

Zhao, Chen, Martin M Thuo, and Xinyu Liu. 2015. "Corrigendum: A microfluidic paper-based electrochemical biosensor array for multiplexed detection of metabolic." *Science and Technology of Advanced Materials* no. 16:049501.

Zhou, Fang, Min Lu, Wei Wang, Zhi-Ping Bian, Jian-Rong Zhang, and Jun-Jie Zhu. 2010. "Electrochemical immunosensor for simultaneous detection of dual cardiac markers based on a poly(dimethylsiloxane)-gold nanoparticles composite microfluidic chip: A proof of principle." *Clinical Chemistry* no. 56 (11):1701–1707. doi: 10.1373/clinchem.2010.147256.

Zhu, Liang, Xiaoxue Liu, Jiao Yang, Yongcheng He, and Yingchun Li. 2020. "Application of multiplex microfluidic electrochemical sensors in monitoring hematological tumor biomarkers." *Analytical Chemistry* no. 92 (17):11981–11986. doi: 10.1021/acs.analchem.0c02430.

Zupančič, Uroš, Pawan Jolly, Pedro Estrela, Despina Moschou, and Donald E. Ingber. 2021. "Graphene enabled low-noise surface chemistry for multiplexed sepsis biomarker detection in whole blood." *Advanced Functional Materials* no. 31: 2010638. doi: 10.1002/adfm.202010638.0960374000

16 Commercialization of Microfluidic Point-of-Care Diagnostic Devices

Pushpesh Ranjan, Mohd. Abubakar Sadique, Arpana Parihar, Chetna Dhand, Alka Mishra, and Raju Khan

CONTENTS

16.1 INTRODUCTION

Over the past decades, biosensor-based diagnostics in a wide range of applications have gained much attention due to their several advantages. Numerous biosensor-based devices were developed for the diagnosis of different kinds of analyte. These are categorized as electrochemical, optical, and mass-based biosensors [1]. Meanwhile, microfluidics-based devices attracted more attention over the other biosensor-based devices because of their numerous advantages and simpler applications [2, 3]. Microfluidics devices offer automation and high-throughput screening and can operate at a low volume of samples. These devices have the potential to reduce the time slot and cost for analytical devices, mainly in clinical, food, and environmental analysis [4]. Microfluidics is widely known in academia and it is rapidly gaining a position in industry mainly for the development of new approaches and devices for

DOI: 10.1201/9781003033479-16

clinical diagnostic applications. Traditional biosensor devices for clinical applications are relatively expensive, large, and difficult to handle, which limits their use in point-of-care (POC) diagnostic applications. Moreover, microfluidics-based devices are an emerging platform that provide the multiplexing and automation of laboratory equipment, and diagnostics devices in biosensor application [3, 5]. Over the several biosensor-based devices, only a few were commercialized in which microfluidics are popular for the diagnostics approach for POC application. Conventional laboratory equipment is costly, requiring a large volume of biological specimens and reagents, well-trained personnel, and analysis takes a long time which results in a costly diagnostics platform. In the current scenario, there is need for a POC diagnostic device platform which not only diagnoses different kinds of biomarker with high specificity and sensitivity, but it should also be cost effective, miniaturized, easy to use, and provide rapid results within a few seconds to minutes. Microfluidics devices have advantages over other conventional techniques in that they are cost effective and require less volume (~10 μL) of the biological sample without any pretreatment and produce rapid results. In addition, they can diagnose multiplex biomarkers in a single device platform. They are disposable and one-time-use devices. They can integrate with other techniques, like electrochemical and optical, to produce fast and accurate results [6, 7]. For instance, Oliveira et al. fabricated a disposable microfluidic chip integrated with the electrochemical technique for the diagnosis of breast cancer biosensor (Figure 16.1) [8]. In addition, microfluidics technology integrates with a portable device like smartphones, which facilitate the application in healthcare diagnosis [9–11].

A biosensor is categorized into five types, such as imaging, biochemical, immune, hybrid, and molecular biosensors. These biosensors are applicable for the detection of biological analytes, such as cells, glucose, protein, amino acid, antibody, antigen, DNA, and other biomarkers in different clinical specimens like blood, serum, saliva, nasal-swab, sweat, urine, etc. [12]. In addition, it is also applicable for environmental analysis as well as food, soil, and water in different sources [13, 14]. Microfluidics-based biosensor systems are classified according to analytical method and sensing mode, including on-chip imaging devices, biochemical parameter sensors, immunoassays, and molecular diagnosis. However, there are still several challenges that are necessary to overcome in the integration and design of smartphone-based microfluidics biosensor systems. Furthermore, excessive accessories attached to the smartphone device may affect convenience and flexibility along with a loss of accuracy in the diagnostics process of the smartphone-based microfluidics biosensors [10, 11].

Moreover, the e-health POC diagnostic device plays a major role in providing quick and real-time data [15]. Recent trends in the miniaturization of microfluidics-based biosensors convert biosensors into flexible and wearable ones. They are effectively monitoring the specific analyte in biological samples without any adverse effect. Figure 16.2 represents miniaturized microfluidics-based systems [16].

Mostly, the wearable devices available in the consumer market provide information related to human "metabolites and/or disease" biomarkers, apart from glucose detection in diabetic patients, even in this, the glycemic levels are taken "under-the-skin" and not "on-the-skin." Recently, it has been well studied that human sweat has

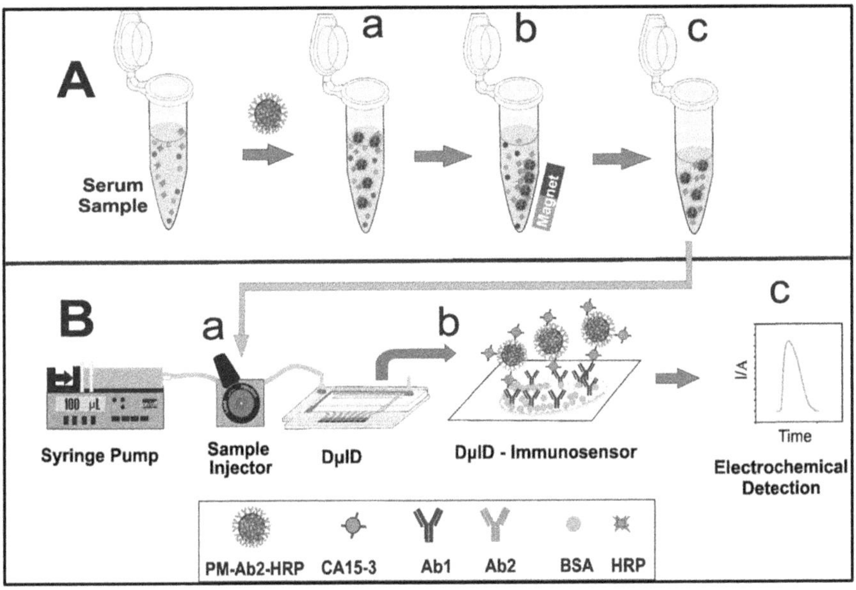

FIGURE 16.1 Disposable microfluidic immunoassay device for sensitive breast cancer biomarker detection. Reprint with permission from [8] R. A. G. de Oliveira, E. M. Materon, M. E. Melendez, A. L. Carvalho, R. C. Faria, Disposable Microfluidic Immunoarray Device for Sensitive Breast Cancer Biomarker Detection, ACS Appl. Mater. Interfaces 2017, 9, 27433–27440.

molecules and other biomarkers (e.g., ions), which makes it an important human fluid that contains crucial medical information via a noninvasive approach at a molecular level. A proper collection of sweat from the top of the skin can effectively convey the required information to the wearable system, this can be done with help from microfluidic systems with ease. The fabrication has become convenient and simple, from tedious photolithography or chemical etching to three-dimensional (3D) printing methods. Cost effectiveness, the flexibility of design, and lower time consumption are a few advantages along with cheap hybrid material for enhancing the mechanical properties and texture control as per the application requirement [17]. Possible wearable microfluidic biosensor platforms are represented in Figure 16.3 [18].

16.2 FABRICATION OF MICROFLUIDIC DEVICE

Microfluidics devices are made of different types of materials such as silicon, glass, and plastic, but mainly of paper or polymeric materials (PDMS) and have been developed for a range of applications. It has a wide range of geometries from simple to complex. On this chip, a microchannel of varying sizes intersects with different dimensions. These microchannels facilitate the mixing, flow, and sorting of samples, cell growth, cell and particle encapsulation, etc. There are many more designs and applications which can be developed but the choice of the microfluidic chip

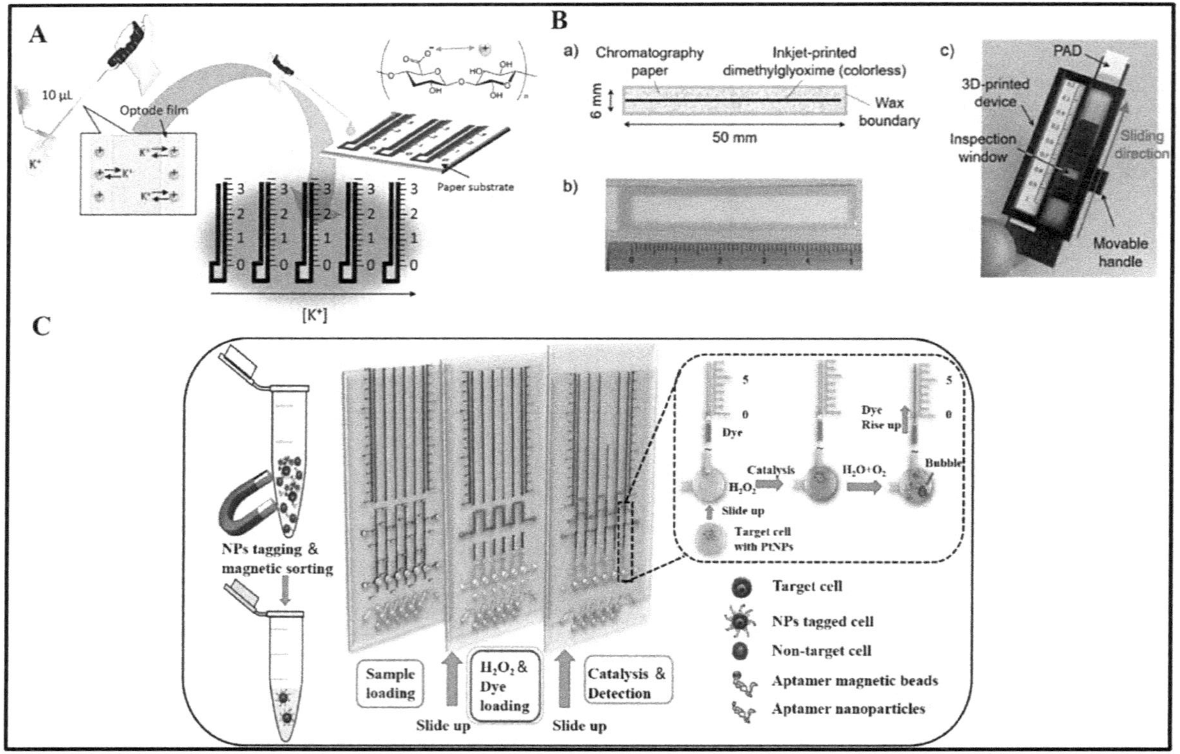

FIGURE 16.2 Microfluidics platform-based miniaturized biosensing systems. Reprint with permission from [16] D. Liu, J. Wang, L. Wu, Y. Huang, Y. Zhang, M. Zhu, Y. Wang, Z. Zhu, C. Yang, Trends in miniaturized biosensors for point-of-care testing, Trends in Analytical Chemistry, TrAC Trends in Analytical Chemistry 2020, 122, 115701. DOI: https://doi.org/10.1016/j.trac.2019.115701

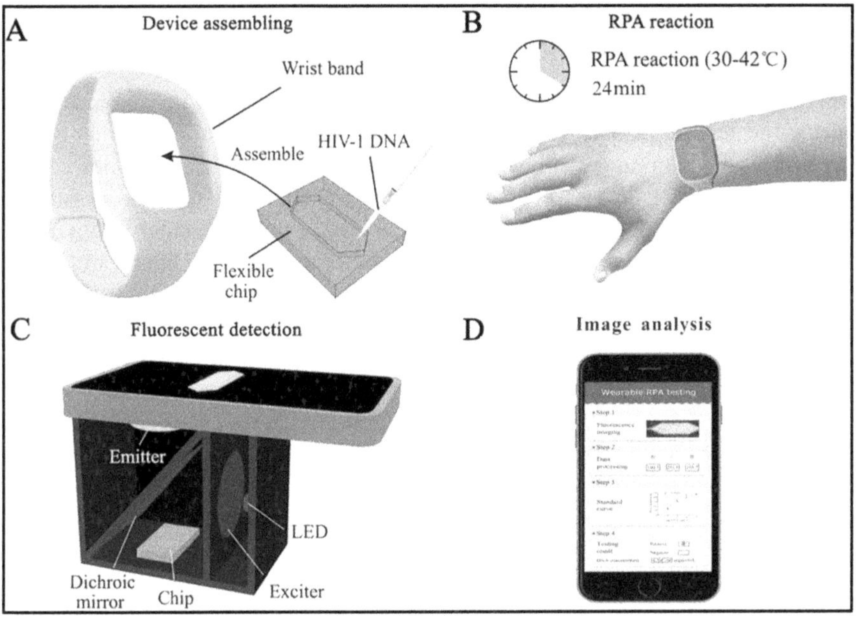

FIGURE 16.3 Wearable microfluidics device for rapid detection of HIV-1 DNA using recombinase polymerase amplification. Reprint with permission from [18] M. Kong, Z. Li, J. Wu, J. Hu, Y. Sheng, D. Wu, Y. Lin, M. Li, X. Wang, S. Wang, A wearable microfluidic device for rapid detection of HIV-1 DNA using recombinase polymerase amplification, Talanta 2019, 205, 120155.

(type of material, geometry, and the dimension of the channel, etc.) is crucial to the experimental setup. During the fabrication of the microfluidic chip, it is important to consider the choice of material which offers different capabilities. For a long time, more favorable PDMS has been utilized by researchers and industry to fabricate the chips. PDMS is a versatile, cheap, transparent material, and it is quick and easy to fabricate, ideally suited to prototyping and low volume applications and can be easily bonded to the glass. In addition, working with PDMS does not require a costly cleanroom. However, due to potential contamination of microfabrication equipment, many microelectromechanical systems (MEMS) avoid it because it can have some adverse effects on cells [6, 19, 20]. In its favor, PDMS is optically transparent, biocompatible, and gas permeable, which makes it suitable for a wide range of biological applications. However, PDMS has some disadvantages and as a result, plastic chips are gaining popularity, particularly as material properties have improved in recent years, offering greater optical quality and multi-layer bonding. However, glass chips are quite difficult to fabricate, and thus, they require well-trained personnel for fabrication. It was also observed that they do not offer the detection of those samples which are insoluble, and this makes it a limitation in production as a microfluidic chip. On the other hand, paper-based microfluidic devices gained much attention in recent years, due to their improvement as they are easy to fabricate, cost effective, and enable real-time monitoring, easier fluid handling, and analysis [21, 22].

16.3 FUTURE DEVELOPMENT OF MICROFLUIDICS-BASED DEVICES

The further development of microfluidics devices for diagnostic applications is expected to feature in our everyday life. The development of such sensors can be easy by keeping in mind the following points – (i) the applicability of the device, like being home-based or field-based, etc. (ii) type of bio-analyte like serum, plasma, or any other biological specimen, (iii) type of analysis, either molecular or any other, (iv) level of training to personnel for operation, etc. The developed biosensors should be well tested on various validation parameters like specificity, sensitivity, low limit of detection, long-term stability, etc. [23].

Microfluidics is the fully functioning integrated advanced technology-based POC diagnostic technique which provides better performance to real clinical samples. It includes some steps to the development of microfluidics devices such as sample collection, sample pretreatment, improvement of long-term stability of device and reagent, working with complex sample specimens. 3D printing of PDMS has been reported using the stereolithography approach. However, certain drawbacks, such as lack of chemical stability, deformation under pressure, and adsorption of small hydrophobic molecules have hindered its industrial-scale utilization. Moreover, the manual molding, cleaning, and bonding processes complicate mass production [24–26].

16.4 PATHWAY TO COMMERCIALIZATION

In the development of microfluidics-based devices, industrial and academic effort should result in the fabrication of an advanced device. Academicians should focus on the fabrication of integrated products with specific applications. Moreover, industry and marketing experts play a major role in the commercialization of the product. They should discuss the benefits of products in the market and provide positive feedback about the products which successfully meet the requirements of the users. So, the combined effort of academicians, industry, and marketing experts plays a leading role in the commercialization of the product in the market [27].

New advances in nanotechnology and microfluidics have allowed the development of advanced nano-biosensors for POC application. For the commercialization of POC devices, several criteria must be followed before diagnosis. On this pathway, a variety of challenges, such as clinical validation and risk assessment, as safety and efficacy must be followed. The US Food and Drug Administration (FDA) has classified the almost 1700 different types of devices and segmented with 16 medical specialty panels. The FDA announced its rules and regulations which should be strictly followed before and after the commercialization of devices. As in the premarket stage, the detailed specification of products such as design, production, labeling, promotion, manufacturing, the testing of the regulated product as well as the processing, content, and evaluation or approval of submission, inspection, and enforcement policies should be submitted for the initial review of the product. In the next stage, as the post market stage, the company should provide the details of its product, tracking system, device malfunctioning, risks of serious injury or death, and registering the establishment where devices are produced or distributed [28]. Before the commercialization of sensor-based devices the degree of regulation as an assessment of the risk level of the device should be evaluated. It is illegal to sell a sensor device without any appropriate regulatory approval.

The code of Federal Regulations can be used to categorize the device into three classes based on the risk level, as low, medium, and high, in which a device in Class I is subjected to least regulatory control, generally for low-risk devices. In Class II devices, they are subjected to a moderate level of risk, but they are designed to perform the activity without causing injury or harmful effects to the end user. Moreover, Class III devices are subjected to a high-risk level that needs premarket approval from the regulatory panel by providing appropriate data to the physician in the treatment of patients suffering from the disease. After considering all reviews and verification of devices, they receive the FDA approval to launch their device in the market. Figure 16.4 represents the pathway to approval of microfluidics devices before launching in the market [29, 30].

16.5 MICROFLUIDICS DEVICE MARKET

The value of biosensors in the global market is billions of US dollars. The overall biosensor market is expected to grow from US$21.0 billion in 2019 to US$31.5 billion by 2024, with an expected compound annual growth rate (CAGR) of 8.3% during this time [31]. However, the microfluidics market expected the value to grow from

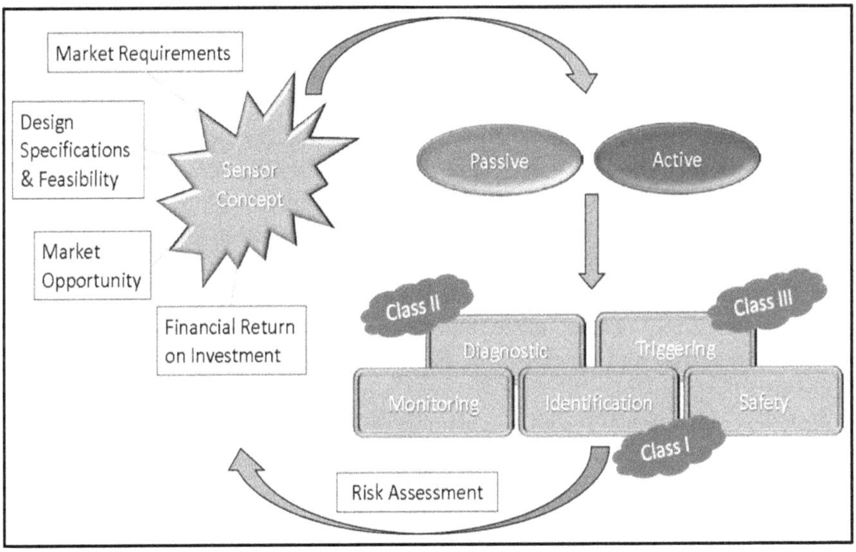

FIGURE 16.4 Represents the pathway to approval of microfluidics device before launching in the market. Reprint with permission from [29] R. A. Mazzocchi, Medical Sensors – Defining a Pathway to Commercialization, ACS Sens. 2016, 1, 1167–1170.

US$3.3 billion in 2019 to US$6.1 billion by 2024, with an expected CAGR of 12.7%. This high growth rate is largely due to recent advances in biotechnology, including gene sequencing and in vitro diagnostics [32].

16.6 MICROFLUIDICS-BASED POINT-OF-CARE DIAGNOSTICS

In the past decade, several microfluidics-based POC devices were developed and later commercialized by their respective developers. The commercialized diagnostics devices meet the requirements of the user and hence benefited the end user. Several POC devices which were commercialized are used for the diagnosis of different target analytes [33]. For instance, Abbott Laboratories launched the i-STAT product, a microfluidics and electrochemical integrated device for the diagnosis of analytes in blood specimens. The i-STAT product can quantify different analytes like metabolites, electrolytes, and gases. In addition, it has also the capability to perform immunoassays. Another product launch by OPKO Inc. acquired Claros Diagnostics Inc. to complement its in-vitro diagnostics platform. Similarly, Claros has developed and commercialized a microfluidic portable device that reads a credit-card-sized disposable chip containing a blood specimen that tests for multiplex in urology and infectious disease. OPKO Inc. used this technology in conjunction with its specific biomarkers, such as antibody-based immunoassays for the diagnosis of Parkinson's or Alzheimer's diseases. In developing countries, rather than the technical novelty, practical and marketable devices are needed to address major clinical problems. Daktari Diagnostics Inc. has developed a microfluidics-based device in which the sample preparation along with their analysis was done in a single device. This device has better diagnostics performance with a microfluidic differential counter. It can quantify CD4+ or CD8+ immune cells in blood samples. This device has high accuracy compared to conventional flow cytometry, which results in reduced cost in the diagnosis of HIV [34].

Another company that has positioned itself for the developing world markets is Diagnostics for All (DFA). Its paper-based microfluidics devices utilize capillary forces to direct the movement of fluidics and enable multiplexed assays with built-in capabilities, which is an improvement over lateral-flow assay. DFA aims to deliver low-cost medical diagnostics, veterinary tests, and environmental monitoring devices to resource-poor countries. Although DFA serves a non-profit mission, it also generates revenue through licensing agreements to supplement funding for further research and development. Some selected POC devices are shown in Figure 16.5 [34].

With the completion of the human genome project in 2003 and the advent of next-generation sequencing, microfluidics technology has been used to increase automation and decrease turnaround times in genomics. Key companies in the field of microfluidic genotyping include Illumina Inc. and Fluidigm Corp. In 2013, Illumina acquired Advanced Liquid Logic Inc. to gain access to its digital microfluidics platform. Their electrowetting technology manipulates discrete droplets in a microfluidics device without pumps, valves, or channels; therefore, these devices have the potential to offer readily scalable solutions. Fluidigm was found to market the

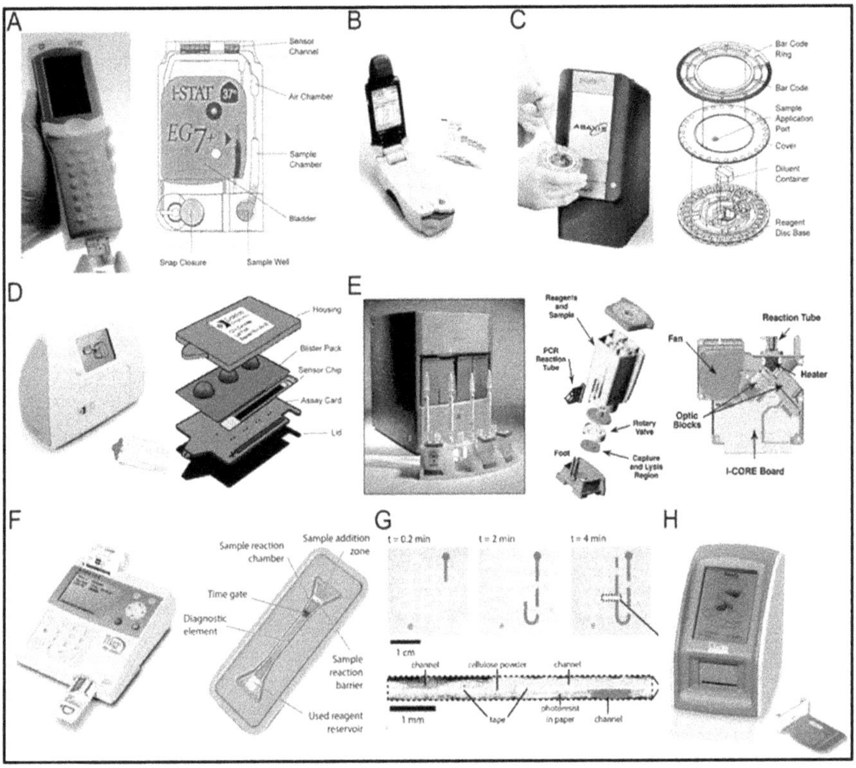

FIGURE 16.5 Images and schematics of selected microfluidics-based POC tests: (A) i-STAT (Abbott). (B) Epocal. (C) Abaxis. (D) Dakari Diagnostics. (E) Cepheid. (F) Biosite. (G) Diagnostics for All. (H) Claros Diagnostics. Reprint with permission from [34] C. D. Chin, V. Linder, S. K. Sia, Commercialization of microfluidic point-of-care diagnostic devices, Lab Chip, 2012, 12, 2118–2134.

integrated fluidic circuit (IFC) based on a pneumatic rubber valve developed in the laboratory of Stephen Quake. With this technology, Fluidigm became the first company to commercialize a digital PCR, and it held its initial public offering in 2011. To facilitate sample preparation further for next-generation sequencing, RainDance Technologies Inc. developed a single-molecule picodroplet system for digital PCR. Each picodroplet is loaded with a uniform quantity of genomic DNA and primers; therefore, this system enhances reproducibility and enables the targeting of specific regions of the genome. Another company with single-molecule expertise is Sphere Fluidics Ltd., whose microfluidic system can perform high-throughput analyses of single cells to produce its genetic, proteomic, and transcriptomic profiles in picoliter volume droplets. These picodroplets are compatible with PCR machines and next-generation sequencers and can also be used in applications such as drug discovery and biomarker identification [35]. Several other microfluidics-based devices which were commercialized are listed in Table 16.1 with their origin and developer.

TABLE 16.1
Enlisting of Microfluidics Devices Products, their Manufacturer, and Application

Company name	Products	Application	Year	Country
Advanced Liquid Logic, Inc.	NeoPrep Library Prep system	Next-generation sequencing	1998/ 2013	US
Agilent Technologies	2100 Bioanalyzer	Human disease, genomics	1999	US
Abaxis	Piccolo Express	Real-time blood chemistry diagnostics	1989	US
Alere (formerly Invesmess)	-	HIV/AIDS diagnosis	2001	-
Bio-Rad	Experion RNA HighSens Chips	RNA analysis	1952	US
Biosite (Alere)	-	Cardiovascular disease, drug, a waterborne parasite	1988	-
bioMérieux	VIDAS	Infectious disease, cardiovascular disease	1963	France
Caliper Life Sciences (Perkin Elmer)	LabChip systems	Diagnostics, molecular testing	1995/ 2011	US
Cellix	VenaFlux	Cell analysis cell culture	2006	Ireland
Cepheid	Xpert and GeneXpert systems	Diagnostics	1996	US
Claros Diagnostics (OPKO)	Prostate specific antigen (total PSA) test, 4K score prostate cancer test	POC diagnostics	2005/ 2011	US
Danaher Corporation	Original equipment's manufacturer	Life science diagnostics	1969	US
Daktari	-	HIV/AIDS	2008	-
Dolomite Microfluidics	Multiflux	Microfluidics device	2005	UK
Elveflow	-	Cell biology	-	France
Fluidigm Corporation	Biomarke HD system, C1 system, EP1 system	Genotyping and sequencing	1999	US
Fluigent	EZ DROP, DROP SEQ	-	-	France
Gyros AB	GyrolabxP workstation, GyrolabBioaffy CDs, Gyrolab mixing CD	Immunoassay, biomarkers monitoring, drug analysis	2000	Sweden
i-STATS Corp. (ABBOTT)	i-STAT systems	POC diagnostics	1983/ 2004	US

(Continued)

TABLE 16.1 (CONTINUED)
Enlisting of Microfluidics Devices Products, their Manufacturer, and Application

Company name	Products	Application	Year	Country
Life Technologies Corporation (Thermo Fisher)	TaqMan Assay	Genotyping, diagnostics, drug discovery	1983/ 2013	US
Micronit Microfluidics	Microreactors, micromixers, droplet generators, Chip Electrophoresis	Microfluidics device	1999	Netherlands
Microfluidic ChipShop	-	POC	2002	Germany
RainDance Technologies	RainDrop system, ThenderStorm system	Genotyping and sequencing	2004	US
Roche Diagnostics	Genome sequencer FLX system, light cycler system, CedexHiRes	Genotyping, microarray analysis, cell analysis,	1896	US & Switzerland
Sphere Fluidics Ltd	Pico-Gen Picodroplet Formation chip	Human disease, drug discovery, biomarkers analyses	2010	UK
thinXXS Microtechnology Acquired by IDEX corporation	-	Lateral flow assay, immunoassay, molecular assay, next-generation sequencing	1988	Germany/USA

16.7 MICROFLUIDICS DEVICE MARKET, COMPANY PROFILES

There has been tremendous research in the field of microfluidics, especially when it comes to the products being commercialized and in use for various applications. Microfluidics has the potential to replace conventional devices; many companies have invested and see a beneficial marketplace for such devices in the future as well. Some of the examples of such commercialized products for various applications have been described in Table 16.1.

16.8 OVERCOMING CHALLENGES TO COMMERCIALIZATION

Microfluidics devices were utilized for diagnosis in the early stage of a specific disease in human beings as well as other living beings. For device commercialization, some of the following challenging factors are pointed out, which before delayed or affected the commercialization of the device [36].

Nature of sample: Microfluidics devices have the potential to diagnose the target analyte in the fluid sample. But they face a serious problem when the target analyte is less soluble or insoluble in a liquid medium. So, it prevents the flow of the insoluble analyte through the channel which shows poor application in the sensor field. This drawback limits their application and further affects their modification into device form and limits their commercialization.

Cost effectiveness: The cost of the device is one of the major factors which slows down the commercialization of the product. The product should be intended to be cheaper in terms of production cost, packaging, and transportation. For this purpose, materials used in the production of the device should be cheaper and would get appropriate results. In addition, the fabricated device should have long-term stability. The technique used for packaging and integration should be cheap as it enhances the overall cost of the product [37, 38].

Specificity and sensitivity: The microfluidic device used for the diagnosis of target analytes should be specific and highly sensitive with high accuracy and should give results rapidly. In addition, it should be designed in a manner that the result generated through the device could be easily interpreted and can be used by a layman [39].

Market return and growth: The major challenges in the commercialization of products involve customer acceptance and market adaptation. In microfluidics platforms, academic publications of proof-of-concept devices are abundant, but the diffusion of this technology to consumer products has been limited over the past two decades due to lower customer demand, development, and validation of market need. Although microfluidics is a promising laboratory tool, the technology is still seeking the best applications. To overcome the restrictions to commercialization, innovators of microfluidics devices need to focus on two main areas, standardization and integration, that currently lack sufficient attention [39].

16.9 CONCLUDING REMARKS AND FUTURE PERSPECTIVES

To aid in the advancement, microfluidics devices play an essential role in the diagnostic platform. They are made of silicon, glass, plastic, or polymeric materials with channels of micrometers in size. Such channels facilitate the sampling and diagnosis process. There are many more designs and applications which can be developed, but the choice of the microfluidic chip (material type, channel geometries, channel dimensions, etc.) is crucial to the experimental setup. Microfluidics has the potential to diagnose multiple analytes in the fluid sample. Most microfluidics-based devices provide a better sensing platform over existing technologies. Moreover, they suffer from the serious problem of the detection of the insoluble analyte. However, this problem could be resolved by the integration of other techniques with the microfluidics device. The academic effort in the development of advanced microfluidics devices that provide fully integrated devices with a selective application should also aim to minimize the cost and look into miniaturizing devices. Industrial partners may lead to the successful launch of the device in the market. Industrial partners in collaboration with marketing experts actively collect the feedback of market and end user requirements and expectations. Since technological advancement, microfluidics has turned into a paper-based analytical device, lab-on-a-chip, and organ-on-a-chip platforms. Moreover, 3D printing technology provides the platform for rapid and

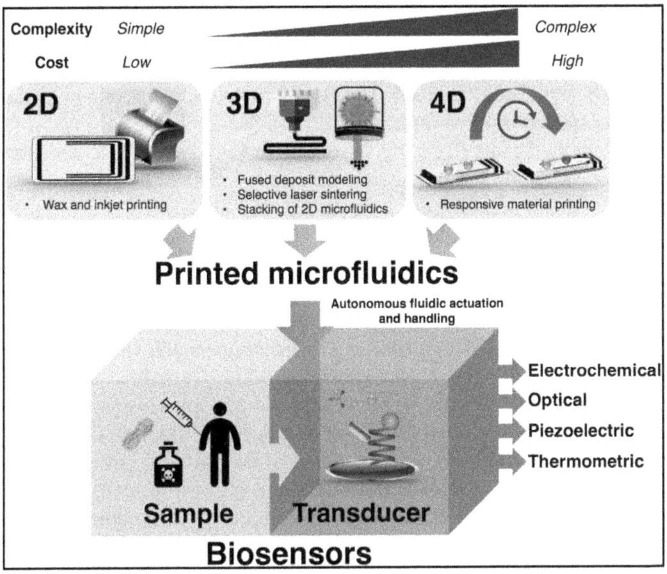

FIGURE 16.6 A summary of printed microfluidics and its biomedical applications. Reprint with permission from [41] J. F. C. Loo, A. H. P. Ho, A. P. F. Turner, W. C. Mak, Integrated Printed Microfluidic Biosensors, Trends in Biotechnology 2019, 37, 10, 1104-1120.

cheap production of the device, which leads to the development of customizing new microfluidics products (Figure 16.6) [40, 41].

Further research is required to develop and explore the potential of microfluidics devices, which may take the leading role over the existing devices in diagnostics applications. They should be cost effective, rapid, specific, and highly sensitive so that they meet the market requirement and benefit the end user in the future.

ACKNOWLEDGMENTS

The authors thank Dr Avanish Kumar Srivastava, Director, Council of Scientific and Industrial Research-Advanced Materials and Processes Research Institute (CSIR-AMPRI) Bhopal, India, for his interest and encouragement in this work. Pushpesh Ranjan is thankful to the CSIR, India, for the award of SRF. Raju Khan would like to acknowledge Science and Engineering Research Board (SERB) for providing funds in the form of the IPA/2020/000130 project.

REFERENCES

1. Goode, J.A., Rushworth, J.V.H. and Millner, P.A., 2015. Biosensor regeneration: A review of common techniques and outcomes. *Langmuir*, 31(23), 6267–6276.
2. Pandey, C.M., Augustine, S., Kumar, S., Kumar, S., Nara, S., Srivastava, S. and Malhotra, B.D., 2018. Microfluidics based point-of-care diagnostics. *Biotechnology Journal*, 13(1), 1700047.

3. Song, Y., Lin, B., Tian, T., Xu, X., Wang, W., Ruan, Q., Guo, J., Zhu, Z. and Yang, C., 2018. Recent progress in microfluidics-based biosensing. *Analytical Chemistry*, 91(1), 388–404.
4. Srinivasan, B. and Tung, S., 2015. Development and applications of portable biosensors. *Journal of Laboratory Automation*, 20(4), 365–389.
5. Liao, Z., Wang, J., Zhang, P., Zhang, Y., Miao, Y., Gao, S., Deng, Y. and Geng, L., 2018. Recent advances in microfluidic chip integrated electronic biosensors for multiplexed detection. *Biosensors and Bioelectronics*, 121, 272–280.
6. Rosen, Y. and Gurman, P., 2010. MEMS and microfluidics for diagnostics devices. *Current Pharmaceutical Biotechnology*, 11(4), 366–375.
7. Uliana, C.V., Peverari, C.R., Afonso, A.S., Cominetti, M.R. and Faria, R.C., 2018. Fully disposable microfluidic electrochemical device for detection of estrogen receptor alpha breast cancer biomarker. *Biosensors and Bioelectronics*, 99, 156–162.
8. De Oliveira, R.A., Materon, E.M., Melendez, M.E., Carvalho, A.L. and Faria, R.C., 2017. Disposable microfluidic immunoarray device for sensitive breast cancer biomarker detection. *ACS Applied Materials & Interfaces*, 9(33), 27433–27440.
9. Zarei, M., 2017. Portable biosensing devices for point-of-care diagnostics: Recent developments and applications. *TrAC Trends in Analytical Chemistry*, 91, 26–41.
10. Xu, D., Huang, X., Guo, J. and Ma, X., 2018. Automatic smartphone-based microfluidic biosensor system at the point of care. *Biosensors and Bioelectronics*, 110, 78–88.
11. Kim, D.W., Jeong, K.Y. and Yoon, H.C., 2020. Smartphone-based medical diagnostics with microfluidic devices. In *Smartphone Based Medical Diagnostics* (103–128). Academic Press.
12. Nasseri, B., Soleimani, N., Rabiee, N., Kalbasi, A., Karimi, M. and Hamblin, M.R., 2018. Point-of-care microfluidic devices for pathogen detection. *Biosensors and Bioelectronics*, 117, 112–128.
13. Kung, C.T., Hou, C.Y., Wang, Y.N. and Fu, L.M., 2019. Microfluidic paper-based analytical devices for environmental analysis of soil, air, ecology and river water. *Sensors and Actuators B: Chemical*, 301, 126855.
14. Liu, J., Jasim, I., Shen, Z., Zhao, L., Dweik, M., Zhang, S. and Almasri, M., 2019. A microfluidic based biosensor for rapid detection of Salmonella in food products. *Plos One*, 14(5), e0216873.
15. Christodouleas, D.C., Kaur, B. and Chorti, P., 2018. From point-of-care testing to eHealth diagnostic devices (eDiagnostics). *ACS Central Science*, 4(12), 1600–1616.
16. Liu, D., Wang, J., Wu, L., Huang, Y., Zhang, Y., Zhu, M., Wang, Y., Zhu, Z. and Yang, C., 2020. Trends in miniaturized biosensors for point-of-care testing. *TrAC Trends in Analytical Chemistry*, 122, 115701.
17. Padash, M., Enz, C. and Carrara, S., 2020. Microfluidics by additive manufacturing for wearable biosensors: A review. *Sensors*, 20(15), 4236.
18. Kong, M., Li, Z., Wu, J., Hu, J., Sheng, Y., Wu, D., Lin, Y., Li, M., Wang, X. and Wang, S., 2019. A wearable microfluidic device for rapid detection of HIV-1 DNA using recombinase polymerase amplification. *Talanta*, 205, 120155.
19. Nge, P.N., Rogers, C.I. and Woolley, A.T., 2013. Advances in microfluidic materials, functions, integration, and applications. *Chemical Reviews*, 113(4), 2550–2583.
20. Tsao, C.W., 2016. Polymer microfluidics: Simple, low-cost fabrication process bridging academic lab research to commercialized production. *Micromachines*, 7(12), 225.
21. Akyazi, T., Basabe-Desmonts, L. and Benito-Lopez, F., 2018. Review on microfluidic paper-based analytical devices towards commercialisation. Analytica Chimica Acta, 1001, 1–17.
22. Cate, D.M., Adkins, J.A., Mettakoonpitak, J. and Henry, C.S., 2015. Recent developments in paper-based microfluidic devices. *Analytical Chemistry*, 87(1), 19–41.

23. Lazar, I.M., 2015. Microfluidic devices in diagnostics: What does the future hold?. *Bioanalysis*, 7(20), 2677–2680.
24. Convery, N. and Gadegaard, N., 2019. 30 years of microfluidics. *Micro and Nano Engineering*, 2, 76–91.
25. Au, A.K., Huynh, W., Horowitz, L.F. and Folch, A., 2016. 3D-printed microfluidics. *Angewandte Chemie International Edition*, 55(12), 3862–3881.
26. Yeh, E.C., Fu, C.C., Hu, L., Thakur, R., Feng, J. and Lee, L.P., 2017. Self-powered integrated microfluidic point-of-care low-cost enabling (SIMPLE) chip. *Science Advances*, 3(3), e1501645.
27. Lin, C.T. and Wang, S.M., 2005. Biosensor commercialization strategy-a theoretical approach. *Frontiers in Bioscience: A Journal and Virtual Library*, 10, 99–106.
28. Web reference: Overview of device regulation and Postmarket requirements (devices). https://www.fda.gov/medical-devices/device-advice-comprehensive-regulatory-assistance
29. Mazzocchi, R.A., 2016. Medical sensors–defining a pathway to commercialization. *ACS Sensors*, 1(10), 1167–1170.
30. McGrath, M.J. and Scanaill, C.N., 2013. Regulations and standards: Considerations for sensor technologies. In *Sensor Technologies* (115–135). Apress.
31. Web reference: Biosensors market by type (sensor patch and embedded device), product (wearable and nonwearable), technology (electrochemical and optical), application (POC, home diagnostics, research lab, food & beverages), and geography - global forecast to 2024. https://www.marketsandmarkets.com/Market-Reports/biosensors-market-798.html
32. Web reference: Microfluidic components market by product (valve, solenoid valve, check valve, nozzle, tubing, micropump, microneedle, shuttle valve), industry (automotive, aerospace & defense, healthcare, consumer electronics, oil & gas) - global forecast to 2024. https://www.marketsandmarkets.com/Market-Reports/microfluidic-components-market-223516809.html
33. Lifesciences, C., Coulter, B., Biosciences, A., Nanosciences, B., Elmer, P., Biosciences, B. and Electron, T., 2006. Microfluidics in commercial applications; an industry perspective. *Lab on a Chip*, 6, 1118–1121.
34. Chin, C.D., Linder, V. and Sia, S.K., 2012. Commercialization of microfluidic point-of-care diagnostic devices. *Lab on a Chip*, 12(12), 2118–2134.
35. Volpatti, L.R. and Yetisen, A.K., 2014. Commercialization of microfluidic devices. *Trends in Biotechnology*, 32(7), 347–350.
36. Chiu, D.T., Demello, A.J., Di Carlo, D., Doyle, P.S., Hansen, C., Maceiczyk, R.M. and Wootton, R.C., 2017. Small but perfectly formed? Successes, challenges, and opportunities for microfluidics in the chemical and biological sciences. *Chem*, 2(2), 201–223.
37. Sanjay, S.T., Fu, G., Dou, M., Xu, F., Liu, R., Qi, H. and Li, X., 2015. Biomarker detection for disease diagnosis using cost-effective microfluidic platforms. *Analyst*, 140(21), 7062–7081.
38. Faustino, V., Catarino, S.O., Lima, R. and Minas, G., 2016. Biomedical microfluidic devices by using low-cost fabrication techniques: A review. *Journal of Biomechanics*, 49(11), 2280–2292.
39. Webb, D.P., Knauf, B., Liu, C., Hutt, D. and Conway, P., 2009. Productionisation issues for commercialisation of microfluidic based devices. *Sensor Review*, 29(4), 349–354.
40. Kadimisetty, K., Song, J., Doto, A.M., Hwang, Y., Peng, J., Mauk, M.G., Bushman, F.D., Gross, R., Jarvis, J.N. and Liu, C., 2018. Fully 3D printed integrated reactor array for point-of-care molecular diagnostics. *Biosensors and Bioelectronics*, 109, 156–163.
41. Loo, J.F., Ho, A.H., Turner, A.P. and Mak, W.C., 2019. Integrated printed microfluidic biosensors. *Trends in Biotechnology*, 37(10), 1104–1120.

Index

Milton Keynes UK
Ingram Content Group UK Ltd.
UKHW021842301123
433521UK00003B/22